Combinatorial Scientific Computing

Chapman & Hall/CRC
Computational Science Series

SERIES EDITOR

Horst Simon
Deputy Director
Lawrence Berkeley National Laboratory
Berkeley, California, U.S.A.

AIMS AND SCOPE

This series aims to capture new developments and applications in the field of computational science through the publication of a broad range of textbooks, reference works, and handbooks. Books in this series will provide introductory as well as advanced material on mathematical, statistical, and computational methods and techniques, and will present researchers with the latest theories and experimentation. The scope of the series includes, but is not limited to, titles in the areas of scientific computing, parallel and distributed computing, high performance computing, grid computing, cluster computing, heterogeneous computing, quantum computing, and their applications in scientific disciplines such as astrophysics, aeronautics, biology, chemistry, climate modeling, combustion, cosmology, earthquake prediction, imaging, materials, neuroscience, oil exploration, and weather forecasting.

PUBLISHED TITLES

Combinatorial Scientific Computing

Edited by

Uwe Naumann
RWTH Aachen University
Germany

Olaf Schenk
University of Basel
Switzerland

CRC Press
Taylor & Francis Group
Boca Raton London New York

CRC Press is an imprint of the
Taylor & Francis Group, an **informa** business
A CHAPMAN & HALL BOOK

CRC Press
Taylor & Francis Group
6000 Broken Sound Parkway NW, Suite 300
Boca Raton, FL 33487-2742

First issued in paperback 2019

ISBN-13: 978-1-4398-2735-2 (hbk)
ISBN-13: 978-0-367-38175-2 (pbk)

Library of Congress Cataloging-in-Publication Data

Combinatorial scientific computing / [edited by] Uwe Naumann, Olaf Schenk.
    p. cm. -- (Chapman & Hall/CRC computational science ; 12)
    Includes bibliographical references and index.
    ISBN 978-1-4398-2735-2 (hardback)
    1. Computer programming. 2. Science--Data processing. 3. Combinatorial analysis. I. Naumann, Uwe, 1969- II. Schenk, Olaf, 1967-

QA76.6.C6275 2012
511'.6--dc23   2011044663

Visit the Taylor & Francis Web site at
http://www.taylorandfrancis.com

and the CRC Press Web site at
http://www.crcpress.com

Contents

10 Unstructured Mesh Generation   257

Jonathan Richard Shewchuk

11 3D Delaunay Mesh Generation   299

Klaus Gärtner, Hang Si, Alexander Rand, and Noel Walkington

Foreword

The ongoing era of high-performance computing is filled with enormous potential for scientific simulation, but also with daunting challenges. Architectures for high-performance computing may have thousands of processors and complex memory hierarchies paired with a relatively poor interconnecting network performance. Due to the advances being made in computational science and engineering, the applications that run on these machines involve complex multiscale or multiphase physics, adaptive meshes and/or sophisticated numerical methods. A key challenge for scientific computing is obtaining high performance for these advanced applications on such complicated computers and, thus, to enable scientific simulations on a scale heretofore impossible.

A typical model in computational science is expressed using the language of continuous mathematics, such as partial differential equations and linear algebra, but techniques from discrete or combinatorial mathematics also play an important role in solving these models efficiently. Several discrete combinatorial problems and data structures, such as graph and hypergraph partitioning, supernodes and elimination trees, vertex and edge reordering, vertex and edge coloring, and bipartite graph matching, arise in these contexts. As an example, parallel partitioning tools can be used to ease the task of distributing the computational workload across the processors. The computation of such problems can be represented as a composition of graphs and multilevel graph problems that have to be mapped to different microprocessors.

The aim of this book on *Combinatorial Scientific Computing* is to address these challenges by setting the stage for accelerated development and deployment of fundamental enabling technologies in high performance scientific computing. Special focus is put on load balancing and parallelization on high-performance computers, large-scale optimization, algorithmic (also: automatic) differentiation of numerical simulation code, sparse matrix software tools, and combinatorial challenges and applications in large-scale social networks. These seemingly disparate areas are unified by a common set of abstractions and algorithms based on combinatorics, graphs, and hypergraphs that represent the main focus of this book.

The book resulted from a seminar held at Schloss Dagstuhl, Germany, in February 2009, on "Combinatorial Scientific Computing," organized by Uwe Naumann (RWTH Aachen University), Olaf Schenk (University of Basel), Horst D. Simon (Lawrence Berkeley National Laboratory), and Sivan Toledo (Tel Aviv University). The Dagstuhl seminar focused on new combinatorial models and techniques for load balancing and scheduling, on combinatorial algorithms in scientific computing applications with increasingly embracing complex geometries, adaptivity and multiscale methods, and on emerging computational applications, e.g., from computational social science, and combinatorial challenges in computational biology. The book includes a number of chapters contributed by participants of the Dagstuhl seminar, and several additional chapters were invited to span the breadth of combinatorial scientific computing applications.

Uwe Nauman, Olaf Schenk

Editors

Uwe Naumann has been an associate professor for Computer Science at RWTH Aachen University, Germany, since 2004. He received his Master's and Doctoral degrees in Mathematics from the Technical University of Dresden, Germany, in 1996 and 1999, respectively. Before coming to Aachen, he held postdoctoral research and lecturer positions at INRIA Sophia-Antipolis, France, at the University of Hertfordshire, United Kingdom, and at Argonne National Laboratory, United States.

Naumann's research interests focus on algorithmic differentiation, combinatorial graph algorithms, high-performance scientific computing, and the application of corresponding methods to real-world problems in computational science, engineering, and finance. In particular, he has been working on the automatic generation of first- and higher-order adjoint numerical simulation programs and on related discrete problems for many years. He has published more than 80 papers in peer-reviewed scientific journals and conference proceedings. His involvement in a number of derivative code compiler development projects resulted in text book titled *The Art of Differentiating Computer Programs* and published by SIAM. Moreover, he is a coeditor of the proceedings of three international workshops on Algorithmic Differentiation held in Nice, France in 2002, in Chicago, IL (USA) in 2004, and in Bonn, Germany in 2008 and published by Springer.

Naumann is a principal investigator of the Aachen Institute for Advanced Study in Computational Engineering Science (AICES)—a graduate school established in 2006 in the frame of the German Excellence Initiative—and a member of the Jülich-Aachen Research Aliance. As a member of the Numerical Algorithms Group Ltd., Oxford, United Kingdom, he is working on the integration of algorithmic differentiation into numerical algorithms. He has cochaired both the First Dagstuhl Seminar on CSC held in February 2009 and the Fifth SIAM Workshop on CSC held in Darmstadt in May 2011.

Olaf Schenk is affiliated with the Department of Mathematics and Computer Science at the University of Basel, Switzerland. He obtained a diploma degree in mathematics from the University of Karlsruhe (KIT), a PhD degree in scientific computing from ETH Zurich, and his habilitation in computer science from the University of Basel in 2010. His research interests concern algorithmic and architectural problems in the field of computational mathematics, scientific computing and high-performance computing. He has published more than 60 peer-reviewed book chapters, journal articles, and conference contributions in the area of parallel and manycore algorithms for computational science. He received a highly-competitive IBM Faculty Award on Cell Processors for Biomedical Hyperthermia Applications in 2008 and was one of the finalists of the International Itanium Award in 2009 in the area of computational intensive applications. He served on the program committee of several international supercomputing conferences and workshops such as the International Conference for High Performance Computing, Networking, Storage and Analysis (SC), the International Conference on High Performance Computing (HiPC), and the Swiss Speedup Workshop Series on High-Performance Computing, and cochaired the First Dagstuhl Seminar on CSC held in February 2009. He is an IEEE Senior Member, and a SIAM Member. He is currently the principal investigator for several research projects funded by Swiss or European research funding agencies or industrial research labs, among others from Intel and NVIDIA. As of January 2012, he will join the University of Lugano (USI) and the Swiss Center of Supercomputing (CSCS) as an associate professor for High-Performance Computing in Computational Science.

Contributors

Peter Arbenz Department of Computer Science, ETH Zürich, CAB H89, Universitätsstrasse 6, CH-8092 Zürich, Switzerland
E-mail: arbenz@inf.ethz.ch

Haim Avron Business Analytics and Mathematical Sciences, IBM T.J. Watson Research Center, P.O. Box 218, Yorktown Heights, NY 10598, USA
E-mail: haim.avron@gmail.com

David Bader College of Computing, Georgia Tech, Atlanta, GA 30332, USA
E-mail: bader@cc.gatech.edu

Costas Bekas Computational Sciences Group, IBM Research, Zurich Research Laboratory, Säumerstrasse 4, CH-8803 Rüschlikon, Switzerland
E-mail: bek@zurich.ibm.com

Rob Bisseling Department of Mathematics, Utrecht University, PO Box 80010, 3508 TA Utrecht, The Netherlands
E-mail: R.H.Bisseling@uu.nl

Erik Boman Scalable Algorithms Department, Sandia National Laboratories, Albuquerque, NM 87185-1318, USA
E-mail: egboman@sandia.gov

Ümit V. Çatalyürek Biomedical Informatics and Electrical & Computer Engineering, The Ohio State University 3172B Graves Hall, 333 W. 10th Ave, Columbus, OH 43210, USA
E-mail: catalyurek.1@osu.edu

Cédric Chevalier Scalable Algorithms Department, Sandia National Laboratories, Albuquerque, NM 87185-1318, USA
E-mail: ccheval@sandia.gov

Alessandro Curioni Computational Sciences Group, IBM Research, Zurich Research Laboratory, Säumerstrasse 4, CH-8803 Rüschlikon, Switzerland
E-mail: cur@zurich.ibm.com

Karen Devine Scalable Algorithms Department, Sandia National Laboratories, Albuquerque, NM 87185-1318, USA
E-mail: kddevin@sandia.gov

Iain Duff Rutherford Appleton Laboratory, Chilton, Didcot, Oxfordshire, OX11 0QX, United Kingdom
E-mail: iain.duff@stfc.ac.uk

Bas Fagginger Auer Department of Mathematics, Utrecht University, PO Box 80010, 3508 TA Utrecht, The Netherlands
E-mail: B.O.FaggingerAuer@uu.nl

Cyril Flaig Department of Computer Science, ETH Zürich, CAB H89, Universitätsstrasse 6, CH-8092 Zürich, Switzerland
E-mail: cflaig@inf.ethz.ch

Klaus Gärtner Numerical Mathematics and Scientific Computing, Weierstrass Institute for Applied Analysis and Stochastics, Mohrenstrasse 39, 10117 Berlin, Germany
E-mail: Klaus.Gaertner@wias-berlin.de

Andreas Griewank Humboldt-Universität zu Berlin, Institut für Mathematik, Unter den Linden 6 D-10099 Berlin, Germany
E-mail: griewank@mathematik.hu-berlin.de

Bruce Hendrickson Discrete Algorithms & Math Department, Sandia National Laboratories, Albuquerque, NM 87185-1318, USA
E-mail: bahendr@sandia.gov

Yifan Hu AT&T Labs Research, 180 Park Avenue, Florham Park, New Jersey, NJ 07932, USA
E-mail: yifanhu@research.att.com

Johannes Huber Department of Mathematics and Computer Science, Universität Basel, Rheinsprung 21, CH-4051 Basel, Switzerland
E-mail: johannes.huber@unibas.ch

Rainald Löhner Department of Computational and Data Sciences, College of Science, George Mason University, 4400 University Dr., MSN 6A2, Fairfax, VA 22030, USA
E-mail: rlohner@gmu.edu

Kamesh Madduri Computational Research Division, Lawrence Berkeley National Laboratory, 1 Cyclotron Rd, Berkeley, CA 94720, USA
E-mail: kmadduri@lbl.gov

Michael Mahoney Computer Science Department, 353 Serra Mall, Stanford University, Stanford, CA 94305-9025, USA
E-mail: mmahoney@cs.stanford.edu

Ralph Müller Institute for Biomechanics, ETH Zürich, HCI E 355.1, 8093 Zürich, Switzerland
E-mail: ram@ethz.ch

Uwe Naumann LuFG Informatik 12: Software and Tools for Computational Engineering, RWTH Aachen University, D-52062 Aachen, Germany
E-mail: naumann@stce.rwth-aachen.de

François Pellegrini Institut Polytechnique de Bordeaux, LaBRI & INRIA Bordeaux – Sud-Ouest, 351, cours de la Libération F-33405 Talence cedex, France
E-mail: francois.pellegrini@labri.fr

Alex Pothen Department of Computer Science, Purdue University, 305 N. University Street, West Lafayette, IN 47907-2107, USA
E-mail: apothen@purdue.edu

Alexander Rand Institute for Computational Engineering and Sciences, University of Texas-Austin, 1 University Station C0200, Austin, TX 78712, USA
E-mail: arand@ices.utexas.edu

Ahmed Sameh Department of Computer Science, Purdue University, 305 N. University Street, West Lafayette, IN 47907-2107, USA
E-mail: sameh@cs.purdue.edu

Madan Sathe Department of Mathematics and Computer Science, Klingelbergstrasse 50, CH-4056 Basel, Switzerland
E-mail: Madan.Sathe@unibas.ch

Olaf Schenk Department of Mathematics and Computer Science, Klingelbergstrasse 50, CH-4056 Basel, Switzerland
E-mail: Olaf.Schenk@unibas.ch

Jonathan Shewchuk Computer Science Division, University of California at Berkeley, Berkeley, CA 94720-1776, USA
E-mail: jrs@cs.berkeley.edu

Hang Si Numerical Mathematics and Scientific Computing, Weierstrass Institute for Applied Analysis and Stochastics, Mohrenstrasse 39, 10117 Berlin, Germany
E-mail: si@wias-berlin.de

Daniel Spielman Department of Computer Science, Yale University, PO Box 208285, New Haven, CT 06520-8285, USA
E-mail: spielman@cs.yale.edu

Sivan Toledo School of Computer Science, Tel-Aviv University, Tel-Aviv 69978, Israel
E-mail: stoledo@tau.ac.il

Bora Uçar Centre National de la Recherche Scientifique and Laboratoire de l'Informatique du Parallélisme, (UMR CNRS-ENS Lyon-INRIA-UCBL), Université de Lyon, 46, Allée d'Italie, ENS Lyon, F-69364, Lyon, France
E-mail: bora.ucar@ens-lyon.fr

Jean Utke Argonne National Laboratory, MCS Bldg. 240/1152, 9700 South Cass Avenue, Argonne, IL 60439-4844, USA
E-mail: utke@mcs.anl.gov

Tristan van Leeuwen Faculty of Civil Engineering and Geosciences, Department of Geotechnology, Stevinweg 1, 2628 CN Delft, The Netherlands
E-mail: T.vanLeeuwen@tudelft.nl

Gerrit Hendrik van Lenthe Institute for Biomechanics, ETH Zürich, HCI E 355.1, 8093 Zürich, Switzerland
E-mail: vanlenthe@ethz.ch

Ebadollah Varnik LuFG Informatik 12: Software and Tools for Computational Engineering, RWTH Aachen University, D-52062 Aachen, Germany
E-mail: varnik@stce.rwth-aachen.de

Andreas Wächter Business Analytics and Mathematical Sciences, IBM T.J. Watson Research Center, P.O. Box 218, Yorktown Heights, NY 10598, USA
E-mail: andreasw@us.ibm.com

Noel Walkington Department of Mathematical Sciences, Carnegie Mellon University, 5000 Forbes Ave, Pittsburgh, PA 15213, USA
E-mail: noelw@cmu.edu

Andrea Walther Institut für Mathematik, Universität Paderborn, 33095 Paderborn, Germany
E-mail: andrea.walther@uni-paderborn.de

Andreas Wirth Institute for Biomechanics, ETH Zürich, HCI E 355.1, 8093 Zürich, Switzerland
E-mail: awi@ethz.ch

Albert-Jan Yzelman Department of Mathematics, Utrecht University, PO Box 80010, 3508 TA Utrecht, The Netherlands
E-mail: A.N.Yzelman@uu.nl

Chapter 1

Combinatorial Scientific Computing: Past Successes, Current Opportunities, Future Challenges

Bruce Hendrickson

Sandia National Labs

Alex Pothen

Purdue University

1.1 Introduction

Combinatorial techniques have become essential tools across the landscape of computational science, and some of the combinatorial ideas undergirding these tools are discussed in the following eighteen chapters in this book. The history of what we now call Combinatorial Scientific Computing (CSC) has involved the steady accretion of new domains where combinatorial theory and algorithms have been used to enable computational science and engineering. The 1970s and 1980s witnessed the flourishing of graph algorithms in sparse direct methods. The 1990s saw the growth of combinatorial algorithms as key enablers of parallel computing. In recent years, graph algorithms have become essential to automatic differentiation, and play a pivotal role in computational biology. Discrete algorithms also play important roles in mesh generation, computational chemistry, performance optimization, and many other fields.

In this chapter, we provide a brief overview of the historical roots of combinatorial scientific computing, the formation of the CSC community in the last decade, and provide our (of course, limited) perspective on likely future directions for CSC. We had provided a longer exposition on our view of Combinatorial Scientific Computing a few years ago [1]. It is our belief that the coming years will see still more applications of combinatorial scientific computing. Computational science remains a young, and rapidly changing discipline. As it evolves, new challenges will arise which discrete algorithms will be able to help solve. Our predictions will certainly be incomplete, and probably wrong—certainly many past trends would have been difficult to foresee. But we hope that by generating curiosity and interest in new CSC applications, our predictions become self-fulfilling.

We also believe that new applications will enrich the CSC community, leading to new combinatorial problems, abstractions, algorithms, and analysis. As has been the case in the past, we expect this work to have broader impact on the combinatorial theory and algorithms communities.

Since CSC is so pervasive within computational science, we believe that the best way to anticipate the future of CSC is to explore general trends within scientific computing. This is the approach we will take in this paper.

1.2 The CSC Community

1.2.1 The Roots of the CSC Community

Sparse Matrix Computations. One root of the CSC community comes from research in sparse matrix computations. Harry M. Markowitz, the Economics Nobel laureate in 1990, writes about working with sparse matrices in modeling industrial capabilities using linear programming at the RAND Corporation in the 1950s [2].

> Our models strained the computer capabilities of the day. I observed that most of the coefficients in our matrices were zero; i.e., the nonzeros were "sparse" in the matrix, and that typically the triangular matrices associated with the forward and back solution provided by Gaussian elimination would remain sparse if pivot ele-

ments were chosen with care. William Orchard-Hayes programmed the first sparse matrix code. ... Sparse matrix techniques are now standard in large linear programming codes.

The *Markowitz scheme* [3] for pivot selection in factoring sparse unsymmetric matrices is still in use today, but research in sparse matrix computations grew gradually in the 1960s before blossoming in the 1970s. Seymour Parter [4] introduced a graph model for Gaussian elimination on sparse matrices in 1961. A number of early sparse matrix conferences were held at I.B.M. from 1969 under the leadership of Ralph Willoughby [5, 6], and a conference was held at Oxford in 1971 [7]. Collections of articles presented at two of these meetings were published as books, and these were followed by further meetings and book collections. Rose [8], in a PhD thesis written at Harvard, showed that the adjacency graph of the Cholesky factor obtained by Gaussian elimination of a sparse symmetric matrix was a chordal graph. George's thesis at Stanford (1971) and Duff's thesis at Oxford (1972) were early contributions to sparse matrix algorithms and their implementations. Duff's comprehensive 1977 survey on sparse matrix research was influential on early researchers in the field. The text book by George and Liu [9], the Waterloo Sparspak, sparse matrix routines in the Harwell library (HSL), and the Yale Sparse Matrix Package (YSMP) enabled the widespread use of sparse matrix computations in many areas of computational science and engineering. The book, *Graph Theory and Sparse Matrix Computation* (edited by George, Gilbert, and Liu), which resulted from a special year on linear algebra held at the Institute for Mathematics and its Applications (IMA) at the University of Minnesota in 1991, provides a snapshot of research current at that time [10]. Much more detail is available in the slides that accompanied Iain Duff's talk on the Development and History of Sparse Direct Methods at the SIAM Conference on Applied Linear Algebra in 2009.[1]

Derivative Computation. Another root of the CSC community comes in the context of derivative computation in optimization. Bauer [11] showed in 1974 how a computational graph could be associated with the evaluation of a mathematical function in terms of intrinsic mathematical operations such as addition, multiplication, and elementary mathematical functions (trigonometric functions, transcendentals, etc.), with vertices corresponding to independent, intermediate, and dependent variables, and with an edge joining an operand of an intrinsic operation to its result. This computational graph is now used in Automatic Differentiation (AD) to compute the derivative of a function specified by a program by applying the chain rule of differentiation along paths between dependent variables and independent variables [12].

Yet another early use of combinatorial methods arose in computing Jacobians and Hessians efficiently with the fewest function evaluations (when finite differencing is used), or with the fewest passes through the computational graph (when AD is used). The essential idea is that of structural or-

---

[1]Slides are available at www.siam.org/meetings/la09/talks/duff.pdf

thogonality, i.e., when two columns of the Jacobian have nonzeros in disjoint row indices, the nonzeros in both columns can be evaluated simultaneously by differencing along both column directions. Furthermore, this observation can be extended to a subset of columns with the property that for each row, at most one of the columns in the subset has a nonzero in it. This observation was used by Curtis, Powell, and Reid in 1974 to compute sparse Jacobians; it was then extended to Hessians, where symmetry could be exploited in addition to sparsity, by Powell and Toint. Coleman and Moré [13] observed that the problem of minimizing the number of function evaluations for the Jacobian could be modeled as a vertex coloring problem on a derived graph, thus establishing that the minimization problem was NP-hard, and proposed a greedy heuristic approach with a number of vertex ordering algorithms. Coleman and various coauthors also extended the graph model to Hessian matrices, to relaxed versions of structural orthogonality (leading to a tree-based computation of the nonzeros beginning with the leaves in the tree, and proceeding iteratively after removing the leaves computed), and bidirectional evaluations of Jacobians where both columns and rows are used. Griwank and Reese [14] identified a vertex elimination problem in computing a Jacobian or Hessian with the least number of operations by transforming the computational graph by eliminating intermediate variables. Variants of this problem play an important role in reducing the number of operations and storage needed to compute derivative matrices.

Statistical Physics. Since the 1930s, statistical physicists have been using combinatorial models as simplified representations of complex physical systems (see [15] for a survey). The paradigm for this work is the Ising spin model, which attempts to represent the salient features of ferromagnetism. In the Ising model, atoms are assumed to be located at the points of a two- or three-dimensional lattice, and they only interact with atoms at neighboring lattice points. Each atom has a spin, and the energy is a function of the spins of neighboring atoms. The goal is to understand the characteristics of low energy configurations of spins, subject to various external and boundary conditions. This seemingly simple problem leads to difficult combinatorial optimization problems with rich graph properties. Key combinatorial kernels include counting the number of perfect matchings in a graph, and their connection to matrix Pfaffians. Several Nobel Prizes in physics have been awarded for work that has deep combinatorial underpinnings.

The success of the Ising model in describing the essence of some magnetic phenomena led to the development of a plethora of combinatorial models for other problems. These include protein folding, quantum systems, material science, and orbital mechanics. The statistical physics community is often interested in the asymptotic statistics of optimal solutions to their combinatorial models. This is somewhat different from the traditional focus in CSC on algorithmic performance to enable scientific modeling and simulation. But the two communities are closely aligned in spirit and in interests, and we foresee significant opportunities for beneficial interactions in the future.

Parallel Computing. Partial differential equations model many physical phenomena in the sciences and engineering, and their solutions are often computed through discretizations on finite difference or finite element meshes. Complex geometries and efficiencies in computation often require the use of unstructured or block-structured meshes. The computational intensity of these simulations necessitates the use of parallel computing. To make effective use of a parallel computer, the problem must be decomposed into subtasks, the subtasks must be assigned to processors, and the resulting communication and computation must be carefully orchestrated. The goals of these steps are to ensure that the work is equally distributed among the processors while keeping the communication costs low, so as to ensure that the parallel computation completes in the least possible time. Algorithms for solving this load-balancing problem use graph models of the computation to compute good partitions and mappings.

Partitioning methods for parallel computing benefited from earlier work on graph partitioning for sparse direct methods. Graph partitioning provides a mechanism for ordering sparse matrices through a divide and conquer paradigm called nested dissection, pioneered by Alan George [16] in 1973. George's work was generalized to sparse and irregular graphs by Lipton, Rose and Tarjan [17], and led to theory and algorithms for finding vertex and edge separators that divide a graph into two subgraphs of with balanced sizes. By recursive application of graph partitioning, the load balancing problem in parallel computation and the problem of computing fill-reducing orderings for sparse matrices could be solved. This work led to the development of spectral graph partitioning algorithms [18] as well as multi-level graph partitioning [19] in the 1990's.

1.2.2   Organization of the CSC Community

Although the many subcommunities mentioned above were employing combinatorial mathematics, algorithms and software in the context of scientific computing, they remained isolated from each other because they contributed to different fields in scientific computing and computational science and engineering, attending different conferences. By 2000, a group of thirty international researchers began efforts to formally organize a community. An email discussion in 2002 settled on the name *Combinatorial Scientific Computing* (CSC) to describe the research undertaken by members of this community. An email subscription server was created for the CSC community[2], and a number of minisymposia on CSC were organized at SIAM and ICIAM conferences during this period.

Five CSC workshops have been organized thus far, beginning with a SIAM workshop on CSC in February 2004, colocated with SIAM Conference on Parallel Processing, in San Francisco, CA. The organizers of this meeting were

---

[2]Subscribe at `https://lists.purdue.edu/mailman/listinfo/csc`

John Gilbert, Horst Simon, Sivan Toledo, and the authors. A special volume of the Electronic Transactions on Numerical Analysis (ETNA) consisting primarily of papers presented at CSC04 was published in 2005 (available as Volume 21 at `www.emis.de/journals/ETNA`). The second CSC Workshop was held in Europe, at CERFACS, Toulouse in 2005, with Iain Duff and Alex Pothen serving as cochairs of the meeting. The following two CSC workshops were organized as SIAM Workshops in 2007 (colocated with SIAM Conference on Computational Science and Engineering at Costa Mesa, CA) and 2009 (colocated with SIAM Conference on Applied Linear Algebra at Seaside, CA). Erik Boman, Sivan Toledo, and the authors served as cochairs of these workshops. The Dagstuhl Seminar on CSC, which served as the stimulus for the papers in this volume, was organized in January 2009. CSC11, the most recent SIAM Workshop on CSC, was held in May 2011 in Darmstadt, Germany, colocated with the SIAM Conference on Optimization, and Uwe Naumann and Ali Pınar were the cochairs of the workshop.

Currently, Paul Hovland, Rob Bisscling, and the authors serve as members of the Steering committee for the CSC community. One measure that indicates that CSC is thriving as a community is that the the responsibility of organizing the biennial SIAM CSC Workshops has been passed from the original group of organizers to the "next generation" in the CSC community.

1.3   Current Opportunities

From the disparate foundations discussed above, combinatorial scientific computing has emerged to be recognized as a key enabler for a wide array of scientific areas. The cross-fertilization of disciplines that has been encouraged by the formation of the CSC community has accelerated progress in both algorithm development and applications. In this section, we briefly sketch some of these areas of current activity. Many are discussed in more detail in the following eighteen chapters in this book, but several are not.

The research areas represented by the chapters in this book give an approximate snap shot of current research in the CSC community. Seven chapters, by far the largest number, describe sparse matrix computations for solving linear systems of equations and computing preconditioners; four chapters discuss combinatorial aspects of Algorithmic (or Automatic) Differentiation; two chapters discuss mesh generation and geometric algorithms; one chapter describes tools for graph visualization; spectral graph theory is the topic of one chapter; combinatorial problems from computational fluid dynamics is the subject of another; and two chapters discuss emerging application areas such as large scale data analysis, social networks, and computational genomics and phylogenetics.

The diversity of applications of CSC can be daunting, but this seeming complexity masks significant underlying commonality. Several algorithmic themes are at the core of CSC, and recur throughout the chapters of this book. Graph and hypergraph algorithms are central to many (perhaps most) CSC activities. Geometric algorithms and algebraic methods on graphs also appear prominently. CSC researchers span a range of activities including modeling, algorithm development, theoretical analysis, application and empirical performance analysis. Parallel and architecturally-informed algorithms also play a prominent role.

One area of high CSC activity and impact is the development and deployment of techniques to support parallel computing. Since the early work of Simon [20], graph partitioning has been used as a tool for dividing a parallel computation among processors to balance the work while minimizing the communication. Sophisticated multilevel heuristics have been developed for this problem, which have subsequently been adapted for a range of other combinatorial optimization problems. The underlying graph model has been enhanced in a variety of ways, most notably by the generalization to hypergraphs by Çatalyürek and Aykanat [21]. These models are surveyed by Hendrickson and Kolda [22]. More detail on parallel computing and partitioning can be found among the six chapters that discuss these topics in this book.

A second recurring combinatorial kernel in parallel computing is graph coloring. Consider a graph in which vertices represent entities that require updating and edges indicate conflicts that preclude simultaneous updating of the adjacent vertices. One example of such a problem arises in mesh refinement [23]. In this setting, it is useful to identify a maximal subset of vertices, no two of which are adjacent to each other. All of the vertices in this subset can be updated in parallel, with no concern about conflicts. More generally, one could color all the vertices in such a way that no pair of adjacent vertices has the same color. The fewer colors required, the fewer the number of phases necessary to update all the entities.

As has been discussed earlier, graph coloring also arises in optimization as a tool for exploiting sparsity structure in the computation of derivative matrices. Several variations of graph coloring, including distance-k vertex coloring (where a vertex must receive a color distinct from every neighbor within a distance k from it), star coloring, acyclic coloring, partial coloring (where only a subset of the vertices need to be colored), edge coloring, and hypergraph coloring, arise as graph models of derivative matrix computation problems. Efficient heuristic algorithms have been developed for these problems, and these algorithms often yield nearly optimal colorings for many classes of graphs and real-life instances of irregular graphs, since they are close to lower bounds which can be computed for the problem. A comprehensive survey of coloring and the related problems of vertex ordering and matrix recovery for derivative computation has been provided in [24].

As sketched in Section 1.2, the CSC community has deep roots in sparse direct methods. Graph algorithms are central to the exploitation of sparse structure in the factorization of various classes of matrices, and this continues to be an active area of research as detailed elsewhere in this book. Iterative methods for solving linear systems also rely upon combinatorial algorithms in various ways. Incomplete factorization preconditioners use graph algorithms to reorder the matrix. Algebraic multigrid approaches employ graph matchings and colorings in the construction of coarse representations. Support theory preconditioners make extensive use of graph embeddings in the construction and analysis of preconditioners. Six chapters in this book discuss sparse matrix computations associated with solving linear systems of equations, and an additional chapter discusses combinatorial preconditioners in greater depth.

A key step in many computational problems is the construction of a discrete representation of the geometry of the system being modeled. The mesh might represent the interior of a combustion chamber for a fluid simulation, the components of an automobile for virtual crash tests, or a tokamak for fusion modeling. The mesh must decompose the overall geometry into a set of well-shaped polyhedra (usually tetrahedra or octahedra). Geometric and combinatorial algorithms are central to this task. More detail can be found in two chapters that discuss mesh generation and geometric algorithms.

Combinatorial algorithms also play a prominent role in scientific applications that are not covered in much depth in this book. One of these is modern biology. String algorithms are central to the assembly of genomic data from shotgun sequencing methods, to the analysis of genomic data, and to searching gene and protein databases for close matches to a new sequence [25]. Dynamic programming on the entire sequences is forbiddingly expensive, and clever heuristic algorithms to identify promising substrings to perform exact dynamic programming on are the basis of search techniques such as BLAST, FASTA, PatternHunter, and others. Suffix trees and suffix arrays are critical for solving genome wide assembly problems as well as comparative studies in metagenomics [26]. Phylogenetics, the construction of evolutionary trees from sequence data, relies on multiple sequence alignments and rich combinatorics to reduce the search space of potential trees. Population genomics includes the reconstruction of genotypes from haplotype data (the number of variations in a specific allele in a pair of chromosomes, which can be 0, 1, or 2), and leads to problems in chordal graphs, as discussed in Section 1.4.4. Biological networks such as gene regulatory networks, protein interaction networks, and metabolic networks lead to a collection of graph problems. Random graph models for generating networks help in understanding the evolutionary relationships of these networks from different organisms. Discovering the modular organization (the decomposition of the network into topological clusters and the overlaps among the clusters) of these networks aids in understanding the biology of the processes of life and help in systematically identifying drug targets.

Another important CSC application area that is not covered in detail in this book is chemistry. The word *graph* as used in CSC was first coined by Cayley in his study of molecular structures, and graph concepts continue to be important in thinking about molecules and materials. The graphs of molecules are used today to characterize potential drug designs. Graph and geometric algorithms are also used in material science to describe complex chemical bond or proximity structure in granular and other complex materials [27, 28, 29].

1.4 Future Challenges

1.4.1 Trends in High Performance Architectures

For the past two decades, computer performance has improved steadily due to faster clock speeds. But power consumption and heat generation are proportional to clock speeds, and these factors are constraining further speedups. Instead, future performance advances will have to come from the increasing use of on-chip parallelism. Current generations of processors are multicore, and the number of cores on a chip is expected to grow quickly as feature sizes continue to shrink exponentially. Effective use of current and future generations of on-chip multiprocessors will require algorithmic ideas that are closely informed by architectural constraints. The CSC community can play an important role in algorithmic design and implementation, and also in impacting the development of future architectures.

Several different designs of multicore processors are competing for market share right now, so it is difficult to foresee the details of future market leaders. But it is likely that future processors will contain multiple types of cores. These heterogeneous multiprocessors will require careful decomposition of the work amongst the different core types to optimize the utilization of each resource. The resulting scheduling and resource allocation problems will benefit from new load balancing models. The parallel computing community has largely avoided heterogeneity in the past, but will not be able to do so in the future.

Communication between cores on a chip is much more efficient than off-chip communication, and so algorithmic choices that are appropriate for our traditional cluster environments may no longer be optimal. For instance, shared-memory algorithms will be more viable on-chip. Shared variables can be used to store bounds in a branch-and-bound computation, or to facilitate partner selection in a matching algorithm. Lower-latency, more finely grained communication will support more fine-grained and asynchronous parallel algorithms.

Multicore processors are likely to provide some multithreading capability to tolerate memory latencies. Work on the Cray XMT [30] has showcased the potential for latency tolerant computers to run graph algorithms with high efficiency while using very different notions of parallelism than traditional message passing codes. The development of combinatorial algorithms that perform well on future processors will be a rich area of research problems.

These multicore processors will be the nodes of our future large parallel machines. For optimal performance this will likely require applications to exploit multiple layers of parallelism—message-passing between nodes but some other model within a node. The extreme scale machines of the future will have hundreds of thousands or millions of cores. To use such machines effectively we will need to improve our models and methods for load balancing, and devise algorithms that require fewer synchronizations. The cost of global collective communication operations on so many cores will penalize synchronous applications. In addition, slight deviations in core performance due to operating system jitter can induce imbalance that will hinder the performance of highly synchronous algorithms. The cost of global collective communication operations will also penalize synchronous applications. Thus, there will be a need for greater asynchrony in key scientific computing algorithms. With so many components, failures will be common occurrences, so fault tolerant and resilient algorithms will be important. In all of these areas, the CSC community will play an important role.

A critical issue with future leadership-class supercomputers will be power consumption [31, 32]. Current petascale machines consume multiple megawatts of power, and deliver fewer than 300 Mflop/sec/Watt. Clearly, dramatic improvements in efficiency are required for exascale computing. Current technology roadmaps indicate that the power consumption in large machines of the future will be dominated by data movement, not by computation. Thus, there will be a need for algorithms that use memory hierarchies efficiently to reduce the overall movement and power consumption. This area of algorithmic analysis is currently immature—the standard measure of algorithmic complexity is based upon computational operations, not data movement. But the theoretical computer science community has generated interesting models for out-of-core algorithms, or cache-oblivious algorithms which provide new ways of thinking about memory hierarchies [33]. As it has always done, the CSC community can serve as a bridge between theoretical computer science and computational applications and bring some of these ideas to the service of scientific computing.

A very different architectural trend is evident in the emergence of cloud computing—highly distributed machines providing computing as a service. It is too early to assess the likely impact of clouds on scientific computing, but their growing importance in business and data analytics suggests that they will be an important feature of the future computing ecosystem. CSC researchers can contribute to improving the efficiency of clouds and also to studying the utility of such platforms for scientific applications.

1.4.2 Trends in Traditional Applications

Computational modeling and simulation are widely recognized as essential components of the scientific process. Computation allows for the study of phenomena that would be dangerous, expensive, or even impossible to study experimentally. It also allows for the exploration of detailed and complex phenomena that are beyond the reach of theory. The forefront of computational science continues to advance via new mathematical models, algorithms and computer architectures.

As they grow in fidelity and sophistication, state-of-the-art simulations increasingly involve complex geometries, multiple temporal and spatial scales, and multiple physical phenomena. These capabilities rely on unstructured and adaptive meshes, advanced linear and non-linear solvers, and complex coupling between models within the simulation. Design optimization and uncertainty quantification are playing a growing role above and beyond the traditional phenomenological modeling.

Combinatorial scientific computing techniques are essential enablers of all this sophistication. Unstructured mesh generation employs sophisticated geometric and graph theoretic algorithms. Numerical solvers rely upon graph algorithms for reordering and multigrid coarsening. Multiscale and multiphysics problems make use of advanced load balancing techniques that are built upon CSC technologies. Optimization and uncertainty quantification use graph algorithms to exploit structure in computing derivatives.

As discussed in a subsequent chapter derivative computation is of growing importance in computational science for design, optimization and uncertainty quantification. Automatic differentiation (AD) relies on combinatorial models to represent transformations of the computational graph representation of the computation of a mathematical function, through vertex, edge, or face elimination, to compute the derivative with the least amount of work or storage. In the adjoint or reverse mode of AD (where the derivatives are computed by applying the chain rule from the dependent to independent variables), which is often more efficient than the forward mode (computation from the independent to dependent variables), one challenge is to reduce the large storage required by placing checkpoints at chosen steps during the computation, and doing forward computations from them. Choosing the number and placement of these checkpoints to reduce the storage required while keeping the additional computations entailed by the checkpoints is an interesting combinatorial problem. When the functions involved recursively call functions within themselves, the problem becomes richer still.

All of these CSC underpinnings are essential to scientific computing and will become even more so as scientific computing continues to progress. There will undoubtedly be new nuances and variants of existing problems that will require new ideas within existing areas of CSC. But of even greater interest to this current paper are broader changes in the relationship between science and computation. We discuss several of these in the subsequent sections.

1.4.3 Emerging Applications

1.4.3.1 Data-Centric Scientific Computing

High-throughput experimental facilities are transforming many areas of science including biology, astronomy, earth science, and high energy physics [34]. All of these fields are struggling to make optimal use of a flood of new data. In biology, novel experimental methodologies are generating data that is qualitatively different from before, e.g., sequence data, gene and protein expression data from tissue micro-arrays, functional magnetic resonance imaging, multichannel flow cytometry and more. In other fields, the data are qualitatively similar to what came before, but orders of magnitude more voluminous. New models and abstractions are needed to address this growth in data quantity and diversity. And new computing approaches will be required to extract maximal scientific insight from the available data.

Extracting useful insight from this data is a preeminent scientific challenge. The internet community has shown how the combination of vision, clever algorithms, and diverse data can make data tremendously valuable. It is a certainty that large scientific data sets will enable answers to scientific questions that have not yet been posed.

Advanced techniques for exploring large data sets will require computational tools that are quite different from those that used in traditional scientific computing applications. Mature capabilities already exist for fundamental operations like indexing, range queries, and searching. More advanced analytics will be application-specific, and will involve various data abstractions. Graphs are increasingly popular as a tool for describing entities and their relationships, so graph algorithms will likely be an important component of data-centric scientific computing (see Section 1.4.3.2). New geometric algorithms will likely play a role in the analysis of complex spatial data in geoscience and other applications.

1.4.3.2 Computing in the Social Sciences

An emerging area of opportunity for CSC is the study of networks that arise in natural or human systems. The connectivity structure of links in the world wide web is central to search ranking algorithms. Communication links between people are reflected in blog postings, email traffic, instant messaging, and other social media. These large, complex networks are of keen interest in the social sciences, but techniques of extracting insight from them remain immature. Two recent books that discuss network science from this perspective have been written by Newman [35] and Easley and Kleinberg [36].

As happened recently with biology, the availability of large data sets is beginning to transform social sciences from qualitative to quantitative disciplines. Not long ago, the study of social interactions required direct observation, interviews and surveys. While richly informative for social modeling, public health, and other applications, these labor-intensive tasks limited the scale of the the communities being studied. But these days a well-connected social scientist can make use of the huge volume of data generated by online social networks, chat rooms, emails, instant messages, and more. In a recent study, Leskovec and Horvitz studied data from a month of Microsoft's instant-messaging traffic. The data encompassed 240 million people and 30 billion interactions [37]. Tellingly, Leskovec and Horvitz generated some summary statistics of the data, but did no real analysis. In another recent study, Michel et al. studied millions of books digitized by Google to analyze changes in language usage and various social trends [38]. As social science on this scale is in its infancy, no one knows what questions to ask, let alone how best to answer them.

Internet data is not only much larger than traditional social science data, but it is qualitatively different. Internet data is less highly curated, and largely uncontrolled. It comes from natural, unconstrained human interaction which is attractive, but also greatly complicates its analysis. New graph algorithms and implementations will be needed to address emerging questions in social science. The structure of a social network is quite different from that of a finite element mesh, so existing parallel algorithms may not be appropriate. As with other areas of computational science, the future will emerge from the intersection of what is scientifically interesting and what is computationally feasible.

1.4.4　Biological Applications

Combinatorial techniques have played an essential role in the development of modern biology, with strings, networks, and trees playing prominent roles. Since these topics are not discussed in detail elsewhere in this book, in this subsection we provide a brief introduction to several important combinatorial problems in biology.

1.4.4.1　Population Genomics

A phylogenetic tree represents the evolutionary history of a set of biological sequences (e.g., proteins) from different organisms, individuals from the same species, etc. Given a matrix of sequences, with columns corresponding to characters, and rows to specific organisms or individuals, the state of a character for an individual is the value of the corresponding matrix element. The problem is to construct an unrooted tree with the given set of sequences at the leaves, generating new sequences at the internal nodes such that the subtree induced by every character-state pair is a subtree (and hence connected). Such a tree has the virtue of not displaying homoplasy, i.e., a state change in a character occurring at more than one place in the tree. The perfect phylogeny problem is to determine if a tree with these properties exists, and to construct such a tree if there is one.

When a fixed number k (with $k > 2$) states are permissible for a character, the perfect phylogeny problem can be posed as a chordal graph completion problem on a k-partite graph. Each character in the sequence matrix corresponds to a vertex part in the k-partite graph, and each character-state pair corresponds to a vertex in this sequence intersection graph. Each sequence is represented as a clique consisting of exactly one vertex from each vertex part.

A legal edge in this graph is an edge that joins two vertices from different vertex parts, whereas an illegal edge joins two vertices in the same vertex part. Buneman [39] proved in 1974 that a perfect phylogeny tree exists if and only if the sequence intersection graph can be made a chordal graph by adding only legal edges.

Recently Gusfield [40] has considered the perfect phylogeny problem with missing data, where the goal is to determine if missing entries can be completed in the sequence matrix so as to obtain a perfect phylogeny. The solution involves computing minimal separators in the sequence intersection graph, and employs a clique tree representation of a chordal completion.

Several genome-scale phylogeny problems remain to be formulated and solved in this area of population genetics with important consequences for understanding genetic similarities and differences in human populations (the International HapMap project). Since researchers in sparse matrix algorithms have made extensive use of chordal graph theory, clique tree representations of chordal graphs, and computing minimal separators in non-chordal graphs [41, 42], this is an example of an area of research in biology where the CSC community might be able to make significant contributions.

1.4.4.2 Computational Systems Biology

Systems biology studies how large sets of genes and proteins within a cell interact with each other to accomplish the manifold functions of the cell. High-throughput large-scale experimental methodologies have been developed to characterize the interactions, and graphs are used to represent and study them. Examples are protein-protein interaction networks and gene regulatory networks, which are known for humans and model organisms such as yeast and the worm *C. elegans*. The graphs occurring in the biological context have modified power-law degree distributions, vertices have small average distances in the graph,

Topology based graph clustering methods have been used to understand the biological role or function of newly discovered genes and proteins from the roles of proteins or genes in their cluster. The frequency with which motifs, which are subgraphs with specific structures, usually of small size due to the computational complexity, occur in a large graph have been investigated through search algorithms. Network alignment is the problem of aligning several networks of a type in a group of related organisms to study how the networks could have evolved. Currently most of these networks are studied as static objects due to the nature of the data that is gathered. How the networks change due to changing conditions within the cell, or due to disease progression, requires the graphs to be dynamic, with edges or vertices appearing or being deleted, or changing their weights. Furthermore, kinetics of their interactions could be described by differential equation models. Most biological systems that are known to such detail are small-scale networks, but as these data become available, more sophisticated combinatorial and algebraic methods would be needed to study them.

Comparing biological networks across organisms leads to the study of graph alignment problems, where the goal is to identify how a subgraph in one network is present, with modifications, in another network. These lead to integer programming formulations, and often require the use of matching algorithms to solve them. Much remains to be done to develop effective algorithms for large network alignment problems.

1.4.4.3   Next Generation Sequencing

As of February 2011, the genomes of more than 2700 individuals have been sequenced, and tens of thousands of others are in the process of being sequenced. The discovery that our specific genetic makeup could lead to different outcomes to a drug necessitates the ability to sequence individual genomes at low costs, and the era of personalized medicine. New sequencing methodologies (next generation sequencing or NGS) that lower costs are being actively developed, and these rely on shotgun sequencing where the genomes are cut into small fragments, which are sequenced, and then assembled based on overlaps and known distances between the fragments. The assembly problems are more massive in the NGS methods, since the fragments are smaller, and the coverage of the genome by fragments is larger, to ensure that the sequences can be assembled correctly. Many new algorithms remain to be discovered and implemented efficiently to solve this challenging problem. Furthermore, new sequencing technologies such as single cell sequencing will soon be available, resulting in new algorithmic problems to be solved.

This is a one instance of a recurring theme in bioinformatics and computational biology. A large number of novel, high-throughput experimental methodologies are being developed that result in massive amounts of multidimensional, error-plagued data that need to be analyzed to discover features of interest to life scientists. An example of downstream analysis of high-dimensional flow cytometry data from leukemias analyzed with techniques from graph matching is in [43]. Often, measurements are taken from the samples at multiple times in varying conditions, and one needs to register features in a sample from one time interval to the next. Researchers in CSC working on these problems need to understand the experimental context well, and need to develop algorithms that can effectively deal with the errors in the data.

1.5   Conclusions

From a diverse set of antecedents, combinatorial scientific computing has emerged as a recognized and essential component of a broad range of scientific disciplines. The range of problems and applications is growing rapidly, and we anticipate further growth in the future due to several trends in science and technology.

First, as parallelism becomes a ubiquitous aspect of all computing platforms, the strong role that CSC has played in facilitating parallelism will impact many more computations. And the unrelenting increase in complexity of leadership class parallel machines will also demand new combinatorial models and techniques for load balancing and scheduling. Second, as traditional scientific computing applications grow in sophistication, they are increasingly embracing complex geometries, adaptivity and multiscale methods. All of these changes require sophisticated combinatorial algorithms to manage mesh generation, and to get optimal performance out of memory hierarchies. Third, emerging computational applications are rich in combinatorial opportunities. Among these applications are data-centric science, computational social science, and new challenges in biology.

For all these reasons, we believe that the breadth and depth of activity in combinatorial scientific computing will continue to grow briskly.

Acknowledgments

Bruce Hendrickson's work was performed at Sandia National Labs, a multiprogram laboratory operated by Sandia Corporation, a Lockheed-Martin Company, for the U.S. DOE undercontract number DE-AC-94AL85000. Alex Pothen's work was supported by the grant DE-FC02-08ER25864 for the CSCAPES Institute, funded under the Scientific Discovery through Advanced Computing program of the U.S. Department of Energy, and by the U.S. National Science Foundation through grant CCF-0830645.

Bibliography

[1] Hendrickson, B. A. and Pothen, A., "Combinatorial Scientific Computing: The enabling power of discrete algorithms in computational science," *Lecture Notes in Computer Science*, Vol. 4395, Springer Verlag, New York, 2007, pp. 260–280.

[2] Markowitz, H. M., "Biosketch," Lex Prix Nobel: Nobel Lectures, edited by T. Frälingsmyr, Nobel Foundation, Stockholm, 1991.

[3] Markowitz, H. M., "The elimination form of the inverse and its application to linear programming," *Management Science*, Vol. 3, 1959, pp. 255–269.

[4] Parter, S. V., "The use of linear graphs in Gaussian elimination," *SIAM Review*, Vol. 3, 1961, pp. 119–130.

[5] Willoughby, R. A., "Proceedings of the Symposium on Sparse Matrices and Their Applications," Tech. Rep. Report RAI (No. 11707), IBM, Yorktown Heights, NY, 1969.

[6] Rose, D. J. and Willoughby, R. A., *Sparse Matrices and Their Applications*, Plenum Press, New York, 1972, Proceedings of a Symposium at the IBM Research Center, New York, Sep. 9-10, 1971.

[7] Reid, J. K., *Large Sparse Sets of Linear Equations*, Academic Press, London, 1971.

[8] Rose, D. J., "A graph-theoretic study of the numerical solution of sparse positive definite systems of linear equations," *Graph Theory and Computing*, edited by R. C. Read, Academic Press, New York, 1972, pp. 183–217.

[9] George, A. and Liu, J. W. H., *Computer Solution of Large Sparse Positive Definite Systems*, Prentice-Hall, New York, 1981.

[10] George, A., Gilbert, J. R., and Liu, J. W. H., editors, *Graph Theory and Sparse Matrix Computation*, Springer-Verlag, 1993, IMA Volumes in Applied Mathematics, Volume 56.

[11] Bauer, F. L., "Computational graphs and rounding error," *SIAM Journal on Numerical Analysis*, Vol. 11, 1974, pp. 87–96.

[12] Griewank, A. and Walther, A., *Evaluating Derivatives: Principles and Techniques of Algorithmic Differentiation*, Society for Industrial and Applied Mathematics, 2nd ed., 2008.

[13] Coleman, T. F. and Moré, J. J., "Estimation of sparse Jacobian matrices and graph coloring problems," *SIAM Journal on Numerical Analysis*, Vol. 20, 1983, pp. 187–209.

[14] Griewank, A. and Reese, S., "On the Calculation of Jacobian Matrices by the Markowitz Rule," *Automatic Differentiation of Algorithms: Theory, Implementation, and Application*, edited by A. Griewank and G. F. Corliss, SIAM, Philadelphia, PA, 1991, pp. 126–135.

[15] Hartmann, A. and Rieger, H., *Optimization Algorithms in Physics*, Wiley & Sons, New York, 2003.

[16] George, A., "Nested Dissection of a regular finite element mesh," *SIAM Journal on Numerical Analysis*, Vol. 10, 1973, pp. 345–363.

[17] Lipton, R. J., Rose, D. J., and Tarjan, R. E., "Generalized Nested Dissection," *SIAM Journal on Numerical Analysis*, Vol. 16, 1979, pp. 346–358.

[18] Pothen, A., Simon, H. D., and Liou, K.-P., "Partitioning sparse matrices with eigenvectors of graphs," *SIAM Journal on Matrix Analysis and its Applications*, Vol. 11, 1990, pp. 430–452.

[19] Hendrickson, B. A. and Leland, R., "A multilevel algorithm for partitioning graphs," *Proceedings of Supercomputing*, 1995.

[20] Simon, H. D., "Partitioning of unstructured problems for parallel processing," *Proc. Conference on Parallel Methods on Large Scale Structural Analysis and Physics Applications*, Pergammon Press, Oxford, UK, 1991.

[21] Çatalyürek, Ü. and Aykanat, C., "Decomposing Irregularly Sparse Matrices for Parallel Matrix-Vector Multiplication," *Lecture Notes in Computer Science 1117*, Springer-Verlag, New York, 1996, pp. 75–86, Proc. Irregular'96.

[22] Hendrickson, B. A. and Kolda, T. G., "Graph Partitioning Models for Parallel Computing," *Parallel Comput.*, Vol. 26, 2000, pp. 1519–1534.

[23] Jones, M. T. and Plassmann, P. E., "A Parallel Graph Coloring Heuristic," *SIAM J. Sci. Comput.*, Vol. 14, No. 3, 1993, pp. 654–669.

[24] Gebremedhin, A., Manne, F., and Pothen, A., "What color is your Jacobian? Graph coloring for computing derivatives," *SIAM Review*, Vol. 47, No. 4, 2005, pp. 629–705.

[25] Gusfield, D., *Algorithms on Strings, Trees, and Sequences*, Cambridge University Press, Cambridge, UK, 1997.

[26] Aluru, S., editor, *Handbook of Computational Molecular Biology*, Chapman & Hall, Boca Raton, FL, 2005.

[27] Garcia-Domenech, R., Gálvez, J., de Julián-Ortiz, J., and Pogliani, L., "Some new trends in chemical graph theory," *J. Molecular Graphics and Modelling*, Vol. 19, February 2001, pp. 60–69.

[28] Martin, S., Roe, D., and Faulon, J.-L., "Predicting protein–protein interactions using signature products," *Bioinformatics*, Vol. 21, No. 2, August 2004, pp. 218–226.

[29] Thorpe, M. F., Lei, M., Rader, A. J., Jacobs, D. J., and Kuhn, L. A., "Protein flexibility and dynamics using constraint theory," *Chem. Rev.*, Vol. 108, No. 3, February 2008, pp. 1127–1169.

[30] Lumsdaine, A., Gregor, D., Hendrickson, B., and Berry, J., "Challenges in Parallel Graph Processing," *Parallel Processing Letters*, Vol. 17, No. 1, 2007, pp. 5–20.

[31] Kogge, P., "The Tops in Flops," *IEEE Spectrum*, February 2011, pp. 50–54.

[32] Kogge (editor), P., "ExaScale Computing Study: Technology Challenges in Achieving Exascale Systems," Tech. Rep. TR-2008-13, University of Notre Dame, CSE Dept., 2008.

[33] Frigo, M., Leiserson, C. E., Prokop, H., and Ramachandran, S., "Cache–Oblivious Algorithms," *Proc. IEEE Symp. Foundations of Computer Science*, IEEE, 1999.

[34] Hey, T., Tansley, S., and Tolle, K., editors, *The Fourth Paradigm: Data–Intensive Scientific Discovery*, Microsoft Research, Redmond, WA, 2009.

[35] Newman, M. E. J., *Networks: An Introduction*, Oxford University Press, 2010.

[36] Easely, D. and Kleinberg, J., *Networks, Crowds, and Markets: Reasoning about a Highly Connected World*, Cambridge University Press, Cambridge, UK, 2010.

[37] Leskovec, J. and Horvitz, E., "Planetary-Scale Views on a Large Instant-Messaging Network," *Proc. World Wide Web Conference*, Beijing, China, April 2008.

[38] Michel, J.-B., Shen, Y. K., Aiden, A. P., Veres, A., Gray, M. K., The Google Books Team, Pickett, J. P., Hoiberg, D., Clancy, D., Norvig, P., Orwant, J., Pinker, S., Nowak, M. A., and Aiden, E. L., "Quantitative Analysis of Culture Using Millions of Digitized Books," *Science*, Vol. 331, No. 6014, 2011, pp. 176–182.

[39] Buneman, P., "A characterisation of rigid circuit graphs," *Discrete Mathematics*, Vol. 9, 1974, pp. 205–212.

[40] Gusfield, D., "The Perfect phylogeny problem with missing data: Solutions via integer-programming and chordal graph theory," *Journal of Computational Biology*, Vol. 17, 2010, pp. 383–399.

[41] Blair, J. R. S. and Peyton, B., "An introduction to chordal graphs and clique trees," *Graph Theory and Sparse Matrix Computation, IMA Volumes in Mathematics and its Applications*, Vol. 56, Springer Verlag, New York, 1993, pp. 1–30.

[42] Peyton, B. W., Pothen, A., and Yuan, X., "Partitioning a chordal graph into transitive subgraphs for parallel sparse triangular solution," *Linear Algebra and its Applications*, Vol. 192, 1993, pp. 329–354.

[43] Azad, A., Langguth, J., Fang, Y., Qi, A., and Pothen, A., "Identifying rare cell populations in comparative flow cytometry," *Proceedings of the Workshop on Algorithms in Bioinformatics, Lecture Notes in Bioinformatics*, Vol. 6293, Springer Verlag, New York, 2010, pp. 162–175.

Chapter 2

Combinatorial Problems in Solving Linear Systems

Iain Duff

Rutherford Appleton Laboratory and CERFACS

Bora Uçar

CNRS and ENS Lyon

2.1 Introduction

In this short review paper, we examine the interplay between the solution of sparse linear systems and combinatorics. Most of this strong association comes from the identification of sparse matrices with graphs so that most algorithms dealing with sparse matrices have a close or exact analogue to an algorithm on a graph. We examine these analogues both in the case of the direct solution of sparse linear equations and their solution by iterative methods, particularly focusing on preconditioning.

Two surveys on combinatorial scientific computing have already been carried out. Hendrickson and Pothen [1] focus on the enabling role of combinatorial algorithms in scientific computing, highlighting a broad range of applications: parallel computing; mesh generation; sparse linear system solution; automatic differentiation for optimization; statistical physics; computational chemistry; bioinformatics; information processing. Bollhöfer and Schenk [2] give an overview of combinatorial aspects of LU factorization. In a spirit similar to the two preceding surveys, Heath, Ng, and Peyton [3] survey parallel algorithms for sparse Cholesky factorization by discussing issues related to the parallelization of the major steps of direct solvers. A recent book by Brualdi and Cvetković [4] covers standard matrix computations where the combinatorial tools are brought to the forefront, and graphs are used to explain standard matrix computations. The contents include matrix powers and their description using directed graphs; graph-theoretical definition of the determinant of a matrix; and the interpretation of matrix inverses and linear system solution. Brualdi and Ryser [5] and Brualdi [6, 7] include a higher level of combinatorial analysis and many linear algebraic concepts beyond the solution of linear systems.

We cover linear system solution with both direct and iterative methods. We try to keep the discussion simple and provide details of some fundamental problems and methods; there are a great many beautiful results on combinatorial problems in linear algebra and reviewing them all would fill a book rather than a short survey paper. Often we review or cite the paper or papers that are at the origin of a particular method. The field has evolved in many ways and many developments have taken place since this original research. We try to provide newer references and software which define the current state-of-the-art. In some cases, survey papers of the highest quality are available, and we list some of these as pointers for readers who wish to explore these areas more fully. All the papers in our reference list are cited and almost all of them are commented on in the text of the paper. In fact we feel that the extensive bibliography is a very useful feature of this review and suggest that the reader may look at these references for further enlightenment on topics of particular interest.

We have intentionally avoided covering subareas that are addressed by other papers in this volume, for example graph partitioning, sparse matrix-vector multiplication, coloring problems, automatic differentiation.

In Section 2.2, we provide basic definitions from graph theory that are used throughout the paper. Some further definitions are deferred to the relevant sections. We start discussing combinatorial problems in direct solvers by a gentle introduction to the elimination process and its relationship to a suitably defined graph in Section 2.3. This section is structured around the main techniques that constitute the essential components of modern direct solvers. Section 2.3.4 covers a special permutation of sparse matrices, known as the block triangular form, which reformulates the solution of a large linear system in terms of the solution on smaller subsystems, thus giving us benefits if the solution scheme is superlinear in the order of the system, which is usually the case. Section 2.4 covers some other combinatorial problems that arise in iterative methods. In this section, we mainly discuss the issues that arise due to the use of preconditioning techniques. We finish with some concluding remarks in Section 2.5.

2.2 Basics

In this section, we collect some elementary terms and definitions in graph theory to be used later. Most of these terms and definitions in direct methods can be found in in [8, 9, 10]. For a more algorithmic treatment of graph theory, we refer the reader to [11].

A *graph* G is a pair (V, E), where V is a finite set, called the vertex or node set, and E is a binary relation on V, called the edge set. There are three standard graph models that are widely used in combinatorial scientific computing. In an *undirected graph* $G = (V, E)$ the edges are unordered pairs of vertices, $\{u, v\} \in E$ for $u, v \in V$ and $u \neq v$. In a *directed graph* $G = (V, E)$, the edges are ordered pairs of vertices, that is, (u, v) and (v, u) are two different edges. A *bipartite graph* $G = (U \cup V, E)$ consists of two disjoint vertex sets U and V where for each edge $(u, v) \in E$ we have $u \in U$ and $v \in V$.

An edge (u, v) is said to be *incident on* the vertices u and v. For any vertex u, the vertices in the set $\text{adj}(u) = \{v : (u, v) \in E\}$ are called the *neighbours* of u. The *degree* of a vertex is the number of edges incident on it. A *path* p of length k is a sequence of vertices $\langle v_0, v_1, \ldots, v_k \rangle$ where $(v_{i-1}, v_i) \in E$ for $i = 1, \ldots, k$. A *cycle* is a path that starts and ends at the same vertex. The two end points v_0 and v_k are said to be connected by the path p, and the vertex v_k is said to be reachable from v_0. An undirected graph is said to be *connected* if every pair of vertices is connected by a path. A directed graph is said to be *strongly connected* if every pair of vertices are reachable from each other. The *subgraph* $H = (W, F)$ of a given graph $G = (V, E)$ is a graph such that $W \subseteq V$ and $F \subseteq W \times W \cap E$. Such an H is called an *induced subgraph*, if $F = W \times W \cap E$ and a *spanning subgraph* if $W = V$. A *tree* is a connected graph without cycles. A *spanning tree* of a connected graph is a spanning subgraph which is also a tree.

Given a sparse square matrix A of order n, one can associate any of the three standard graph models described above. Formally one can associate the following three graphs. The first one is the bipartite graph $G_B = (V_R \cup V_C, E)$, where the vertex sets V_R and V_C correspond to the rows and columns of A, respectively, and the edge set E corresponds to the set of nonzeros of the matrix A so that $(i, j) \in E$ iff $a_{ij} \neq 0$. The second one is the directed graph

$G_D = (V, E)$, where each vertex corresponds to a row and the respective column of A, and the edge set E corresponds to the set of nonzeros of A so that $(i, j) \in E$ iff $a_{ij} \neq 0$. The third one is the undirected graph $G_U = (V, E)$ which is defined for a pattern symmetric matrix A (that is $a_{ij} \neq 0$ whenever $a_{ji} \neq 0$), where each vertex corresponds to a row and the respective column of A, and the edge set E corresponds to the set of nonzeros so that $(i, j) \in E$ iff $a_{ij} \neq 0$ and $a_{ji} \neq 0$. We note that among these three alternatives, only the bipartite graph G_B can represent a rectangular matrix.

The three graph models discussed above suffice for the material that we cover in this paper. For completeness, we also mention three hypergraph models that are used in modelling sparse matrices. We do not cover the combinatorial scientific computing problems for which hypergraph models are used. A hypergraph $\mathcal{H}$ is a pair (V, H) where V is a finite set, called the vertex or node set, and H is a family of subsets of V, called the hyperedge set. Each hyperedge is therefore a set of vertices. Clearly, an undirected graph is a hypergraph where each hyperedge has two vertices. There are three known hypergraph models for sparse matrices. In the column-net hypergraph model [12] $\mathcal{H}_\mathcal{R} = (V_\mathcal{R}, H_\mathcal{C})$ of a sparse matrix matrix A, there exists one vertex $v_i \in V_\mathcal{R}$ for each row r_i, and there exists one hyperedge $h_j \in H_\mathcal{C}$ for each column c_j. In this model, $v_i \in h_j$ if and only if $a_{ij} \neq 0$. In the row-net hypergraph model [12] $\mathcal{H}_\mathcal{C} = (V_\mathcal{C}, H_\mathcal{R})$ of matrix A, there exists one vertex $v_j \in V_\mathcal{C}$ for each column c_j, and there exists one hyperedge $h_i \in H_\mathcal{R}$ for each row r_i. In this model, $v_j \in h_i$ if and only if $a_{ij} \neq 0$. In the fine-grain model [13] (see also [14]), $\mathcal{H}_z = (V_z, H_{\mathcal{RC}})$ of an $m \times n$ sparse matrix matrix A, there exists one vertex $v_{ij} \in V_z$ corresponding to each nonzero a_{ij} in matrix A. For each row and for each column there exists a hyperedge such that $H_{\mathcal{RC}}$ contains m row-hyperedges $r_1, \ldots, r_m$ and n column-hyperedges $c_1, \ldots, c_n$ where $v_{ij} \in r_i$ and $v_{ij} \in c_j$ if and only if $a_{ij} \neq 0$.

A *matching* in a graph is a set of edges such that no two are incident on the same vertex. In this paper, we will be mostly interested in matchings in bipartite graphs. A matching in the bipartite graph of a matrix A corresponds to a set of nonzeros in A no two of which are in the same row or column. An *independent set* in an undirected graph is a set of vertices no two of which are adjacent. An independent set in the undirected graph of a matrix corresponds to a square principal submatrix whose nonzeros can only be on the diagonal. A *clique* is a set of mutually adjacent vertices. A clique in the undirected graph of a matrix corresponds to a dense square principal submatrix, assuming a zero free diagonal.

2.3   Direct Methods

We start by describing the LU decomposition, sometimes called Gaussian elimination, of a nonsingular, square sparse matrix A of order n. Although there are many variations, the basic point-wise LU decomposition proceeds in $n-1$ steps, where at step $k = 1, 2, \ldots, n-1$, the formulae

$$a_{ij}^{(k+1)} \leftarrow a_{ij}^{(k)} - \left(a_{ik}^{(k)} / a_{kk}^{(k)} \right) a_{kj}^{(k)} , \quad \text{for } i, j > k \tag{2.1}$$

are used to create zeros below the diagonal entry in column k. Matrices of the form $A^{(k)} = \{a_{ij}^{(k)}\}$ of order $n - k + 1$ are called reduced matrices. This process leads to an upper triangular matrix U. Here, each updated entry $a_{ij}^{(k+1)}$ overwrites $a_{ij}^{(k)}$, and the multipliers $l_{ik} = a_{ik}^{(k)} / a_{kk}^{(k)}$ may overwrite $a_{ik}^{(k)}$ resulting in the decomposition $A = LU$ stored in-place. Here L is a unit lower triangular matrix, and U is an upper triangular matrix. In order for this method run to completion, the inequalities $a_{kk}^{(k)} \neq 0$ should hold. These updated diagonal entries are called *pivots* and the operation performed using the above formulae is referred to as eliminating the variable x_k from the subsequent equations.

Suppose at step k, either of the matrix entries $a_{ik}^{(k)}$ or $a_{kj}^{(k)}$ is zero. Then there would be no update to $a_{ij}^{(k)}$. On the other hand, if both are nonzero, then $a_{ij}^{(k+1)}$ becomes nonzero even if it was previously zero (accidental cancellations due to existing values are not considered as zeros, rather they are held as if they were nonzero). Now consider the first elimination step on a symmetric matrix characterized by an undirected graph. If a_{i1} is nonzero we zero out that entry. Suppose that a_{1j} is also nonzero for some $j > 1$, then we will have a nonzero value at a_{ij} after the application of the above formulae. Consider now the undirected graph $G_U(V, E)$ of A. As $a_{i1} \neq 0$ and $a_{1j} \neq 0$, we have the edges $(1, i)$ and $(1, j)$ in E. After the elimination, the new nonzero a_{ij} will thus correspond to the edge (i, j) in the graph. Since the vertex 1 does not concern us any further (due to the condition $i, j > k$ in the formulae above), we can remove the vertex 1 from the graph, thereby obtaining the graph of the reduced matrix $A^{(1)}$ of size $(n - 1) \times (n - 1)$. In other words, step k of the elimination process on the matrix $A^{(k-1)}$ corresponds to removing the kth vertex from the graph and adding edges between all the neighbours of vertex k that were not connected before.

Algorithm 1 Elimination process in the graph

$G_U(V, E) \leftarrow$ undirected graph of A
for $k = 1 : n - 1$ **do**
$\quad V \leftarrow V - \{k\}$ {remove vertex k}
$\quad E \leftarrow E - \{(k, \ell) : \ell \in \text{adj}(k)\} \cup \{(x, y) : x \in \text{adj}(k) \text{ and } y \in \text{adj}(k)\}$

$$
\begin{array}{cccc}
\times & \times & \times & \times \\
\times & \times &   &   \\
\times &   & \times &   \\
\times &   &   & \times
\end{array}
$$

Original matrix

$$
\begin{array}{cccc}
\times &   &   & \times \\
       & \times &   & \times \\
       &   & \times & \times \\
\times & \times & \times & \times
\end{array}
$$

Reordered matrix

FIGURE 2.1: Ordering affects the sparsity during elimination.

This relation between Gaussian elimination on A and the vertex elimination process on the graph of A, shown in Algorithm 1, was first observed by Parter [15]. Although it looks innocent and trivial[1], this relation was the starting point for much of what follows in the following subsections.

The following discussion is intentionally simplified. We refer the reader to [8] and [9] for more rigorous treatment.

2.3.1   Labelling or Ordering

Consider the elimination process on the matrices shown in Figure 2.1. The original ordering of the matrix A is shown on the left. The elimination process on this matrix will lead to nonzeros in the factors that are zeros in the matrix A. These new nonzeros are called *fill-in*. Indeed, the resulting matrix of factors will be full. On the other hand, the ordering obtained by permuting the first row and column to the end (shown on the right) will not create any fill-in. As is clearly seen from this simple example, the ordering of the eliminations affects the cost of the computation and the storage requirements of the factors.

Ordering the elimination operations corresponds to choosing the pivots among combinatorially many alternatives. It is therefore important to define and find the best ordering in an efficient way. There are many different ordering methods; most of them attempt to reduce the fill-in. Minimizing the fill-in is an NP-complete problem. This was first conjectured to be true in 1976 by Rose, Tarjan, and Lueker [16] in terms of the elimination process on undirected graphs. Then Rose and Tarjan [17] proved in 1978 that finding an elimination ordering on a directed graph that gives minimum fill-in is NP-complete (there was apparently a glitch in the proof which was rectified by Gilbert [18] two years later). Finally, Yannakakis [19] proved the NP-completeness of the minimum fill-in problem on undirected graphs in 1981.

Heuristic ordering methods to reduce the fill-in predate these complexity results by about two decades. The first of these ordering methods is due to Markowitz [20]. At the beginning of the kth elimination step, a nonzero entry $a_{ij}^{(k)}$ in the reduced matrix is chosen to reduce the fill-in, and the chosen entry is permuted to the diagonal, thus defining the kth pivot. The criterion for

---

[1] In 2000 at Los Alamos, Seymour V. Parter told Michele Benzi that such were the reactions he had received from the referees on his paper.

choosing an entry $a_{ij}^{(k)}$ is to select the entry to minimize the product of the number of other entries in its row and the number of other entries in its column. Markowitz's paper deals with nonsymmetric matrices. The selection criterion was later adapted to symmetric matrices by Tinney and Walker [21] (they do not cite Markowitz's paper but state that their method might be used already). Tinney and Walker's method of choosing a diagonal entry as the pivot at step k, referred to as S2 in their paper, can be seen more elegantly during the elimination process on the graph, as noted by Rose [22]. Here, the vertex with the minimum degree in the current graph is eliminated. In other words, instead of eliminating vertex k at step k of the Algorithm 1, a vertex with minimum degree is selected as the pivot and labelled as k. Due to this correspondence, Rose renamed the method S2 of Tinney and Walker as the minimum degree algorithm.

There have been many improvements over the basic minimum degree algorithm, reducing both the run time and the space complexity. Probably the most striking result is that the method can be implemented in the same amount of space used to represent the original graph with a few additional arrays of size n. This is surprising as the degrees changes dynamically, fill-in normally occurs throughout the execution of the algorithm, and to be able to select a vertex of minimum degree, the elimination process should somehow be simulated. The methods used to achieve this goal are described in [23, 24, 25]. The survey by George and Liu [26] lists, inter alia, the following improvements and algorithmic follow-ups: *mass eliminations* [27], where it is shown that, for finite-element problems, after a minimum degree vertex is eliminated a subset of adjacent vertices can be eliminated next, together at the same time; *indistinguishable nodes* [9], where it is shown that two adjacent nodes having the same adjacency can be merged and treated as one; *incomplete degree update* [28], where it is shown that if the adjacency set of a vertex becomes a subset of the adjacency set of another one, then the degree of the first vertex does not need to be updated before the second one has been eliminated; *element absorption* [29], where based on a compact representation of elimination graphs, redundant structures (cliques being subsets of other cliques) are detected and removed; *multiple elimination* [30], where it was shown that once a vertex v is eliminated, if there is a vertex with the same degree that is not adjacent to the eliminated vertex, then that vertex can be eliminated before updating the degree of the vertices in adj(v), that is the degree updates can be postponed; *external degree* [30], where instead of the true degree of a vertex, the number of adjacent and indistinguishable nodes is used as a selection criteria. Some further improvements include the use of compressed graphs [31], where the indistinguishable nodes are detected even before the elimination process and the graph is reduced, and the extensions of the concept of the external degree [32, 33]. The *approximate minimum degree* as described in [34] is shown to be more accurate than previous degree approximations and leads to almost always faster execution with an ordering often as good as or better than minimum degree. Assuming a linear space implementation, the run-time complexity of the minimum degree (and multiple minimum degree) and the approximate minimum degree ordering heuristics are shown to be, respectively, $\mathcal{O}(n^2 m)$ and $\mathcal{O}(nm)$ for a graph of n vertices and m edges [35].

A crucial issue with the minimum degree algorithm is that ties arise while selecting the minimum degree vertex [36]. It is still of interest, though a little daunting, to develop a tie-breaking strategy and beat the current fill-reducing algorithms.

As mentioned above, the minimum degree based approaches order the matrix by selecting pivots using the degree of a vertex without any reference to later steps of the elimination. For this reason, the general class of such approaches are called *local strategies*. Another class, called *global strategies*, permute the matrix in a global sense so as to confine the fill-in within certain parts of the permuted matrix. A widely used and cited algorithm is by Cuthill and McKee [37]. A structurally symmetric matrix A is said to have *bandwidth* $2m+1$, if m is the smallest integer such that $a_{ij} = 0$, whenever $|i-j| > m$. If no interchanges are performed during elimination, fill-in occurs only within the band. The algorithm is referred to as CM and is usually based on a breadth-first search algorithm. George [38] found that reversing the ordering found by the CM algorithm effectively always reduces the total storage requirement and the arithmetic operations when using a variant of the band-based factorization algorithm (a rigorous treatment and analysis of these two algorithms is given in [39]). This algorithm is called reverse Cuthill-McKee and often referred to as RCM.

Another global approach that received and continues to receive considerable attention is called the *nested dissection* method, proposed by George [40] and baptized by Birkhoff (acknowledged in George's paper). The central concept is a *vertex separator* in a graph: that is a set of vertices whose removal leaves the remaining graph disconnected. In matrix terms, such a separator corresponds to a set of rows and columns whose removal yields a block diagonal matrix after suitable permutation. Permuting the rows and columns corresponding to the separator vertices last, and each connected component of the remaining graph consecutively results in the *doubly bordered block diagonal* form

$$\begin{pmatrix} A_{11} & & A_{1S} \\ & A_{22} & A_{2S} \\ A_{S1} & A_{S2} & A_{SS} \end{pmatrix}. \tag{2.2}$$

The blocks A_{11} and A_{22} can be further dissected using the vertex separator of the corresponding graphs and can themselves be permuted into the above form, resulting in a nested dissection ordering. Given such an ordering, it is evident that fill-ins are confined to the blocks shown in the form.

A significant property of nested-dissection based orderings is that they yield asymptotically optimal fill-in and operation counts for certain types of problems. It was shown in [40] that, for a matrix corresponding to a regular finite-element mesh of size $q \times q$, the fill-in and operation count using a nested dissection ordering are $\mathcal{O}(q^2 \log_2 q)$ and $\mathcal{O}(q^3)$, respectively. For a three dimensional mesh of size $q \times q \times q$, the bounds are $\mathcal{O}(q^6)$ for the operation count and $\mathcal{O}(q^4)$ for the fill-in, see also [41] for a detailed analysis. George [40] shows the asymptotic optimality of the operation count but not the fill-in. The asymptotic results were settled thoroughly in [42]. Further developments for finite-element meshes include automatic generation of nested dissection on square meshes with $q = 2^l - 1$, with l integer, in [43], and methods for arbitrary square meshes and irregular shaped regions in [36]. In [44], a heuristic is presented to perform nested dissection on general sparse symmetric matrices. The nested dissection approach was then generalized so that it yields the same bounds for systems defined on planar and almost planar graphs [45]. The generalization essentially addresses all $n \times n$ systems of linear equations whose graph has a bounded separator of size $n^{1/2}$. The results of [46] are used to obtain separators and bounds on separator sizes on planar graphs, and therefore the asymptotic optimality results apply to planar or almost planar graphs, a general class of graphs which includes two dimensional finite-element meshes. Gilbert and Tarjan [47] combine and extend the work in [44] and [45] to develop algorithms that are easier to implement than the earlier alternatives and have smaller constant factors. In [47] asymptotic optimality results are demonstrated on planar graphs, two-dimensional finite-element graphs, graphs of bounded genus, and graphs of bounded degree with $n^{1/2}$-separators (note that without the bounded degree condition, the algorithm can be shown not to achieve the bound on fill-in).

It is not much of a surprise that hybrid fill-reducing ordering methods (combining the minimum degree and nested dissection heuristics) have been developed. We note that the essential ideas can be seen already in [21]. Tinney and Walker suggest that if there is a natural decomposition of the underlying network, in the sense of (2.2), then it may be advantageous to run the minimum degree algorithm on each subnetwork. The first formalization of the hybrid ordering approach was, however, presented in [48]. In this work, the hybrid ordering method is applied to finite-element meshes, where first a few steps of nested dissection are applied before ordering all entries (a precise recommendation is not given). The remaining entries are ordered using bandwidth minimization methods such as CM and RCM. Liu [49] uses a similar idea on general symmetric sparse matrices. Probably this is the first paper where minimum degree based algorithms are called bottom-up approaches and the separator based, nested dissection algorithms are called top-down approaches, thus defining the current terminology. Liu terminates the nested dissection earlier (up to 5 levels of dissections are applied; but this depends on the size of the graphs and there is no precise recommendation for a general problem), and then orders the remaining vertices with minimum degree, including the separator vertices in the degree counts but ignoring them during vertex selection (this method is known as *constrained minimum degree* or *minimum degree with constraints*). The merits of the proposed algorithm are listed as a reduced sensitivity to the initial ordering of the matrix and an ordering algorithm more appropriate for parallel factorization.

As we stated before, a nested dissection ordering gives asymptotically optimal storage and operation counts for square grids. For rectangular grids, however, this is not the case. It has been shown that for rectangular grids with a large aspect ratio, nested dissection is inferior to the minimum degree ordering [50], and even to the natural ordering [51]. The main issue, as stated by Ashcraft and Liu [50], is that the ordering of the separator vertices found at different dissection steps is important. For rectangular grids, Bhat et al. [51] propose to order the separator vertices in the natural ordering to minimize the profile after partitioning the rectangular grid into square sub-grids each of which is ordered using nested dissection. Ashcraft and Liu [50] develop this idea to propose a family of ordering algorithms. In these algorithms, a partitioning of the graph is first found by using either the elimination tree or by recursive application of graph bisection [52]. Then each part is ordered using constrained minimum degree. The Schur complement of the domains is formed symbolically and used to reorder the separator vertices using multiple minimum degree. Hendrickson and Rothberg [53] (concurrently with Ashcraft and Liu) and Schulze [54] develop similar algorithms. Ashcraft and Liu find multi-sectors instead of performing recursive bisection; Hendrickson and Rothberg and Schulze use multilevel graph partitioning; Schulze proposes an elegant coarsening algorithm.

Not surprisingly, the current state-of-the-art in fill-reducing ordering methods is based on hybrid ordering approaches of the kind outlined above. The efficiency of these methods is due to developments in graph partitioning methods such as efficient algorithms for computing eigenvectors to use in partitioning graphs [55, 56]; and the genesis of the multilevel paradigm [57, 58] which enables better use of vertex-move-based iterative refinement algorithms [59, 60]. These developments are neatly incorporated in graph partitioning and ordering software packages such as Chaco [61], MeTiS [62], SCOTCH [63], and WGPP [64]. These libraries usually have a certain threshold (according to F. Pellegrini, around 200 vertices seems to be a common choice) to terminate the dissection process and to switch to a variant of the minimum degree algorithm.

2.3.2   Matching and Scaling

As discussed before, for the elimination process to succeed, the pivots $a_{kk}^{(k)}$ should be nonzero. This can be achieved by searching for a nonzero in the reduced matrix and permuting the rows and columns to place that entry on the diagonal. Such permutations are called *pivoting* and guarantee that an $a_{kk}^{(k)} \neq 0$ can be found for all k so long as the original matrix is nonsingular. In *partial pivoting*, the search is restricted to the kth column. A general technique used to control the growth factor is to search the column for a maximum entry or to accept an entry as pivot so long as it passes certain numerical tests. These pivoting operations are detrimental to the fill-reducing orderings discussed in the previous section, as those ordering methods assume that the actual numerical elimination will follow the ordering produced by the symbolic elimination process on the graph.

Suppose that the diagonal entries of the matrix are all nonzero. Assuming no exact cancellation, all pivots will be nonzero when they are taken from the diagonal. Notice that any symmetric permutation of the original matrix keeps the set of diagonal entries the same, and hence the fill-reducing orderings of the previous section are applicable in this case. Our purpose in this section is to summarize the methods to find permutations that yield such diagonals.

As is clear, we are searching for n nonzeros in an $n \times n$ matrix A no two of which are in the same row or column. As we mentioned earlier, this corresponds to finding a perfect matching in the bipartite graph representation of A. The existence of such a set of n nonzeros (i.e., a perfect matching in the bipartite graph) is guaranteed to exist if, for $k = 1, 2, \ldots, n$ any k distinct columns have nonzeros in at least k distinct rows—a result shown by P. Hall [65].

A few definitions are in order before describing bipartite matching algorithms. Given a matching $\mathcal{M}$, a vertex is said to be *matched* if there is an edge in the matching incident on the vertex, and to be *unmatched* otherwise. An $\mathcal{M}$-*alternating path* is a path whose edges are alternately in $\mathcal{M}$ and not in $\mathcal{M}$. An alternating path is called an *augmenting path*, if it starts and ends at unmatched vertices. The *cardinality of a matching* is the number of edges in it. We will be mostly interested in matchings of maximum cardinality. Given a bipartite graph G and a matching $\mathcal{M}$, a necessary and sufficient condition for $\mathcal{M}$ to be of maximum cardinality is that there is no $\mathcal{M}$-augmenting path in G [66, Theorem 1]—for the curious reader the second theorem of Berge gives a similar condition for minimum vertex covers. Given a matching $\mathcal{M}$ on the bipartite graph of a square matrix A, one can create a permutation matrix M such that $m_{ji} = 1$ iff row i and column j are matched in $\mathcal{M}$. Then, the matrix AM has a zero-free diagonal. It is therefore convenient to abuse the notation and refer to a matching as a permutation matrix.

The essence of bipartite cardinality matching algorithms is to start with an empty matching and then to augment it until no further augmentations are possible. The existing algorithms mostly differ in the way the augmenting paths are found and the way the augmentations are performed. In [67] a breadth-first search is started from an unmatched row vertex to reach an unmatched column vertex. The time complexity is $\mathcal{O}(n\tau)$, where τ is the number of nonzeros in the matrix. The algorithm in [68, 69], known as MC21, uses depth-first search where, before continuing the depth-first search with an arbitrary neighbour of the current vertex, all its adjacency set is scanned to see if there is an unmatched vertex. This is called a cheap assignment and helps reduce the run time. The time complexity is $\mathcal{O}(n\tau)$, but it is observed to run usually much faster than that bound. Depth-first search is also used in [70] with a complexity of again $\mathcal{O}(n\tau)$. Hopcroft and Karp [71] find a maximal set of shortest augmenting paths using breadth-first search and perform the associated augmentations at the same time. With a detailed analysis of the possible length and number of such augmentations, they demonstrate a complexity of $\mathcal{O}(\sqrt{n}\tau)$. Building upon the work of Hopcroft and Karp, Alt et al. [72] judiciously combine depth- and breadth-first searches to further reduce the complexity to $\mathcal{O}(\min\{\sqrt{n}\tau, n^{1.5}\sqrt{\tau/\log n}\})$.

Not all matrices have perfect matchings. Those that have a perfect matching are referred to as structurally nonsingular, or structurally full rank, whereas those that do not have a perfect matching are referred to as structurally singular, or structurally rank deficient. The maximum cardinality of a matching is referred to as the structural rank which is at least as large as the numerical rank.

Although permuting a matching to the diagonal guarantees existence of the pivots, it does not say anything about their magnitudes. In order to control the growth factor, it may still be necessary to perform pivoting during the course of the elimination. It is known that for diagonally dominant matrices, pivoting on the grounds of numerical stability is not necessary. Therefore, if we find a matching which guarantees diagonal dominance after a permutation and probably some scaling, we can avoid numerical pivoting. Unfortunately not all matrices can be permuted to such a form but trying to do so is likely to reduce the need for numerical pivoting. These were the motivating ideas in [73] and [74], where such an attempt is formulated in terms of *maximum weighted bipartite matchings*.

In matrix terms, Olschowka and Neumaier [74] and Duff and Koster [73] find a permutation matrix (and hence a perfect matching) M such that the product of the diagonal of the permuted matrix, $\prod \text{diag}(AM)$, is maximum (in magnitude) among all permutations. Although the product form of the variables is intimidating, a simple transformation by changing each entry of the matrix to the logarithm of its magnitude reduces the problem to the well known maximum weighted bipartite matching problem. In particular, maximizing $\prod \text{diag}(AM)$ is equivalent to maximizing the diagonal sum given by $\sum \text{diag}(\hat{C}M)$ for $\hat{C} = (\hat{c}_{ij})$ where

$$\hat{c}_{ij} = \begin{cases} \log|a_{ij}|, & \text{if } a_{ij} \neq 0 \\ -\infty, & \text{otherwise} , \end{cases}$$

or, to minimizing the diagonal sum given by $\sum \text{diag}(CM)$ for $C = (c_{ij})$ where

$$c_{ij} = \begin{cases} \log\max_i |a_{ij}| - \log|a_{ij}|, & \text{if } a_{ij} \neq 0 \\ \infty, & \text{otherwise} . \end{cases}$$

The literature on the minimum weighted matching problem is much larger than that on the cardinality matching problem. A recent book lists 21 algorithms [75, p. 121] from years 1946 to 2001 and provides codes or a link to codes of eight of them (http://www.assignmentproblems.com/). The best strongly polynomial time algorithm is by Fredman and Tarjan [76] and runs in $\mathcal{O}(n(\tau + n \log n))$. In addition to those 21 algorithms in the abovementioned book, we include Duff and Koster's implementation of the matching algorithm, now known as MC64 and available as an HSL subroutine (http://www.hsl.rl.ac.uk/). MC64 was initially designed for square matrices, but the latest version extends the algorithm to rectangular matrices. The running time of MC64 for the maximum weighted matching problem is $\mathcal{O}(n(\tau + n)\log n)$. MC64 also provides algorithms for a few more bipartite matching problems. There are also recent efforts which aim to develop practical parallel algorithms for the weighted matching problem. In [77] and [78] parallel algorithms for maximum weighted bipartite matching are proposed. Although these algorithms are still being investigated, one cannot expect them to be entirely successful for all problems (given that depth-first search is inherently sequential [79] and certain variations of breadth-first search are also inherently sequential [80]); nevertheless, there are many cases where each algorithm delivers solutions in quite reasonable time with quite reasonable speed-ups. There are also efforts in designing parallel approximate matching algorithms [81, 82, 83].

We say a few words about the pioneering work of Kuhn in maximum weighted weighted matchings [70] for two reasons. Firstly, his paper was selected as the best paper of Naval Research Logistics in its first 50 years and was reproduced as [84] (there is a delightful history of the paper by Kuhn himself [85]). Secondly, it forms the basis for the algorithms [73, 74] that combine matchings with matrix scaling for better numerical properties during elimination.

By linear programming duality, it is known [70] that M is a maximum weighted matching if and only if there exist dual variables u_i and v_j with

$$\begin{cases} u_i + v_j \leq c_{ij} & \text{for } (i,j) \in E \setminus M \\ u_i + v_j = c_{ij} & \text{for } (i,j) \in M \end{cases}$$

For such u_i and v_j, setting

$$D_1 = \text{diag}(e^{u_i}) \quad \text{and} \quad D_2 = \text{diag}(e^{v_j}/\max_i |a_{ij}|)$$

scales the matrix so that $D_1 A M D_2$ has all ones on the diagonal and all other entries are less than or equal to one, see [73, 74]. Off-diagonal entries can be one in magnitude (this can happen for example when there is more than one maximum weighted matching), but otherwise the combined effect is such that the resulting matrix has larger entries on the diagonal. If the given matrix was obtained from a strongly diagonally dominant matrix or a symmetric positive definite matrix by permutations, the maximum product matching recovers the original diagonal [74]. Therefore, it is believed that the combination of the matching and the associated scaling would yield a set of good pivots. This was experimentally observed but has never been proved.

2.3.3 Elimination Tree and the Multifrontal Method

In this section, we describe arguably the most important graphical representation for sparse matrices: the elimination tree. We discuss its properties, construction and complexity, and illustrate the flexibility of this model. We then consider one of the most important elimination-tree based class of direct methods: the multifrontal method. We indicate how we can modify the tree for greater efficiency of the multifrontal method and show how it is used in a parallel and out-of-core context.

2.3.3.1 Elimination Tree

The elimination tree is a graphical model that represents the storage and computational requirements of sparse matrix factorization. The name elimination tree was first used in [86] principally as a computational tree to guide the factorization process. The term was also used by Jess and Kees [87], who again used the elimination tree to represent the computational dependencies in order to exploit parallelism. We note that Jess and Kees used the elimination tree of a triangulated graph. That is, they consider the graph that includes the fill-in edges obtained using a fill-reducing ordering. The formalized definitions of the elimination tree and the use of the data structures resulting from it for efficient factorization and solution are given by Schreiber [88]. Liu considers the elimination tree in some detail and provides a detailed discussion on it in [89] and [90]. The latter paper is a comprehensive survey of the elimination tree structure. Here we provide a few properties of the elimination tree and comment on its computation, mostly following the exposition in [90].

The elimination tree essentially relates to the factorization of pattern symmetric matrices. However, some modern solvers such as MUMPS [91] extend the use of the elimination tree to unsymmetric systems by using a tree based on the structure of the matrix $|A| + |A^T|$. This extends the benefits of the efficient symmetric symbolic factorization to the unsymmetric case, and so the following discussion can apply to general LU factorization. Note that there is a rich literature on the use of directed graphs and alternative tree models for the unsymmetric case which we do not cover—the reader is referred to [32, 92, 93, 94, 95] for the use of directed graph models in sparse factorization and to [96, 97] for alternative tree models for unsymmetric matrices.

We first give a few definitions before listing some important properties of the elimination tree. A *depth-first search* (DFS) of a graph starts with an initial node v, marks the node as visited and then recursively visits an unvisited vertex which is adjacent to v. If there is no unvisited vertex adjacent to v, the algorithm backtracks to the vertex which had initiated the search on v. The edges that are traversed when going to an unvisited vertex form a tree, which is called a *depth-first search tree* [98]. Let A be a symmetric positive definite matrix having the factorization LL^T. Let G_F represent the graph of the filled in matrix, i.e., the undirected graph of $L + L^T$. Then the elimination tree $T = (V, E)$ is a depth-first search tree of the undirected graph G_F. To us, this last statement summarizes most of the structural information relating to the factorization process. Therefore we discuss this correspondence a little more. As discussed in Section 2.2, the ith vertex of G_F corresponds to the ith row/column of $L + L^T$. If we run the DFS starting from the vertex n of G_F and, during the recursive calls from a vertex to its unvisited neighbours, we choose the one with the highest index first, we will obtain the elimination tree as its search tree. We note that if the matrix is reducible the first call from the vertex n will not explore all the vertices; when this happens, the highest numbered unvisited vertex should be taken as the starting vertex of another DFS. If the matrix is reducible, this will give the forest of elimination trees, one tree for each irreducible block.

We now give the most significant characteristics of the elimination tree. First, the elimination tree is a spanning tree of the graph corresponding to the filled in matrix. Second, the elimination tree can be constructed by making an edge from each vertex $j = 1, \ldots, n-1$ to the first nonzero l_{ij} in column j so that vertex i is the parent of vertex j. As there are no *cross edges* (a cross edge is an edge of the graph but not the DFS tree and it connects vertices which do not have an ancestor-descendant relationship in the DFS tree) with respect to a DFS tree of an undirected graph, the edges of G_F that are not in the tree T are *back edges*, i.e., they are from a vertex v to another vertex in the unique path joining v to the root (see [98] and [11, Section 23.3]). Combining this with the fill-path theorem of Rose et al. [16, Lemma 4, p. 270], we have that for $i > j$, the entry l_{ij} is nonzero if and only if there exists a path $v_j, v_{p1}, \ldots, v_{pt}, v_i$ in the graph of A such that the vertices $v_j, v_{p1}, \ldots, v_{pt}$ are all in the subtree of the elimination tree T rooted at node v_j. Another important property characterizing the fill-in is that $l_{ij} \neq 0$, if and only if the vertex v_j is an ancestor of some vertex v_k in the elimination tree T, where $a_{ik} \neq 0$ (see [89, Theorem 2.4]).

As is evident from the previous discussion, the elimination tree depends on the ordering of elimination operations. In particular, a node (that is the associated variable) can only be eliminated after all of its descendants have been eliminated—otherwise the structure is lost and the tree no longer corresponds to the depth-first search tree of the filled-in matrix anticipated at the beginning. A *topological ordering* of the nodes of a tree refers to an ordering in which each node is ordered before its parent. It was known earlier (see, e.g., [99] and comments in [89, p. 133]) that all topological orderings of the elimination tree are equivalent in terms of fill-in and computation; in particular any postorderings of the tree are equivalent. Liu [100] investigates a larger class of equivalent orderings obtained by tree restructuring operations, refereed to as *tree rotations*. Liu combines a result of Rose [22, Corollary 4, p. 198] and one of his own [89, Theorem 2.4, p. 132] to note that for any node v in the tree there is an equivalent ordering in which the nodes adjacent (in the graph of A) to the nodes in the subtree rooted at v are numbered last. Using a result of Schreiber [88, Proposition 5, p. 260], Liu identifies a node y as eligible for rotation if the following is true for all ancestors z (in the elimination tree) of y but not for the node y itself: the ancestors of z forms a clique in the filled graph and are connected (in the original graph, not in the tree) to the subtree rooted at z. Then, the nodes are partitioned into three sets and numbered in the following order: those that are not ancestors of the selected node while retaining their relative order; those that are ancestors of the selected node but not connected to (in the original graph, not in the tree) the subtree rooted at the selected node; then finally the remaining nodes. The final ordering will correspond to a new elimination tree in which the subtree rooted at the selected node would be closer to the root of the new tree.

Liu [89] (also in the survey [90]) provides a detailed description of the computation of the elimination tree. Firstly, an algorithm is given in which the structure of row i of L is computed using only A and the parent pointers set for the first $i-1$ nodes. This algorithm processes the rows of A in turn and runs in time proportional to the number of nonzeros in L. At row i, the pattern of the ith row of L is generated; for this the latest property of the elimination tree that we mention above (see [89, Theorem 2.4]) is used. Then for for each k for which $l_{ik} \neq 0$ and the parent k is not set, the parent of k is set to be i. In order to reduce the time and space complexity, Liu observes that parent pointers can be set using the graph of A and repeated applications of set operations for the disjoint set union problem [101]. A relatively simple implementation using these operations reduces the time complexity to $\mathcal{O}(\tau \alpha(\tau, n))$, where τ is the number of entries in A and $\alpha(\tau, n)$ is the two parameter variation of the inverse of the Ackermann function—$\alpha(\tau, n)$ can be safely assumed to be less than four.

2.3.3.2 Multifrontal Method

Based on techniques developed for finite-element analysis by Irons [102] and Speelpenning [103], Duff and Reid proposed the multifrontal method for symmetric [29] and unsymmetric [104] systems of equations. Liu [105] provides a good overview of the multifrontal method for symmetric positive definite matrices.

The essence of a frontal method is that all elimination operations are performed on dense submatrices (called *frontal matrices*) so that these operations can be performed very efficiently on any computer often by using the Level 3 BLAS [106]. The frontal matrix can be partitioned into a block two by two matrix where all variables from the (1,1) block can be eliminated (the variables are called fully summed) but the Schur complement formed by the elimination of these on the (2,2) block cannot be eliminated until later in the factorization. This Schur complement is often called a *contribution block*.

The multifrontal method uses an *assembly tree* to generalize the notion of frontal matrices and to allow any fill-reducing ordering. At each node of the assembly tree, a dense frontal matrix is constructed from parts of the original matrix and the contribution blocks of the children—this process is called the *assembly* operation. Then, the fully summed variables are eliminated (factorization of the (1,1) block takes place) and the resulting contribution block is passed to the parent node for assembly into the frontal matrix at that node. Clearly, one can only perform the eliminations at a node when all the contributions have been received from its children. Liu [105] defines a tree (associated with LL^T factorization of a matrix A) as an assembly tree if for any node j and its parent p, the off-diagonal structure of the jth column of L is a subset of the structure of the pth column of L and $p > j$. It is easy to see that the elimination tree satisfies this condition and that there are other tree structures which will exhibit the same property. In this setting, the contributions from j (or any descendant of it in the assembly tree) can be accommodated in the frontal matrix of p and hence the generality offered by the assembly tree might be used for reducing the memory requirements [105, p. 98].

Since the elimination tree is defined with one variable (row/column) per node, it only allows one elimination per node and the (1,1) block would be of order one. Therefore, there would be insufficient computation at a node for efficient implementation. It is thus advantageous to combine or amalgamate nodes of the elimination tree. The amalgamation can be restricted to avoid any additional fill-in. That is, two nodes of the elimination tree are amalgamated only if the corresponding columns of the L factor have the same structure below the diagonal. As even this may not give a large enough (1,1) block, a threshold based amalgamation strategy can be used in which the columns to be amalgamated are allowed to have a certain number of discrepancies in their patterns, introducing logical zeros. Duff and Reid do this in their original paper [29] where they amalgamate nodes so that a minimum number of eliminations are performed at each node of the resulting tree. That is, they make the (1,1) blocks at least of a user-defined order. Ashcraft and Grimes [107] investigate the effect of this relaxation of the amalgamation and provide new algorithms. Firstly, the notion of a fundamental supernode is defined. A fundamental supernode is a maximal chain $(n_1, n_2, \ldots, n_p)$ of nodes in the tree such that each n_i is the only child of n_{i+1}, for $i = 1, \ldots, p-1$, and the associated column structures are perfectly nested. Then, the fundamental supernodes are visited in an ordering given by a postorder, and the effect of merging children with a parent node on the number of logical zeros created is taken into account to amalgamate nodes. In [108], an efficient algorithm which determines the fundamental supernodes in time proportional to the number of nonzeros in the original matrix (avoiding the symbolic factorization altogether) is presented.

The simple tree structure of the computations helps to identify a number of combinatorial problems, usually relating to scheduling and task mapping for efficient memory use, in out-of-core solution and in parallel computing contexts. In the rest of this section, we discuss some of the issues relating to efficient memory use and parallelization.

In a multifrontal method, the pattern of memory use permits an easy extension to out-of-core execution. In these methods, memory is divided into two parts. In the static part, the computed factors of the frontal matrices are stored. This part can be moved to secondary storage. The second part, called the active memory, contains the frontal matrix being factorized and a stack of contributions from the children of still uneliminated nodes. Liu [109] minimizes the size of the active memory by rearranging the children of each node (hence creating an equivalent ordering) in order to minimize the peak active memory for processing the whole tree. In a series of papers, Agullo et al. [110, 111] and Guermouche and L'Excellent [112] develop these ideas of Liu [109] concerning the assembly and partial assembly of the children and their ordering. In these papers, algorithmic models that differentiate between the concerns of I/O volume and the peak memory size are developed, and a new reorganization of the computations within the context of an out-of-core multifrontal method are presented.

George et al. [113] propose the *subtree-to-subcube* mapping to reduce the communication overhead in parallel sparse Cholesky factorization on hypercubes. This mapping mostly addresses parallelization of the factorization of matrices arising from a nested dissection based ordering of regular meshes. The essential idea is to start from the root and to assign the nodes of the amalgamated elimination tree in a round robin-like fashion along the chains of nodes and to divide the processors according to the subtrees in the elimination tree and then recursively applying the idea in each subtree. The method reduces the communication cost but can lead to load imbalance if the elimination tree is not balanced. Geist and Ng [114] improve upon the subtree-to-subcube mapping to alleviate the load imbalance problem. Given an arbitrary tree, Geist and Ng find the smallest set of subtrees such that this set can be partitioned among the processors while attaining a load balance threshold supplied by the user (in the experiments an imbalance as high as 95% is allowed). A breadth-first search of the tree is performed to search for such a set of subtrees. The remaining nodes of the tree are partitioned in a round robin fashion. Improving upon the previous two algorithms, Pothen and Sun [115] propose the *proportional mapping* algorithm. They observe that the remaining nodes after the subtree mapping can be assigned to the processors in order to reduce the communication overhead that arises because of the round robin scheme. The essential idea is to map the nodes (they use a clique tree representation of the filled-in graph and therefore nodes are cliques of the nodes of the original graph) that lie on the paths from already assigned subtrees to the root onto the processors that are associated with those subtrees. A first-fit decreasing bin-packing heuristic is used to select a processor among the candidates. In a different framework, Gilbert and Schreiber [116] address the parallelization on a massively parallel, fine-grained architecture (using a virtual processor per each entry of L). In this work, a submatrix corresponding to supernodes is treated as a dense submatrix and is factorized in a square grid of processors. In order to facilitate parallelism among independent, dense submatrices, a two-dimensional bin-packing is performed. It is interesting to note the relevance of this work to current massively parallel architectures. Amestoy et al. [117, 118, 119] generalize and improve the heuristics in [114, 115] by taking memory scalability issues into account and by incorporating dynamic load balancing decisions for which some preprocessing is done in the analysis phase.

Many of the known results on out-of-core factorization methods are surveyed in a recent thesis by Agullo [120]. The thesis surveys some out-of-core solvers including [121, 122, 123, 124, 125], provides NP-completeness results for the problem of minimizing I/O volume in certain variations of factorization methods, and also develops polynomial time algorithms for some other variations. Reid and Scott discuss the design issues for HSL_MA77, a robust, state-of-the-art, out-of-core multifrontal solver for symmetric positive-definite systems [126] and for symmetric indefinite systems [127].

2.3.4 Block Triangular Form

Consider a permutation of a square, nonsingular sparse matrix that yields a block upper triangular form (BTF):

$$
A = \begin{pmatrix}
A_{11} & * & * & * \\
O & A_{22} & * & * \\
\vdots & \vdots & \ddots & * \\
O & O & \cdots & A_{pp}
\end{pmatrix},
$$

where each block on the diagonal is square and nonsingular and the nonzeros are confined to the block upper triangular part of the permuted matrix. If a permutation to this form is used when solving the linear system, the whole system can be solved as a sequence of subproblems, each involving a solution with one of the blocks on the diagonal.

The algorithms to obtain the BTF proceed in two steps, see e.g., [41, 128] and [129]. First, a maximum cardinality matching on the bipartite graph representation is found, see [68, 69]. In the case of a structurally full-rank matrix, this would be a perfect matching. Then, the matrix is nonsymmetrically permuted so that the matching entries are on the main diagonal. The directed graph of this matrix is then constructed, and its strongly connected components are found [98] which define the blocks on the diagonal. Efficient and very compact implementations in Fortran are provided in [130, 131].

The block structure of the BTF is unique, apart from possible renumbering of the blocks or possible orderings within blocks, as shown in [132, 133, 134]. In other words, the same block structure would be obtained from any perfect matching. We note that any such matching contains nonzeros that are only in the diagonal blocks of the target BTF.

The BTF form is generalized to rectangular and unsymmetric, structurally rank deficient matrices by Pothen [135] and Pothen and Fan [136] following the work of Dulmage and Mendelsohn [128, 133, 134]. According to this generalization any matrix has the following form

$$\begin{pmatrix} A_H & * & * \\ O & A_S & * \\ O & O & A_V \end{pmatrix},$$

where A_H is underdetermined (horizontal), A_S is square, and A_V is overdetermined (vertical). Each row of A_H is matched to a column in A_H, but there are unmatched columns in A_H; each row and column of A_S are matched; each column of A_V is matched to a row in A_V, but there are unmatched rows in A_V. Furthermore, Pothen and Fan [136] and Dulmage and Mendelsohn [128] give a finer structural characterization. The underdetermined matrix A_H can be permuted into block diagonal form, each block being underdetermined. The square block A_S can be permuted into upper BTF with square diagonal blocks, as discussed before. The overdetermined block A_V can be permuted into block diagonal form, with each block being overdetermined. Again, the fine permutation is unique [135], ignoring permutations within each fine block. The permutation to the generalized BTF is performed in three steps. In the first step, a maximum cardinality matching is found, not necessarily a perfect matching. Then each row that reaches an unmatched column through alternating paths (these rows are all matched, otherwise the matching is not of maximum cardinality) are put into the horizontal block, along with any column vertex in those paths. Then, a corresponding process is run to detect the columns and rows of the vertical block. Finally, the previous algorithm is run on the remaining full rank square block to detect its fine structure. Pothen [135] proves the essential uniqueness of the BTF for rectangular and structurally singular square matrices (see also [133, 134]).

In recent work, we have presented a few observations on the BTF of symmetric matrices [137]. Firstly, the blocks A_H and A_V are transposes of each other. That is, the set of rows and the set of columns that define the horizontal block are equal to the set of columns and the set of rows that define the vertical block, respectively. Secondly, a fine block of the square submatrix A_S is such that either the set of its row indices is equal to the set of its column indices, or they are totally disjoint and there is another square block equal to the transpose of the block.

2.4   Iterative Methods

A different but equally well known family of methods used for solving linear systems of the form

$$Ax = b \tag{2.3}$$

starts with an initial guess and, by successive approximations, obtains a solution with a desired accuracy. The computations proceed in iterations (hence the name *iterative methods*), where a set of linear vector operations and usually one or two sparse matrix-vector multiplication operations take place at each iteration. As one can easily guess, the presence of sparse matrices raises a number of combinatorial problems. Indeed, there is a beautiful interaction between the parallelization of a certain class of iterative methods and combinatorial optimization which revolves around graph and hypergraph partitioning. As this is the subject of another survey [138], we do not cover this issue. Instead, we refer the reader to [12, 139, 140, 141] for this interaction and to [142] for different aspects of parallelization (including a survey of earlier studies, the effect of granularity on parallelism, and algorithm design issues).

There are many aspects to iterative methods; here we restrict the discussion to some preconditioning techniques because of their combinatorial ingredients. *Preconditioning* refers to transforming the linear system (2.3) to another one which is easier to solve. A *preconditioner* is a matrix enabling such a transformation. Suppose that M is a nonsingular matrix which is a good approximation to A in the sense that M^{-1} well approximates A^{-1}, then the solution of the system $M^{-1}Ax = M^{-1}b$ may be much easier than the solution of the original system. There are alternative formulations as to how to apply the preconditioner: from the left, from the right, or both from left and right. For our purposes in this section, those formulations do not make any difference, and we refer the reader to a survey by Benzi [143] that thoroughly covers most of the developments up to 2002. We refer to some other surveys when necessary.

Among various preconditioning techniques, we briefly discuss preconditioners based on incomplete factorization, on algebraic multigrid, and support graph preconditioners. We choose these preconditioners because of the strong presence of combinatorial issues and because of their maturity. We do not discuss sparse approximate inverse preconditioning [144] nor its factored form [145], although the choice of pattern for the approximate inverse is a combinatorial issue (see [146] and the references therein), and sparse matrix ordering has a significant effect on the factored form of approximate inverses (see for example [147]). We also do not include a recent family of banded preconditioners [148] which raises various combinatorial problems that are currently being investigated (see the relevant chapter in this book).

2.4.1 Preconditioners Based on Incomplete Factorization

As discussed in the previous section, fill-in usually occurs during the LU decomposition of a matrix A. By selectively dropping the entries computed during the decomposition, one can obtain an *incomplete LU factorization* (ILU) of A with a lower and an upper triangular matrix $\hat{L}$ and $\hat{U}$. The matrix $M = \hat{L}\hat{U}$, then can approximate the matrix A and hence can be used as a preconditioner. Benzi traces incomplete factorization methods back to the 1950s and 1960s by citing papers by Buleev and Varga dating, respectively, 1959 and 1960, but credits Meijerink and van der Vorst [149] for recognizing the potential of incomplete factorization as a preconditioner for the conjugate gradient method.

As just mentioned, the essence of incomplete factorization is to drop entries in the course of the elimination process. Current methods either discard entries according to their position, value, or with a combination of both criteria.

Consider a pattern $\mathcal{S} \subseteq \{1, \ldots, n\} \times \{1, \ldots, n\}$ and perform the elimination process as in (2.1) but allow fill-in if the position is in $\mathcal{S}$ using the formulae

$$
a_{ij}^{(k+1)} \leftarrow
\begin{cases}
a_{ij}^{(k)} - \left(a_{ik}^{(k)} / a_{kk}^{(k)} \right) a_{kj}^{(k)}, & \text{if } (i,j) \in \mathcal{S} \\
a_{ij}^{(k)}, & \text{otherwise}
\end{cases}
\tag{2.4}
$$

for each major elimination step k ($k = 1, \ldots, n$) and $i, j > k$. Often, $\mathcal{S}$ is set to the nonzero pattern of A, in which case one obtains ILU(0), a *no-fill* ILU factorization. This was used in [149] within the context of incomplete Cholesky factorization.

A generalization was formalized by Gustafsson [150] (see in [151, p. 259]). Gustafsson develops the notion of level of fill-in and drops fill-in entries according to this criterion. The initial level of fill-in for a_{ij} is defined as

$$
lev_{ij} \leftarrow
\begin{cases}
0 & \text{if } a_{ij} \neq 0 \text{ or } i = j, \\
\infty & \text{otherwise},
\end{cases}
$$

and for each major elimination step k ($k = 1, \ldots, n$) during the elimination (2.4), the level of fill-in is updated using the formula

$$
lev_{ij} = \min\{lev_{ij}, lev_{ik} + lev_{kj} + 1\}.
$$

Given an initial choice of drop level ℓ, ILU(ℓ) drops entries whose level is larger than ℓ. Observe that the level of fill-in is a static value that can be computed by following the elimination process on graphs.

There have been many improvements upon the basic two incomplete factorization methods discussed above, resulting in almost always better preconditioners. However, these two methods are still quite useful (and effective for certain problems) because of the structural properties of the computed factors, as we shall see later when discussing the parallel computation of incomplete factorization preconditioners.

For example, Saad [152] develops a dual threshold strategy ILUT(τ, p), where a fill-in entry is dropped if its value is smaller than τ, and at most p fill-ins per row are allowed. For more on the variations and the properties of preconditioners based on incomplete factorization, we refer the reader to [143, 153, 154].

2.4.1.1 Orderings and Their Effects

As for their complete factorization counterparts, incomplete factorization preconditioners are sensitive to the ordering of the elimination. Recall from

Section 2.3 that, for a complete factorization, the ordering affects both the fill-in and stability of the factorization. For an incomplete factorization, in addition to these two effects, the ordering of the eliminations also affects the convergence of the iterative method. This last issue, although demonstrated by many, has yet to be understood in a fully satisfactory way. Benzi [143] cites 23 papers between the years 1989 and 2002 that experimentally investigate the effects of ordering on incomplete factorization preconditioners.

The first comparative study of the effect of ordering on incomplete Cholesky factorization was performed by Duff and Meurant [155]. The paper shows that, contrary to what was conjectured in [156], the number of iterations of the conjugate gradient method is not related to the number of fill-ins discarded but is almost directly related to the norm of the residual matrix $R = A - \bar{L}\bar{L}^T$. Chow and Saad [157] show that for more general problems the norm of the preconditioned residual $(\bar{L}\bar{U})^{-1}R$ is also important.

The general consensus of the experimental papers, starting with [155], including [158], strongly favour the use of RCM. Bridson and Tang [159] prove a structural result (using only the connectivity information on the graph of A) as to why RCM yields successful orderings for incomplete factorization preconditioning. One of the results showing why RCM works for IC(0) is based on $(\bar{L}\bar{L}^T)^{-1}$ being fully dense if and only if each column of $\bar{L}$ has a nonzero below the diagonal. Any ordering yielding such a structure is called a reversed graph traversal in [159] and RCM is shown to yield such a structure. We note that for the complete factorization case such characterizations were used before; for example the irreducibility characterization of A in terms of the structure of L (see [160] and [161]). The other result of [159] is based on the intuition that if the structures of L^{-1} and $\bar{L}^{-1}$ coincide, then the incomplete factor returned by IC(0) could be a good approximation. It is then shown that reversing an ordering that can be found by a graph search procedure that visits, at each step, a node that is adjacent to the most recently visited node (allowing backtracking) will order A so that the above condition holds. RCM does not yield such an ordering in general, but a close variant always will.

As an alternative to using the ordering methods originally designed for complete factorization, orderings specially designed for incomplete factorization have also been developed. In [162], a *minimum discarded fill* ordering (MDF) is proposed. The algorithm is considered as the numerical analogue of the minimum deficiency ordering (scheme S3 of [21]), and it corresponds to ILU(ℓ). The basic idea is to eliminate the node with the minimum discarded fill at each stage of the incomplete elimination in an attempt to minimize the Frobenius norm of the matrix of discarded elements. The method has been developed in [163] yielding two variants, both of which still are fairly expensive. D'Azevedo et al. deserve the credit for giving the static characterization of the factors in ILU(ℓ) in terms of the graph of the original matrix. In [164], two sets of ordering methods which use the values of the matrix elements are proposed. The first one is a variation of RCM where ties are broken according to the numerical values of the matrix entries corresponding to edges between the vertex and the already ordered vertices. The second one is based on minimum spanning trees, where at each node the least heavy edge is chosen to produce an ordering of the nodes. These algorithms use heuristics based on a theorem [164, Theorem 1] (the proof refers to the results of [149, 163]) relating the element l_{ij} to the entries along the path joining vertices i and j in the original graph of an M-matrix.

In recent studies, such as [73, 165, 166], nonsymmetric orderings that permute rows and columns differently are used to permute large entries to the diagonal before computing an incomplete preconditioner. Other more recent work that uses similar ideas includes [167] where pivoting is performed during incomplete elimination; [168] where fill-reducing ordering methods are interleaved with the elimination; and [169] where weighted matching-like algorithms are applied to detect a diagonally dominant square submatrix, which is then approximately factorized. Its approximate Schur complement is then constructed, on which the algorithm is applied recursively.

Blocking methods for the complete factorization are adapted to the incomplete factorization as well. The aim here is to speed up the computations as for complete factorization and to have more effective preconditioners (in terms of their effect on the convergence rate). A significant issue is that in certain incomplete factorization methods, the structure of the incomplete factors are only revealed during the elimination process. Ng et al. [170] present a technique for the incomplete Cholesky factorization that starts with the supernodal structure of the complete factors. If standard dropping techniques are applied to individual columns, the pre-computed supernodal structure is usually lost. In order to retain the supernodal structure as much as possible, Ng et al. either drop the set of nonzeros of a row in the current set of columns (the supernode) or retain that set. In order to obtain sparser incomplete factors, they subdivide each supernode so that more rows can be dropped.

In [171] and [172] blocking operations are relaxed in such a way that the supernodes are not exact, but are allowed to incur some fill-in. In the first step, the set of exact supernodes are found. Then, in [172], a compressed matrix is created from the exact supernodes, and the cosine-similarity between nodes or supernodes are computed to allow some inexact supernodes. In [171], inexact amalgamations are performed between the parents and children in the assembly tree with a threshold measuring the inexactness of the supernodes.

Another set of blocking approaches are presented in [173, 174], explicitly for preconditioning purposes. Here, a large number of small dense blocks are found and permuted to the diagonal. The initial intention of these methods was to obtain block diagonal preconditioners, but the resulting orderings are found to be useful for point incomplete factorizations as well, see [158]. The blocking methods are fast (in general run in $\mathcal{O}(n+\tau)$ time although the current version finds a maximum product matching with MC64 as a preprocessor) and are provided in the PABLO library (http://www.math.temple.edu/~daffi/software/pablo/).

Many of the current state-of-the-art variations of ILU methods are provided in ILUPACK [175]. Other efforts include PETSc [176], IFPACK [177], and ITSOL (http://www-users.cs.umn.edu/~saad/software/ITSOL/index.html).

Benzi et al. [158] ask the following questions; answers to which will shed light into the effect of orderings on incomplete factorization preconditioners: (i) why does the choice of the initial node and the ordering within level sets affect the performance of (reverse) Cuthill-McKee? (ii) why does ILU(0) with a minimum degree ordering not suffer from the instability that occurs when a natural ordering is used, for the model problem or similar ones?

2.4.1.2 Parallelization

The parallel computation of ILU preconditioners is often implemented in two steps. Firstly, the matrix is partitioned into blocks to create a high level parallelism where the ILU of the interior blocks can be performed independently. Secondly, dependencies between blocks are identified and sequential bottlenecks reduced to increase the parallelism.

The basic algorithm for no-fill ILU can be found in [154, p. 398]. A parallel algorithm for a threshold based ILU preconditioner is given in [178]. In this work, after the initial partitioning and ILUT elimination of interior nodes, an independent set of boundary nodes is found using Luby's algorithm [179]. After elimination of these nodes, which can be done in parallel, fill-in edges are determined and added between the remaining nodes. Another independent set is then found and eliminated. The process is continued until all nodes have been eliminated. Hysom and Pothen [180] develop a parallel algorithm for a level-based ILU. They order each subdomain locally, ordering the interface nodes of each domain after the interior nodes. Then, a graph of subdomains is constructed that represents interactions between the subdomains. If two subdomains intersect, ordering one before the other introduces a directed edge from the first domain to the second. Considering these directions, Hysom and Pothen color the vertices of the subdomain graph to reduce the length of directed paths in this graph. The color classes can again be found using Luby's algorithm. Hysom and Pothen impose constraints on the fill-in that can be obtained from a pure ILU(ℓ) factorization. This helps improve the parallelization. Their paper presents an improvement to the scheme outlined above and provides a fill-in theorem for the incomplete factorization.

2.4.2 Support Graph Preconditioners

Combinatorial structures have been used to construct and analyse preconditioners. For example, Rose [181] defines the R-regular splitting (matrix splitting is a form of preconditioning, see [182] for splitting methods) of singular M-matrices. Starting from a given choice of diagonal blocks, Rose reorders the blocks so that the vertices in a cycle (guaranteed to exist) are ordered consecutively. This ordering guarantees the convergence of any given block regular splitting for singular M-matrices. This was not true without this simple combinatorial tinkering of the given choice of diagonal blocks.

A more recent combinatorial preconditioner and a set of tools used in designing and proving the effectiveness of the constructed preconditioner is based on work by Vaidya [183] (we refer the reader to Chapter 3 for a more thorough discussion). Although Vaidya's manuscript is not published, his main theorem and the associated preconditioners are given in the thesis of his student Joshi [184, Chapter 5]. In this work, preconditioners for symmetric, positive definite, diagonally dominant matrices are constructed using a maximum weighted spanning tree of the associated undirected graph (the edge weights are equal to the absolute values of the corresponding matrix entries). In other words, some off-diagonal entries of the given matrix are dropped to obtain the preconditioner. In Joshi's thesis there is a condition on which entries of A to drop: an edge can be dropped if one can associate a single path in the graph of the preconditioner matrix such that all edges in this path have a weight at least as large as the weight of the dropped edge. A maximum weighted spanning tree satisfies this condition. Any matrix containing that spanning tree and some additional edges also satisfies this condition. Joshi demonstrates the development on two dimensional regular grids. First, he separates the boundary nodes from the internal ones by removing all the edges between the boundary nodes and the internal ones but keeping the edges between boundary nodes. Then, he constructs a spanning tree of the internal nodes, and finally joins the boundary to this tree with a single edge (one of those removed previously).

The proof that such structures give effective preconditioners uses two *graph embedding* notions (Joshi uses the term embedding just to mean the representation of a grid by a graph). For simplicity, consider two graphs H and G defined on the same set of vertices. The embedding of H into G is a set of paths of G, such that each edge in H is associated with a single path in G. The *congestion* of an edge of G is the sum of the weights of such paths that pass through that edge, and the *dilation* of an edge of H is the length of the associated path in G. The maximum congestion of an edge of G and the maximum dilation of an edge of H define, respectively, the congestion and the dilation of the embedding. Vaidya's main result as stated by Joshi says that the condition number of the preconditioned system is less than the product of the congestion and the dilation of the embedding. We note that these graph embedding notions are also known in parallel computing literature and are used in studying the interconnection networks (see for example [185, p. 39]).

The basic support-tree preconditioners and the graph embedding tools used in bounding the condition number of the preconditioned system were extended and generalized by Miller and his students Gremban and Guattery [186, 187, 188]. The extensions by Miller and Gremban include projecting the matrix onto a larger space and building support trees using Steiner trees (a Steiner tree forms a spanning tree of a graph with possibly additional vertices and edges). Vertex separators that are used to partition the underlying graph are defined as the additional nodes. The leaves of the constructed support-tree correspond to the nodes of the underlying graph, and the internal nodes correspond to vertex separators. This form of preconditioner is demonstrated to be more amenable to parallelization than the original support-tree preconditioners [186, Section 3.4]. Reif [189] also develops Vaidya's preconditioners and reduces the bounds on the condition number of the preconditioned matrix by using a weighted decomposition, i.e., by partial embedding of the edges into multiple paths. Similar decompositions are used and defined more clearly by Guattery [188] based on Gremban's thesis. Guattery uses the embedding tools to analyse incomplete Cholesky factorization as well.

As seen from the references cited in the above paragraph, the earlier papers on support tree preconditioners are not published in widely accessible journals, except the rather theoretical paper by Reif [189]. The preconditioners, the related tools that are used to analyse them, and the potential of the theory as a means to analyse preconditioners are presented by Bern et al. [190]. This paper collects many results and refines them, at the same time extending the techniques to analyse modified incomplete Cholesky factorization (the dropped entries are added to the diagonal entry, keeping the row sums of the original matrix and the preconditioner matrix the same). The main tool that is used in bounding the condition numbers of the preconditioned matrix is called the splitting lemma. Suppose we use a preconditioner B for the matrix A, where A is symmetric and diagonally dominant with nonnegative off-diagonal entries, and B is symmetric positive semidefinite, and both A and B have zero row sums. One way to bound the maximum eigenvalue of $B^{-1}A$ is to split A and B into m parts as

$$A = A_1 + \cdots + A_m \quad \text{and} \quad B = B_1 + \cdots + B_m \quad ,$$

where each A_k and B_k are symmetric positive semidefinite. Proving $\tau B_k - A_k$ is positive semidefinite for all k gives a bound on $\lambda_{max}(B^{-1}A)$. A similar technique is used to bound $\lambda_{min}(B^{-1}A)$ so that the condition number of the preconditioned system given by $\lambda_{max}(B^{-1}A)/\lambda_{min}(B^{-1}A)$ can be bounded. The relation with the graph embedding concept is that each A_k represents an edge in the graph of A and each B_k represents the associated path in the graph of B. Let A_k and B_k correspond to the edge (i,j), i.e., to the nonzero $a_{ij} = a_{ji}$. The matrix A_k contains the weight of the edge a_{ij} in its entirety, whereas the matrix B_k contains fractions of the weights of the edges along the path associated with the corresponding edge so that the sum of the weights of the same edge in different paths add up to the weight of the edge in B. For example in Vaidya's construction, if b_{ij} represents the weight of an edge b in the graph of B, and if the edge b appears in paths associated with some a_{ij}, then each such a_{ij} contributes a_{ij} divided by the congestion of b to b_{ij}. The diagonal entries of A_k and B_k are set in such a way that the row sums are zero. The congestion due to the edge a_{ij} represented by A_k in the path represented by B_k is $|a_{ij}|/b_k$, where b_k is the minimum magnitude of a nonzero off-diagonal entry in B_k. The dilation d_k is the length of the associated path, hence one less than the number of non-null rows or columns of B_k. These two numbers can be used to show that $d_k|a_{ij}|/b_k B_k - A_k$ is positive semidefinite. The bound on λ_{max} is therefore $\max_{ij} d_k(|a_{ij}|/b_k)$.

Bern et al. [190] use the above tools to analyse Vaidya's maximum-weight spanning tree preconditioners. In this analysis, B's underlying graph is a maximum-weight spanning tree T. Loose asymptotic bounds for congestion and dilation can be computed as follows. Suppose there are m edges in the graph of A, then B will be split using m paths, each defined uniquely in T. If one allocates $1/m$ of the weight of each edge of T to each path, then the maximum congestion due to an edge a_{ij} in the associated path would be $\frac{a_{ij}}{a_{ij}/m} = m$. Furthermore, the dilation can be at most $n - 1$, one less than the number of vertices. Hence, $\mathcal{O}(mn)$ is a loose upper bound on the maximum eigenvalue λ_{max} of the preconditioned system. The analysis for λ_{min} is easier: λ_{min} is at least 1, as each edge of B is already in A. Therefore, the preconditioned system has a condition number bound of $\mathcal{O}(mn)$. Another class of preconditioners that are based on maximum-weight spanning trees is also proposed by Vaidya and analysed in [190]. These preconditioners are built by partitioning the vertices of a maximum-weight spanning tree T into t connected parts, and then enriching the edge set of T by adding the maximum weighted edge between any pair of parts (if two parts are already connected by an edge of T, nothing is done for these two). With this kind of preconditioner, the preconditioned system is shown to have a condition number of $\mathcal{O}(n^2/t^2)$ [190].

Later, Boman and Hendrickson [191] generalized these embedding tools to develop more widely applicable algebraic tools. Specifically, the tools can now address symmetric positive semidefinite matrices. The insights gained with this generalization are used to analyse the block Jacobi preconditioner and have enabled the development of new preconditioners [192]. Additionally, the support theory techniques have been extended to include all diagonally dominant matrices [191].

The work by Chen and Toledo [193] presents an easily accessible description of Vaidya's preconditioners and their implementation. They report mixed results with Vaidya's preconditioners. On certain problems (2D and 3D with diagonal dominance) remarkable performance is obtained but on some others (3D general) the performance is poorer than standard preconditioners based on incomplete factorization. Sophisticated algorithmic approaches which aim at constructing spanning trees yielding a provably better condition number for the preconditioned matrix, and also provably better graph decompositions are given in [194, 195, 196]. The developments in these papers lead to nearly linear time algorithms for solving a certain class of linear systems. Another very recent study is by Koutis, another student of Miller. Koutis [197] proposes preconditioners for Laplacian matrices of planar graphs. Koutis develops the preconditioners by aggregating graph-based preconditioners of very small subsystems. Furthermore, Steiner tree preconditioners [186] are extended and algebraic multigrid preconditioners are cast in terms of Steiner trees, yielding combinatorial implications for the algebraic preconditioners.

The research on support graph preconditioners is very active. It seems that much of the results are theoretical, focusing only on some particular classes of linear systems. Therefore, there is much to do in this relatively young intersection of combinatorics and linear system solution. For example, except for a few comments in [191], nothing is said for nonsymmetric matrices, not even for pattern symmetric ones. Although we have very little experience with support graph preconditioning methods, we think that they can help understand the effects of ordering methods for incomplete factorization preconditioners discussed in the previous subsection.

Support-graph preconditioners are available in the TAUCS library of iterative solvers http://www.tau.ac.il/~stoledo/taucs/ and in PETSc [176].

2.4.3 Algebraic Multigrid Preconditioning

Algebraic multigrid preconditioners approximate a given matrix by a series of smaller matrices. Simply put, the system of equations is coarsened to a much smaller system, a system is solved and refined at each level to obtain a correction to the original solution. There are a number of combinatorial issues regarding efficient parallelization of multigrid solvers, see Chow at el. [198]. Here, we will look at the coarsening operation which incorporates a number of combinatorial techniques. For a survey on algebraic multigrid, we refer the reader to [199].

In general, there are three coarsening approaches used in algebraic multigrid: classical coarsening (see, e.g., [200]), aggregation based coarsening (see, e.g., [201]), and graph matching (this is a relatively new method described in [202]).

In classical coarsening approaches of the type given in [200], the grid points are classified into coarse or fine points. The coarse points are used to define the coarser grid. In order to restrict the size of the coarser grid, such points are restricted to be a maximal independent set. As we have seen before, this can be achieved using Luby's algorithm [179]. Two modifications of Luby's algorithm are presented in [203] for the coarsening operation in the algebraic multigrid context. The modifications include directing Luby's algorithm to choose points that have a higher number of influenced points (that is, those that are connected to the chosen points by heavy weights) and removing certain points before running the algorithm.

In aggregation based coarsening [201], an aggregate is defined as a root point and its immediate neighbours for which a certain condition in the magnitudes of the coefficient between the neighbours and the root point is satisfied. A constraint on a root point is that the aggregate defined around it cannot be adjacent to another root point. Therefore, a maximal independent set in the square of the graph (a graph formed by adding edges between any two vertices that are connected by a path of length 2 in the original graph) of the fine grid is found to define the roots of the aggregates again using Luby's algorithm. The exposition suggests that Luby's algorithm is run on the square graph. The graph coloring heuristics, see e.g., [204, 205], can be modified and used to reduce the space requirements by avoiding the construction of the square graph (similar applications of the distance-k graph coloring heuristics can also boost the performance of some other aspects of multigrid solvers [206] as well as the coarsening [207]).

In the matching based coarsening [202], the coarse grids are defined using simple graph matching heuristics. In this work, a matching is found on the graph of a fine grid, and the matched vertices are reduced to a single vertex in the coarser grid. The matching is of the cardinality matching type, but does not aim at maximizing the cardinality. An investigation in the paper shows that if the original matrix is an M-matrix, so are the coarser matrices.

Current state-of-the-art multigrid solvers include ML [208] and Boomer-AMG [209]. PETSc [176] provide interfaces for a number of multigrid pre-conditioners. For a much larger list of multigrid solvers and related multilevel ones see `http://www.mgnet.org/mgnet-codes.html`.

2.5   Conclusions

In this review, we have been rather eclectic in our choice of topics to illustrate the symbiotic relationship between combinatorics and sparse linear algebra. This is in part because other papers in this volume address specific subareas and in part because of our own interest and expertise. Although space and energy prevent us from going into significant detail, we have given a substantial number of references that should easily quench the thirst of anyone eager to dig more deeply.

We have discussed graph search algorithms in the spirit of depth- and breadth-first search methods; both weighted and unweighted bipartite matchings; spanning trees; and graph embedding concepts. We believe that these are the most important and useful tools of the trade, and hence by having some level of acquaintance with these concepts, a computational scientist will be able to start understanding many of the issues that arise in solving sparse linear systems and be able to see how combinatorial approaches can be used to solve them.

We hope that we have communicated to the reader that combinatorial optimization and graph theory play a dominant role in sparse linear system solution. This is a delightful combination as the discrete and continuous worlds often seem so far apart, yet the synergy created by the interaction of the two leads to developments and advances in both worlds. Much of the combinatorial material that we have discussed is fairly elementary and indeed most would be covered in the context of an undergraduate level discrete mathematics course, or a senior-level algorithms course. We view this very positively as it means that these basic techniques are accessible to many people. However, the way these elementary techniques are applied requires substantial conceptualization, both in casting a problem in combinatorial terms and in restructuring computational methods to accommodate the combinatorial results.

Acknowledgments

We thank our colleagues whose works enabled much of the issues covered in this paper. We also thank Patrick R. Amestoy and Jean-Yves L'Excellent for their contribution to the presented material. We also thank Ümit V. Çatalyürek for his comments on an earlier version of the paper.

Bibliography

[1] Hendrickson, B. and Pothen, A., "Combinatorial scientific computing: The enabling power of discrete algorithms in computational science," *High Performance Computing for Computational Science—VECPAR 2006*, edited by M. Dayde, M. L. M. Palma, L. G. A. Coutinho, E. Pacitti, and J. C. Lopes, Vol. 4395 of *Lecture Notes in Computer Science*, 2007, pp. 260–280.

[2] Bollhöfer, M. and Schenk, O., "Combinatorial aspects in sparse elimination methods," *GAMM Mitteilungen*, Vol. 29, 2006, pp. 342–367.

[3] Heath, M. T., Ng, E., and Peyton, B. W., "Parallel algorithms for sparse linear systems," *SIAM Review*, Vol. 33, No. 3, 1991, pp. 420–460.

[4] Brualdi, R. A. and Cvetković, D. M., *A Combinatorial Approach to Matrix Theory and its Applications*, Chapman & Hall/CRC Press, Boca Raton, 2009.

[5] Brualdi, R. A. and Ryser, H. J., *Combinatorial Matrix Theory*, Vol. 39 of *Encyclopedia of Mathematics and its Applications*, Cambridge University Press, Cambridge, UK; New York, US, Melbourne, AU, 1991.

[6] Brualdi, R. A., "The symbiotic relationship of combinatorics and matrix theory," *Linear Algebra and its Applications*, Vol. 162–164, 1992, pp. 65–105.

[7] Brualdi, R. A., *Combinatorial Matrix Classes*, Vol. 108 of *Encyclopedia of Mathematics and its Applications*, Cambridge University Press, Cambridge, UK; New York, USA, 2006.

[8] Duff, I. S., Erisman, A. M., and Reid, J. K., *Direct Methods for Sparse Matrices*, Oxford University Press, London, 1986.

[9] George, A. and Liu, J. W. H., *Computer Solution of Large Sparse Positive Definite Systems*, Prentice Hall, Englewood Cliffs, N.J., 1981.

[10] Davis, T. A., *Direct Methods for Sparse Linear Systems*, No. 2 in Fundamentals of Algorithms, Society for Industrial and Applied Mathematics, Philadelphia, PA, USA, 2006.

[11] Cormen, T. H., Leiserson, C. E., and Rivest, R. L., *Introduction to Algorithms*, MIT Press, Cambridge, MA, USA, 1st ed., 1990.

[12] Çatalyürek, Ü. V. and Aykanat, C., "Hypergraph-partitioning-based decomposition for parallel sparse-matrix vector multiplication," *IEEE Transactions on Parallel and Distributed Systems*, Vol. 10, No. 7, Jul 1999, pp. 673–693.

[13] Çatalyürek, Ü. V. and Aykanat, C., "A fine-grain hypergraph model for 2D decomposition of sparse matrices," *Proceedings of the 15th International Parallel and Distributed Processing Symposium (IPDPS*, San Francisco, CA, 2001.

[14] Çatalyürek, U. V., Aykanat, C., and Uçar, B., "On two-dimensional sparse matrix partitioning: Models, methods, and a recipe," *SIAM Journal on Scientific Computing*, Vol. 32, No. 2, 2010, pp. 656–683.

[15] Parter, S., "The use of linear graphs in Gauss elimination," *SIAM Review*, Vol. 3, No. 2, 1961, pp. 119–130.

[16] Rose, D. J., Tarjan, R. E., and Lueker, G. S., "Algorithmic aspects of vertex elimination on graphs," *SIAM Journal on Computing*, Vol. 5, No. 2, 1976, pp. 266–283.

[17] Rose, D. J. and Tarjan, R. E., "Algorithmic aspects of vertex elimination in directed graphs," *SIAM Journal on Applied Mathematics*, Vol. 34, No. 1, 1978, pp. 176–197.

[18] Gilbert, J. R., "A note on the NP-completeness of vertex elimination on directed graphs," *SIAM Journal on Algebraic and Discrete Methods*, Vol. 1, No. 3, 1980, pp. 292–294.

[19] Yannakakis, M., "Computing the minimum fill-in is NP-complete," *SIAM Journal on Algebraic and Discrete Methods*, Vol. 2, No. 1, 1981, pp. 77–79.

[20] Markowitz, H. M., "The elimination form of the inverse and its application to linear programming," *Management Science*, Vol. 3, 1957, pp. 255–269.

[21] Tinney, W. F. and Walker, J. W., "Direct solutions of sparse network equations by optimally ordered triangular factorization," *Proceedings of the IEEE*, Vol. 55, No. 11, Nov. 1967, pp. 1801–1809.

[22] Rose, D. J., "A graph-theoretic study of the numerical solution of sparse positive definite systems of linear equations," *Graph Theory and Computing*, edited by R. C. Read, Academic Press, New York, USA, 1972, pp. 183–217.

[23] Duff, I. S. and Reid, J. K., "MA27–A set of Fortran subroutines for solving sparse symmetric sets of linear equations," Tech. Rep. AERE R10533, HMSO, London, UK, 1982.

[24] George, A. and Liu, J. W. H., "A fast implementation of the minimum degree algorithm using quotient graphs," *ACM Transactions on Mathematical Software*, Vol. 6, No. 3, 1980, pp. 337–358.

[25] George, A. and Liu, J. W. H., "A minimal storage implementation of the minimum degree algorithm," *SIAM Journal on Numerical Analysis*, Vol. 17, No. 2, 1980, pp. 282–299.

[26] George, A. and Liu, J. W. H., "The evolution of the minimum degree ordering algorithm," *SIAM Review*, Vol. 31, No. 1, 1989, pp. 1–19.

[27] George, A. and McIntyre, D. R., "On the application of the minimum degree algorithm to finite element systems," *SIAM Journal on Numerical Analysis*, Vol. 15, No. 1, 1978, pp. 90–112.

[28] Eisenstat, S. C., Gursky, M. C., Schultz, M. H., and Sherman, A. H., "The Yale sparse matrix package I: The symmetric codes," *International Journal for Numerical Methods in Engineering*, Vol. 18, 1982, pp. 1145–1151.

[29] Duff, I. S. and Reid, J. K., "The multifrontal solution of indefinite sparse symmetric linear equations," *ACM Transactions on Mathematical Software*, Vol. 9, 1983, pp. 302–325.

[30] Liu, J. W. H., "Modification of the minimum-degree algorithm by multiple elimination," *ACM Transactions on Mathematical Software*, Vol. 11, No. 2, 1985, pp. 141–153.

[31] Ashcraft, C., "Compressed graphs and the minimum degree algorithm," *SIAM Journal on Scientific Computing*, Vol. 16, 1995, pp. 1404–1411.

[32] Davis, T. A. and Duff, I. S., "An Unsymmetric-Pattern Multifrontal Method for Sparse LU Factorization," *SIAM Journal on Matrix Analysis and Applications*, Vol. 18, No. 1, 1997, pp. 140–158.

[33] Gilbert, J. R., Moler, C., and Schreiber, R., "Sparse matrices in MAT-LAB: Design and implementation," *SIAM Journal on Matrix Analysis and Applications*, Vol. 13, 1992, pp. 333–356.

[34] Amestoy, P. R., Davis, T. A., and Duff, I. S., "An approximate minimum degree ordering algorithm," *SIAM Journal on Matrix Analysis and Applications*, Vol. 17, No. 4, 1996, pp. 886–905.

[35] Heggernes, P., Eisenstat, S. C., Kumfert, G., and Pothen, A., "The computational complexity of the minimum degree algorithm," *Proceedings of NIK 2001—14th Norwegian Computer Science Conference*, Tromsø, Norway, 2001, pp. 98–109.

[36] Duff, I. S., Erisman, A. M., and Reid, J. K., "On George's nested dissection method," *SIAM Journal on Numerical Analysis*, Vol. 13, No. 5, 1976, pp. 686–695.

[37] Cuthill, E. and McKee, J., "Reducing the bandwidth of sparse symmetric matrices," *Proceedings of the 24th national conference*, ACM, New York, USA, 1969, pp. 157–172.

[38] George, A. J., *Computer Implementation of the Finite Element Method*, Ph.D. thesis, Stanford University, Stanford, CA, USA, 1971.

[39] Liu, W.-H. and Sherman, A. H., "Comparative analysis of the Cuthill-McKee and the Reverse Cuthill–McKee ordering algorithms for sparse matrices," *SIAM Journal on Numerical Analysis*, Vol. 13, No. 2, 1976, pp. 198–213.

[40] George, A., "Nested dissection of a regular finite element mesh," *SIAM Journal on Numerical Analysis*, Vol. 10, No. 2, 1973, pp. 345–363.

[41] Duff, I. S., *Analysis of Sparse Systems*, Ph.D. thesis, Oxford University, England, 1972.

[42] Hoffman, A. J., Martin, M. S., and Rose, D. J., "Complexity bounds for regular finite difference and finite element grids," *SIAM Journal on Numerical Analysis*, Vol. 10, No. 2, 1973, pp. 364–369.

[43] Rose, D. J. and Whitten, G. F., "Automatic nested dissection," *ACM 74: Proceedings of the 1974 annual conference*, ACM, New York, USA, 1974, pp. 82–88.

[44] George, A. and Liu, J. W. H., "An automatic nested dissection algorithm for irregular finite element problems," *SIAM Journal on Numerical Analysis*, Vol. 15, No. 5, 1978, pp. 1053–1069.

[45] Lipton, R. J., Rose, D. J., and Tarjan, R. E., "Generalized nested dissection," *SIAM Journal on Numerical Analysis*, Vol. 16, No. 2, 1979, pp. 346–358.

[46] Lipton, R. J. and Tarjan, R. E., "A separator theorem for planar graphs," *SIAM Journal on Applied Mathematics*, Vol. 36, 1979, pp. 177–189.

[47] Gilbert, J. R. and Tarjan, R. E., "The analysis of a nested dissection algorithm," *Numerische Mathematik*, Vol. 50, No. 4, 1987, pp. 377–404.

[48] George, A., Poole, W. G., and Voigt, R. G., "Incomplete nested dissection for solving n by n grid problems," *SIAM Journal on Numerical Analysis*, Vol. 15, No. 4, 1978, pp. 662–673.

[49] Liu, J. W. H., "The minimum degree ordering with constraints," *SIAM Journal on Scientific and Statistical Computing*, Vol. 10, No. 6, 1989, pp. 1136–1145.

[50] Ashcraft, C. and Liu, J. W. H., "Robust ordering of sparse matrices using multisection," *SIAM Journal on Matrix Analysis and Applications*, Vol. 19, No. 3, 1998, pp. 816–832.

[51] Bhat, M. V., Habashi, W. G., Liu, J. W. H., Nguyen, V. N., and Peeters, M. F., "A note on nested dissection for rectangular grids," *SIAM Journal on Matrix Analysis and Applications*, Vol. 14, No. 1, 1993, pp. 253–258.

[52] Ashcraft, C. and Liu, J. W. H., "A partition improvement algorithm for generalized nested dissection," Tech. Rep. BCSTECH-94-020, Boeing Computer Services, Seattle, WA, USA, 1996.

[53] Hendrickson, B. and Rothberg, E., "Improving the run time and quality of nested dissection ordering," *SIAM Journal on Scientific Computing*, Vol. 20, No. 2, 1998, pp. 468–489.

[54] Schulze, J., "Towards a tighter coupling of bottom-up and top-down sparse matrix ordering methods," *BIT Numerical Mathematics*, Vol. 41, No. 4, 2001, pp. 800–841.

[55] Barnard, S. and Simon, H. D., "A fast multilevel implementation of recursive spectral bisection for partitioning unstructured problems," *Concurrency: Practice and Experience*, Vol. 6, 1994, pp. 101–117.

[56] Pothen, A., Simon, H. D., and Liou, K.-P., "Partitioning sparse matrices with eigenvectors of graphs," *SIAM Journal on Matrix Analysis and Applications*, Vol. 11, No. 3, 1990, pp. 430–452.

[57] Bui, T. N. and Jones, C., "A heuristic for reducing fill-in in sparse matrix factorization," *6th SIAM Conference on Parallel Processing for Scientific Computing*, Norfolk, VA, USA, 1993, pp. 445–452.

[58] Hendrickson, B. and Leland, R., "A multilevel algorithm for partitioning graphs," *Supercomputing '95: Proceedings of the 1995 ACM/IEEE conference on Supercomputing (CDROM)*, ACM, New York, USA, 1995, p. 28.

[59] Fiduccia, C. M. and Mattheyses, R. M., "A linear-time heuristic for improving network partitions," *DAC '82: Proceedings of the 19th Conference on Design Automation*, IEEE Press, Piscataway, NJ, USA, 1982, pp. 175–181.

[60] Kernighan, B. W. and Lin, S., "An efficient heuristic procedure for partitioning graphs," *The Bell System Technical Journal*, Vol. 49, Feb. 1970, pp. 291–307.

[61] Hendrickson, B. and Leland, R., *The Chaco user's guide, version 2.0*, Sandia National Laboratories, Alburquerque, NM, 87185, 1995.

[62] Karypis, G. and Kumar, V., *MeTiS: A software package for partitioning unstructured graphs, partitioning meshes, and computing fill-reducing orderings of sparse matrices version 4.0*, University of Minnesota, Department of Comp. Sci. and Eng., Army HPC Research Center, Minneapolis, MN, USA, 1998.

[63] Pellegrini, F., *SCOTCH 5.1 User's Guide*, Laboratoire Bordelais de Recherche en Informatique (LaBRI), 2008.

[64] Gupta, A., "Fast and effective algorithms for graph partitioning and sparse matrix ordering," Tech. Rep. RC 20496 (90799), IBM Research Division, T. J. Watson Research Center, Yorktown Heights, NY, USA, 1996.

[65] Hall, P., "On representatives of subsets," *Journal of the London Mathematical Society*, Vol. s1-10, No. 37, 1935, pp. 26–30.

[66] Berge, C., "Two theorems in graph theory," *Proceedings of the National Academy of Sciences of the USA*, Vol. 43, 1957, pp. 842–844.

[67] Hall, Jr., M., "An Algorithm for Distinct Representatives," *The American Mathematical Monthly*, Vol. 63, No. 10, 1956, pp. 716–717.

[68] Duff, I. S., "Algorithm 575: Permutations for a zero-free diagonal," *ACM Transactions on Mathematical Software*, Vol. 7, No. 3, 1981, pp. 387–390.

[69] Duff, I. S., "On algorithms for obtaining a maximum transversal," *ACM Transactions on Mathematical Software*, Vol. 7, No. 3, 1981, pp. 315–330.

[70] Kuhn, H. W., "The Hungarian method for the assignment problem," *Naval Research Logistics Quarterly*, Vol. 2, No. 1-2, 1955, pp. 83–97.

[71] Hopcroft, J. E. and Karp, R. M., "An $n^{5/2}$ algorithm for maximum matchings in bipartite graphs," *SIAM Journal on Computing*, Vol. 2, No. 4, 1973, pp. 225–231.

[72] Alt, H., Blum, N., Mehlhorn, K., and Paul, M., "Computing a maximum cardinality matching in a bipartite graph in time $\mathcal{O}(n^{1.5}\sqrt{m/\log n})$," *Information Processing Letters*, Vol. 37, No. 4, 1991, pp. 237–240.

[73] Duff, I. S. and Koster, J., "On algorithms for permuting large entries to the diagonal of a sparse matrix," *SIAM Journal on Matrix Analysis and Applications*, Vol. 22, 2001, pp. 973–996.

[74] Olschowka, M. and Neumaier, A., "A new pivoting strategy for Gaussian elimination," *Linear Algebra and Its Applications*, Vol. 240, 1996, pp. 131–151.

[75] Burkard, R., Dell'Amico, M., and Martello, S., *Assignment Problems*, SIAM, Philadelphia, PA, USA, 2009.

[76] Fredman, M. L. and Tarjan, R. E., "Fibonacci heaps and their uses in improved network optimization algorithms," *J. ACM*, Vol. 34, No. 3, 1987, pp. 596–615.

[77] Riedy, J. and Demmel, J., "Parallel weighted bipartite matching and applications," Presentation at SIAM 11th Conference on Parallel Processing for Scientific Computing (PP04), San Francisco, CA, USA, February 2004.

[78] Duff, I. S., Ruiz, D., and Uçar, B., "Computing a class of bipartite matchings in parallel," Presentation at SIAM 13th Conference on Parallel Processing for Scientific Computing (PP08), Atlanta, GA, USA, March 2008.

[79] Reif, J. H., "Depth-first search is inherently sequential," *Information Processing Letters*, Vol. 20, No. 5, 1985, pp. 229–234.

[80] Greenlaw, R., "A model classifying algorithms as inherently sequential with applications to graph searching," *Information and Computation*, Vol. 97, No. 2, 1992, pp. 133–149.

[81] Halappanavar, M., *Algorithms for vertex-weighted matching in graphs*, Ph.D. thesis, Old Dominion University, Norfolk, VA, USA, 2008.

[82] Manne, F. and Bisseling, R. H., "A parallel approximation algorithm for the weighted maximum matching problem," *Parallel Processing and Applied Mathematics*, edited by R. Wyrzykowski, K. Karczewski, J. Dongarra, and J. Wasniewski, Vol. 4967 of *Lecture Notes in Computer Science*, 2008, pp. 708–717.

[83] Pothen, A., "Graph matchings in combinatorial scientific computing (Vertex-weighted and parallel edge-weighted)," Presentation at Dagstuhl Seminar on Combinatorial Scientific Computing (09061), February 2009.

[84] Kuhn, H. W., "The Hungarian method for the assignment problem," *Naval Research Logistics*, Vol. 52, No. 1, 2005, pp. 7–21.

[85] Kuhn, H. W., "Statement for Naval Research Logistics," *Naval Research Logistics*, Vol. 52, No. 1, 2005, pp. 6.

[86] Duff, I. S., "Full matrix techniques in sparse Gaussian elimination," *Proceedings of 1981 Dundee Biennal Conference on Numerical Analysis*, edited by G. A. Watson, Vol. 912 of *Lecture Notes in Mathematics*, 1982, pp. 71–84.

[87] Jess, J. A. G. and Kees, H. G. M., "A data structure for parallel L/U decomposition," *IEEE Transactions on Computers*, Vol. 31, No. 3, 1982, pp. 231–239.

[88] Schreiber, R., "A new implementation of sparse Gaussian elimination," *ACM Transactions on Mathematical Software*, Vol. 8, No. 3, 1982, pp. 256–276.

[89] Liu, J. W. H., "A compact row storage scheme for Cholesky factors using elimination trees," *ACM Transactions on Mathematical Software*, Vol. 12, No. 2, 1986, pp. 127–148.

[90] Liu, J. W. H., "The role of elimination trees in sparse factorization," *SIAM Journal on Matrix Analysis and Applications*, Vol. 11, No. 1, 1990, pp. 134–172.

[91] Amestoy, P. R., Duff, I. S., and L'Excellent, J.-Y., "Multifrontal parallel distributed symmetric and unsymmetric solvers," *Computer methods in applied mechanics and engineering*, Vol. 184, No. 2–4, 2000, pp. 501–520.

[92] Demmel, J. W., Eisenstat, S. C., Gilbert, J. R., Li, X. S., and Liu, J. W. H., "A supernodal approach to sparse partial pivoting," *SIAM Journal on Matrix Analysis and Applications*, Vol. 20, No. 3, 1999, pp. 720–755.

[93] Gilbert, J. R. and Liu, J. W. H., "Elimination structures for unsymmetric sparse LU factors," *SIAM Journal on Matrix Analysis and Applications*, Vol. 14, No. 2, 1993, pp. 334–352.

[94] Gilbert, J. R. and Peierls, T., "Sparse partial pivoting in time proportional to arithmetic operations," *SIAM Journal on Scientific and Statistical Computing*, Vol. 9, No. 5, 1988, pp. 862–874.

[95] Gupta, A., "Improved symbolic and numerical factorization algorithms for unsymmetric sparse matrices," *SIAM Journal on Matrix Analysis and Applications*, Vol. 24, No. 2, 2002, pp. 529–552.

[96] Eisenstat, S. C. and Liu, J. W. H., "The theory of elimination trees for sparse unsymmetric matrices," *SIAM Journal on Matrix Analysis and Applications*, Vol. 26, No. 3, 2005, pp. 686–705.

[97] Eisenstat, S. C. and Liu, J. W. H., "A tree-based dataflow model for the unsymmetric multifrontal method," *Electronic Transactions on Numerical Analysis*, Vol. 21, 2005, pp. 1–19.

[98] Tarjan, R. E., "Depth-first search and linear graph algorithms," *SIAM Journal on Computing*, Vol. 1, No. 2, 1972, pp. 146–160.

[99] Duff, I. S. and Reid, J. K., "A note on the work involved in no-fill sparse matrix factorization," *IMA Journal on Numerical Analysis*, Vol. 1, 1983, pp. 37–40.

[100] Liu, J. W. H., "Equivalent sparse matrix reordering by elimination tree rotations," *SIAM Journal on Scientific and Statistical Computing*, Vol. 9, No. 3, 1988, pp. 424–444.

[101] Tarjan, R. E., *Data Structures and Network Algorithms*, Vol. 44 of *CBMS-NSF Regional Conference Series in Applied Mathematics*, SIAM, Philadelphia, PA, USA, 1983.

[102] Irons, B. M., "A frontal solution program for finite-element analysis," *International Journal for Numerical Methods in Engineering*, Vol. 2, No. 1, 1970, pp. 5–32.

[103] Speelpenning, B., "The generalized element method," Tech. Rep. UIUCDCS-R-78-946, Department of Computer Science, University of Illinois at Urbana-Champaign, IL, 1978.

[104] Duff, I. S. and Reid, J. K., "The multifrontal solution of unsymmetric sets of linear equations," *SIAM Journal on Scientific and Statistical Computing*, Vol. 5, No. 3, 1984, pp. 633–641.

[105] Liu, J. W. H., "The multifrontal method for sparse matrix solution: Theory and practice," *SIAM Review*, Vol. 34, No. 1, 1992, pp. 82–109.

[106] Dongarra, J. J., Du Croz, J., Duff, I. S., and Hammarling, S., "A set of level 3 Basic Linear Algebra Subprograms." *ACM Transactions on Mathematical Software*, Vol. 16, 1990, pp. 1–17.

[107] Ashcraft, C. and Grimes, R., "The influence of relaxed supernode partitions on the multifrontal method," *ACM Transactions on Mathematical Software*, Vol. 15, No. 4, 1989, pp. 291–309.

[108] Liu, J. W. H., Ng, E. G., and Peyton, B. W., "On finding supernodes for sparse matrix computations," *SIAM Journal on Matrix Analysis and Applications*, Vol. 14, No. 1, 1993, pp. 242–252.

[109] Liu, J. W. H., "On the storage requirement in the out-of-core multifrontal method for sparse factorization," *ACM Transactions on Mathematical Software*, Vol. 12, No. 3, 1986, pp. 249–264.

[110] Agullo, E., Guermouche, A., and L'Excellent, J.-Y., "Reducing the I/O volume in an out-of-core sparse multifrontal solver," *High Performance Computing – HiPC2007; 14th International Conference*, edited by S. Aluru, M. Parashar, R. Badrinath, and V. K. Prasanna, Vol. 4873 of *Lecture Notes in Computer Science*, 2007, pp. 260–280.

[111] Agullo, E., Guermouche, A., and L'Excellent, J.-Y., "A parallel out-of-core multifrontal method: Storage of factors on disk and analysis of models for an out-of-core active memory," *Parallel Computing*, Vol. 34, No. 6-8, 2008, pp. 296–317.

[112] Guermouche, A. and L'Excellent, J.-Y., "Constructing memory-minimizing schedules for multifrontal methods," *ACM Transactions on Mathematical Software*, Vol. 32, No. 1, 2006, pp. 17–32.

[113] George, A., Liu, J. W. H., and Ng, E., "Communication results for parallel sparse Cholesky factorization on a hypercube," *Parallel Computing*, Vol. 10, No. 3, 1989, pp. 287–298.

[114] Geist, G. A. and Ng, E. G., "Task scheduling for parallel sparse Cholesky factorization," *International Journal of Parallel Programming*, Vol. 18, No. 4, 1989, pp. 291–314.

[115] Pothen, A. and Sun, C., "A mapping algorithm for parallel sparse Cholesky factorization," *SIAM Journal on Scientific Computing*, Vol. 14, No. 5, 1993, pp. 1253–1257.

[116] Gilbert, J. R. and Schreiber, R., "Highly parallel sparse Cholesky factorization," *SIAM Journal on Scientific and Statistical Computing*, Vol. 13, No. 5, 1992, pp. 1151–1172.

[117] Amestoy, P. R., Duff, I. S., L'Excellent, J.-Y., and Koster, J., "A fully asynchronous multifrontal solver using distributed dynamic scheduling," *SIAM Journal on Matrix Analysis and Applications*, Vol. 23, No. 1, 2001, pp. 15–41.

[118] Amestoy, P. R., Duff, I. S., and Vömel, C., "Task scheduling in an asynchronous distributed memory multifrontal solver," *SIAM Journal on Matrix Analysis and Applications*, Vol. 26, No. 2, 2004, pp. 544–565.

[119] Amestoy, P. R., Guermouche, A., L'Excellent, J.-Y., and Pralet, S., "Hybrid scheduling for the parallel solution of linear systems," *Parallel Computing*, Vol. 32, No. 2, 2006, pp. 136–156.

[120] Agullo, E., *On the out-of-core factorization of large sparse matrices*, Ph.D. thesis, Ecole Normale Supérieure de Lyon, Lyon, France, 2008.

[121] Dobrian, F., *External Memory Algorithms for Factoring Sparse Matrices*, Ph.D. thesis, Old Dominion University, Norfolk, VA, USA, 2001.

[122] Gilbert, J. R. and Toledo, S., "High-performance out-of-core sparse LU factorization," *9th SIAM Conference on Parallel Processing for Scientific Computing (CDROM)*, 1999, p. 10.

[123] Meshar, O., Irony, D., and Toledo, S., "An out-of-core sparse symmetric-indefinite factorization method," *ACM Transactions on Mathematical Software*, Vol. 32, No. 3, 2006, pp. 445–471.

[124] Rotkin, V. and Toledo, S., "The design and implementation of a new out-of-core sparse Cholesky factorization method," *ACM Transactions on Mathematical Software*, Vol. 30, No. 1, 2004, pp. 19–46.

[125] Toledo, S. and Uchitel, A., "A supernodal out-of-core sparse Gaussian-elimination method," *7th International Conference on Parallel Processing and Applied Mathematics (PPAM 2007)*,, edited by R. Wyrzykowski, K. Karczewski, J. Dongarra, and J. Wasniewski, Vol. 4967 of *Lecture Notes in Computer Science*, Springer-Verlag, Berlin Heidelberg, 2008, pp. 728–737.

[126] Reid, J. K. and Scott, J. A., "An out-of-core sparse Cholesky solver," Tech. Rep. RAL-TR-2006-013, Computational Sciences and Engineering Department, Rutherford Appleton Laboratory, Oxon, OX11 0QX, England, 2006.

[127] Reid, J. K. and Scott, J. A., "An efficient out-of-core sparse symmetric indefinite direct solver," Tech. Rep. RAL-TR-2008-024, Computational Sciences and Engineering Department, Rutherford Appleton Laboratory, Oxon, OX11 0QX, England, December 2008.

[128] Dulmage, A. L. and Mendelsohn, N. S., "Two algorithms for bipartite graphs," *SIAM Journal on Applied Mathematics*, Vol. 11, No. 1, 1963, pp. 183–194.

[129] Gustavson, F. G., "Finding the block lower-triangular form of a sparse matrix," *Sparse Matrix Computations*, edited by J. R. Bunch and D. J. Rose, Academic Press, New York and London, 1976, pp. 275–289.

[130] Duff, I. S. and Reid, J. K., "Algorithm 529: Permutations to block triangular form," *ACM Transactions on Mathematical Software*, Vol. 4, No. 2, 1978, pp. 189–192.

[131] Duff, I. S. and Reid, J. K., "An implementation of Tarjan's algorithm for the block triangularization of a matrix," *ACM Transactions on Mathematical Software*, Vol. 4, No. 2, 1978, pp. 137–147.

[132] Duff, I. S., "On permutations to block triangular form," *Journal of the Institute of Mathematics and its Applications*, Vol. 19, No. 3, 1977, pp. 339–342.

[133] Dulmage, A. L. and Mendelsohn, N. S., "Coverings of bipartite graphs," *Canadian Journal of Mathematics*, Vol. 10, 1958, pp. 517–534.

[134] Dulmage, A. L. and Mendelsohn, N. S., "A structure theory of bipartite graphs of finite exterior dimension," *Trans. Roy. Soc. Can. Sec. III*, Vol. 53, 1959, pp. 1–13.

[135] Pothen, A., *Sparse Null Bases and Marriage Theorems*, Ph.D. thesis, Department of Computer Science, Cornell University, Ithaca, NY, 1984.

[136] Pothen, A. and Fan, C.-J., "Computing the block triangular form of a sparse matrix," *ACM Transactions on Mathematical Software*, Vol. 16, 1990, pp. 303–324.

[137] Duff, I. S. and Uçar, B., "On the block triangular form of symmetric matrices," Tech. Rep. TR/PA/08/26, CERFACS, Toulouse, France, 2008.

[138] Bisseling, R. H., "Combinatorial problems in high-performance computing," Presentation at Dagstuhl Seminar on Combinatorial Scientific Computing (09061).

[139] Devine, K. D., Boman, E. G., and Karypis, G., "Partitioning and load balancing for emerging parallel applications and architectures," *Frontiers of Scientific Computing*, edited by M. Heroux, A. Raghavan, and H. Simon, SIAM, Philadelphia, 2006.

[140] Hendrickson, B. and Kolda, T. G., "Graph partitioning models for parallel computing," *Parallel Computing*, Vol. 26, No. 12, 2000, pp. 1519–1534.

[141] Uçar, B. and Aykanat, C., "Partitioning sparse matrices for parallel preconditioned iterative methods," *SIAM Journal on Scientific Computing*, Vol. 29, No. 4, 2007, pp. 1683–1709.

[142] Duff, I. S. and van der Vorst, H. A., "Developments and trends in the parallel solution of linear systems," *Parallel Computing*, Vol. 25, No. 13–14, 1999, pp. 1931–1970.

[143] Benzi, M., "Preconditioning techniques for large linear systems: A survey," *Journal of Computational Physics*, Vol. 182, No. 2, 2002, pp. 418–477.

[144] Grote, M. J. and Huckle, T., "Parallel preconditioning with sparse approximate inverses," *SIAM Journal on Scientific Computing*, Vol. 18, No. 3, 1997, pp. 838–853.

[145] Benzi, M., Meyer, C. D., and Tůma, M., "A sparse approximate inverse preconditioner for the conjugate gradient method," *SIAM Journal on Scientific Computing*, Vol. 17, No. 5, 1996, pp. 1135–1149.

[146] Chow, E., "A priori sparsity patterns for parallel sparse approximate inverse preconditioners," *SIAM Journal on Scientific Computing*, Vol. 21, No. 5, 2000, pp. 1804–1822.

[147] Benzi, M. and Tůma, M., "Orderings for factorized sparse approximate inverse preconditioners," *SIAM Journal on Scientific Computing*, Vol. 21, No. 5, 2000, pp. 1851–1868.

[148] Manguoglu, M., Sameh, A., and Schenk, O., "PSPIKE: Parallel sparse linear system solver," *In Proc. Euro-Par 2009 Parallel Processing*, 2009, pp. 797–808.

[149] Meijerink, J. A. and van der Vorst, H. A., "An iterative solution method for linear systems of which the coefficient matrix is a symmetric M-matrix," *Mathematics of Computation*, Vol. 31, No. 137, 1977, pp. 148–162.

[150] Gustafsson, I., "A class of first order factorization methods," *BIT Numerical Mathematics*, Vol. 18, No. 2, 1978, pp. 142–156.

[151] Axelsson, O., *Iterative solution methods*, Cambridge University Press, Cambridge, 1994.

[152] Saad, Y., "ILUT: A dual threshold incomplete LU factorization," *Numerical Linear Algebra with Applications*, Vol. 1, No. 4, 1994, pp. 387–402.

[153] Meurant, G. A., *Computer Solution of Large Linear Systems*, Vol. 28 of *Studies in Mathematics and Its Applications*, North-Holland, Amsterdam, Netherlands, 1999.

[154] Saad, Y., *Iterative Methods for Sparse Linear Systems*, SIAM, Philadelphia, 2nd ed., 2003.

[155] Duff, I. S. and Meurant, G. A., "The effect of ordering on preconditioned conjugate gradients," *BIT*, Vol. 29, No. 4, 1989, pp. 635–657.

[156] Simon, H. D., "Incomplete LU preconditioners for conjugate-gradient-type iterative methods," *Proceedings of the 1985 Reservoir Simulation Symposium*, Dallas, February 1985, pp. 387–396.

[157] Chow, E. and Saad, Y., "Experimental study of ILU preconditioners for indefinite matrices," *Journal of Computational and Applied Mathematics*, Vol. 86, No. 2, 1997, pp. 387–414.

[158] Benzi, M., Szyld, D. B., and van Duin, A., "Orderings for incomplete factorization preconditioning of nonsymmetric problems," *SIAM Journal on Scientific Computing*, Vol. 20, No. 5, 1999, pp. 1652–1670.

[159] Bridson, R. and Tang, W.-P., "A structural diagnosis of some IC orderings," *SIAM Journal on Scientific Computing*, Vol. 22, No. 5, 2001, pp. 1527–1532.

[160] Willoughby, R. A., "A characterization of matrix irreducibility," *Numerische Methoden bei Graphentheoretischen und Kombinatorischen Problemen*, edited by L. Collatz, G. Meinardus, and H. Werner, Vol. 29 of *International Series of Numerical Mathematics*, Birkhäuser Verlag, Basel-Stuttgart, 1975, pp. 131–143.

[161] Duff, I. S., Erisman, A. M., Gear, C. W., and Reid, J. K., "Sparsity structure and Gaussian elimination," *SIGNUM Newsletter*, Vol. 23, 1988, pp. 2–8.

[162] D'Azevedo, E. F., Forsyth, P. A., and Tang, W.-P., "Ordering methods for preconditioned conjugate gradient methods applied to unstructured grid problems," *SIAM Journal on Matrix Analysis and Applications*, Vol. 13, No. 3, 1992, pp. 944–961.

[163] D'Azevedo, E. F., Forsyth, P. A., and Tang, W.-P., "Towards a cost-effective ILU preconditioner with high level fill," *BIT Numerical Mathematics*, Vol. 32, No. 3, 1992, pp. 442–463.

[164] Clift, S. S. and Tang, W.-P., "Weighted graph based ordering techniques for preconditioned conjugate gradient methods," *BIT Numerical Mathematics*, Vol. 35, No. 1, 1995, pp. 30–47.

[165] Benzi, M., Haws, J. C., and Tůma, M., "Preconditioning highly indefinite and nonsymmetric matrices," *SIAM Journal on Scientific Computing*, Vol. 22, No. 4, 2000, pp. 1333–1353.

[166] Duff, I. S. and Koster, J., "The design and use of algorithms for permuting large entries to the diagonal of sparse matrices," *SIAM Journal on Matrix Analysis and Applications*, Vol. 20, No. 4, 1999, pp. 889–901.

[167] Bollhöfer, M., "A robust ILU with pivoting based on monitoring the growth of the inverse factors," *Linear Algebra and its Applications*, Vol. 338, No. 1–3, 2001, pp. 201–218.

[168] Lee, I., Raghavan, P., and Ng, E. G., "Effective preconditioning through ordering interleaved with incomplete factorization," *SIAM Journal on Matrix Analysis and Applications*, Vol. 27, No. 4, 2006, pp. 1069–1088.

[169] Saad, Y., "Multilevel ILU with reorderings for diagonal dominance," *SIAM Journal on Scientific Computing*, Vol. 27, No. 3, 2005, pp. 1032–1057.

[170] Ng, E. G., Peyton, B. W., and Raghavan, P., "A blocked incomplete Cholesky preconditioner for hierarchical-memory computers," *Iterative methods in scientific computation IV*, edited by D. R. Kincaid and A. C. Elster, Vol. 5 of *IMACS Series in Computational Applied Mathematics*, IMACS, New Brunswick, NJ, USA, 1999, pp. 211–222.

[171] Hénon, P., Ramet, P., and Roman, J., "On finding approximate supernodes for an efficient block-ILU(k) factorization," *Parallel Computing*, Vol. 34, No. 6–8, 2008, pp. 345–362.

[172] Saad, Y., "Finding exact and approximate block structures for ILU preconditioning," *SIAM Journal on Scientific Computing*, Vol. 24, No. 4, 2003, pp. 1107–1123.

[173] Fritzsche, D., Frommer, A., and Szyld, D. B., "Extensions of certain graph-based algorithms for preconditioning," *SIAM Journal on Scientific Computing*, Vol. 29, No. 5, 2007, pp. 2144–2161.

[174] O'Neil, J. and Szyld, D. B., "A block ordering method for sparse matrices," *SIAM Journal on Scientific and Statistical Computing*, Vol. 11, No. 5, 1990, pp. 811–823.

[175] Bollhöfer, M., Saad, Y., and Schenk, O., "ILUPACK," `http:// www-public.tu-bs.de/~bolle/ilupack/` (last access: March, 2009).

[176] Balay, S., Buschelman, K., Gropp, W. D., Kaushik, D., Knepley, M. G., McInnes, L. C., Smith, B. F., and Zhang, H., "PETSc Web page," `\hbox{http://www.mcs.anl.gov/petsc}`, (last access: March, 2009), 2001.

[177] Sala, M. and Heroux, M., "Robust algebraic preconditioners with IFPACK 3.0," Tech. Rep. SAND-0662, Sandia National Laboratories, Alburquerque, NM, USA, 2005.

[178] Karypis, G. and Kumar, V., "Parallel threshold-based ILU factorization," *Supercomputing '97: Proceedings of the 1997 ACM/IEEE conference on Supercomputing (CDROM)*, ACM, New York, NY, USA, 1997, pp. 1–24.

[179] Luby, M., "A simple parallel algorithm for the maximal independent set problem," *SIAM Journal on Computing*, Vol. 15, No. 4, 1986, pp. 1036–1053.

[180] Hysom, D. and Pothen, A., "A scalable parallel algorithm for incomplete factor preconditioning," *SIAM Journal on Scientific Computing*, Vol. 22, No. 6, 2001, pp. 2194–2215.

[181] Rose, D. J., "Convergent regular splittings for singular M-matrices," *SIAM Journal on Algebraic and Discrete Methods*, Vol. 5, No. 1, 1984, pp. 133–144.

[182] Varga, R. S., *Matrix Iterative Analysis*, Springer, Berlin, Heidelberg, New York, 2nd ed., 2000.

[183] Vaidya, P. M., "Solving linear equations with symmetric diagonally dominant matrices by constructing good preconditioners," Unpublished manuscript presented at the IMA Workshop on Graph Theory and Sparse Matrix Computation, Minneapolis, MN, USA.

[184] Joshi, A., *Topics in Optimization and Sparse Linear Systems*, Ph.D. thesis, Department of Computer Science, University of Illinois Urbana-Champaign, Urbana, IL, USA, December 1996.

[185] Kumar, V., Grama, A., Gupta, A., and Karypis, G., *Introduction to Parallel Computing: Desing and Analysis of Algorithms*, The Benjamin/Cummings Publishing Company, Inc., Redwood, CA, USA, 1994.

[186] Gremban, K. D., *Combinatorial Preconditioners for Sparse, Symmetric, Diagonally Dominant Linear Systems*, Ph.D. thesis, School of Computer Science, Carnegie Mellon University, Pittsburgh, PA, USA, 1996.

[187] Gremban, K. D., Miller, G. L., and Zagha, M., "Performance evaluation of a parallel preconditioner," *9th International Parallel Processing Symposium*, IEEE, Santa Barbara, April 1995, pp. 65–69.

[188] Guattery, S., "Graph embedding techniques for bounding condition numbers of incomplete factor preconditioners," Tech. Rep. ICASE Report No.97-47, Institute for Computer Applications in Science and Engineering, NASA Langley Research Center, Hampton, VA, USA, 1997.

[189] Reif, J. H., "Efficient approximate solution of sparse linear systems," *Computers and Mathematics with Applications*, Vol. 36, No. 9, November 1998, pp. 37–58.

[190] Bern, M., Gilbert, J. R., Hendrickson, B., Nguyen, N., and Toledo, S., "Support-graph preconditioners," *SIAM Journal on Matrix Analysis and Applications*, Vol. 27, No. 4, 2006, pp. 930–951.

[191] Boman, E. G. and Hendrickson, B., "Support theory for preconditioning," *SIAM Journal on Matrix Analysis and Applications*, Vol. 25, No. 3, 2003, pp. 694–717.

[192] Boman, E. G., Chen, D., Hendrickson, B., and Toledo, S., "Maximum-weight-basis preconditioners," *Numerical Linear Algebra with Applications*, Vol. 11, No. 8-9, 2004, pp. 695–721.

[193] Chen, D. and Toledo, S., "Vaidya's preconditioners: Implementation and experimental study," *Electronic Transactions on Numerical Analysis*, Vol. 16, 2003, pp. 30–49.

[194] Elkin, M., Emek, Y., Spielman, D. A., and Teng, S.-H., "Lower-stretch spanning trees," *SIAM Journal on Computing*, Vol. 38, No. 2, 2008, pp. 608–628.

[195] Spielman, D. A. and Teng, S.-H., "Solving sparse, symmetric, diagonally dominant linear systems in time $\mathcal{O}(m^{1.31})$," *44th Annual IEEE Symposium on Foundations of Computer Science*, IEEE, 2003, pp. 416–427.

[196] Spielman, D. A. and Teng, S.-H., "Nearly-linear time algorithms for graph partitioning, graph sparsification, and solving linear systems," *STOC'04: Proceedings of the 36th annual ACM symposium on Theory of computing*, ACM, New York, NY, USA, 2004, pp. 81–90.

[197] Koutis, I., *Combinatorial and algebraic tools for multigrid algorithms*, Ph.D. thesis, Carnegie Mellon University, Pittsburgh, May 2007.

[198] Chow, E., Falgout, R. D., Hu, J. J., Tuminaro, R. S., and Yang, U. M., "A survey of parallelization techniques for multigrid solvers," *Parallel Processing for Scientific Computing*, edited by M. A. Heroux, P. Raghavan, and H. D. Simon, Vol. 20 of *Software, Environments, and Tools*, chap. 10, SIAM, 2006, pp. 179–201.

[199] Stüben, K., "A review of algebraic multigrid," *Journal of Computational and Applied Mathematics*, Vol. 128, No. 1–2, 2001, pp. 281–309.

[200] Ruge, J. W. and Stüben, K., "Algebraic multigrid," *Multigrid Methods*, edited by S. F. McCormick, chap. 4, SIAM, Philadelphia, PA, 1987, pp. 73–130.

[201] Tuminaro, R. S. and Tong, C., "Parallel smoothed aggregation multigrid: Aggregation strategies on massively parallel machines," *Supercomputing '00: Proceedings of the 2000 ACM/IEEE conference on Supercomputing (CDROM)*, IEEE Computer Society, Washington, DC, USA, 2000, p. 5.

[202] Kim, H., Xu, J., and Zikatanov, L., "A multigrid method based on graph matching for convection-diffusion equations," *Numerical Linear Algebra with Applications*, Vol. 10, No. 1–2, 2003, pp. 181–195.

[203] de Sterck, H., Yang, U. M., and Heys, J. J., "Reducing complexity in parallel algebraic multigrid preconditioners," *SIAM Journal on Matrix Analysis and Applications*, Vol. 27, No. 4, 2006, pp. 1019–1039.

[204] Bozdağ, D., Çatalyürek, Ü. V., Gebremedhin, A. H., Manne, F., Boman, E. G., and Özgüner, F., "A parallel distance-2 graph coloring algorithm for distributed memory computers," *Proceedings of 2005 International Conference on High Performance Computing and Communications (HPCC-05)*, edited by L. T. Yang, O. F. Rana, B. Di Martino, and J. Dongarra, Vol. 3726 of *Lecture Notes in Computer Science*, 2005, pp. 796–806.

[205] Gebremedhin, A. H., Manne, F., and Pothen, A., "What color is your Jacobian? Graph coloring for computing derivatives," *SIAM Review*, Vol. 47, No. 4, 2005, pp. 629–705.

[206] de Sterck, H., Falgout, R. D., Nolting, J. W., and Yang, U. M., "Distance-two interpolation for parallel algebraic multigrid," *Numerical Linear Algebra with Applications*, Vol. 15, No. 2–3, MAR-APR 2008, pp. 115–139.

[207] Alber, D. M. and Olson, L. N., "Parallel coarse-grid selection," *Numerical Linear Algebra with Applications*, Vol. 14, No. 8, 2007, pp. 611–643.

[208] Gee, M. W., Siefert, C. M., Hu, J. J., Tuminaro, R. S., and Sala, M. G., "ML 5.0 smoothed aggregation user's guide," Tech. Rep. SAND2006-2649, Sandia National Laboratories, 2006.

[209] Henson, V. E. and Yang, U. M., "BoomerAMG: A parallel algebraic multigrid solver and preconditioner," *Applied Numerical Mathematics*, Vol. 41, No. 1, 2002, pp. 155–177.

Chapter 3

Combinatorial Preconditioners[1]

Sivan Toledo

Tel Aviv University

Haim Avron

IBM T.J. Watson Research Center

3.1 Introduction

The Conjugate Gradient (CG) method is an iterative algorithm for solving linear systems of equations $Ax = b$, where A is symmetric and positive definite. The convergence of the method depends on the spectrum of A; when its eigenvalues are clustered, the method converges rapidly. In partic-

---

[1]This research was supported in part by an IBM Faculty Partnership Award and by grant 1045/09 from the Israel Science Foundation (founded by the Israel Academy of Sciences and Humanities).

ular, CG converges to within a fixed tolerance in $O(\sqrt{\kappa})$ iterations, where $\kappa = \kappa(A) = \lambda_{\max}(A)/\lambda_{\min}(A)$ is the spectral condition number of A. This upper bound is tight when eigenvalues are poorly distributed (i.e., they are not clustered). For example, if there are only two distinct eigenvalues, CG converges in two iterations.

When the spectrum of A is not clustered, a *preconditioner* can accelerate convergence. The Preconditioned Conjugate Gradients (PCG) method implicitly applies the CG iteration to the linear system $(B^{-1/2}AB^{-1/2})(B^{1/2}x) = B^{-1/2}b$ using a clever transformation that only requires applications of A and B^{-1} in every iteration; B also needs to be symmetric positive definite. The convergence of PCG is determined by the spectrum of $(B^{-1/2}AB^{-1/2})$, which is the same as the spectrum of $B^{-1}A$. If a representation of B^{-1} can be constructed quickly and applied quickly, and if $B^{-1}A$ has a clustered spectrum, the method is very effective. There are also variants of PCG that require only one of A and B to be positive definite [1], and variants that allow them to be singular, under some technical conditions on their null spaces [2].

Combinatorial preconditioning is a technique that relies on graph algorithms to construct effective preconditioners. The simplest applications of combinatorial preconditioning targets symmetric diagonally-dominant matrices with non-positive offdiagonals, a class of matrices that are isomorphic to weighted undirected graphs. The coefficient matrix A is viewed as its isomorphic graph G_A. A specialized graph algorithm constructs another graph G_B such that the isomorphic matrix B is a good preconditioner for A. The graph algorithm aims to achieve the same two goals: the inverse of B should be easy to apply, and the spectrum of $B^{-1}A$ should be clustered. It turns out that the spectrum of $B^{-1}A$ can be bounded in terms of properties of the graphs G_A and G_B; in particular, the quality of embeddings of G_A in G_B (and sometimes vice versa) plays a fundamental role in these spectral bounds.

Combinatorial preconditioners can also be constructed for many other classes of symmetric positive semidefinite matrices, including M-matrices (a class that includes all symmetric diagonally-dominant matrices) and matrices that arise from finite-elements discretizations of scalar elliptic partial differential equations. One important application area of combinatorial preconditioners is max-flow problems; interior-point algorithms for max-flow problems require the solution of many linear systems with coefficient matrices that are symmetric diagonally-dominant or M matrices.

This chapter focuses on explaining the relationship between the spectrum of $B^{-1}A$ and quantitative properties of embeddings of the two graphs. The set of mathematical tools that are used in the analysis of this relationship is called *support theory* [3]. The last section briefly surveys algorithms that construct combinatorial preconditioners. The literature describes combinatorial preconditioners that are practical to implement and are known to work well, as well as algorithms that appear to be more theoretical (that is, they provide strong theoretical performance guarantees but are complex and have not been implemented yet).

We omit most proofs from this chapter; some are trivial, and the others appear in the paper cited in the statement of the theorem or lemma. There are two new results in this chapter. Lemma 3.4.10 is a variation of the Symmetric Support Lemma for bounding the trace. Theorem 3.4.11 is a combinatorial bound on the trace of the preconditioned matrix. Both generalize results from [4].

3.2 Symmetric Diagonally-Dominant Matrices and Graphs

We begin by exploring the structure of diagonally-dominant matrices and their relation to graphs.

3.2.1 Incidence Factorizations of Diagonally-Dominantt Matrices

Definition 3.2.1 *A square matrix $A \in \mathbb{R}^{n \times n}$ is called* diagonally-dominant *if for every $i = 1, 2, \ldots n$ we have $A_{ii} \geq \sum_{i \neq j} |A_{ij}|$.*

Symmetric diagonally dominant matrices have symmetric factorizations $A = UU^T$ such that each column of U has at most two nonzeros, and all nonzeros in each column have the same absolute values [5]. We now establish a notation for such columns.

Let $1 \leq i, j \leq n$, $i \neq j$. The length-n *positive edge vector* denoted $\langle i, -j \rangle$ and the *negative edge vector* $\langle i, j \rangle$ are defined by

$$\langle i, -j \rangle_k = \begin{cases} +1 & k = i \\ -1 & k = j \\ 0 & \text{otherwise,} \end{cases} \quad \text{and} \quad \langle i, j \rangle_k = \begin{cases} +1 & k = i \\ +1 & k = j \\ 0 & \text{otherwise.} \end{cases}$$

The reason for the assignment of signs to edge vectors will become apparent later. A *vertex vector* $\langle i \rangle$ is the unit vector

$$\langle i \rangle_k = \begin{cases} +1 & k = i \\ 0 & \text{otherwise.} \end{cases}$$

For example, for $n = 4$ we have

$$\langle 1, -3 \rangle = \begin{bmatrix} 1 \\ 0 \\ -1 \\ 0 \end{bmatrix}, \quad \langle 1, 4 \rangle = \begin{bmatrix} 1 \\ 0 \\ 0 \\ 1 \end{bmatrix}, \quad \text{and} \quad \langle 4 \rangle = \begin{bmatrix} 0 \\ 0 \\ 0 \\ 1 \end{bmatrix}.$$

A symmetric diagonally dominant matrix can always be expressed as a sum of outer products of edge and vertex vectors, and therefore, as a symmetric product of a matrix whose columns are edge and vertex vectors.

Lemma 3.2.2 *([5]) Let $A \in \mathbb{R}^{n \times n}$ be a diagonally dominant symmetric matrix. We can decompose A as follows*

$$
A = \sum_{\substack{i<j \\ A_{ij}>0}} |A_{ij}| \langle i,j \rangle \langle i,j \rangle^T
$$

$$
+ \sum_{\substack{i<j \\ A_{ij}<0}} |A_{ij}| \langle i,-j \rangle \langle i,-j \rangle^T
$$

$$
+ \sum_{i=1}^{n} \left(A_{ii} - \sum_{\substack{j=1 \\ j \neq i}}^{n} |A_{ij}| \right) \langle i \rangle \langle i \rangle^T
$$

For example,

$$
\begin{bmatrix} 5 & 0 & -2 & 3 \\ 0 & 0 & 0 & 0 \\ -2 & 0 & 2 & 0 \\ 3 & 0 & 0 & 4 \end{bmatrix} = 2 \begin{bmatrix} 1 & 0 & -1 & 0 \\ 0 & 0 & 0 & 0 \\ -1 & 0 & 1 & 0 \\ 0 & 0 & 0 & 0 \end{bmatrix}
$$

$$
+3 \begin{bmatrix} 1 & 0 & 0 & 1 \\ 0 & 0 & 0 & 0 \\ 0 & 0 & 0 & 0 \\ 1 & 0 & 0 & 1 \end{bmatrix} + \begin{bmatrix} 0 & 0 & 0 & 0 \\ 0 & 0 & 0 & 0 \\ 0 & 0 & 0 & 0 \\ 0 & 0 & 0 & 1 \end{bmatrix}
$$

$$
= 2 \langle 1,-3 \rangle \langle 1,-3 \rangle^T + 2 \langle 1,4 \rangle \langle 1,4 \rangle^T + \langle 4 \rangle \langle 4 \rangle^T .
$$

Matrix decompositions of this form play a prominent role in combinatorial preconditioners, so we give them a name:

Definition 3.2.3 *A matrix whose columns are scaled edge and vertex vectors (that is, vectors of the forms $c \langle i,-j \rangle$, $c \langle i,j \rangle$, and $c \langle i \rangle$) is called an* incidence *matrix. A factorization $A = UU^T$ where U is an incidence matrix is called an* incidence factorization. *An incidence factorization with no zero columns, with at most one vertex vector for each index i, with at most one edge vector for each index pair i,j, and whose positive edge vectors are all of the form $c \langle \min(i,j), -\max(i,j) \rangle$ is called a* canonical incidence factorization.

Lemma 3.2.4 *Let $A \in \mathbb{R}^{n \times n}$ be a diagonally dominant symmetric matrix. Then A has an incidence factorization $A = UU^T$, and a unique canonical incidence factorization.*

3.2.2 Graphs and Their Laplacian Matrices

We now define the connection between undirected graphs and diagonally-dominant symmetric matrices.

Definition 3.2.5 *Let* $G = (\{1, 2, \ldots n\}, E, c, d)$ *be a weighted undirected graph on the vertex set* $\{1, 2, \ldots, n\}$ *with no self loops and no parallel edges (i.e., at most one edge between two vertices), and with weight functions* $c : E \to \mathbb{R} \setminus \{0\}$ *and* $d : \{1, \ldots, n\} \to \mathbb{R}_+ \cup \{0\}$. *That is, the edge set consists of unordered pairs of unequal integers* (i, j) *such that* $1 \leq i, j \leq n$. *The* Laplacian *of* G *is the matrix* $A \in \mathbb{R}^{n \times n}$ *such that*

$$
A_{ij} = \begin{cases} d(i) + \sum_{(i,k) \in E} |c(i, k)| & i = j \\ -c(i, j) & (i, j) \in E \\ 0 & \text{otherwise.} \end{cases}
$$

A vertex i *such that* $d(i) > 0$ *is called a* strictly dominant vertex. *If* $c = 1$, *the graph is not considered weighted. If* $c > 0$ *and is not always 1, the graph is* weighted. *If some weights are negative, the graph is* signed.

Lemma 3.2.6 *The Laplacians of the graphs defined in Definition 3.2.5 are symmetric and diagonally dominant. Furthermore, these graphs are isomorphic to symmetric diagonally-dominant matrices under this Laplacian mapping.*

In algorithms, given an explicit representation of a diagonally-dominant matrix A, we can easily compute an explicit representation of an incidence factor U (including the canonical incidence factor if desired). Sparse matrices are often represented by a data structure that stores a compressed array of nonzero entries for each row or each column of the matrix. Each entry in a row (column) array stores the column index (row index) of the nonzero, and the value of the nonzero. From such a representation of A we can easily construct a sparse representation of U by columns. We traverse each row of A, creating a column of U for each nonzero in the upper (or lower) part of A. During the traversal, we can also compute all the $d(i)$'s. The conversion works even if only the upper or lower part of A is represented explicitly.

We can use the explicit representation of A as an implicit representation of U, with each off-diagonal nonzero of A representing an edge-vector column of U. If A has no strictly-dominant rows, that is all. If A has strictly dominant rows, we need to compute their weights using a one-pass traversal of A.

3.3 Support Theory

Support theory is a set of tools that aim to bound the generalized eigenvalues λ that satisfy $Ax = \lambda Bx$ from above and below. If B is nonsingular, these eigenvalues are also the eigenvalues of $B^{-1}A$, but the generalized representation allows us to derive bounds that also apply to singular matrices.

Definition 3.3.1 *Let A and B be n-by-n complex matrices. We say that a scalar λ is a finite generalized eigenvalue of the matrix pencil (pair) (A, B) if there is a vector $v \neq 0$ such that $Av = \lambda Bv$ and $Bv \neq 0$. We say that ∞ is an infinite generalized eigenvalue of (A, B) if there exists a vector $v \neq 0$ such that $Bv = 0$ but $Av \neq 0$. Note that ∞ is an eigenvalue of (A, B) if and only if 0 is an eigenvalue of (B, A). The finite and infinite eigenvalues of a pencil are determined eigenvalues (the eigenvector uniquely determines the eigenvalue). If both $Av = Bv = 0$ for a vector $v \neq 0$, we say that v is an indeterminate eigenvector, because $Av = \lambda Bv$ for any scalar λ.*

The tools of support theory rely on symmetric factorizations $A = UU^T$ and $B = VV^T$; this is why incidence factorizations are useful. In fact, the algebraic tools of support theory are particularly easy to apply when U and V are incidence matrices.

3.3.1 From Generalized Eigenvalues to Singular Values

If $A = UU^T$ then $\Lambda(A) = \Sigma^2(U^T)$, where $\Lambda(A)$ is the set of eigenvalues of A and $\Sigma(U^T)$ is the set of singular values of U^T, and Σ^2 is the set of the squares of the singular values. The following lemma extends this trivial result to generalized eigenvalues.

Lemma 3.3.2 *([6]) Let $A = UU^T$ and $B = VV^T$ with $\text{null}(B) = \text{null}(A) = \mathbb{S}$. We have*

$$\Lambda(A, B) = \Sigma^2 \left(V^+ U \right)$$

and

$$\Lambda(A, B) = \Sigma^{-2} \left(U^+ V \right) .$$

In these expressions, $\Sigma(\cdot)$ is the set of nonzero singular values of the matrix within the parentheses, Σ^ℓ denotes the same singular values to the ℓth power, and V^+ denotes the Moore-Penrose pseudoinverse of V.

The lemma characterizes all the generalized eigenvalues of the pair (A, B), but for large matrices, it is not particularly useful. Even if U and V are highly structured (e.g., they are incidence matrices), U^+ and V^+ are usually not structured and are expensive to compute. The next section shows that if we lower our expectations a bit and only try to bound Λ from above and below, then we do not need the pseudo-inverses.

3.3.2 The Symmetric Support Lemma

In the previous section, we have seen that the singular values of $V^+ U$ provide complete information on the generalized eigenvalues of (A, B). This product is important, so we give it a notation: $W_{\text{opt}} = V^+U$. If $\text{null}(V) \subseteq \text{null}(U)$, we have

$$
\begin{aligned}
VW_{\text{opt}} &= VV^+U \\
&= U \; .
\end{aligned}
$$

It turns out that any W such that $VW = U$ provides some information on the generalized eigenvalues of (A, B).

Lemma 3.3.3 *([3]) (**Symmetric Support Lemma**) Let $A = UU^T$ and let $B = VV^T$, and assume that $\text{null}(B) \subseteq \text{null}(A)$. Then*

$$
\max\left\{\lambda \mid Ax = \lambda Bx, Bx \neq 0\right\} = \min\left\{\|W\|_2^2 \mid U = VW\right\} \; .
$$

This lemma is fundamental to support theory and preconditioning, because it is often possible to prove that W such that $U = VW$ exists and to give a-priori bounds on its norm.

3.3.3 Norm Bounds

The Symmetric Product Support Lemma bounds generalized eigenvalues in terms of the 2-norm of some matrix W such that $U = VW$. Even if we have a simple way to construct such a W, we still cannot easily derive a corresponding bound on the spectrum from the Symmetric Support Lemma. The difficulty is that there is no simple closed form expression for the 2-norm of a matrix, since it is not related to the entries of W in a simple way. It is equivalent to the largest singular value, but this must usually be computed numerically.

Fortunately, there are simple (and also some not-so-simple) functions of the elements of the matrix that yield useful bounds on its 2-norm. The following bounds are standard and are well known and widely used (see [7, Fact 9.8.10.ix] and [7, Fact 9.8.15]).

Lemma 3.3.4 *The 2-norm of $W \in \mathbb{C}^{k \times m}$ is bounded by*

$$
\|W\|_2^2 \;\leq\; \|W\|_F^2 = \sum_{i=1}^{k}\sum_{i=1}^{m} W_{ij}^2 \; , \tag{3.1}
$$

$$
\|W\|_2^2 \;\leq\; \|W\|_1 \|W\|_\infty = \left(\max_{j=1}^{m}\sum_{i=1}^{k}|W_{ij}|\right)\left(\max_{i=1}^{k}\sum_{j=1}^{m}|W_{ij}|\right) \; . \tag{3.2}
$$

The next two bounds are standard and well known; they follow directly from $\|WW^T\|_2 = \|W\|_2^2$ and from the fact that $\|S\|_1 = \|S\|_\infty$ for a symmetric S.

Lemma 3.3.5 *The 2-norm of $W \in \mathbb{C}^{k \times m}$ is bounded by*

$$\|W\|_2^2 \ \leq \ \|WW^T\|_1 = \|WW^T\|_\infty \,, \tag{3.3}$$

$$\|W\|_2^2 \ \leq \ \|W^TW\|_1 = \|W^TW\|_\infty \,. \tag{3.4}$$

The following bounds are more specialized. They all exploit the sparsity of W to obtain bounds that are usually tighter than the bounds given so far.

Lemma 3.3.6 *([8]) The 2-norm of $W \in \mathbb{C}^{k \times m}$ is bounded by*

$$\|W\|_2^2 \ \leq \ \max_j \sum_{i:W_{i,j}\neq 0} \|W_{i,:}\|_2^2 = \max_j \sum_{i:W_{i,j}\neq 0} \sum_{c=1}^m W_{i,c}^2 \,, \tag{3.5}$$

$$\|W\|_2^2 \ \leq \ \max_i \sum_{j:W_{i,j}\neq 0} \|W_{:,j}\|_2^2 = \max_i \sum_{j:W_{i,j}\neq 0} \sum_{r=1}^k W_{r,j}^2 \,. \tag{3.6}$$

The bounds in this lemma are a refinement of the bound $\|W\|_2^2 \leq \|W\|_F^2$. The Frobenius norm, which bounds the 2-norm, sums the squares of *all* the elements of W. The bounds (3.5) and (3.6) sum only the squares in some of the rows or some of the columns, unless the matrix has a row or a column with no zeros.

There are similar refinements of the bound $\|W\|_2^2 \leq \|W\|_1 \|W\|_\infty$.

Lemma 3.3.7 *([8]) The 2-norm of $W \in \mathbb{C}^{k \times m}$ is bounded by*

$$\|W\|_2^2 \ \leq \ \max_j \sum_{i:W_{i,j}\neq 0} |W_{i,j}| \cdot \left(\sum_{c=1}^m |W_{i,c}| \right) \,, \tag{3.7}$$

$$\|W\|_2^2 \ \leq \ \max_i \sum_{j:W_{i,j}\neq 0} |W_{i,j}| \cdot \left(\sum_{r=1}^k |W_{r,j}| \right) \,. \tag{3.8}$$

3.3.4   Support Numbers

Support numbers generalize the notion of the maximal eigenvalue of a matrix pencil.

Definition 3.3.8 *A matrix B dominates a matrix A if for any vector x we have $x^T(B - A)x \geq 0$. We denote domination by $B \succeq A$.*

Definition 3.3.9 *The* support number *for a matrix pencil (A, B) is*

$$\sigma(A, B) = \min \{t \,|\, \tau B \succeq A, \text{ for all } \tau \geq t\} \,.$$

If B is symmetric positive definite and A is symmetric, then the support number is always finite, because $x^T Bx / x^T x$ is bounded from below by $\min \Lambda(B) > 0$ and $x^T Ax / x^T x$ is bounded from above by $\max \Lambda(A)$, which is finite. In other cases, there may not be any t satisfying the formula; in such cases, we say that $\sigma(A, B) = \infty$.

Example 1 *Suppose that $x \in \text{null}(B)$ and that A is positive definite. Then for any $\tau > 0$ we have $x^T(\tau B - A)x = -x^T Ax < 0$. Therefore, $\sigma(A, B) = \infty$.*

Example 2 *If B is not positive semidefinite, then there is some x for which $x^T Bx < 0$. This implies that for any A and for any large enough τ, $x^T(\tau B - A)x < 0$. Therefore, $\sigma(A, B) = \infty$.*

The next result, like the Symmetric Support Lemma, bounds generalized eigenvalues.

Theorem 3.3.10 *([3]) Let A and B be symmetric, let B also be positive semidefinite. If $\text{null}(B) \subseteq \text{null}(A)$, then*

$$\sigma(A, B) = \max\left\{\lambda | Ax = \lambda Bx, Bx \neq 0\right\} .$$

A primary motivation for support numbers is to bound (spectral) condition numbers. For symmetric matrices, the relation $\kappa(A) = \lambda_{\max}(A)/\lambda_{\min}(A)$ holds. Let $\kappa(A, B)$ denote the condition number of the matrix pencil (A, B), that is, $\kappa(B^{-1}A)$ when B is non-singular.

Theorem 3.3.11 *([3]) When A and B are symmetric positive definite, then $\kappa(A, B) = \sigma(A, B)\sigma(B, A)$.*

A common strategy in support theory is to bound condition numbers by bounding both $\sigma(A, B)$ and $\sigma(B, A)$. Typically, bounding one of them is easy while bounding the other is hard.

Applications usually do not solve singular systems. Nonetheless, it is often convenient to analyze preconditioners in the singular context. For example, finite element systems are often singular until boundary conditions are imposed, so we can build B from the singular part of A and then impose the same boundary constraints on both matrices.

3.3.5 Splitting

Support numbers are convenient for algebraic manipulation. One of their most powerful properties is that they allow us to split complicated matrices into simpler pieces (matrices) and analyze these separately. Let $A = A_1 + A_2 + \cdots A_q$, and similarly, $B = B_1 + B_2 + \cdots + B_q$. We can then match up pairs (A_i, B_i) and consider the support number for each such pencil separately.

Lemma 3.3.12 *([9]) Let $A = A_1 + A_2 + \cdots A_q$, and similarly, $B = B_1 + B_2 + \cdots + B_q$, where all A_i and B_i are symmetric and positive semidefinite. Then*

$$\sigma(A, B) \leq \max_i \sigma(A_i, B_i)$$

Proof Let $\sigma = \sigma(A, B)$, let $\sigma_i = \sigma(A_i, B_i)$, and let $\sigma_{max} = \max_i \sigma_i$. Then for any x

$$
\begin{aligned}
x^T(\sigma_{max} B - A)x &= x^T \left(\sigma_{max} \sum_i B_i - \sum_i A_i \right) x \\
&= \sum_i x^T (\sigma_{max} B_i - A_i) x \\
&\geq \sum_i x^T (\sigma_i B_i - A_i) x \\
&\geq 0 .
\end{aligned}
$$

Therefore, $\sigma \leq \sigma_{max}$.

The splitting lemma is quite general, and can be used in many ways. In practice we want to break both A and B into simpler matrices that we know how to analyze. The term "simpler" can mean sparser, or lower rank, and so on. Upper bound obtained from the splitting lemma might be loose. In order to get a good upper bound on the support number, the splitting must be chosen carefully. Poor splitting gives poor bounds. In the following example, the splitting bound is not tight, but is also not particularly loose.

Example 3 *Let*

$$
A = \begin{bmatrix} 3 & -2 \\ -2 & 2 \end{bmatrix} = \begin{bmatrix} 2 & -2 \\ -2 & 2 \end{bmatrix} + \begin{bmatrix} 1 & 0 \\ 0 & 0 \end{bmatrix} = A_1 + A_2 ,
$$

and

$$
B = \begin{bmatrix} 2 & -1 \\ -1 & 2 \end{bmatrix} = \begin{bmatrix} 1 & -1 \\ -1 & 1 \end{bmatrix} + \begin{bmatrix} 1 & 0 \\ 0 & 1 \end{bmatrix} = B_1 + B_2 .
$$

Then the Splitting Lemma says $\sigma(A, B) \leq \max \{\sigma(A_1, B_1), \sigma(A_2, B_2)\}$. *It is easy to verify that* $\sigma(A_1, B_1) = 2$ *and that* $\sigma(A_2, B_2) = 1$; *hence* $\sigma(A, B) \leq 2$. *Note that* B_1 *can not support* A_2, *so correct pairing of the terms in A and B is essential. The exact support number is* $\sigma(A, B) = \lambda_{max}(A, B) = 1.557$.

3.4 Embeddings and Combinatorial Support Bounds

To bound $\sigma(A, B)$ using the Symmetric Support Lemma, we need to factor A and B into $A = UU^T$ and $B = VV^T$, and we need to find a W such that $U = VW$. We have seen that if A and B are diagonally dominant, then there is an almost trivial way to factor A and B such that U and V are about as sparse as A and B. But how do we find a W such that $U = VW$? In this section, we show that when A and B are weighted (but not signed) Laplacians, we can construct such a W using an embedding of the edges of G_A into paths in G_B. Furthermore, when W is constructed from an embedding, the bounds on $\|W\|_2$ can be interpreted as combinatorial bounds on the quality of the embedding.

3.4.1 Defining W Using Path Embeddings

We start with the construction of a matrix W, such that $U = VW$.

Lemma 3.4.1 *Let* $(i_1, i_2, \ldots, i_\ell)$ *be a sequence of integers between 1 and n, such that* $i_j \neq i_{j+1}$ *for* $j = 1, \ldots \ell - 1$. *Then*

$$\langle i_1, -i_\ell \rangle = \sum_{j=1}^{\ell-1} \langle i_j, -i_{j+1} \rangle ,$$

where all the edge vectors are length n.

To see why this lemma is important, consider the role of a column of W. Suppose that the columns of U and V are all positive edge vectors. Denote column c of U by

$$U_{:,c} = \langle \min(i_1, i_\ell), -\max(i_1, i_\ell) \rangle = (-1)^{i_1 > i_\ell} \langle i_1, -i_\ell \rangle ,$$

where the $(-1)^{i_1 > i_\ell}$ evaluates to -1 if $i_1 > i_\ell$ and to 1 otherwise. This column corresponds to the edge (i_1, i_ℓ) in G_{UU^T}. Now let $(i_1, i_2, \ldots, i_\ell)$ be a simple path in G_{VV^T} (a simple path is a sequence of vertices $(i_1, i_2, \ldots, i_\ell)$ such that (i_j, i_{j+1}) is an edge in the graph for $1 \leq j < \ell$ and such that any vertex appears at most once on the path).

Let $r_1, r_2, \ldots, r_{\ell-1}$ be the columns of V that corresponds to the edges of the path $(i_1, i_2, \ldots, i_\ell)$, in order. That is, $V_{:,r_1} = \langle \min(i_1, i_2), -\max(i_1, i_2) \rangle$, $V_{:,r_2} = \langle \min(i_2, i_3), -\max(i_2, i_3) \rangle$, and so on. By the lemma,

$$
\begin{aligned}
U_{:,c} &= (-1)^{i_1 > i_\ell} \langle i_1, -i_\ell \rangle \\
&= (-1)^{i_1 > i_\ell} \sum_{j=1}^{\ell-1} \langle i_j, -i_{j+1} \rangle \\
&= (-1)^{i_1 > i_\ell} \sum_{j=1}^{\ell-1} (-1)^{i_j > i_{j+1}} V_{:,r_j} .
\end{aligned}
$$

It follows that if we define $W_{:,c}$ to be

$$W_{r,c} = \begin{cases} (-1)^{i_1 > i_\ell}(-1)^{i_j > i_{j+1}} & r = r_j \text{ for some } 1 \leq j < \ell \\ 0 & \text{otherwise,} \end{cases}$$

then we have

$$U_{:,c} = VW_{:,c} = \sum_{r=1}^{k} V_{:,r} W_{r,c} .$$

We can construct all the columns of W in this way, so that W satisfies $U = VW$.

A path of edge vectors that ends in a vertex vector supports the vertex vector associated with the first vertex of the path.

Lemma 3.4.2 *Let $(i_1, i_2, \ldots, i_\ell)$ be a sequence of integers between 1 and n, such that $i_j \neq i_{j+1}$ for $j = 1, \ldots \ell - 1$. Then*

$$\langle i_1 \rangle = \langle i_\ell \rangle + \sum_{j=1}^{\ell-1} \langle i_j, -i_{j+1} \rangle \; ,$$

where all the edge and vertex vectors are length n.

The following theorem generalizes these ideas to scaled positive edge vectors and to scaled vertex vectors. The theorem also states how to construct all the columns of W. The theorem summarizes results in [3, 10].

Theorem 3.4.3 *Let A and B be weighted (unsigned) Laplacians and let U and V be their canonical incidence factors. Let π be a path embedding of the edges and strictly-dominant vertices of G_A into G_B, such that for an edge (i_1, i_ℓ) in G_A, $i_1 < i_\ell$, we have*

$$\pi(i_1, i_\ell) = (i_1, i_2, \ldots, i_\ell)$$

for some simple path $(i_1, i_2, \ldots, i_\ell)$ in G_B, and such that for a strictly-dominant i_1 in G_A,

$$\pi(i_1) = (i_1, i_2, \ldots, i_\ell)$$

for some simple path $(i_1, i_2, \ldots, i_\ell)$ in G_B that ends in a strictly-dominant vertex i_ℓ in G_B. Denote by $c_V(i_j, i_{j+1})$ the index of the column of V that is a scaling of $\langle i_j, -i_{j+1} \rangle$. That is,

$$V_{:\,, c_V(i_j, i_{j+1})} = \sqrt{-B_{i_j, i_{j+1}}} \, \langle \min(i_j, i_{j+1}), -\max(i_j, i_{j+1}) \rangle \; .$$

Similarly, denote by $c_V(i_j)$ the index of the column of V that is a scaling of $\langle i_j \rangle$,

$$V_{:\,, c_V(i_j)} = \sqrt{B_{i_j, i_j} - \sum_{\substack{i_k=1 \\ i_k \neq i_j}}^{n} \left| B_{i_k, i_j} \right|} \, \langle i_j \rangle \; ,$$

and similarly for U.

We define a matrix W as follows. For a column index $c_U(i_1, i_\ell)$ with $i_1 < i_\ell$ we define

$$W_{r, c_U(i_1, i_\ell)} = \begin{cases} (-1)^{i_j > i_{j+1}} \sqrt{A_{i_1, i_\ell}/B_{i_j, i_{j+1}}} & \text{if } r = c_V(i_j, i_{j+1}) \text{ for} \\ & \text{some edge } (i_j, i_{j+1}) \text{ in } \pi(i_1, i_\ell) \\ 0 & \text{otherwise.} \end{cases}$$

For a column index $c_U(i_1)$, we define

$$
W_{r,c_U(i_1)} = \begin{cases}
\sqrt{\dfrac{A_{i_1,i_1} - \sum_{j \neq i_1} |A_{i_1,j}|}{B_{i_\ell,i_\ell} - \sum_{j \neq i_\ell} |B_{i_\ell,j}|}} & \text{if } r = c_V(i_\ell) \\[2em]
(-1)^{i_j > i_{j+1}} \sqrt{\dfrac{A_{i_1,i_1} - \sum_{k \neq i_1} |A_{i_1,k}|}{|B_{i_j,i_{j+1}}|}} & \text{if } r = c_V(i_j, i_{j+1}) \text{ for} \\
& \text{some edge } (i_j, i_{j+1}) \text{ in } \pi(i_1) \\[1em]
0 & \text{otherwise.}
\end{cases}
$$

Then $U = VW$.

Proof For scaled edge-vector columns in U we have

$$
\begin{aligned}
VW_{:,c_U(i_1,i_\ell)} &= \sum_r V_{:,r} W_{r,c_U(i_1,i_\ell)} \\
&= \sum_{\substack{r = c_V(i_j,i_{j+1}) \\ \text{for some edge} \\ (i_j,i_{j+1}) \text{ in } \pi(i_1,i_\ell)}} V_{:,r} W_{r,c_U(i_1,i_\ell)} \\
&= \sum_{j=1}^{\ell-1} \sqrt{|B_{i_j,i_{j+1}}|} \langle \min(i_j,i_{j+1}), -\max(i_j,i_{j+1}) \rangle (-1)^{i_j > i_{j+1}} \sqrt{\frac{A_{i_1,i_\ell}}{B_{i_j,i_{j+1}}}} \\
&= \sqrt{|A_{i_1,i_\ell}|} \sum_{j=1}^{\ell-1} \langle i_j, -i_{j+1} \rangle \\
&= U_{:,c_U(i_1,i_\ell)} \, .
\end{aligned}
$$

For scaled vertex-vector columns in U, we have

$$
\begin{aligned}
VW_{:,c_U(i_1)} &= \sum_r V_{:,r} W_{r,c_U(i_1)} \\
&= V_{:,c_V(i_\ell)} W_{c_V(i_\ell),c_U(i_1,i_\ell)} + \sum_{\substack{r = c_V(i_j,i_{j+1}) \\ \text{for some edge} \\ (i_j,i_{j+1}) \text{ in } \pi(i_1)}} V_{:,r} W_{r,c_U(i_1)} \\
&= \sqrt{B_{i_\ell,i_\ell} - \sum_{j \neq i_\ell} |B_{i_k,i_\ell}|} \, \langle i_\ell \rangle \sqrt{\frac{A_{i_1,i_1} - \sum_{j \neq i_1} |A_{i_1,j}|}{B_{i_\ell,i_\ell} - \sum_{j \neq i_\ell} |B_{i_\ell,j}|}} \\
&\quad + \sum_{j=1}^{\ell-1} \sqrt{|B_{i_j,i_{j+1}}|} \, \langle \min(i_j,i_{j+1}), -\max(i_j,i_{j+1}) \rangle \\
&\qquad \cdot (-1)^{i_j > i_{j+1}} \sqrt{\frac{A_{i_1,i_1} - \sum_{j \neq i_1} |A_{i_1,j}|}{|B_{i_j,i_{j+1}}|}}
\end{aligned}
$$

$$= \sqrt{A_{i_1,i_1} - \sum_{j \neq i_1} |A_{i_1,j}|} \langle i_\ell \rangle + \sqrt{A_{i_1,i_1} - \sum_{j \neq i_1} |A_{i_1,j}|} \sum_{j=1}^{\ell-1} \langle i_j, -i_{j+1} \rangle$$

$$= \sqrt{A_{i_1,i_1} - \sum_{j \neq i_1} |A_{i_1,j}|} \langle i_1 \rangle$$

$$= U_{:,c_U(i_1)} \, .$$

The generalization of this theorem to signed Laplacians is more complex, because a path from i_1 to i_ℓ supports an edge (i_1, i_ℓ) only if the parity of positive edges in the path and in the edge is the same. In addition, a cycle with an odd number of positive edges spans all the vertex vectors of the path. For details, see [5].

Theorem 3.4.3 plays a fundamental role in many applications of support theory. A path embedding π that can be used to construct W exists if and only if the graphs of A and B are related in a specific way, which the next lemma specifies.

Lemma 3.4.4 *Let $A = UU^T$ and $B = VV^T$ be weighted (but not signed) Laplacians with arbitrary symmetric-product factorizations. The following conditions are necessary for the equation $U = VW$ to hold for some matrix W (by Theorem 3.4.3, these conditions are also sufficient).*

1. *For each edge (i, j) in G_A, either i and j are in the same connected component in G_B, or the two components of G_B that contain i and j both include a strictly-dominant vertex.*

2. *For each strictly-dominant vertex i in G_A, the component of G_B that contains i includes a strictly-dominant vertex.*

Proof Suppose for contradiction that one of the conditions is not satisfied, but that there is a W that satisfies $U = VW$. Without loss of generality, we assume that the vertices are ordered such that vertices that belong to a connected component in G_B are consecutive. Under that assumption,

$$V = \begin{bmatrix} V_1 & & & \\ & V_2 & & \\ & & \ddots & \\ & & & V_k \end{bmatrix},$$

and

$$B = \begin{bmatrix} B_1 & & & \\ & B_2 & & \\ & & \ddots & \\ & & & B_k \end{bmatrix} = \begin{bmatrix} V_1 V_1^T & & & \\ & V_2 V_2^T & & \\ & & \ddots & \\ & & & V_k V_k^T \end{bmatrix}.$$

The blocks of V are possibly rectangular, whereas the nonzero blocks of B are all diagonal and square.

We now prove the necessity of the first condition. Suppose for some edge (i, j) in G_A, i and j belong to different connected components of G_B (without loss of generality, to the first two components), and that one of the components (w.l.o.g. the first) does not have a strictly-dominant vertex. Because this component does not have a strictly-dominant vertex, the row sums in $V_1 V_1^T$ are exactly zero. Therefore, $V_1 V_1^T \vec{1} = \vec{0}$, so V_1 must be rank deficient.

Since (i, j) is in G_A, the vector $\langle i, -j \rangle$ is in the column space of the canonical incidence factor of A, and therefore in the column space of any U such that $A = UU^T$. If $U = VW$, then the vector $\langle i, -j \rangle$ must also be in the column space of V, so for some x

$$
\langle i, -j \rangle = V x =
\begin{bmatrix}
V_1 & & & \\
& V_2 & & \\
& & \ddots & \\
& & & V_k
\end{bmatrix}
\begin{bmatrix}
x_1 \\
x_2 \\
\vdots \\
x_k
\end{bmatrix}
=
\begin{bmatrix}
V_1 x_1 \\
V_2 x_2 \\
\vdots \\
V_k x_k
\end{bmatrix}.
$$

Therefore, $V_1 x_1$ is a vertex vector. By Lemma 3.4.2, if V_1 spans a vertex vector, it spans all the vertex vectors associated with the vertices of the connected component. This implies that V_1 is full rank, a contradiction.

The necessity of the second condition follows from a similar argument. Suppose that vertex i is strictly dominant in G_A and that it belongs to a connected component in G_B (w.l.o.g. the first) that does not have a vertex that is strictly dominant in G_B. This implies that for some y

$$
\langle i \rangle = V y =
\begin{bmatrix}
V_1 & & & \\
& V_2 & & \\
& & \ddots & \\
& & & V_k
\end{bmatrix}
\begin{bmatrix}
y_1 \\
y_2 \\
\vdots \\
y_k
\end{bmatrix}
=
\begin{bmatrix}
V_1 y_1 \\
V_2 y_2 \\
\vdots \\
V_k y_k
\end{bmatrix}.
$$

Again $V_1 y_1$ is a vertex vector, so V_1 must be full rank, but it cannot be full rank because $V_1 V_1^T$ has zero row sums.

Not every W such that $U = VW$ corresponds to a path embedding, even if U and V are the canonical incidence factors of A and B. In particular, a column of W can correspond to a linear combination of multiple paths. Also, even if W does correspond to a path embedding, the paths are not necessarily simple. A linear combination of scaled positive edge vectors that correspond to a simple cycle can be identically zero, so the coefficients of such linear combinations can be added to W without affecting the product VW. However, it seems that adding cycles to a path embedding cannot reduce the 2-norm of W, so cycles are unlikely to improve support bounds.

3.4.2 Combinatorial Support Bounds

To bound $\sigma(A, B)$ using the Symmetric Support Lemma, we factor A into $A = UU^T$, B into $B = VV^T$, find a matrix W such that $U = VW$, and bound the 2-norm of W from above. We have seen how to factor A and B (if they are weighted Laplacians) and how to construct an appropriate W from an embedding of G_B in G_A. We now show how to use combinatorial metrics of the path embeddings to bound $\|W\|_2$.

Bounding the 2-norm directly is hard, because the 2-norm is not related in a simple way to the entries of W. But the 2-norm can be bounded using simpler norms, such as the Frobenius norm, the ∞-norm, and the 1-norm (see Section 3.3.3). These simpler norms have natural and useful combinatorial interpretations when W represents a path embedding.

To keep the notation and the definition simple, we now assume that A and B are weighted Laplacians with zero row sums. We will show later how to deal with positive row sums. We also assume that W corresponds to a path embedding π. The following definitions provide a combinatorial interpretation of these bounds.

Definition 3.4.5 *The* weighted dilation *of an edge (i_1, i_2) of G_A in an path embedding π of G_A into G_B is*

$$dilation_\pi(i_1, i_2) = \sum_{\substack{(j_1, j_2) \\ (j_1, j_2) \in \pi(i_1, i_2)}} \sqrt{\frac{A_{i_1, i_2}}{B_{j_1, j_2}}} .$$

The weighted congestion *of an edge (j_1, j_2) of G_B is*

$$congestion_\pi(j_1, j_2) = \sum_{\substack{(i_1, i_2) \\ (j_1, j_2) \in \pi(i_1, i_2)}} \sqrt{\frac{A_{i_1, i_2}}{B_{j_1, j_2}}} .$$

The weighted stretch *of an edge of G_A is*

$$stretch_\pi(i_1, i_2) = \sum_{\substack{(j_1, j_2) \\ (j_1, j_2) \in \pi(i_1, i_2)}} \frac{A_{i_1, i_2}}{B_{j_1, j_2}} .$$

The weighted crowding *of an edge in G_B is*

$$crowding_\pi(j_1, j_2) = \sum_{\substack{(i_1, i_2) \\ (j_1, j_2) \in \pi(i_1, i_2)}} \frac{A_{i_1, i_2}}{B_{j_1, j_2}} .$$

Note that stretch is a summation of the squares of the quantities that constitute dilation, and similarly for crowding and congestion. Papers in the support-preconditioning literature are not consistent in their definition of these terms, so check the definitions carefully when you consult a paper that deals with congestion, dilation, and similar terms.

Lemma 3.4.6 *([3, 11]) Let A and B be weighted Laplacians with zero row sums, and let π be a path embedding of G_A into G_B. Then*

$$\sigma(A, B) \ \leq \ \sum_{(i_1, i_2) \in G_A} stretch_\pi(i_1, i_2)$$

$$\sigma(A, B) \ \leq \ \sum_{(j_1, j_2) \in G_B} crowding_\pi(j_1, j_2)$$

$$\sigma(A, B) \ \leq \ \left(\max_{(i_1, i_2) \in G_A} dilation_\pi(i_1, i_2) \right)$$
$$\cdot \left(\max_{(j_1, j_2) \in G_B} congestion_\pi(j_1, j_2) \right) .$$

We now describe one way to deal with matrices with some positive row sums. In many applications, the preconditioner B is not given, but rather constructed. One simple way to deal with positive row sums in A is to define $\pi(i_1) = (i_1)$. That is, vertex vectors in the canonical incidence factor of A are mapped into the same vertex vectors in the incidence factor of B. In other words, we construct B to have exactly the same row sums as A. With such a construction, the rows of W that correspond to vertex vectors in U are columns of the identity. The same idea is also used in other families of preconditioners, like those based on incomplete Cholesky factorizations. The next lemma gives bounds based on this strategy (it is essentially a special case of Lemma 2.5 in [10]).

Lemma 3.4.7 *Let A and B be weighted Laplacians with the same row sums, let π be a path embedding of G_A into G_B, and let ℓ be the number of rows with positive row sums in A and B. Then*

$$\sigma(A, B) \ \leq \ \ell + \sum_{(i_1, i_2) \in G_A} stretch_\pi(i_1, i_2)$$

$$\sigma(A, B) \ \leq \ \left(\max \left\{ 1, \max_{(i_1, i_2) \in G_A} dilation_\pi(i_1, i_2) \right\} \right)$$
$$\cdot \left(\max \left\{ 1, \max_{(j_1, j_2) \in G_B} congestion_\pi(j_1, j_2) \right\} \right) .$$

Proof Under the hypothesis of the lemma, the rows and columns of W can be permuted into a block matrix

$$W = \left(\begin{array}{cc} W_Z & 0 \\ 0 & I_{\ell \times \ell} \end{array} \right) ,$$

where W_Z represents the path embedding of the edges of G_A into paths in G_B. The bounds follow from the structure of W and from the proof of the previous lemma.

The sparse bounds on the 2-norm of a matrix lead to tighter combinatorial bounds.

Lemma 3.4.8 *Let A and B be weighted Laplacians with zero row sums, and let π be a path embedding of G_A into G_B. Then*

$$\sigma(A,B) \leq \max_{(j_1,j_2)\in G_B} \sum_{\substack{(i_1,i_2)\in G_A \\ (j_1,j_2)\in \pi(i_1,i_2)}} stretch_\pi(i_1,i_2) \,,$$

$$\sigma(A,B) \leq \max_{(i_1,i_2)\in G_A} \sum_{\substack{(j_1,j_2)\in G_A \\ (j_1,j_2)\in \pi(i_1,i_2)}} crowding_\pi(j_1,j_2) \,.$$

We can derive similar bounds for the other sparse 2-norm bounds.

3.4.3   Subset Preconditioners

To obtain a bound on $\kappa(A,B)$, we need a bound on both $\sigma(A,B)$ and $\sigma(B,A)$. But in one common case, bounding $\sigma(B,A)$ is trivial. Many support preconditioners construct G_B to be a subgraph of G_A, with the same weights. That is, V is constructed to have a subset of the columns in U. If we denote by $\bar{V}$ the set of columns of U that are *not* in V, we have

$$\begin{aligned} B &= VV^T \\ A &= UU^T \\ &= VV^T + \bar{V}\bar{V}^T \\ &= B + \bar{V}\bar{V}^T \,. \end{aligned}$$

This immediately implies $x^T A x \geq x^T B x$ for any x, so $\lambda_{\min}(A,B) \geq 1$.

3.4.4   Combinatorial Trace Bounds

The Preconditioned Conjugate Gradients (PCG) algorithm requires $\Theta(\sqrt{\kappa(A,B)})$ iterations only when the generalized eigenvalues are distributed poorly between $\lambda_{\min}(A,B)$ and $\lambda_{\max}(A,B)$. It turns out that a bound on $\text{trace}(A,B) = \sum \lambda_i(A,B)$ also yields a bound on the number of iterations, and in some cases this bound is sharper than the $O(\sqrt{\kappa(A,B)})$ bound.

Lemma 3.4.9 *([4]) The Preconditioned Conjugate Algorithm converges to within a fixed tolerance in*

$$O\left(\sqrt[3]{\frac{\text{trace}(A,B)}{\lambda_{\min}(A,B)}} \right)$$

iterations.

A variation of the Symmetric Support Lemma bounds the trace, and this leads to a combinatorial support bound.

Lemma 3.4.10 *Let* $A = UU^T \in \mathbb{R}^{n \times n}$ *and let* $B = VV^T \in \mathbb{R}^{n \times n}$, *and assume that* $\text{null}(B) = \text{null}(A)$. *Then*

$$\text{trace}(A, B) = \min\left\{\|W\|_F^2 \mid U = VW\right\}.$$

Proof Let W be a matrix such that $U = VW$. For every $x \notin \text{null}(B)$ we can write

$$\frac{x^T A x}{x^T B x} = \frac{x^T VWW^T V^T x}{x^T VV^T x}$$
$$= \frac{y^T WW^T y}{y^T y},$$

where $y = V^T x$. Let $S \subseteq \mathbb{R}^n$ be a subspace orthogonal to $\text{null}(B)$ of dimension k, and define $T_S = \left\{V^T x \mid x \in S\right\}$. Because S is orthogonal to $\text{null}(B)$ we have $\dim(T_S) = k$. The sets $\left\{x^T A x / x^T B x \mid x \in S\right\}$ and $\left\{y^T WW^T y \mid y \in T_S\right\}$ are identical so their minima are equal as well. The group of subspaces $\left\{T_S \mid \dim(S) = k, \ S \perp \text{null}(B)\right\}$ is a subset of all subspaces of dimension k, therefore

$$\max_{\substack{\dim(S) = k \\ S \perp \text{null}(B)}} \min_{\substack{x \in S \\ x \neq 0}} \frac{x^T UU^T x}{x^T VV^T x} = \max_{\substack{\dim(S) = k \\ S \perp \text{null}(B)}} \min_{\substack{y \in T_S \\ y \neq 0}} \frac{y^T WW^T y}{y^T y}$$

$$\leq \max_{\dim(T)=k} \min_{\substack{y \in T \\ y \neq 0}} \frac{y^T WW^T y}{y^T y}.$$

According to the Courant-Fischer Minimax Theorem,

$$\lambda_{n-k+1}(WW^T) = \max_{\dim(T)=k} \min_{\substack{y \in T \\ y \neq 0}} \frac{y^T WW^T y}{y^T y}$$

and by the generalization of Courant-Fischer in [12],

$$\lambda_{n-k+1}(UU^T, VV^T) = \max_{\substack{\dim(S) = k \\ V \perp \text{null}(B)}} \min_{x \in S} \frac{x^T UU^T x}{x^T VV^T x}.$$

Therefore, for $k = 1, \ldots, \text{rank}(B)$ we have $\lambda_{n-k+1}(A, B) \leq \lambda_{n-k+1}(WW^T)$ so $\text{trace}(A, B) \leq \text{trace}(WW^T) = \|W\|_F^2$. This shows that $\text{trace}(A, B) \leq \min\left\{\|W\|_F^2 \mid U = VW\right\}$.

According to Lemma 3.3.2 the minimum is attainable at $W = V^+U$.

To the best of our knowledge, Lemma 3.4.10 is new. It was inspired by a specialized bound on the trace from [4]. The next theorem generalizes the result form [4]. The proof is trivial given Definition 3.4.5 and Lemma 3.4.10.

Theorem 3.4.11 *Let A and B be weighted Laplacians with the same row sums, let π be a path embedding of G_A into G_B. Then*

$$\text{trace}(A, B) \leq \sum_{(i_1, i_2) \in G_A} stretch_\pi(i_1, i_2) \,,$$

$$\text{trace}(A, B) \leq \sum_{(j_1, j_2) \in G_B} crowding_\pi(j_1, j_2) \,.$$

3.5 Combinatorial Preconditioners

Early research on combinatorial preconditioners focused on symmetric diagonally-dominant matrices. The earliest graph algorithm to construct a preconditioner was proposed by Vaidya [13] (see also [10]). He proposed to use a so-called augmented maximum spanning tree of a weighted Laplacian as a preconditioner. This is a subset preconditioner that drops some of the edges in G_A while maintaining the weights of the remaining edges. When B^{-1} is applied using a sparse Cholesky factorization, this construction leads to a total solution time of $O(n^{7/4})$ for Laplacians with a bounded degree and $O(n^{6/5})$ when the graph is planar. For regular unweighted meshes in 2 and 3 dimensions, special constructions are even more effective [14]. Vaidya also proposed to use recursion to apply B^{-1}, but without a rigorous analysis; in this scheme, sparse Gaussian elimination steps are performed on B as long as the reduced matrix has a row with only two nonzeros. At that point, the reduced system is solved by constructing a graph preconditioner. Vaidya's preconditioners are quite effective in practice [15]. A generalization to complex matrices proved effective in handling a certain kind of ill conditioning [16].

Vaidya's bounds used a congestion-dilation product. The research on subset graph preconditioners continued with an observation that the sum of the stretch can also yield a spectral bound, and that so-called *low-stretch trees* would give better worst-case bounds than the maximum-spanning trees that Vaidya used [17]. Constructing low-stretch trees is more complicated than constructing maximum spanning trees. When the utility of low-stretch trees was discovered, one algorithm for constructing them was known [18]; better algorithms were discovered later [19]. Low-stretch trees can be used to build preconditioners that can solve any Laplacian system with m nonzeros in $O(m^{4/3})$, up to some additional polylogarithmic factors [20, 4]. By employing recursion, the theoretical running time can be reduced to close to linear in

m [21]. An experimental comparison of simplified versions of these sophisticated algorithms to Vaidya's algorithm did not yield a conclusive result [22]. Heuristic subset graph preconditioners have also been proposed [23].

Gremban and Miller proposed a class of combinatorial preconditioners called *support-tree* preconditioners [24, 9]. Their algorithms construct a weighted tree whose leaves are the vertices of G_A. Therefore, the graph G_T of the preconditioner T has additional vertices. They show that applying T^{-1} to extensions of residuals is equivalent to using a preconditioner B that is the Schur complement of T with respect to the original vertices. Bounding the condition number $\kappa(A, B)$ is more difficult than bounding the condition number of subset preconditioners, because the Schur complement is dense. More effective versions of this strategy have been proposed later [25, 26, 27].

Efforts to generalize these constructions to matrices that are not weighted Laplacians followed several paths. Gremban showed how to transform a linear system whose coefficient matrix is a signed Laplacian to a linear system of twice the size whose matrix is a weighted Laplacian. The coefficient matrix is a 2-by-2 block matrix with diagonal blocks with the same sparsity pattern as the original matrix A and with identity off-diagonal blocks. A different approach is to extend Vaidya's construction to signed graphs [5]. The class of symmetric matrices with a symmetric factorization $A = UU^T$ where columns of U have at most 2 nonzeros contains not only signed graphs, but also gain graphs, which are not diagonally dominant [28]. In paricular, this class contains symmetric positive semidefinite M matrices. It turns out that matrices in this class can be scaled to diagonal dominance, which allows graph preconditioners to be applied to them [29] (this reduction requires not only that a factorization $A = UU^T$ with at most 2 nonzeros exists, but that it be known).

Linear systems of equations with symmetric positive semidefinite M-matrices and symmetric diagonally-dominant matrices arise in interior-point solvers for max-flow problems. Combinatorial preconditioners are very effective in this application, both theoretically [29] and in practice [23].

The matrices that arise in finite-element discretizations of elliptic partial differential equations (PDEs) are positive semi-definite, but in general they are not diagonally dominant. However, when the PDE is scalar (e.g., describes a problem in electrostatics), the matrices can sometimes be approximated by diagonally dominant matrices. In this scheme, the coefficient matrix A is first approximated by a diagonally-dominant matrix D, and then G_D is used to construct the graph G_B of the preconditioner B. For large matrices of this class, the first step is expensive, but because finite-element matrices have a natural representation as a sum of very sparse matrices, the diagonally-dominant approximation can be constructed for each term in the sum separately. There are at least three ways to construct these approximations: during the finite-element discretization process [30], algebraically [6], and geometrically [31]. A slightly modified construction that can accommodate terms that do not have a close diagonally-dominant approximation works well in practice [6].

Another approach for constructing combinatorial preconditioners to finite element problems is to rely on a graph that describes the relations between neighboring elements. This graph is the dual of the finite-element mesh; elements in the mesh are the vertices of the graph. Once the graph is constructed, it can be sparsified much like subset preconditioners. This approach, which is applicable to vector problems like linear elasticity, was proposed in [32]; this paper also showed how to construct the dual graph algebraically and how to construct the finite-element problem that corresponds to the sparsified dual graph. The first effective preconditioner of this class was proposed in [33]. It is not yet known how to weigh the edges of the dual graph effectively, which limits the applicability of this method. However, in applications where there is no need to weigh the edges, the method is effective [34].

Bibliography

[1] Ashby, S. F., Manteuffel, T. A., and Saylor, P. E., "A taxonomy for conjugate gradient methods," *SIAM J. Numer. Anal.*, Vol. 27, No. 6, 1990, pp. 1542–1568.

[2] Chen, D. and Toledo, S., "Combinatorial characterization of the null spaces of symmetric H-matrices," *Linear Algebra and its Applications*, Vol. 392, 2004, pp. 71–90.

[3] Boman, E. G. and Hendrickson, B., "Support Theory for Preconditioning," *SIAM J. Matrix Anal. Appl.*, Vol. 25, No. 3, 2003, pp. 694–717.

[4] Spielman, D. A. and Woo, J., "A Note on Preconditioning by Low-Stretch Spanning Trees," Manuscript available online at http://arxiv.org/abs/0903.2816.

[5] Boman, E. G., Chen, D., Hendrickson, B., and Toledo, S., "Maximum-weight-basis Preconditioners," *Numerical Linear Algebra with Applications*, Vol. 11, 2004, pp. 695–721.

[6] Avron, H., Chen, D., Shklarski, G., and Toledo, S., "Combinatorial Preconditioners for Scalar Elliptic Finite-Element Problems," *SIAM Journal on Matrix Analysis and Applications*, Vol. 31, No. 2, 2009, pp. 694–720.

[7] Bernstein, D. S., *Matrix Mathematics: Theory, Facts, and Formulas with Applications to Linear Systems Theory*, Princeton University Press, Princeton, NJ: 2005.

[8] Chen, D., Gilbert, J. R., and Toledo, S., "Obtaining bounds on the two norm of a matrix from the splitting lemma," *Electronic Transactions on Numerical Analysis*, Vol. 21, 2005, pp. 28–46.

[9] Gremban, K. D., *Combinatorial Preconditioners for Sparse, Symmetric, Diagonally Dominant Linear Systems*, Ph.D. thesis, School of Computer Science, Carnegie Mellon University, Oct. 1996, Available as Technical Report CMU-CS-96-123.

[10] Bern, M., Gilbert, J. R., Hendrickson, B., Nguyen, N., and Toledo, S., "Support-Graph Preconditioners," *SIAM Journal on Matrix Analysis and Applications*, Vol. 27, 2006, pp. 930–951.

[11] Spielman, D. A. and Teng, S.-H., "Nearly-Linear Time Algorithms for Preconditioning and Solving Symmetric, Diagonally Dominant Linear Systems," Unpublished manuscript available online at http://arxiv.org/abs/cs/0607105.

[12] Avron, H., Ng, E., and Toledo, S., "Using Perturbed QR Factorizations to Solve Linear Least-Squares Problems," *SIAM Journal on Matrix Analysis and Applications*, Vol. 31, No. 2, 2009, pp. 674–693.

[13] Vaidya, P. M., "Solving linear equations with symmetric diagonally dominant matrices by constructing good preconditioners," Unpublished manuscript. A talk based on this manuscript was presented at the IMA Workshop on Graph Theory and Sparse Matrix Computations, Minneapolis, October 1991.

[14] Joshi, A., *Topics in Optimization and Sparse Linear Systems*, Ph.D. thesis, Department of Computer Science, University of Illinois at Urbana-Champaign, 1997.

[15] Chen, D. and Toledo, S., "Vaidya's Preconditioners: Implementation and Experimental Study," *Electronic Transactions on Numerical Analysis*, Vol. 16, 2003, pp. 30–49.

[16] Howle, V. E. and Vavasis, S. A., "An Iterative Method for Solving Complex-Symmetric Systems Arising in Electrical Power Modeling," *SIAM Journal on Matrix Analysis and Applications*, Vol. 26, No. 4, 2005, pp. 1150–1178.

[17] Boman, E. G. and Hendrickson, B., "On Spanning Tree Preconditioners," Unpublished manuscript, Sandia National Laboratories, Albuquerque, NM.

[18] Alon, N., Karp, R. M., Peleg, D., and West, D., "A Graph-Theoretic Game and Its Application to the k-Server Problem," *SIAM Journal on Computing*, Vol. 24, 1995, pp. 78–100.

[19] Elkin, M., Emek, Y., Spielman, D. A., and Teng, S.-H., "Lower-stretch spanning trees," *Proceedings of the 37th Annual ACM Symposium on Theory of Computing (STOC)*, ACM Press, Baltimore, MD, 2005, pp. 494–503.

[20] Spielman, D. A. and Teng, S.-H., "Solving Sparse, Symmetric, Diagonally-Dominant Linear Systems in Time $0(m^{1.31})$," *Proceedings of the 44th Annual IEEE Symposium on Foundations of Computer Science*, Oct. 2003, pp. 416–427.

[21] Spielman, D. A. and Teng, S.-H., "Nearly-linear time algorithms for graph partitioning, graph sparsification, and solving linear systems," *STOC '04: Proceedings of the thirty-sixth annual ACM symposium on Theory of computing*, ACM Press, New York, NY, USA, 2004, pp. 81–90.

[22] Unger, U., *An Experimental Evaluation of Combinatorial Preconditioners*, Master's thesis, Tel-Aviv University, July 2007.

[23] Frangioni, A. and Gentile, C., "New Preconditioners for KKT systems of network flow problems," *SIAM Journal on Optimization*, Vol. 14, 2004, pp. 894–913.

[24] Gremban, K. D., Miller, G. L., and Zagha, M., "Performance evaluation of a new parallel preconditioner," *Proceedings of the 9th International Parallel Processing Symposium*, IEEE Computer Society, 1995, pp. 65–69, A longer version is available as Technical Report CMU-CS-94-205, Carnegie-Mellon University, Pittsburgh, PA.

[25] Maggs, B. M., Miller, G. L., Parekh, O., Ravi, R., and Woo, S. L. M., "Finding effective support-tree preconditioners," *SPAA '05: Proceedings of the seventeenth annual ACM symposium on Parallelism in algorithms and architectures*, ACM Press, Las Vegas, NV, 2005, pp. 176–185.

[26] Koutis, I. and Miller, G. L., "A linear work, $O(n^{1/6})$ time, parallel algorithm for solving planar Laplacians," *Proceedings of the Eighteenth Annual ACM-SIAM Symposium on Discrete Algorithms, SODA 2007*, New Orleans, LA, USA, January 7-9, 2007, edited by N. Bansal, K. Pruhs, and C. Stein, SIAM, 2007, pp. 1002–1011.

[27] Koutis, I. and Miller, G. L., "Graph partitioning into isolated, high conductance clusters: theory, computation and applications to preconditioning," *SPAA '08: Proceedings of the twentieth annual Symposium on Parallelism in Algorithms and Architectures*, ACM, New York, USA, 2008, pp. 137–145.

[28] Boman, E. G., Chen, D., Parekh, O., and Toledo, S., "On the factor-width and symmetric H-matrices," *Numerical Linear Algebra with Applications*, Vol. 405, 2005, pp. 239–248.

[29] Daitch, S. I. and Spielman, D. A., "Faster approximate lossy generalized flow via interior point algorithms," *STOC '08: Proceedings of the 40th annual ACM Symposium on Theory of Computing*, ACM, New York, USA, 2008, pp. 451–460.

[30] Boman, E. G., Hendrickson, B., and Vavasis, S., "Solving Elliptic Finite Element Systems in Near-Linear Time with Support Preconditioners," *SIAM Journal on Numerical Analysis*, Vol. 46, No. 6, 2008, pp. 3264–3284.

[31] Wang, M. and Sarin, V., "Parallel Support Graph Preconditioners," *High Performance Computing - HiPC 2006*, edited by Y. Robert, M. Parashar, R. Badrinath, and V. K. Prasanna, Vol. 4297, chap. 39, Springer, Berlin Heidelberg, 2006, pp. 387–398.

[32] Shklarski, G. and Toledo, S., "Rigidity in Finite-Element Matrices: Sufficient Conditions for the Rigidity of Structures and Substructures," *SIAM Journal on Matrix Analysis and Applications*, Vol. 30, No. 1, 2008, pp. 7–40.

[33] Daitch, S. I. and Spielman, D. A., "Support-Graph Preconditioners for 2-Dimensional Trusses," available at http://arxiv.org/abs/cs/0703119.

[34] Shklarski, G. and Toledo, S., "Computing the null space of finite element problems," *Computer Methods in Applied Mechanics and Engineering*, Vol. 198, No. 37-40, August 2009, pp. 3084–3095.

Chapter 4

A Scalable Hybrid Linear Solver Based on Combinatorial Algorithms

Madan Sathe, Olaf Schenk

University of Basel

Bora Uçar

CNRS and ENS Lyon

Ahmed Sameh

Purdue University

4.1   Introduction

The availability of large-scale computing platforms comprising of tens of thousands of multicore processors motivates the need for the next generation of highly scalable sparse linear system solvers. These solvers must optimize parallel performance, processor (serial) performance, as well as memory requirements, while being robust across broad classes of applications and systems. In this chapter, we present a hybrid parallel solver that combines the desirable characteristics of direct methods (robustness) and effective iterative solvers (low computational cost), while alleviating their drawbacks (memory requirements, lack of robustness). The hybrid solver is based on the general sparse direct solver PARDISO [1], and a class of Spike factorization [2, 3, 4, 5, 6, 7, 8, 9, 10, 11] solvers. The resulting algorithm, called PSPIKE, is as robust as direct solvers, more reliable than classical preconditioned Krylov-subspace methods, and much more scalable than direct sparse solvers. We discuss several combinatorial problems that arise in the design of this hybrid solver, present algorithms to solve these combinatorial problems, and demonstrate their impact on a large-scale three-dimensional PDE-constrained optimization problem.

4.2   PSPIKE—A Scalable Hybrid Linear Solver

Most of the parallel linear solvers run either on distributed memory machines by using only one compute core per node [12, 13, 14], or on shared memory machines using multiple cores [1]. A very good scalability of these solvers has been shown for a small number of compute cores or compute nodes. Unfortunately, to solve a system with tens of millions of unknowns one requires much higher number of compute cores/nodes, and there are no candidate solvers that can utilize thousands of cores simultaneously. A promising approach to solve large systems in a reasonable time is to combine a distributed-memory implementation with a shared-memory implementation to achieve a much better scalability. For instance, such an attempt has been recently proposed by Gupta et al. [15] based on the direct solver WSMP, where symmetric positive definite systems on up to 16, 384 compute cores have been solved.

To obtain such a scalability for indefinite systems, hybrid solvers—combining a direct with an iterative solver scheme—are very promising candidates for current and upcoming hardware architectures. To understand the ingredients of this kind of hybrid solvers, we will describe the main steps of PSPIKE [16, 17], which is a combination of an iterative and direct solver. Iterative solvers are well suited for distributed systems, but it is also well known that an iterative solver does not work out-of-the-box for indefinite and ill-conditioned systems. Often, preconditioning greatly improves the convergence and numerical stability of the solver. With this technique, the original matrix is transformed into a matrix that has much better numerical properties, and this preconditioned matrix is fed into an iterative Krylov-subspace solver. The preconditioned solver still converges to a solution which is also valid for the original system, but with a smaller number of iterations expected. In `PSPIKE`, the iterative solver `BiCGStab` [18] is used in combination with the direct solver `PARDISO` [1]. The direct solver `PARDISO` solves the systems on each node based on a shared-memory parallelization, whereas the communication and coordination between the nodes is established by the distributed parallelization of the preconditioned iterative solver.

For a nonsingular $n \times n$ matrix A, in contrast to the $A = LU$ or $A = LDL^T$ factorizations of direct solvers, the general idea of `PSPIKE` is to permute the matrix A into a banded form A^b and then use the factorization $A^b = DS$, where D is a sparse block diagonal matrix, and S is a spike matrix. The diagonal blocks of D are distributed among the nodes of the cluster, and a dense spike matrix is locally stored on each node. This factorization idea is illustrated in Figure 4.1 for three MPI processes. The sparse banded matrix A^b is factorized via a sparse block diagonal matrix D and a spike matrix S. A^b is split into the block diagonal matrices A_1, A_2, A_3 and the coupling block matrices B_1, B_2, C_2, C_3. The size of each coupling block is given by the bandwidth k. Thus, the sparse coupling block matrices of size $k \times k$ cover the nonzero elements which are not already covered by the diagonal blocks A_1, A_2, A_3. D contains the square block diagonal matrices A_1, A_2, A_3, where each block is the input matrix for the direct solver. Each MPI processor owns a diagonal block and performs the LU factorization using a multi-threaded sparse direct linear system solver. The spike matrix S contains dense spike matrices V_1, W_2, V_2, W_3 of bandwidth k. The bottom *tips* V_1^B, V_2^B of size $k \times k$ of V_1, V_2 and the top *tips* W_2^T, W_3^T of size $k \times k$ of W_2, W_3 are emphasized due to their crucial role in the hybrid solver. The diagonal of S is occupied by 1.

Similar to the forward and backward substitution of direct solvers, in `PSPIKE` the block diagonal matrix D and the spike matrix S are used to obtain the solution of the linear equation system $A^b x = f$. In the diagonal solve, the linear system

$$Dg = f \qquad (4.1)$$

has to be solved, where D is the block diagonal matrix, f is the right-hand side, the solution g is the new right-hand side for the spike solve

$$Sx = g, \qquad (4.2)$$

where S is the spike matrix, and x is the solution of the system with the input matrix A^b. The diagonal solve (4.1) can be done in parallel by using the direct solver, because each block in the diagonal matrix D can be LU factorized by

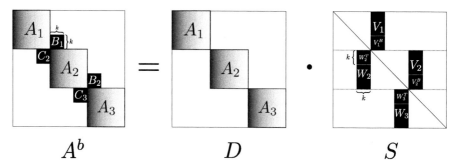

FIGURE 4.1: Factorization of the sparse banded matrix A^b in a block diagonal matrix D and a spike matrix S in the hybrid solver PSPIKE by using three MPI processes.

the direct solver and each diagonal block is independent from other diagonal blocks. Thus we obtain

$$A_i = L_i U_i. \tag{4.3}$$

If the diagonal blocks are singular (or close to being singular), we obtain the LU factorization by diagonal boosting via $L_i U_i = (A_i + \delta A_i)$ in which $\|\delta A_i\| = \mathcal{O}(\varepsilon\|A_i\|)$, where ε is the unit roundoff. For the spike solve (4.2) the right-hand side is generated, but the spike matrix S is still missing. We compute the spike matrix S with $S = D^{-1}A$ by solving, for instance, the following systems in parallel

$$L_i U_i [V_i, W_i] = \left[\left(\begin{smallmatrix} 0 \\ B_i \end{smallmatrix} \right), \left(\begin{smallmatrix} C_i \\ 0 \end{smallmatrix} \right) \right], \tag{4.4}$$

where the first processor computes only V_1 and the last processor p_K computes only W_K. The direct solver PARDISO is able to solve the systems with multiple right-hand sides, here k right-hand sides are given by the coupling block matrix B and k right-hand sides are given by the coupling block matrix C. Also, it is possible to provide to PARDISO the coupling block matrices without explicitly storing the zeros. Then, the solution x can be obtained by solving the system (4.2) directly.

Nevertheless, this step would be computationally very expensive for the linear solver because the spikes V, W are dense and $size(A_i) \times 2k$ elements need to be stored in the spikes, which can be a large memory requirement. In PSPIKE a very efficient approach has been applied. A reduced spike system is generated by computing only the bottom and top tips V^B, W^T of the spikes V, W by

$$L_i U_i \left[V_i^B, W_i^T \right] = \left[\left(\begin{smallmatrix} 0 \\ B_i \end{smallmatrix} \right), \left(\begin{smallmatrix} C_i \\ 0 \end{smallmatrix} \right) \right]. \tag{4.5}$$

PARDISO calculates the bottom and top tips of the spikes very efficiently by taking advantage of the zero blocks in the coupling block matrices. The direct solver permutes the linear system such that the right-hand side can be fed into the solver in a compressed form and returns back the tips of the spikes also in a compressed form. With the help of the reduced spike system we are able to compute the bottom and top solutions x^B, x^T of the original system by

$$
\begin{pmatrix}
I & V_i^B & 0 & 0 \\
W_{i+1}^T & I & 0 & V_{i+1}^T \\
W_{i+1}^B & 0 & I & V_{i+1}^B \\
0 & 0 & W_{i+2}^T & I
\end{pmatrix}
\begin{bmatrix}
\begin{pmatrix}
x_i^B \\
x_{i+1}^T \\
x_{i+1}^B \\
x_{i+2}^T
\end{pmatrix}
\end{bmatrix}
=
\begin{bmatrix}
\begin{pmatrix}
g_i^B \\
g_{i+1}^T \\
g_{i+1}^B \\
g_{i+2}^T
\end{pmatrix}
\end{bmatrix}. \tag{4.6}
$$

To have independent systems of the reduced form

$$
\begin{pmatrix}
I & V_i^B \\
W_{i+1}^T & I
\end{pmatrix}
\begin{bmatrix}
\begin{pmatrix}
x_i^B \\
x_{i+1}^T
\end{pmatrix}
\end{bmatrix}
=
\begin{bmatrix}
\begin{pmatrix}
g_i^B \\
g_{i+1}^T
\end{pmatrix}
\end{bmatrix} \tag{4.7}
$$

on each MPI processor we are ignoring the V_{i+1}^T and W_{i+1}^B matrices (see Equation 4.6), which should have a minor influence on the convergence of the solver.

The top and bottom solutions are now subtracted from the right-hand side of the original system such that the systems can be solved with `PARDISO` in parallel independently from each other,

$$
L_i U_i [x_i] = f_i - B_i x_{i+1}^T - C_i x_{i-1}^B, \tag{4.8}
$$

where the first processor sets $x_0^B = 0$ and the last processor sets $x_{K+1}^T = 0$.

4.2.1   The `PSPIKE` Algorithm

The `PSPIKE` algorithm (see Algorithm 2) consists of three phases: preprocessing (line 3), numerical hybrid factorization (lines 4–9), and solving the linear system (lines 10–13).

4.2.2   Preconditioning

In the preprocessing phase of Algorithm 2 (line 3), it is required to reorder the indefinite and ill-conditioned matrix A into a sparse banded matrix A^b. The input matrix has to be reordered in such a way that the diagonal blocks A_i, and the coupling blocks B_i, C_{i+1} are filled with heavy weighted entries as much as possible. The goal for the reordering procedure is to cover almost all weighted entries by the spike structure and only leave a few elements uncovered. In `PSPIKE`, the preconditioner M is built as

$$
M = K_f Q_m P_{s_1} D_r A D_c P_{s_2} K_f^T, \tag{4.9}
$$

where D_r, D_c are the scaling factors for the original matrix A (if A is symmetric, one prefers $D_c = D_r^T$), P_{s_1}, P_{s_2} are the permutations given by a partitioner/spectral heuristic, Q_m is the row permutation matrix to permute large entries onto the diagonal (note in [19] the column permutation matrix is returned), and K_f is the reordering that aims at maximizing the weights in the coupling block matrices B_i and C_{i+1} for all i. In Figure 4.2, a saddle-point matrix (left) is reordered to the `PSPIKE` structure (right) by applying the reordering techniques to the matrix.

Algorithm 2 Outline of the Hybrid Solver PSPIKE for Solving $Ax = f$

Input: $n \times n$ matrix A stored in compressed sparse row format, right-hand side f, bandwidth k, #right-hand sides, residual tolerance level ϵ;

Output: Solution x;

1: Initialize solver; {*init* PARDISO; *I/O and error handling*}
2: Read matrix A and copy numerical values into sparse matrix data structures; {*init preconditioner; allocate memory for local matrix structures*}
3: Create preconditioner $M = \left[K_f Q_m P_{s_1} D_r A D_c P_{s_2} K_f^T \right]$ by solving several combinatorial problems where D_r, D_c are row and column scaling factors; P_{s_1}, P_{s_2} the permutation matrices of a partitioner/spectral heuristic; Q_m the row permutation matrix by solving the weighted graph matching problem; K_f the symmetric permutation matrix obtained by solving the quadratic knapsack problem; generate reorderings $\Pi_l = [K_f Q_m P_{s_1}]$ and $\Pi_r = \left[P_{s_2} K_f^T \right]$; {*result is a sparse banded matrix A^b (see Figure 4.1)*}
4: Split the matrix A^b into an equal row-block-size and distribute the row-block matrices A_i^r to MPI processor i;
5: Extract the square diagonal block structure A_i and the square coupling block matrices B_i, C_i of size k from the local row block A_i^r {*most of the weighted entries of the input matrix are either covered by the diagonal or coupling blocks; store the diagonal block, the coupling blocks, and the non-covered part separately; processor 0 stores only B and the last processor only C*}
6: Perform the LU factorization of the block diagonal matrices A_i with the direct solver PARDISO in parallel;
7: Solve with PARDISO: $L_i U_i \left[V_i^B, W_i^T \right] = \left[\left(\begin{smallmatrix} 0 \\ B_i \end{smallmatrix} \right), \left(\begin{smallmatrix} C_i \\ 0 \end{smallmatrix} \right) \right]$; {*the dense spike matrix S is generated in a reduced form; processor 0 computes only V^B and the last processor only W^T; main computational advantage of using direct solver PARDISO by computing the bottom and top tips of the spikes V, W*}
8: Send dense matrix W_i^T of size k to predecessor processor $i - 1$ and build the small reduced system $S_i^r = \left(\begin{smallmatrix} I & V_i^B \\ W_{i+1}^T & I \end{smallmatrix} \right)$ of size $2k$; {*Processor 0 is only receiving, the last processor is only sending data; main communication part especially for large k (e.g., $k > 1000$)* }
9: Perform an LU factorization of the reduced system S_i^r;
10: Reorder and scale the right-hand side with $f^r = [\Pi_l D_r f]$ and distribute the RHS to the corresponding processors;
11: Preparation for the parallel matrix-vector multiplication; {*identify entries which are residing on the other processors and construct the data structures for an efficient matrix-vector multiply*}
12: Perform the distributed preconditioned BiCGStab scheme with the parallel matrix-vector multiply and call Algorithm 3 for each solving step $M\bar{x} = z$ with the preconditioner M until convergence criterion $\frac{\|Ax^b - b\|}{\|b\|} < \epsilon$ is met; {*due to the scaling mechanism it could be necessary to check intermediate solutions of the preconditioned system for the original system*}
13: Gather, scale back, and reorder the solution x^b of the preconditioned system to a solution of the original system with $x = D_c \Pi_r x^b$;
14: Deallocate all data structures;

Algorithm 3 Outline of the Call $M\bar{x} = z$ in the Preconditioned `BiCGStab`

Input: Block diagonal matrix M stored in compressed sparse row format, right-hand side z, bandwidth k;

Output: Solution $\bar{x}$;

1: Solve with `PARDISO` $L_i U_i [g_i] = [z_i]$;

2: Partition the solution $g_i = \left(g_i^T, g_i^M, g_i^B\right)$ where the top and bottom parts are of size k; send the solution of the upper part g_i^T to the predecessor processor; {*processor 0 is only receiving; prepare to solve with the reduced system*}

3: Solve the system $\begin{pmatrix} I & V_i^B \\ W_{i+1}^T & I \end{pmatrix} \left[\begin{pmatrix} \bar{x}_i^B \\ \bar{x}_{i+1}^T \end{pmatrix}\right] = \left[\begin{pmatrix} g_i^B \\ g_{i+1}^T \end{pmatrix}\right]$; {*last processor is idle*}

4: Send the solution $\bar{x}_i^B$ to the successor processor; {*last processor is only receiving*}

5: Solve with `PARDISO` $L_i U_i [\bar{x}_i] = \left[z_i - \begin{pmatrix} 0 \\ I^B \end{pmatrix} B_i \bar{x}_{i+1}^T - \begin{pmatrix} I^T \\ 0 \end{pmatrix} C_i \bar{x}_{i-1}^B\right]$; {*processor 0 has no predecessor solution $\bar{x}_{i-1}^B$ and the last processor has no successor solution $\bar{x}_{i+1}^T$; thus set both to 0*}

The scaling factors D_r and D_c that we use ensure that all entries of the matrix are smaller than or equal to 1 in absolute value, which has a strong influence on the robustness and scalability of the hybrid solver. For instance, the scaling factors can be obtained by solving the matching problem with a primal-dual method, where the scaling factors are computed as a by-product of the matching solution, i.e., the dual variables [20, 21], see Section 4.3.2, Chapter 2. The permutations P_{s_1} and P_{s_2} should reorder the matrix in such a way that the entries with large magnitudes are placed close to the diagonal. This structure should engender a sufficient load balance among the processors.

The permutation Q_m that permutes large entries on the diagonal is solved by an exact graph matching. Each row and column represent a vertex in the bipartite graph and each entry of the matrix represents a weighted edge between them. The matching algorithm should find a maximum weighted perfect matching—a matching where every vertex has exactly one incident undirected edge with maximum edge-weight. The solution is coded in a row or column permutation, which ensures that the square block diagonal matrices are nonsingular—at least if A is diagonal dominant. This is crucial because the diagonal blocks need to be factorized by a direct solver.

It is important to note that the performance of the overall preconditioner depends heavily on the numbers of elements in the coupling matrices B_i and C_{i+1}. The permutations P_{s_1}, P_{s_2} and the row permutation Q_m only guarantee that the diagonal blocks have a sufficient diagonal structure. However, if the matrices B_i and C_{i+1} are zero-matrices or only contain a small number of elements, then the overall method boils down to a block Jacobi method. It is well known that a simple block Jacobi preconditioner will have problems with the convergence, especially for unsymmetric and indefinite matrices. Therefore, it

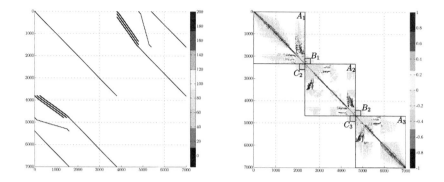

FIGURE 4.2: (See color insert.) Reordering of a saddle-point matrix (left) to the PSPIKE structure (right) by applying the scaling and reordering techniques to the matrix.

is of utmost importance to permute additional *heavy* elements into the B_i and C_{i+1} matrices. The last reordering step (finding the permutation matrix K_f) should permute most of the entries, which are not yet covered by the diagonal blocks, into the coupling blocks.

The next section describes the combinatorial aspects in the preconditioner, and provides solution approaches.

4.3   Combinatorics in the Hybrid Solver PSPIKE

Since the hybrid solver needs a banded structure of the input matrix, we have to apply several reordering techniques to the original matrix to achieve all entries of the matrix being covered, ideally either by the diagonal blocks A_i or by the coupling blocks B_i and C_i. If all entries are confined within the blocks, the Krylov-subspace solver will converge in a few iterations. In practical applications this situation does not occur very often; thus, a realistic goal would be to find a preconditioner such that possibly all heavy-weighted entries are included in the block structures and some small-weighted entries are not.

The algorithms, which return these reordering schemes, must be able to deal with large and sparse matrices. We will use different graph models to represent matrices and use well-developed notions and algorithms from graph theory to devise the reordering algorithms. Three goals have to be satisfied by the algorithms: find the diagonal blocks, establish that each diagonal block is nonsingular, and find heavy coupling blocks. Each goal can be formalized as a combinatorial optimization problem:

- finding the diagonal blocks can be interpreted as a graph partitioning problem (Section 4.3.1)

- guaranteeing (in practice) the nonsingularity of the diagonal blocks can be modeled as a graph matching problem (Section 4.3.2)

- finding the coupling blocks can be represented as a quadratic optimization problem of the form similar to the knapsack and clique problems (Section 4.3.3)

Ideally, the partitioner balances the workload associated with the diagonal blocks while minimizing the entries in the off-diagonal blocks. The problem has many facets, each requiring fast and effective heuristics. The solution of the second combinatorial problem, graph matching, permutes large entries into the diagonal of the matrix and is solvable in cubic complexity. Additionally, the matching algorithm returns back scaling vectors such that entries on the diagonal are 1 or -1, and the off-diagonal entries are between 1 and -1. The scaling of the whole matrix has a strong influence on the robustness and scalability of the hybrid solver. The reordering that is based on the quadratic optimization problem fills the coupling blocks (finds B_i and C_{i+1}) with heavy-weighted entries as much as possible. As the general problem is $\mathcal{NP}$-hard, we aim for fast and effective heuristics.

Because the hybrid solver works in parallel and the matrix will be assembled in a distributed fashion, all algorithms should ideally be able to deal with a distributed matrix as an input and work in parallel for building the preconditioner. Thus, we will give a literature overview with a focus, whenever possible, on the challenging task of hard-to-parallelize combinatorial algorithms and explain the implemented algorithm for each problem. Consequently, we will describe each combinatorial problem in more detail.

4.3.1 Graph Partitioning

The partitioning routines for efficient parallelization of PSPIKE should balance the loads of the processors during each phase of Algorithm 3 and should reduce the communication cost, as in other similar parallelization problems. One significant property is that the partitioning routines should also guarantee that each B_i and C_j is nonempty (recall that the speed of convergence of the method is hampered if those blocks are empty). We have envisioned two alternatives. One of them applies standard graph/hypergraph partitioning techniques to address the standard partitioning objectives and then appends a phase to reorder the parts such that the blocks B_i and C_j are nonempty. The second one orders the matrix in a first phase and then partitions the ordered matrix (always respecting the order found in the first phase). We elaborate on these two approaches below.

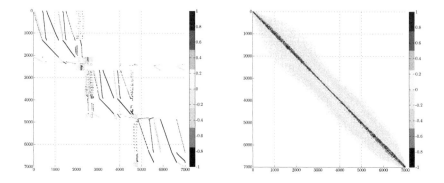

FIGURE 4.3: (See color insert.) Partitioning obtained by the partitioner Mondriaan (left) and the ordering obtained by the spectral method HSL_MC73 (right) for the previous matrix. The second matrix would be partitioned with horizontal and vertical lines while respecting the permutation.

4.3.1.1   Partitioning and Ordering

The standard graph and hypergraph partitioning models can be used to partition the matrix so that most of the nonzeros are on the diagonal blocks and each block has almost equal numbers of rows/columns. Below we describe the graph model. For the hypergraph model we refer the reader to [22].

Recall that an undirected graph $G = (V, E)$ consists of a set of vertices $V = \{v_1, \ldots, v_n\}$ and a set of edges $E \subset V \times V$. Sometimes nonnegative weights and costs are associated with the vertices and edges, respectively, where $w(v_i)$ represents the weight of vertex v_i and $c(v_i, v_j)$ represents the cost of the edge (v_i, v_j). A standard problem is to partition the vertices of such a graph into K parts, for a given integer $K \geq 2$, such that the parts have almost equal weights (defined in terms of the sum of the weights of the vertices lying within a part), and the total weight of the edges having vertices in two different parts is minimized. In general, one uses an imbalance parameter in the problem definition to define an allowable upper bound for the part weights. Formally, the constraint can be written as

$$\sum_{v \in V_k} w(v) \leq \frac{(1+\varepsilon)}{K} \sum_{v \in V} w(v) \qquad \text{for } k = 1, \ldots, K,$$

where ε is the imbalance parameter (here $\varepsilon \geq 0$ so that it defines how much percent part weights can deviate from the perfect balance). The objective function, edge cut, is to minimize and can be written as

$$\sum_{(v_i, v_j) \in E} c(i, j) \qquad \text{for } \pi(v_i) \neq \pi(v_j),$$

where $\pi(v_i)$ is the part which contains the vertex v_i. This problem is $\mathcal{NP}$-hard (for a proof see [23] for the special case of $K = 2$).

If one uses unit vertex weights, each block will have almost equal number of rows/columns and hence one aspect of local PARDISO solves would be balanced. If one uses the matrix entries as costs of the edges so that $c(v_i, v_j) = |a_{ij}|$, then reducing the edge cut will likely have a positive effect for PSPIKE (only a small portion of the magnitudes of the matrix entries lies in the off-diagonal blocks). Alternatively, one can try to reduce the number of matrix entries lying in the off-diagonal blocks by using unit edge costs.

Heuristics and software for graph partitioning. Most of the successful graph partitioning heuristics follow the multilevel paradigm for partitioning the given graph. The multilevel paradigm was proposed in the 1990s [24, 25] and is implemented by many successful graph and hypergraph partitioning tools, including graph partitioning tools Chaco [26], MeTiS [27], and Scotch [28], and hypergraph partitioning tools Mondriaan [29] and Pa-ToH [22, 30].

The multilevel paradigm consists of three phases: coarsening, initial partitioning, and uncoarsening. In the first phase, a multilevel clustering is applied starting from the original graph/hypergraph by adopting various matching/-clustering heuristics until the number of vertices in the coarsened graph/hypergraph falls below a predetermined threshold. The main issue in this step is to match/cluster similar vertices so that the small graphs/hypergraphs capture the essential structure of the original one. Often, the quality of the partitioning depends on the success of the coarsening phase. A common heuristic is known as the heavy-connectivity matching or heavy-edge matching. This matching heuristic visits the vertices in an order and matches each unmatched vertex to a neighboring unmatched one with the heaviest edge. In the second phase, a partition is obtained on the coarsest graph/hypergraph using various heuristics. This step is usually performed by simple and fast greedy heuristics—a quite common one is known as greedy graph/hypergraph growing, in which a breadth-first search-like heuristic is run from a seed vertex until a big enough part is obtained. In the third phase, the partition found in the second phase is successively prolonged back towards the original graph/hypergraph by refining the prolonged partitions on the intermediate level graphs/hypergraphs using various heuristics. A common refinement heuristic is FM, which is a localized iterative improvement method proposed for graph/hypergraph bipartitioning proposed by Fiduccia and Mattheyses [31] as a faster implementation of the KL algorithm proposed by Kernighan and Lin [32].

The K-way partitioning is usually achieved by recursively bisecting the given graph/hypergraph until the desired number of parts is achieved. As intuition suggests, the performance of such a recursive bisection approach deteriorates with the increasing number of parts. This fact is also shown [33]. Therefore, direct K-way algorithms have been and are currently being investigated; see, for example, Chapter 14 and [34].

Although the multilevel paradigm permits fast heuristics, there is still a performance issue of partitioning large graphs/hypergraphs on a single processor. More importantly, partitioning large graphs/hypergraphs (which are quite common in applications) is sometimes impossible on a single processor due to memory limits. Fortunately, there has been some work leading to tool support for parallel graph/hypergraph partitioning, such ParMeTiS [35], PT-Scotch (see Chapter 14), and Zoltan (see Chapter 13).

Ordering. As we want the blocks B_i and C_j to be nonempty, it is necessary to reorder the blocks so that this is guaranteed. Assume for the sake of simplicity that the matrix A is symmetric, each block contains the same set of row and column indices, and we would like again a symmetric ordering (that is, $P_{s_1} = P_{s_2}^T$). Under these assumptions, we would therefore like to order the blocks so that the rows in the ith block have nonzeros only in the columns in the blocks i and $(i+1)$. This is not guaranteed with the standard partitioning tools—see for example the first image in Figure 4.3 where Mondriaan is used as is to partition the matrix into three parts. One therefore needs to modify the output of the standard graph partitioning routines by combining different blocks so that the aforementioned condition holds. There are two alternatives to modify the output of the standard graph partitioning problem to meet the condition.

In the first alternative, one partitions into more parts than the number of processors and then tries to combine blocks (while respecting load balance) to reduce the number of blocks to the number of processors. There are two difficulties associated with this approach.

The first difficulty arises in combining different parts. Consider the $\mathcal{NP}$-complete problem of graph contractibility [36, GT51], which is stated as follows.

INSTANCE: Graphs $G = (V_1, E_1)$ and $H = (V_2, E_2)$

QUESTION: Can a graph isomorphic to H be obtained from G by a sequence of edge contractions, i.e., a sequence in which each step replaces two adjacent vertices u, v by a single vertex w adjacent to exactly those vertices that were previously adjacent to at least one of u and v?

Clearly, our problem includes the case where H is a cycle and G contains a vertex for each part obtained by the graph partitioning routine. If $|V_2| = 3$, then H is a triangle and the problem can be solved in polynomial time [36, GT51]. We did not study whether the problem remains $\mathcal{NP}$-complete when H is a cycle with more than three vertices. We conjecture that it is so.

On top of the difficulty of ordering the blocks by contraction, we would also need to specify a number greater than the number of processors for the standard graph partitioning problem. How one chooses such a number is not a question with an obvious answer.

In the second alternative, one applies the standard graph partitioning routines and then combines the blocks to meet the condition above. This time one obtains fewer blocks than there are processors. The outcome is that one needs to run a parallel direct solver on some of the blocks, rendering the balance issue for PSPIKE more challenging. It is probably preferable this time to minimize the number of contractions so that the resulting graph is a cycle. The problem, being an optimization version of the problem above, seems to be $\mathcal{NP}$-complete.

4.3.1.2   Ordering and Partitioning

In this approach, we first reorder the matrix (symmetrically) so that the nonzeros are around the diagonal, within a small band. Then we partition the ordered matrix by assigning consecutive row/column indices to processors; i.e., we find $K - 1$ indices $1 < i_1 < \cdots < i_{K-1} < n$, where the first processor gets the rows/columns (of the permuted matrix) 1 to $i_1 - 1$, the second processor gets the rows/columns i_1 to $i_2 - 1$, and so on, where the last one gets those between i_{K-1} and n. If the bandwidth is not too large, then this partition can be used to permute the matrix into block tridiagonal form (so that one can find the coupling blocks B_i and C_j to permute the matrix into the form shown in Figure 4.1). If the bandwidth is too large, then it may not be possible to obtain K blocks of reasonable sizes. Assuming $n \gg K$, this last case should not be encountered in practice, as one can guess from the second image in Figure 4.3 where the matrix is ordered first and to be partitioned as discussed above.

Among the two subproblems (ordering and partitioning) of this section, the first one is known as the profile or the bandwidth reduction problem. These two problems are known to be $\mathcal{NP}$-complete (see, respectively, the problems GT42 and GT40 in [36]). The second one is known as the chains-on-chains partitioning problem which is polynomial time solvable. Below we discuss these two problems and solution methods.

Ordering. There are usually two metrics for ordering matrices to have nonzero entries around the diagonal: the profile and the bandwidth (see [37], p. 127). Let A be a pattern symmetric matrix with a zero-free diagonal. Let $b_j = j - \min\{i : a_{ij} \neq 0\}$, i.e., b_j is the distance of the furthest entry in column j to the diagonal. Then, the profile and the bandwidth of A are defined, respectively, as $\sum_j b_j$ and $\max_j b_j$. Finding a permutation so as to minimize the profile is equivalent to the $\mathcal{NP}$-complete optimal linear arrangement problem [36, GT42]. Finding a permutation so as to minimize the bandwidth is also an $\mathcal{NP}$-complete problem [36, GT40]). Therefore, heuristics are used. Fortunately, there are quite successful heuristics available in software libraries.

The most traditional bandwidth reducing heuristic is the reverse Cuthill–McKee (CM) method (see [38] for the original method and [39] for the reversed one). The original method, CM, proceeds in steps. It starts by ordering a node first. Then, at each step the unnumbered neighbors of ordered nodes are ordered by increasing degrees. The reverse CM, RCM, reverses the order found by CM. As is clear, the starting node determines a partial order (based on the distances to the starting node) which has a significant effect on the solution. Quite a bit of work has been done to find a good starting node; see [40].

For the profile reduction problem, there are enhanced variants of the basic ordering scheme described above. These take the form of ordering a node and then choosing the next node to order from those that are at most at distance two from the ordered ones. The first of these important class of algorithms is given by Sloan [41] and has been improved later by a number of authors; see [42, 43, 44, 45] and the references therein. A recent variant [46, 43] of this line of algorithms uses the multilevel approach described above for the standard graph and hypergraph partitioning problems.

Another line of heuristics for profile reduction uses spectral techniques to order the matrix. Based on an earlier work of Fiedler [47] and Juvan and Mohar [48], Barnard et al. [49] observe that ordering the vertices according to the components of the eigenvector corresponding to the second-smallest eigenvalue of the Laplacian (see Chapter 18) of a graph (known as the Fiedler vector) minimizes a quadratic function intimately related to the profile. Kumfert and Pothen [44] and Hu and Scott [43] discuss such ordering methods in detail with a description of the software to do this ordering. Manguoglu et al. [50, 51] describe an efficient algorithm to compute the Fiedler vector in parallel.

The most successful heuristics reportedly [43, 44] use spectral methods and then refine the ordering found locally using a variant of Sloan's algorithm.

Partitioning. The partitioning routines now have to respect the ordering such that each processor gets a contiguous set of row/column indices. There are two goals to attain during the partitioning. Load balance between the processors should be achieved, and the coupling between two consecutive blocks should somehow be reduced. If we ignore the second goal, the first one is solvable in practice efficiently.

Given a linear chain graph on n nodes, each with a positive weight and K processors, the chains-on-chains partitioning problem asks for K sub-chains of the original graph, each to be assigned to a processor such that the total weight of the nodes in all subchains are balanced. Clearly the chain in our case corresponds to indices 1 to n of rows/columns, where each node has a weight equal to the total number of nonzeros in the corresponding row or column. Then partitioning this chain on a chain of K processors assigns almost equal numbers of nonzeros to each processor. The problem has received much attention since the late 1980s, starting with the work of Bokhari [52]. As a result a number of fast algorithms exist; see, for example, [53] for almost linear time algorithms under very realistic assumptions about the weights of nodes.

For a survey of heuristics and exact algorithms for the chains-on-chains partitioning problem, see [53]; for mapping into heterogeneous processors see [54], and see [55] for the same problems where communication costs under various models of communication are also discussed from a more theoretical point of view.

4.3.2 Graph Matching

In order to prevent the nonsingularity of the diagonal blocks, it is of fundamental importance for the solver to permute large entries onto the diagonal of the sparse matrix. To do this, the ordering is modeled as a weighted matching problem in a bipartite graph $G = (U, W, E)$, where $U = \{u_1, \ldots, u_n\}, W = \{w_1, \ldots, w_n\}$, and $E \subset U \times W$.

Each row of the matrix A is represented by a vertex in U and each column by a vertex in W. A nonzero entry c_{ij} of A forms an undirected weighted edge between (i, j). A subset $M \subseteq E$ in G is called a matching if no two edges of M are incident to the same vertex. In a maximum (cardinality) matching M, M contains the largest possible number of edges. An important concept in finding a maximum matching in bipartite graphs is that of an augmenting path. Let M be a matching in G. A path P in G is called M-augmenting if P has odd length, its ends are not covered by M, and its edges are alternatively out of and in M. Then, the symmetric difference $M \Delta P = (M \cup P) \setminus (M \cap P)$ is a matching and $|M \Delta P| = |M| + 1$.

A matching is called perfect if it is a maximum matching and every vertex of U, W is incident to a matching edge. It is not always possible to find a perfect matching. However, if the matrix is of full rank and, consequently, it is nonsingular, and there exists a perfect matching of G. The perfect matching M has cardinality n according to the $n \times n$ matrix A and defines an $n \times n$ permutation matrix $\Pi = (\pi_{ij})$ with $\pi_{ij} = 1$ if $(i, j) \in M$ and a zero entry otherwise. The result of the perfect matching can be obtained by the row permutation ΠA with the matching entries on the main diagonal. Figure 4.4 includes the permuted matrix after the partitioning and ordering methods (notice that the permutations take place within the blocks found by the partitioners). By using a standalone partitioner, the coupling blocks between the diagonal blocks (here, between A_2 and A_3) can be very sparse, which has a negative influence on the convergence of the hybrid solver.

In a perfect weighted matching, possible objectives are to maximize the sum or the product of the weights of the matched edges. It is sometimes helpful to scale the entries of the matrix via logarithmic scaling [56]. Then, the original problem is transformed into a minimum weight matching problem and the problem can be interpreted as the linear sum assignment problem (LSAP) [57, 58]. Thus, solution techniques of the LSAP can be applied to the matching problem as well.

Basically, two exact sequential algorithms are well known and often used. In order to find a maximum matching, the algorithm of Hopcroft and Karp [59] with a worst case complexity of $\mathcal{O}(\sqrt{n}m)$ is implemented; or, in case of perfect weight matching, the Hungarian method [60] with $\mathcal{O}(n(m + n \log n))$, where $m = |E|$ and $n = |W| = |U|$ is used. One of the inherently fastest sequential

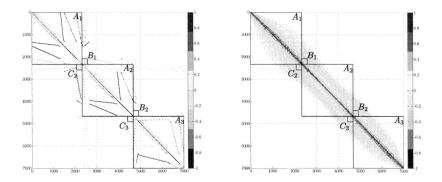

FIGURE 4.4: (See color insert.) Permutation of the weighted graph matching implementation HSL_MC64 applied to the partitioned matrix by Mondriaan (left) versus applied to the spectral method reordered matrix (right).

implementations for sparse matching is a package of the HSL library collection called MC64 [19].

The goal is to obtain a perfect matching with maximum/minimum weight in a parallel manner. In general. partitioning the matrix before doing the parallel matching may have a strong influence on the communication and computation time for the whole matching algorithm. To date, there are mainly three general concepts to solve the matching problem: approximation algorithms, primal-dual methods, and simplex-based algorithms. We will outline three different concepts of existing parallel algorithms.

4.3.2.1   Parallel Approximation Algorithms

While exact algorithms for weighted matching problems are quite expensive to compute and hard to parallelize, approximation algorithms are not. In general, approximation algorithms are very fast, and nevertheless produce very good results even if these results are not optimal. A promising idea is to put the solution of an approximation algorithm to an exact matching algorithm to speed up the process of finding matchings [61].

The quality of an approximation algorithm for the weighted matching problem is measured by its approximation ratio. An approximation algorithm has an approximation ratio of ϕ if it finds a matching with a weight of at least ϕ times the weight of an optimal solution for all graphs.

A $\frac{1}{2}$-approximation algorithm can be obtained by the following greedy strategy. First, sort the weights of the edges in decreasing order. Then, pick up the heaviest edge e, delete e and the incident edges, and repeat this process until the graph is empty. The sorting procedure results in a worst case complexity of $\mathcal{O}(m \log n)$. A linear-time implementation of this algorithm is provided by Preis [62]. From the parallelization point of view the algorithm is inherently sequential. Thus, new parallel algorithms are emerging which are trying to improve the time complexity and quality of the algorithm.

There are several other approaches to achieve a better approximation ratio most likely in linear time (see [63, 64] and the references therein).

Manne et al. [65] present a parallel algorithm for distributed architectures which is very similar to the parallel calculation of the independent set [66]. The main idea is the use of dominating edges—edges that are heavier than their incident edges. In the algorithm, vertices are equally distributed across the processors and each processor will compute dominating edges. If the corresponding vertex of the edge does not reside on the same processor, communication is required to discover if the two vertices are allowed to be matched. The algorithm has been applied to a problem with 5 million edges and $50,000$ vertices, where it scales well with up to 32 processors. However, there are counter examples in which the algorithm works sequentially. The worst case complexity of the algorithm is $\mathcal{O}(m \log m)$. Recently, a parallelization strategy for a greedy matching algorithm has been proposed [67].

4.3.2.2 Simplex-Based Algorithms

A different idea to find a solution for the weighted matching problem is the formalization as a mathematical program. The weighted matching problem can be formalized as a primal linear program (LP)

$$
\begin{aligned}
\min \quad & \sum_{(i,j)\in E} c_{ij} x_{ij} \\
\text{s.t.} \quad & \sum_{i\in U} x_{ij} = 1 \qquad \text{for } j \in W, \\
& \sum_{j\in W} x_{ij} = 1 \qquad \text{for } i \in U, \\
& x_{ij} \geq 0 \qquad \text{for } (i,j) \in E, \\
& x_{ij} = 0 \qquad \text{for } (i,j) \notin E,
\end{aligned}
$$

in which methods such as the simplex algorithm can be applied to the program.

The linear programming duality theorem states that the minimum weight value of the primal LP is equal to the maximum value of the following dual LP:

$$
\begin{aligned}
\max \quad & \sum_{i\in U} u_i + \sum_{j\in W} w_j \\
\text{s.t.} \quad & u_i + w_j \leq c_{ij} \qquad \text{for } (i,j) \in E, \\
& u_i, w_j \in \mathbb{R} \qquad \text{for } i \in U, j \in W.
\end{aligned}
$$

Any basic solution of the primal LP corresponds to a permutation matrix. The most expensive step in the simplex algorithm is the computation of a new edge to be included in the basis solution. One idea for parallelization of the computation would be to search for the edge in parallel where each processor returns one edge from the set of rows. One processor picks an edge from the list of candidates and performs the pivoting sequentially [68]. Another approach is to perform pivoting operations during the search for edges to enter the basis. Additionally, a pivot processor is established which checks the feasibility of the pivot due to the asynchronism of the algorithm [69]. The worst case complexity amounts to $\mathcal{O}(n^3)$.

4.3.2.3 Parallel Primal-Dual Methods

Primal-dual algorithms operate with a pair of an unfeasible primal solution $x_{ij} \in \{0, 1\}, 1 \leq i, j \leq n$, and a feasible dual solution $u_i, w_j, 1 \leq i, j \leq n$, which fulfill the complementary slackness conditions

$$x_{ij}(c_{ij} - u_i - w_j) = 0, \qquad 1 \leq i, j \leq n.$$

We denote $\bar{c}_{ij} := w_{ij} - u_i - w_j$ and call $\bar{c}_{ij}$ reduced costs with respect to the dual solution u_i, w_j. In general, these algorithms are focused on how to obtain a primal and dual solution which fulfills the complementary slackness conditions, and how to implement the update of dual variables.

In the remainder of this section, we will present parallel approaches based on the idea of augmenting path and auction algorithms.

Shortest Augmenting Path. Finding the shortest augmenting path of a vertex requires a transformation of the original undirected weighted bipartite graph G into a directed weighted bipartite graph $\tilde{G} = (U, W, \tilde{E})$, where $\tilde{E}$ is filled with directed edges. If $(i, j) \in M$, the corresponding edge is directed from i to j with weight 0; otherwise, the edge is directed from j to i with a weight equal to the corresponding reduced costs. Then, a free (not matched) vertex $s \in U$ is selected and the shortest path from s to all vertices of $\tilde{G}$ is computed. This problem is known as the single-source shortest path problem. The shortest among all paths from s to some free vertex in W is used to augment the current primal solution by swapping the free and the matched edges. The primal and dual solutions as well as the reduced costs are updated via $\delta = \min\{\bar{c}_{ij} : (i, j) \in E\}$, the currently minimum uncovered reduced costs. After n augmentations, an optimal primal solution is created. The overall complexity of the augmenting path algorithms amounts to $\mathcal{O}(n^3)$. Primal-dual parallel algorithms perform the shortest path computation either by starting from a single vertex and doing the shortest path computation from the single free vertex in parallel, or by starting from multiple free vertices in order to execute multiple augmentation and variable updates in parallel. The former strategy will be outperformed by the latter one in the case of sparse problems because of the increase of the idle time of processors, especially at the end of the algorithm. Recently, Madduri et al. [70] presented a very fast shortest path computation on a special shared-memory architecture for certain graph families and problem sizes.

Auction Algorithms. These algorithms [71, 72, 73] are well suited for implementation on parallel machines and, therefore, are interesting for large-scale weighted matching problems. These algorithms simulate the auction process with bids, bidders, objects, and benefits of an object. U is interpreted as a set of persons i and W as a set of objects j. The weighted edge c_{ij} corresponds to the benefit which a person i obtains by buying the object j for free. Additionally, there is a price p_j, the dual variable, for each object j, and at each iteration unassigned persons bid simultaneously for objects with most benefit ($\max_{j \in W} \{c_{ij} - p_j\}$). Each person i increments the price p_{j_i} by a bidding increment $\gamma_i > 0$. Objects are awarded to the highest bidder. If an object remains without any bids during an iteration, its price and assignment status are left unchanged. This process terminates to a perfect matching in a finite number of iterations. Whether the perfect matching is also a minimum weight matching, highly depends on the bidding increment γ_i. If the increment γ_i is small enough to ensure the bid will be accepted, then the bid will be almost optimal for the bidder. Consequently, the final matching will be almost optimal too [71].

In particular, if the ϵ-complementary slackness condition (ϵ-CS)

$$\max_j \{c_{ij} - p_j\} - \epsilon \leq c_{ij_i} - p_{j_i} \quad \text{for all matched pairs } (i, j_i)$$

holds upon termination, the total benefit of the final matching is within $n\epsilon$ of being optimal. The standard method for choosing the bidding increment γ_i to fulfill the ϵ-CS condition is when

$$\gamma_i = d_i - e_i + \epsilon,$$

where $d_i = \max_{j \in W} \{c_{ij} - p_j\}$ is the best objective value while e_i is the second-best objective value. At the beginning of the algorithm ϵ is initialized with a large value and successively reduced up to a value which is less than the critical value $\frac{1}{n}$. This value is required to ensure that $n\epsilon < 1$ and, finally, to obtain a perfect matching.

We define a search task to be the calculation of the best and second-best objective value. Here we use s to denote the number of searching processors used per person, and p to denote the maximum available number of processors. Shared-memory implementations of the auction algorithm can be divided into three classes according to the parallelization of the search task: Jacobi parallelization, Gauss–Seidel parallelization, and a combination of the two mentioned before, called block Gauss–Seidel parallelization.

In the Jacobi parallelization, the calculation for the bids per person i is assigned to a single processor ($s = 1$). The search tasks are executed in parallel and followed by merging the results of the search tasks of the other processors. As, in our case, if the number of persons exceeds p, then some processors are overloaded with multiple work. However, in the late stages of the method, processors are idle during the bidding phase, thereby reducing efficiency.

In the Gauss–Seidel parallelization, the search task of an unassigned person is performed by p processors. Thus, the set of objects is divided into p pieces and all processors are assigned to the pieces doing the computation for one person in parallel ($s = p$). After the search task has completed, the results are merged together to figure out the best and second best objective values. The main drawback of this method is the large number of iterations taking place for a huge number of persons, such as in our case.

To overcome the drawbacks of the two approaches a hybrid method has been devised. Again, the bid calculation is parallelized as in the original Gauss–Seidel, but the number of searching processors used per bid is $1 < s < p$. With a proper choice of s, this method combines the best features of the other methods and alleviates their drawbacks. Most promising in the case of sparse matching problems are the Jacobi and hybrid methods. It is an open question whether it is possible to transfer auction algorithms to distributed architectures in order to achieve scalability of the matching process.

4.3.3 Quadratic Knapsack Problem

As outlined before, given the blocks $A_1, \ldots, A_K$ of the banded matrix (see Figure 4.1), we would like to find $k \times k$ coupling matrices B_i and C_{i+1}, for $i = 1, \ldots, K - 1$ such that these coupling matrices contain many of the entries (or their magnitudes) residing in the encompassing off-diagonal blocks. To formalize, we define the following densest square submatrix (DSS) problem.

Definition 1 (DSS) *Given an $m \times n$ matrix N and an integer $k \leq \min(m, n)$, find two sets of indices $\mathcal{R} = \{r_1, r_2, \ldots, r_k\}$ and $\mathcal{C} = \{c_1, c_2, \ldots, c_k\}$ such that*

$$\sum_{r \in \mathcal{R}, c \in \mathcal{C}} |N_{rc}|$$

is maximized.

In the above definition of the problem, the objective function relates to maximizing the total weight of the magnitude of the entries in the selected submatrix. An alternative objective function is to try to maximize the number of nonzero entries in the selected submatrix, which is a special case of the above problem in which N is a $\{0, 1\}$ matrix. The $\mathcal{NP}$-hardness of the DSS problem can be shown by a reduction from the balanced complete bipartite subgraph problem (known to be $\mathcal{NP}$-complete [36, GT24]) to the case with N being a $\{0, 1\}$ matrix. The DSS problem with both of the objective functions also arises in manufacturing problems [74].

We will need the following notation. Let $A^{i,i+1}$ denote the off-diagonal block between A_i and A_{i+1}, that is, $A^{i,i+1}$ contains nonzeros between the rows

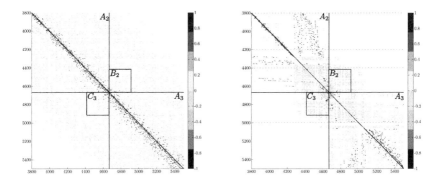

FIGURE 4.5: (See color insert.) Reordered matrix by applying a heuristic to improve the weight of the coupling blocks without reordering (left) and with reordering (right).

in A_i and the columns in A_{i+1}. Similarly, let $A^{i+1,i}$ denote the off-diagonal block between A_{i+1} and A_i, that is, $A^{i+1,i}$ contains nonzeros between the rows in A_{i+1} and the columns in A_i. In the PSPIKE context, the column permutation placing B_i just after A_i should be equivalent to the row permutation aligning C_{i+1} with the upper part of A_{i+1} in order to preserve the diagonal elements of A_{i+1}; similarly the row permutation aligning B_i with the lower part of A_i should be equivalent to the column permutation placing C_{i+1} before A_{i+1} in order to preserve the diagonal elements of A_i. This requires the two $k \times k$ submatrices B_i and C_{i+1} of A, respectively, $A^{i,i+1}$ and $A^{i+1,i}$ to be determined simultaneously. As the row permutation of one of the matrices is the column permutation of the other one, the problem can be stated as a DSS problem with the input matrix being $|A^{i,i+1}| + |(A^{i+1,i})^T|$. See Figure 4.5 for a sample where B_2 and C_3 between blocks A_2 and A_3 are determined simultaneously. Note that if the knapsack problem is not solved, only a few weighted elements can be inside of the coupling blocks.

As the problem in PSPIKE is equivalent to the DSS problem, we will continue with a single matrix formulation, where N replaces $|A^{i,i+1}| + |(A^{i+1,i})^T|$.

4.3.3.1   Heuristics

We present a simple heuristic algorithm for the DSS problem. The heuristic starts with the given matrix N as an initially infeasible solution and discards the row or column with the least contribution to the remaining matrix until a $k \times k$ matrix is obtained. We refer to this heuristic as DeMin. A similar procedure is discussed and analyzed in [75] for undirected graphs.

For the $\{0, 1\}$-matrices, the procedure can be implemented to run in linear time (linear in terms of the number of nonzeros in N). Two important observations are that one does not distinguish any two rows or columns having the same number of nonzeros in the remaining matrix, and that after discarding a row or column, the impacted columns or rows lose only one of their contributions to the remaining submatrix. Therefore, the rows and columns can be put into bins according to their initial degrees. Then, after discarding a row or column, the adjacent columns or rows need to be moved from their current bin to the one which contains rows/columns with one degree less.

For the general case of arbitrary matrix N, an initial sort seems to be necessary to make this algorithm work efficiently. First, rows and columns are sorted according to the sum of their nonzeros (or their magnitude). Then the lightest row or column is discarded, and the positions of the affected columns or rows are updated to keep the list sorted. As the list is nearly sorted after discarding a row or column, the insertion sort may be preferable.

4.3.3.2　An Upper Bound and Evaluation of the Heuristics

The DSS problem with a given matrix N and an integer k can be cast as a quadratic programming problem,

$$\max \quad \tfrac{1}{2}x^T M x - k$$
$$\text{s.t.}$$
$$\sum_{i=1}^{m} x_i = k,$$
$$\sum_{i=m+1}^{m+n} x_i = k,$$
$$x_i \in \{0,1\},$$

with

$$M = \begin{bmatrix} I_m & N \\ N^T & I_n \end{bmatrix}$$

and x_i's being the design variables. The first m variables correspond to the rows, and the remaining n variables correspond to the columns. In the formulation above, $x_i = 1$ if the corresponding row or column is in the densest submatrix, and it is zero otherwise. The first and second constraints guarantee that we choose only k rows and k columns, respectively. In an optimal solution, the two constraints are equally satisfied, in which case the objective function $\tfrac{1}{2}x^T M x - k$ corresponds to the optimum value of a solution to the DSS problem, as $x^T x = 2k$. Note that when N is connected, matrix M is irreducible—this is the reason for having identity matrices embedded in M; multiples of identity matrices could have been added to make M positive definite.

We now write a relaxed version of the problem as

$$\max \quad \tfrac{1}{2}x^T M x - k$$

$$\text{s.t.}$$

$$\sum_{i=1}^{m+n} x_i^2 = 2k,$$

$$0 \le x_i \le 1.$$

We will now use a well-known theorem (see for example [76, p. 176]) to obtain an analytical solution to the relaxed problem. Let A be a real, symmetric, $s \times s$ matrix, and $\lambda_{max}(A)$ be the largest eigenvalue of A. Then, the following relations hold:

$$\lambda_{max}(A) = \max\{x^T A x : x \in \mathbb{R}^s \text{ and } x^T x = 1\}.$$

Once we note the relation to the relaxed problem, we see that

$$k \, \lambda_{max}(M) - k$$

is a solution to the relaxed formulation above, and hence is an upper bound of the DSS problem. This kind of upper bounds is used in other problems as well (see, for example, for the graph bipartitioning problem in [77, Remark 5.1]). The main point in having such an upper bound is to evaluate the heuristics performance with respect to a known result.

Note that the problem then becomes a special form of the quadratic knapsack problem. See [78] for this connection and a survey of upper bounds.

Now list a few results where we compare the heuristics performance with the upper bound stated above for a number of matrices from the University of Florida Sparse Matrix Collection. These matrices satisfy the following properties: $300 \le m, n \le 500$, and $\text{nnz} \ge 2.5m$. At the time of writing, there were a total of 107 matrices at the collection. We tested with $k \in \{5, 10, 15, 20, 50, 75\}$ with the objective functions of (i) maximizing the total weight of the selected $k \times k$ submatrix and (ii) maximizing the number of nonzero entries in the selected submatrix.

As seen in Table 4.1, the performance of the DeMin heuristic for the $\{0, 1\}$ case increases with increasing k. This is in accordance with our expectations, as the proposed DeMin heuristic makes fewer decisions for large k. The performance of the heuristic for the weighted case seems to be deteriorating for larger k. This should be, however, due to the upper bound being too loose for large k values, in which case $k\lambda_{max}(M) - k$ becomes too large to attain.

TABLE 4.1: Performance of the DEMIN heuristic with respect to the upper bound on 107 sparse real-life matrices. Column $\{0,1\}$ contains results for the objective function of maximizing the number of entries, and column W contains results for the objective function of maximizing the weight of the selected submatrix.

k	$\{0,1\}$	W
5	0.70	0.79
10	0.68	0.74
15	0.69	0.61
20	0.67	0.54
50	0.72	0.41
75	0.74	0.36
min	0.45	0.07

4.4   Computational Results in PDE-Constrained Optimization

Let us consider a nonlinear optimization problems of the form

$$\begin{aligned} \min \quad & f(x) \\ \text{s.t.} \quad & c_{\mathbb{E}}(x) = 0, \\ & c_{\mathbb{I}}(x) \geq 0, \\ & x \geq 0, \end{aligned}$$

where the objective function $f : \mathbb{R}^n \to \mathbb{R}$ and the constraints $c_{\mathbb{E}} : \mathbb{R}^n \to \mathbb{R}^p$ and $c_{\mathbb{I}} : \mathbb{R}^n \to \mathbb{R}^q$ are sufficiently smooth. We are particularly interested in large-scale problems such as those where the equality constraints are obtained by discretizing PDEs and the inequality constraints are, for example, restrictions on a set of control and/or state variables.

Here, we consider a nonconvex 3D boundary control problem [79]

$$\begin{aligned} \min_{y,u} \quad & \int_\Omega \Phi(y(x) - y_t(x))dx \;\; + \;\; 10^{-2}\int_{\partial\Omega} u(x)^2 dx \\ \text{s.t.} \quad & {-}\Delta y(x) = 20 && \text{on} \quad \Omega = [0,1]^3, \\ & y(x) \leq 3.2 && \text{on} \quad \Omega, \\ & y(x) = u(x) && \text{on} \quad \partial\Omega, \\ & 1.6 \leq u(x) \leq 2.3 && \text{on} \quad \partial\Omega, \end{aligned}$$

where the target profile $y_t(x)$ is defined as

$$y_t(x) = 2.8 + 40 \cdot x_1(x_1 - 1) \cdot x_2(x_2 - 1) \cdot x_3(x_3 - 1)$$

and $\Phi(x)$ is the Beaton–Tukey penalty function.

The computational efficiency of all PDE-constrained optimization problems strongly depends on the performance of the numerical linear algebra kernel to solve Karush–Kuhn–Tucker (KKT) systems $Ax = f$. Almost 99% of the total optimization time is spent in the solution of such linear systems, which are typically sparse, symmetric, indefinite, and highly ill-conditioned.

The computational results are obtained by running the parallel optimization framework Ipopt [80, 81] combined with PSPIKE on the Cray XE6 from the National Supercomputing Centre CSCS in Switzerland. The cluster has currently 176 nodes, each node consisting of two sockets hosting 12-core Magny-Cours running at 2.1 GHz and up to 16 GB DDR3 main memory. For our performance tests we limit the number of compute cores to 3072. We are solving the 3D boundary control problem (4.10) using finite differences with a 27-point stencil. We discretize the PDE with a grid size of $N = 100$, which leads to a KKT system with $4,300,000$ equations and $62,772,832$ nonzeros. Additionally, we limit the maximum number of Ipopt iterations to 11, to have a comparable setup for the scalability results.

We are running the experiments with 256 MPI processes at maximum, mapping each MPI processor to a socket, and holding the number of OMP threads per socket set to 12 (see Figure 4.6). Almost linear scalability is obtained for this type of matrices by using a bandwidth of 100 in PSPIKE.

TABLE 4.2: Accumulated sequential timings (in seconds) for the preprocessing phase.

Algorithm	Implementation	Timings
Scaling Algorithm	2×2 Symmetric Matching [20, 21]	72s
Spectral Heuristic	HSL_MC73 [43]	167s*
Matching Algorithm	HSL_MC64 [19]	50s
Knapsack Heuristic	similar to DEMIN	40s
		Total 329s

Note: *The spectral heuristic is requested only once for the full optimization process; the other algorithms are called in every Ipopt iteration.

When using more than $1,000$ compute cores, the preprocessing phase, and especially the computational time for computing the spectral reordering, will be the crucial reordering routine to consider so as to preserve the linear scalability (see Table 4.2). A second observation is the increasing number of BiCGStab iterations within the hybrid solver. This can be logically explained, because a fewer number of elements are covered by the banded preconditioner when using more and more cores.

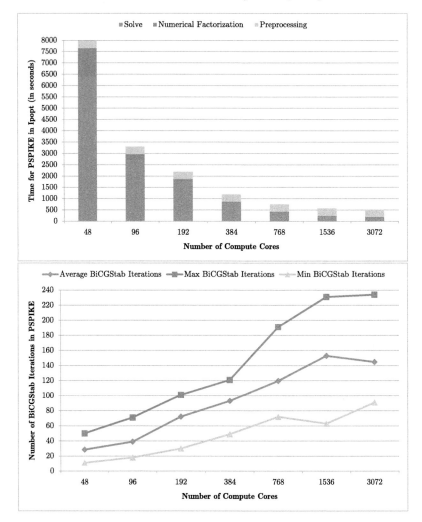

FIGURE 4.6: Computational results for PSPIKE when solving the boundary control problem within 11 `Ipopt` iterations. Top: Timing results for the sequential preprocessing, parallel numerical factorization, and parallel solve phase when using up to 3,072 compute cores. Bottom: maximum, minimum, and average number of `BiCGStab` iterations performed in PSPIKE during the optimization process.

4.5   Conclusions

Hybrid linear solvers—combining an iterative solver with a direct solver—are very promising candidates for achieving higher scalability on high-performance clusters with thousands of cores in order to solve very large and indefinite linear systems of equations. However, Krylov-subspace methods require an effective and efficient preconditioner to improve the robustness and scalability of the hybrid solver. Enabling combinatorial algorithms in the pre-processing phase, shown on the basis of PSPIKE, plays an important role for the scalability of such hybrid solvers on a large number of cores. We discussed several strategies to increase the robustness of these hybrid solvers. All these strategies are related to combinatorial algorithms, such as graph partitioning, graph matchings, and quadratic $\mathcal{NP}$-hard optimization problems.

Bibliography

[1] Schenk, O. and Gärtner, K., "On fast factorization pivoting methods for symmetric indefinite systems," *Elec. Trans. Numer. Anal.*, Vol. 23, No. 1, 2006, pp. 158–179.

[2] Mikkelsen, C. C. K. and Manguoglu, M., "Analysis of the Truncated SPIKE Algorithm," *SIAM Journal on Matrix Analysis and Applications*, Vol. 30, No. 4, 2008, pp. 1500–1519.

[3] Sameh, A. H. and Kuck, D. J., "On Stable Parallel Linear System Solvers," *J. ACM*, Vol. 25, No. 1, 1978, pp. 81–91.

[4] S. C. Chen, D. J. K. and Sameh, A. H., "Practical Parallel Band Triangular System Solvers," *ACM Transactions on Mathematical Software*, Vol. 4, No. 3, 1978, pp. 270–277.

[5] Lawrie, D. H. and Sameh, A. H., "The computation and communication complexity of a parallel banded system solver," *ACM Trans. Math. Softw.*, Vol. 10, No. 2, 1984, pp. 185–195.

[6] Berry, M. W. and Sameh, A., "Multiprocessor schemes for solving block tridiagonal linear systems," *The International Journal of Supercomputer Applications*, Vol. 1, No. 3, 1988, pp. 37–57.

[7] Dongarra, J. J. and Sameh, A. H., "On some parallel banded system solvers," *Parallel Computing*, Vol. 1, No. 3, 1984, pp. 223–235.

[8] Polizzi, E. and Sameh, A. H., "A parallel hybrid banded system solver: the SPIKE algorithm," *Parallel Comput.*, Vol. 32, No. 2, 2006, pp. 177–194.

[9] Polizzi, E. and Sameh, A. H., "SPIKE: A parallel environment for solving banded linear systems," *Computers & Fluids*, Vol. 36, No. 1, 2007, pp. 113–120.

[10] Sameh, A. H. and Sarin, V., "Hybrid parallel linear system solvers," *Inter. J. of Comp. Fluid Dynamics*, Vol. 12, 1999, pp. 213–223.

[11] Naumov, M. and Sameh, A. H., "A tearing-based hybrid parallel banded linear system solver," *J. Comput. Appl. Math.*, Vol. 226, April 2009, pp. 306–318.

[12] Amestoy, P. R., Duff, I. S., and L'Excellent, J. Y., "Multifrontal parallel distributed symmetric and unsymmetric solvers," *Computer Methods in Applied Mechanics and Engineering*, Vol. 184, No. 2–4, 2000, pp. 501–520.

[13] Li, X. S. and Demmel, J. W., "Making sparse Gaussian elimination scalable by static pivoting," *Proceedings of the 1998 ACM/IEEE conference on Supercomputing*, Supercomputing '98, IEEE Computer Society, Washington, DC, USA, 1998, pp. 1–17.

[14] Demmel, J., Grigori, L., and Xiang, H., "CALU: A Communication Optimal LU Factorization Algorithm," Tech. Rep. UCB/EECS-2010-29, EECS Department, University of California, Berkeley, Mar 2010.

[15] Gupta, A., Koric, S., and George, T., "Sparse matrix factorization on massively parallel computers," *Proceedings of the Conference on High Performance Computing Networking, Storage and Analysis*, SC '09, ACM, New York, USA, 2009, pp. 1:1–1:12.

[16] Manguoglu, M., Sameh, A. H., and Schenk, O., "PSPIKE: A Parallel Hybrid Sparse Linear System Solver," *Proceedings of the 15th International Euro-Par Conference on Parallel Processing*, Euro-Par '09, Springer-Verlag, Berlin, Heidelberg, 2009, pp. 797–808.

[17] Schenk, O., Manguoglu, M., Sameh, A. H., Christen, M., and Sathe, M., "Parallel scalable PDE-constrained optimization: antenna identification in hyperthermia cancer treatment planning," *Computer Science—R&D*, Vol. 23, No. 3–4, 2009, pp. 177–183.

[18] van der Vorst, H. A., "BI-CGSTAB: a fast and smoothly converging variant of BI-CG for the solution of nonsymmetric linear systems," *SIAM J. Sci. Stat. Comput.*, Vol. 13, No. 2, 1992, pp. 631–644.

[19] Duff, I. S. and Koster, J., "On Algorithms for Permuting Large Entries to the Diagonal of a Sparse Matrix," *SIAM J. Matrix Anal. Appl*, Vol. 22, 1999, pp. 973–996.

[20] Hagemann, M. and Schenk, O., "Weighted matchings for preconditioning symmetric indefinite linear systems," *SIAM J. Sci. Comput*, Vol. 28, 2006, pp. 403–420.

[21] Schenk, O., Wächter, A., and Hagemann, M., "Matching-based preprocessing algorithms to the solution of saddle-point problems in large-scale nonconvex interior-point optimization," *Comput. Optim. Appl.*, Vol. 36, 2007, pp. 321–341.

[22] Çatalyürek, Ü. V. and Aykanat, C., "Hypergraph-partitioning-based decomposition for parallel sparse-matrix vector multiplication," *IEEE Transactions on Parallel and Distributed Systems*, Vol. 10, No. 7, Jul 1999, pp. 673–693.

[23] Bui, T. N. and Jones, C., "Finding good approximate vertex and edge partitions is NP-hard," *Information Processing Letters*, Vol. 42, 1992, pp. 153–159.

[24] Bui, T. N. and Jones, C., "A heuristic for reducing fill-in in sparse matrix factorization," *6th SIAM Conference on Parallel Processing for Scientific Computing*, Norfolk, VA, USA, 1993, pp. 445–452.

[25] Hendrickson, B. and Leland, R., "A multilevel algorithm for partitioning graphs," *Supercomputing '95: Proceedings of the 1995 ACM/IEEE conference on Supercomputing (CDROM)*, ACM, New York, USA, 1995, p. 28.

[26] Hendrickson, B. and Leland, R., *The Chaco user's guide, version 2.0*, Sandia National Laboratories, Alburquerque, NM, 87185, 1995.

[27] Karypis, G. and Kumar, V., *MeTiS: A software package for partitioning unstructured graphs, partitioning meshes, and computing fill-reducing orderings of sparse matrices version 4.0*, University of Minnesota, Department of Comp. Sci. and Eng., Army HPC Research Center, Minneapolis, MN, 1998.

[28] Pellegrini, F., *SCOTCH 5.1 User's guide*, Laboratoire Bordelais de Recherche en Informatique (LaBRI), 2008.

[29] Vastenhouw, B. and Bisseling, R. H., "A two-dimensional data distribution method for parallel sparse matrix-vector multiplication," *SIAM Review*, Vol. 47, No. 1, 2005, pp. 67–95.

[30] Çatalyürek, Ü. V. and Aykanat, C., *PaToH: A Multilevel Hypergraph Partitioning Tool, Version 3.0*, Bilkent University, Department of Computer Engineering, Ankara, 06533 Turkey. PaToH is available at \hbox{http://bmi.osu.edu/~umit/software.htm}, 1999.

[31] Fiduccia, C. M. and Mattheyses, R. M., "A linear-time heuristic for improving network partitions," *DAC '82: Proceedings of the 19th Conference on Design Automation*, IEEE Press, Piscataway, NJ, USA, 1982, pp. 175–181.

[32] Kernighan, B. W. and Lin, S., "An efficient heuristic procedure for partitioning graphs," *The Bell System Technical Journal*, Vol. 49, Feb. 1970, pp. 291–307.

[33] Simon, H. D. and Teng, S.-H., "How good is recursive bisection?" *SIAM Journal on Scientific Computing*, Vol. 18, No. 5, 1997, pp. 1436–1445.

[34] Aykanat, C., Cambazoglu, B. B., and Uçar, B., "Multi-level direct K-way hypergraph partitioning with multiple constraints and fixed vertices," *Journal of Parallel and Distributed Computing*, Vol. 68, 2008, pp. 609–625.

[35] Karypis, G. and Kumar, V., "Parallel Multilevel series k-Way Partitioning Scheme for Irregular Graphs," *SIAM Review*, Vol. 41, No. 2, 1999, pp. 278–300.

[36] Garey, M. R. and Johnson, D. S., *Computers and Intractability; A Guide to the Theory of NP-Completeness*, W. H. Freeman & Co., New York, USA, 1979.

[37] Davis, T. A., *Direct Methods for Sparse Linear Systems*, No. 2 in Fundamentals of Algorithms, Society for Industrial and Applied Mathematics, Philadelphia, PA, USA, 2006.

[38] Cuthill, E. and McKee, J., "Reducing the bandwidth of sparse symmetric matrices," *Proceedings of the 24th national conference*, ACM, New York, USA, 1969, pp. 157–172.

[39] George, A. J., *Computer implementation of the finite element method*, Ph.D. thesis, Stanford University, Stanford, CA, USA, 1971.

[40] George, A. and Liu, J. W. H., "An Implementation of a Pseudoperipheral Node Finder," *ACM Trans. Math. Softw.*, Vol. 5, September 1979, pp. 284–295.

[41] Sloan, S. W., "An algorithm for profile and wavefront reduction of sparse matrices," *International Journal for Numerical Methods in Engineering*, Vol. 23, No. 2, 1986, pp. 239–251.

[42] Duff, I. S., Reid, J. K., and Scott, J. A., "The use of profile reduction algorithms with a frontal code," *International Journal for Numerical Methods in Engineering*, Vol. 28, No. 11, 1989, pp. 2555–2568.

[43] Hu, Y. and Scott, J., "HSL_MC73: a fast multilevel Fiedler and profile reduction code," Technical Report RAL-TR-2003-036, RAL, 2003.

[44] Kumfert, G. and Pothen, A., "Two improved algorithms for envelope and wavefront reduction," *BIT Numerical Mathematics*, Vol. 37, 1997, pp. 559–590.

[45] Reid, J. K. and Scott, J. A., "Ordering symmetric sparse matrices for small profile and wavefront," *International Journal for Numerical Methods in Engineering*, Vol. 45, No. 12, 1999, pp. 1737–1755.

[46] Hu, Y. F. and Scott, J. A., "A Multilevel Algorithm for Wavefront Reduction," *SIAM J. Sci. Comput.*, Vol. 23, April 2001, pp. 1352–1375.

[47] Fiedler, M., "A property of eigenvectors of nonnegative symmetric matrices and its application to graph theory." *Czechoslovak Mathematical Journal*, Vol. 25, No. 4, 1975, pp. 619–633.

[48] Juvan, M. and Mohar, B., "Optimal linear labelings and eigenvalues of graphs," *Discrete Applied Mathematics*, Vol. 36, No. 2, 1992, pp. 153–168.

[49] Barnard, S. T., Pothen, A., and Simon, H. D., "A spectral algorithm for envelope reduction of sparse matrices," *Proceedings of the 1993 ACM/IEEE conference on Supercomputing*, Supercomputing '93, ACM, New York, USA, 1993, pp. 493–502.

[50] Manguoglu, M., Cox, E., Saied, F., and Sameh, A. H., "TRACEMIN-Fiedler: A Parallel Algorithm for Computing the Fiedler Vector," *VECPAR*, 2010, pp. 449–455.

[51] Manguoglu, M., Koyutürk, M., Sameh, A. H., and Grama, A., "Weighted Matrix Ordering and Parallel Banded Preconditioners for Iterative Linear System Solvers," *SIAM J. Sci. Comput.*, Vol. 32, 2010, pp. 1201–1216.

[52] Bokhari, S. H., "Partitioning Problems in Parallel, Pipeline, and Distributed Computing," *IEEE Trans. Comput.*, Vol. 37, January 1988, pp. 48–57.

[53] Pinar, A. and Aykanat, C., "Fast optimal load balancing algorithms for 1D partitioning," *Journal of Parallel and Distributed Computing*, Vol. 64, No. 8, 2004, pp. 974 – 996.

[54] Pınar, A., Tabak, E. K., and Aykanat, C., "One-dimensional partitioning for heterogeneous systems: Theory and practice," *Journal of Parallel and Distributed Computing*, Vol. 68, No. 11, 2008, pp. 1473–1486.

[55] Benoit, A. and Robert, Y., "Mapping pipeline skeletons onto heterogeneous platforms," *Journal of Parallel and Distributed Computing*, Vol. 68, No. 6, 2008, pp. 790–808.

[56] Olschowka, M. and Neumaier, A., "A New Pivoting Strategy For Gaussian Elimination," *Linear Algebra and Its Application*, Vol. 240, 1996, pp. 131–151.

[57] Burkard, R., Dell'Amico, M., and Martello, S., *Assignment Problems*, Society for Industrial and Applied Mathematics, Philadelphia, PA, USA, 2009.

[58] Avis, D., "A survey of heuristics for the weighted matching problem," *Networks*, Vol. 13, No. 4, 1983, pp. 475–493.

[59] Hopcroft, J. E. and Karp, R. M., "An $n^{5/2}$ algorithm for maximum matching in bipartite graphs," *SIAM J. Computing*, Vol. 2, 1973, pp. 225–231.

[60] Kuhn, H. W., "The Hungarian Method for the assignment problem," *Naval Research Logistics Quarterly*, Vol. 2, 1955, pp. 83–97.

[61] Langguth, J., Manne, F., and Sanders, P., "Heuristic initialization for bipartite matching problems," *J. Exp. Algorithmics*, Vol. 15, 2010, pp. 1–22.

[62] Preis, R., "Linear time 1/2-approximation algorithm for maximum weighted matching in general graphs." *In Proceedings of STACS 1999. Lecture Notes in Computer Science*, Vol. 1563, 1999, pp. 259–269.

[63] Hougardy, S. and Vinkemeier, D. E., "Approximating weighted matchings in parallel." *Information Processing Letters*, Vol. 99, No. 3, 2006, pp. 119–123.

[64] Duan, R. and Pettie, S., "Approximating Maximum Weight Matching in Near-Linear Time," *Foundations of Computer Science, Annual IEEE Symposium on*, Vol. 0, 2010, pp. 673–682.

[65] Manne, F. and Bisseling, R. H., "A Parallel Approximation Algorithm for the Weighted Maximum Matching Problem," *Proceedings Seventh International Conference on Parallel Processing and Applied Mathematics (PPAM 2007)*, Vol. 4967 of *Lecture Notes in Computer Science*, 2008, pp. 708–717.

[66] Luby, M., "A Simple Parallel Algorithm for the Maximal Independent Set Problem," *SIAM J. Comput.*, Vol. 15, No. 4, 1986, pp. 1036–1053.

[67] Patwary, M. A., Bisseling, R. H., and Manne, F., "Parallel Greedy Graph Matching using an Edge Partitioning Approach," *Proceedings of the Fourth ACM SIGPLAN Workshop on High-level Parallel Programming and Applications (HLPP 2010)*, 2010, pp. 45–54.

[68] Miller, D. L., Pekny, J. F., and Thompson, G. L., "Solution of large dense transportation problems using a parallel primal algorithm," *Operations Research Letters*, Vol. 9, No. 5, 1990, pp. 319–324.

[69] Peters, J., "The network simplex method on a multiprocessor," *Networks*, Vol. 20, No. 7, 1990, pp. 845–859.

[70] Madduri, K., Bader, D. A., Berry, J. W., and Crobak, J. R., "Parallel Shortest Path Algorithms for Solving Large-Scale Instances," *Shortest Path Computations: Ninth DIMACS Challenge*, 2008.

[71] Bertsekas, D. P. and Castañon, D. A., "Parallel synchronous and asynchronous implementations of the auction algorithm," *Parallel Computing*, Vol. 17, No. 707–732, 1991.

[72] Wein, J. and Zenios, S., "Massively parallel auction algorithms for the assignment problem," *Proceedings of the 3-rd Symposium on the Frontiers of Massively Parallel Computations*, 1990, pp. 90–99.

[73] Buš, L. and Tvrdik, P., "Towards auction algorithms for large dense assignment problems," *Computational Optimization and Application*, Vol. 43, No. 3, 2009, pp. 411–436.

[74] Dawande, M., Keskinocak, P., Swaminathan, J. M., and Tayur, S., "On bipartite and multipartite clique problems," *Journal of Algorithms*, Vol. 41, No. 2, 2001, pp. 388–403.

[75] Asahiro, Y., Iwama, K., Tamaki, H., and Tokuyama, T., "Greedily finding a dense subgraph," *Journal of Algorithms*, Vol. 34, No. 2, 2000, pp. 203–221.

[76] Horn, R. A. and Johnson, C. R., *Matrix Analysis*, Cambridge University Press, New York, 1985.

[77] Boyd, S. and Vandenberghe, L., *Convex Optimization*, Cambridge University Press, New York, USA, 2004.

[78] Pisinger, D., "The Quadratic Knapsack Problem—A Survey," *Discrete Applied Mathematics*, Vol. 155, 2007, pp. 623–648.

[79] Maurer, H. and Mittelmann, H. D., "Optimization Techniques for Solving Elliptic Control Problems with Control and State Constraints. Part 1: Boundary Control," *Comp. Optim. Applic.*, Vol. 16, 2000, pp. 29–55.

[80] Curtis, F. E., Schenk, O., and Wächter, A., "An Interior-Point Algorithm for Large-Scale Nonlinear Optimization with Inexact Step Computations," *SIAM Journal on Scientific Computing*, Vol. 32, No. 6, 2010, pp. 3447–3475.

[81] Wächter, A. and Biegler, L. T., "On the implementation of an interior-point filter line-search algorithm for large-scale nonlinear programming," *Math. Program.*, Vol. 106, No. 1, 2006, pp. 25–57.

Chapter 5

Combinatorial Problems in Algorithmic Differentiation

Uwe Naumann

RWTH Aachen University

Andrea Walther

Universität Paderborn

5.1 Introduction

Conceptually, in *Algorithmic* (also known as *Automatic*) *Differentiation* (AD) we consider multivariate vector functions $F : \mathbb{R}^n \to \mathbb{R}^m$ mapping a vector of *independent* inputs $\mathbf{x} \in \mathbb{R}^n$ onto a vector of *dependent* outputs $\mathbf{y} \in \mathbb{R}^m$ as $\mathbf{y} = F(\mathbf{x})$. For the purpose of the current chapter the function F is assumed to be twice continuously differentiable in a neighborhood of all arguments at which it is supposed to be evaluated. Hence, its Jacobian

$$\nabla F(\mathbf{x}) = \left(\frac{\partial y_j}{\partial x_i}\right)_{i=1,\dots,n}^{j=1,\dots,m} \in \mathbb{R}^{m \times n}$$

and Hessian

$$\nabla^2 F(\mathbf{x}) = \left(\frac{\partial^2 y_j}{\partial x_i \partial x_k}\right)_{i,k=1,\dots,n}^{j=1,\dots,m} \in \mathbb{R}^{m \times n \times n}$$

exist at all of these points. The Hessian $\nabla^2 F(\mathbf{x}) \equiv H = (h_{j,i,k})_{i,k=1,\dots,n}^{j=1,\dots,m}$ is a symmetric 3-tensor, that is, $h_{j,i,k} = h_{j,k,i}$.

AD is a technique for transforming implementations of such multivariate vector functions as computer programs into code that computes projections of the Jacobian onto $\mathbb{R}^m$ (*forward* or *tangent-linear mode*) or $\mathbb{R}^n$ (*reverse* or *adjoint mode*). Reapplications of forward and reverse modes to first derivative models yield second derivative models and so forth. Higher derivatives can also be calculated by propagating truncated Taylor series. Refer to [1] for details. For an efficient evaluation of the higher derivative tensors the choice of these Taylor polynomials forms a combinatorial problem that is not discussed here.

AD can be implemented very elegantly and robustly by means of operator and function overloading if this technique is supported by the used programming language. AD of large-scale numerical simulations is often performed by preprocessors that transform the source code correspondingly based on the results of various domain-specific program analyses. While source transformation potentially yields more efficient derivative codes the implementation of such compilers is a highly challenging software development task.

Refer to [2] for a discussion of the mathematical foundations of AD. Links to research groups, tool developers, applications, and a comprehensive bibliography can be found on the AD community's web portal http://www.autodiff.org. Two selected tools are discussed in Chapters 6 (the source transformation tool OpenAD for Fortran) and 7 (the overloading tool ADOL-C for C/C++) in the light of combinatorial problems and specific approaches to their solution. Another source transformation tool for C/C++ (dcc) is used in the context of a case study in nonlinear constrained optimization in Chapter 8.

Definition 2 *The* tangent-linear model *of $F : \mathbb{R}^n \to \mathbb{R}^m$ where $\mathbf{y} = F(\mathbf{x})$ is defined as $F^{(1)} : \mathbb{R}^n \times \mathbb{R}^n \to \mathbb{R}^m$ such that*

$$\mathbf{y}^{(1)} = F^{(1)}(\mathbf{x}, \mathbf{x}^{(1)}) \equiv \nabla F(\mathbf{x}) \cdot \mathbf{x}^{(1)} \quad .$$

Example 4 *Tangent-linear code for the simple product reduction*

```
void f(int n, float* x, float& y) {
  y=x[0];
  for (int i=1;i<n;i++)
    y=y*x[i];
}
```

is generated by association of directional derivatives t1_v with all active [3] floating-point variables v followed by augmenting all assignments with their respective tangent-linear models. The derivatives of the passive floating-point variables vanish identically at all points in the domain of F. Both x and y are assumed to be active.

```
void t1_f(int n, float* x, float* t1_x,
                 float& y, float& t1_y) {
  t1_y=t1_x[0];
  y=x[0];
  for (int i=1;i<n;i++) {
    t1_y=t1_y*x[i]+y*t1_x[i];
    y=y*x[i];
  }
}
```

The correctness of this approach follows immediately from the chain rule of differential calculus.

The whole Jacobian can be accumulated with machine accuracy by letting $\mathbf{x}^{(1)}$ range over the Cartesian basis vectors $\mathbf{e}_i \in \mathbb{R}^n$, $i = 1, \ldots, n$, with a computational complexity of $O(n) \cdot OPS(F)$, where $OPS(F)$ denotes the number of arithmetic operations performed by the given implementation of F. This cost of running the tangent-linear model is proportional to and often lower than that of the approximation of the Jacobian's columns by finite differences. Moreover, the improved accuracy is likely to turn out advantageous in the context of numerical algorithms such as Newton's method for the solution of systems of nonlinear equations.

Example 5 *For* float x[n], t1_x[n], y, g[n] *and* t1_x[i]=0 *for* i =0,..., n−1 *the driver program*

```
...
for (int i=0;i<n;i++) {
  t1_x[i]=1;
  t1_f(n,x,t1_x,y,g[i]);
  t1_x[i]=0;
}
...
```

computes the gradient of y *with respect to* x *at the current point.*

Several directional derivatives (e.g., $l \geq 1$) can be propagated in *vector mode*. Therefore the t1_v are allocated as vectors of length l for all active program variables v. The wanted derivatives are computed by loops over the corresponding entries of the directional derivative vectors.

Definition 3 *The adjoint model of* $F : \mathbb{R}^n \to \mathbb{R}^m$ *where* $\mathbf{y} = F(\mathbf{x})$ *is defined as* $F_{(1)} : \mathbb{R}^n \times \mathbb{R}^m \to \mathbb{R}^n$ *such that*

$$\mathbf{x}_{(1)}^T = F_{(1)}(\mathbf{x}, \mathbf{y}_{(1)}) \equiv \mathbf{y}_{(1)}^T \cdot \nabla F(\mathbf{x}) \quad .$$

Example 6 *Nonincremental adjoint code for the product reduction from Example 4 is generated by association of adjoints* a1_v *with all active floating-point variables* v *followed by code for running the adjoint models of all individual assignments in reverse order. All intermediate values of* y *are required in reverse order. One way to make these values available is to store (push) them on a stack during an appropriately augmented run of the original code and to restore (pop) them prior to the execution of the corresponding adjoint code.*

```
void a1_f(int n, float* x, float& a1_x,
                  float& y, float* a1_y) {
    y=x[0];
    for (int i=1;i<n;i++) {
        push(y);
        y=y*x[i];
    }
    for (int i=n-1;i>0;i--) {
        pop(y);
        a1_x[i]=y*a1_y;
        a1_y=x[i]*a1_y;
    }
    a1_x[0]=a1_y;
}
```

AD tools often generate incremental adjoint code *yielding further technical issues that are beyond the scope of this survey article. Some implication are shown in Chapter 8.*

The whole Jacobian can be accumulated with machine accuracy by letting $y_{(1)}$ range over the Cartesian basis vectors e_i, $i = 1, \ldots, m$, in $\mathbb{R}^m$ with a computational complexity of $O(m) \cdot OPS(F)$. The independence of the computational cost from the number of independent variables n is of particular interest in large-scale nonlinear optimization (ref. Chapter 8). Large gradients ($m = 1$) can be computed at a typically small constant multiple of $OPS(F)$. As a function of the adjoint outputs of F the adjoint model requires the reversal of the original data flow. This seemingly minor side note causes most of the technical and some highly relevant combinatorial problems in AD (ref. Section 5.3).

Example 7 *For* float x[n], g[n], y, a1_y *the driver program*

```
...
a1_y=1;
a1_f(n,x,g,y,a1_y);
...
```

computes the gradient of y *with respect to* x *at the current point* x.

Adjoint vector mode allocates the adjoints a1_v of all scalar program variables v as vectors. The local statement-level adjoint models are replaced by corresponding loops over the components of the adjoint vectors.

When considering second derivatives we focus on scalar functions ($m = 1$) for notational simplicity.

Definition 4 *The* second-order tangent-linear model *of* $F : \mathbb{R}^n \to \mathbb{R}$ *where* $y = F(\mathbf{x})$ *is defined as* $F^{(1,2)} : \mathbb{R}^n \times \mathbb{R}^n \times \mathbb{R}^n \to \mathbb{R}$ *such that*

$$y^{(1,2)} = F^{(1,2)}(\mathbf{x}, \mathbf{x}^{(1)}, \mathbf{x}^{(2)}) \equiv (\mathbf{x}^{(2)})^T \cdot \nabla^2 F(\mathbf{x}) \cdot \mathbf{x}^{(1)} \quad .$$

The whole Hessian at point $\mathbf{x}$ can be accumulated with machine accuracy by letting $\mathbf{x}^{(2)}$ and $\mathbf{x}^{(1)}$ range independently over the Cartesian basis vectors in $\mathbb{R}^n$. Even if symmetry is exploited the computational cost is $O(n^2) \cdot OPS(F)$. Again, the computational complexity is proportional to that of finite difference approximation. However, the higher accuracy of the second derivatives is likely to have a positive impact on second-order numerical algorithms such as Newton's method for nonlinear optimization.

Example 8 *Second-order tangent-linear code for the product reduction can be obtained by reapplication of the procedure sketched in Example 4 to the tangent-linear code. We omit a full listing due to space restrictions. The mechanical reapplication of the source transformation procedure does not yield any conceptual surprises.*

Definition 5 *The* second-order adjoint model *of* $F : \mathbb{R}^n \to \mathbb{R}$ *where* $y = F(\mathbf{x})$ *is defined as* $F_{(1)}^{(2)} : \mathbb{R}^n \times \mathbb{R}^n \times \mathbb{R} \to \mathbb{R}^n$ *such that*

$$\mathbf{x}_{(1)}^{(2)} = F_{(1)}^{(2)}(\mathbf{x}, \mathbf{x}^{(2)}, y_{(1)}) \equiv y_{(1)} \cdot \nabla^2 F(\mathbf{x}) \cdot \mathbf{x}^{(2)} \quad .$$

The whole Hessian at point $\mathbf{x}$ can be accumulated with machine accuracy by setting $y_{(1)} = 1$ and by letting $\mathbf{x}^{(2)}$ range over the Cartesian basis vectors in $\mathbb{R}^n$. The computational cost is $O(n) \cdot OPS(F)$. Hessian-vector products for use, for example, in matrix-free implementations of Newton-Krylov methods for nonlinear optimization can be obtained at a typically small constant multiple of the cost of evaluating F.

Example 9 *Second-order adjoint code for the product reduction can be obtained by application of the procedure sketched in Example 4 to the adjoint code derived in Example 6.*

Implementations of AD include libraries based on operator overloading in C++ [4, 5] and Fortran [6] as well as source transformation tools for Fortran [7, 8, 9, 10] and C/C++ [11]. The main advantage of derivative code compilers is the ability to run certain domain-specific program analyses [3, 12] often resulting in more efficient derivative code.

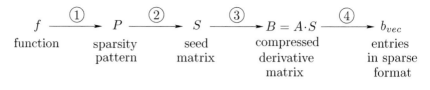

FIGURE 5.1: Computing sparse derivative matrices.

5.2 Compression Techniques

For some applications, e.g., the integration of stiff ODE systems or non-linear optimization using second-order information, the computation of whole Jacobians or Hessians is required. If these matrices are sparse, i.e., contain a dominant number of zeros, the derivative calculation can be considerably accelerated if sparsity is taken into account. For this purpose AD technology provides two approaches.

First, one may employ so-called sparse vectors that exploit the sparsity directly during the derivative calculation. Here, sparse vector based approaches use *sparse data structures*, instead of dense arrays, to execute a fundamental operation in an AD code as implemented for example in the ADIFOR / SparsLinC combination [13]. SparsLinC uses three different data structures to represent sparse vectors (one for vectors with at most one nonzero, a second for vectors with a few scattered nonzeros, and a third for vectors with a contiguous block of nonzeros) and heuristically switches from one representation to another as needed to reduce the large runtime overhead observed in earlier sparse vector based approaches [14].

Second, one may follow the procedure in Figure. 5.1:

1. *Detection of sparsity pattern P:* For several applications the sparsity pattern of the derivative matrix A is known in advance, for example, when using finite difference schemes for the discretization of partial differential equations. If the sparsity pattern is unknown, one may employ AD for its computation, see, e.g., [15, 16]. Ideally this step is performed only once. If the control flow of the program changes, then the sparsity pattern may also change and, hence, has to be recomputed.

2. *Generation of the seed matrix S:* Graph coloring methods are employed for this purpose. We distinguish cases where the derivative matrix A is symmetric or not and whether so-called direct or indirect methods are used. Details will be discussed in the remainder of this section. Again, ideally this step is executed only once.

3. *Evaluation the product $A \cdot S$ of the derivative matrix A and the seed matrix S to obtain the compressed derivative matrix B:* The vector forward mode of AD can be used.

4. *Recovery of the nonzero derivative entries:* For direct methods one just has to identify the correct indices of the entries. For indirect methods, one also has to recompute the values of the entries as the solution of usually linear systems of equations.

Steps 1 and 3 may rely on AD whereas steps 2 and possibly 4 involve numerous combinatorial aspects including graph coloring methods and suitable combinations of the forward and reverse modes.

As an alternative to the row compression illustrated in Figure 5.1 and described above, one may also employ a column compression $B = S \cdot A$ using a suitable seed matrix S and the reverse (vector) mode of AD as well as a combination of both approaches.

For the detection of sparsity using AD, one has two main choices. First, one may employ bit vectors. In this case, the Jacobian is multiplied from the left by n bit vectors, where n is the number of independent variables; each arithmetic operation then corresponds to a logical OR, yielding the overall sparsity pattern of the Jacobian. Since one Jacobian-vector product needs to be performed for each independent variable, the complexity of this approach is $O(n) \cdot OPS(F)$. Bit vector approaches have been implemented for source transformation and operator overloading tools to detect the sparsity pattern of Jacobians. For the sparsity pattern of Hessians no such implementation is available.

Alternatively, one may propagate dependence information through the function evaluation as proposed for example in [2]. In the case of Jacobians, so-called *index domains* $\mathcal{X}_k$ that contain the indices of all independent variables that influence the intermediate value v_k are computed. The corresponding sparsity pattern can be constructed directly from the index domains of the dependent variables y_i, $i = 1, \ldots, m$. The number of operations required for the determination of all index domains is proportional to $OPS(F)$ multiplied by the maximal number of nonzeros in one row of the Jacobian. This last number is equivalent to the maximal number of entries in one index domain $\mathcal{X}_k$. In the case of Hessians, in addition to the *index domains* one may propagate nonlinearity information either in forward mode, see, e.g., [16], or in reverse mode, see, e.g., [2]. Both approaches increase the computational complexity considerably making the efficient detection of sparsity patterns of Hessians a highly relevant part of ongoing research. For example, in Chapter 8 the recursive structure of partial separability is exploited for the computation of conservative estimates of Hessian sparsity patterns. A considerable reduction in the computational cost is observed.

5.2.1 Computation of Sparse Jacobians

Several graphs have been associated with the computation of suitable seed matrices when combined with the forward mode of AD. In this case, the goal is to generate a seed matrix with as few columns as possible, since the complexity of the evaluation of $A \cdot S$ depends linearly on the number of columns of S.

Curtis, Powell, and Reid [17] were the first to propose a *structurally orthogonal* partition of a Jacobian matrix A – a partition of the columns of A in which no two columns in a group share a nonzero in the same row. Combining structurally orthogonal columns into one group, i.e., setting the corresponding entries of the column of the seed matrix representing the group equal to one, gives a seed matrix S where the entries of A can be directly recovered from the compressed representation $B \equiv A \cdot S$.

Coleman and Moré [18] model the associated problem of partitioning the columns of the Jacobian into the fewest possible groups as a distance-1 coloring problem of a suitably defined graph.

Definition 6 *The* column-incidence graph $G_c(V_c, E_c)$ *associated with the function* $F : \mathbb{R}^n \to \mathbb{R}^m$ *is a graph, where* V_c *contains a vertex for each independent variable* $x_1, \dots, x_n$. *Furthermore, there is an edge between the vertices* i *and* j *if the corresponding independent variables* x_i *and* x_j *jointly impact any dependent variable.*

Example 10 *The sparsity pattern and the column-incidence graph for the function* $F : \mathbb{R}^6 \to \mathbb{R}^5$, *where*

$$F_1(\mathbf{x}) = x_1 \cdot \cos(x_5 \cdot x_6), \quad F_2(\mathbf{x}) = x_2 + x_3 + x_4, \quad F_3(\mathbf{x}) = x_3 \cdot x_5,$$
$$F_4(\mathbf{x}) = \exp(x_1 + x_4), \quad F_5(\mathbf{x}) = x_2 + x_6$$

and its distance-1 coloring are given below.

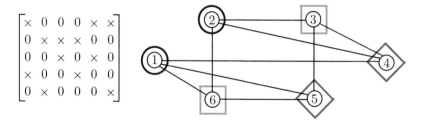

$$\begin{bmatrix} \times & 0 & 0 & 0 & \times & \times \\ 0 & \times & \times & \times & 0 & 0 \\ 0 & 0 & \times & 0 & \times & 0 \\ \times & 0 & 0 & \times & 0 & 0 \\ 0 & \times & 0 & 0 & 0 & \times \end{bmatrix}$$

Three colors and hence three Jacobian-vector products are required to accumulate all twelve Jacobian entries.

As mentioned above, we would like to keep the number p of different colors as small as possible. The absolute minimum for any given graph G is called its chromatic number, and its determination is known to be NP-hard [19]. Various heuristic algorithms have been developed. They generally yield values p that are only slightly larger than the chromatic number itself. A comprehensive introduction of corresponding coloring algorithms together with a comparison of their performance can be found in [20]. All this methodology can be immediately transferred to AD. Moreover, it can also be applied to grouping (coloring) Jacobian rows rather than columns.

If the requirement that the entries of a Jacobian be recovered directly from a compressed representation is relaxed, then the compression can be done much more compactly. In this case, the recovery of the entries involves the solution of usually linear systems of equations. The detection of seed matrices S yielding linear systems that can be solved efficiently and accurately turns out to be challenging.

Newsam and Ramsdell [21] originally proposed generalized Vandermonde matrices

$$S = \left[P_k(\hat{\lambda}_j)\right]_{k=1,\dots,p}^{j=1,\dots,n} \in \mathbb{R}^{n \times p}$$

where the $P_k(\lambda)$ for $k = 1, \dots, p$ denote linearly independent polynomials of degree less than p and the n real abscissas $\hat{\lambda}_j$ may be restricted to the interval $[-1, 1]$ without loss of generality. Exploiting the generalized Vandermonde structure as proposed, for example, in [22] one can reduce the effort required for the solution of the resulting linear systems to grow only quadratically in p_i. This complexity estimate has long been known for the classical Vandermonde matrices [23]. However, Vandermonde matrices are prominent examples for ill-conditioned matrices having a condition number that grows exponentially with the order p_i. If finite difference quotient approximations are applied to approximate the derivatives, this drawback could become very significant. For exact derivatives computed with AD, it was shown by Björck and Pereyra [24] that the algorithm mentioned above yields much better accuracy than could be expected from the general estimates based on conditioning. Nevertheless, the conditioning of the seed matrices is a reason for concern. Several other approaches have been examined and proposed, as, for example, the Pascal seeding by Hossain and Steihaug [25]. However, up to now a good generic choice for the seed matrices in the indirect case is still an open question.

Since the reverse mode of AD provides products of the form $W^T \cdot \nabla F$, one may apply the described approach also to column compression. For this purpose, we define another graph associated with the underlying function F.

Definition 7 *The* row-incidence graph $G_r(V_r, E_r)$ *associated with the function* $F : \mathbb{R}^n \to \mathbb{R}^m$ *is a graph, where* V_r *contains a vertex for each dependent variable* $y_1, \dots, y_m$. *Furthermore, there is an edge between the vertices* i *and* j *if the corresponding dependent variables* y_i *and* y_j *are both impacted by the same independent variable.*

Example 11 *The sparsity pattern and the row-incidence graph for the function given in Example 10 and its distance-1 coloring are given below.*

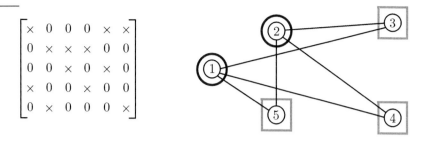

In this case, two colors and hence two vector-Jacobian products are required to accumulate the twelve Jacobian entries.

The number of colors may differ for the column- and row-incidence graphs. One can construct examples, where this difference is quite large. Therefore, it is in general not a-priory clear whether one should prefer row or column compression. Once more, indirect approaches can be applied to column compression yielding usually a smaller number of columns in the corresponding seed matrix. For arrow-shaped matrices a combination of row and column compression is the most efficient method as explained for example in [2]. This observation opens up a wide range of combinatorial problems, especially since the forward mode does not require as much memory as the reverse mode. Therefore, very specific AD aspects come into play.

5.2.2   Computation of Sparse Hessians

In contrast to the compression of sparse Jacobians, for sparse second-order derivative tensors one may exploit additionally the inherent symmetry. This is especially important for the case of scalar-valued functions, i.e., $F : \mathbb{R}^n \to \mathbb{R}$, as well as for adjoint projections of Hessian tensors (see Chapter 8). A compressed representation of the sparse Hessian $\nabla^2 F$ can be computed as

$$H = \nabla^2 F \cdot S$$

for a given seed matrix $S \in \mathbb{R}^{n \times p}$ by p Hessian-vector products. For simplicity, it is always assumed that no element of the diagonal of H vanishes.

Coleman and Moré [26] considered a direct approach and showed that the partitioning problem that occurs in the computation of the Hessian in this case corresponds to a *star coloring* of the adjacency graph of the Hessian.

Definition 8 *The* adjacency graph $G_a(V_a, E_a)$ *associated with the function* $F : \mathbb{R}^n \to \mathbb{R}^m$ *is a graph, where* V_a *contains a vertex for each independent variable* $x_1, \ldots, x_n$. *Furthermore, there is an edge between the vertices* i *and* j *if the corresponding independent variables* x_i *and* x_j *jointly impact in a nonlinear way any dependent variable.*

In a star-colored graph, a subgraph induced by any two color classes, that is, sets of vertices having the same color, represents a collection of stars. The resulting approach for computing a sparse Hessian is illustrated in the next example taken from [20].

Example 12 *The sparsity pattern and the adjacency graph of the Hessian for the function* $F : \mathbb{R}^{10} \to \mathbb{R}$, *where*

$$F(\mathbf{x}) = \sum_{i=1}^{10} x_i^2 + x_1 \cdot x_2 \cdot x_7 + \cos(x_2 + x_3 + x_5) + x_3 \cdot x_4 \cdot x_6 +$$

$$\exp(x_4 \cdot x_{10}) + x_5 \cdot x_6 \cdot x_8 + \sin(x_6 - x_9) + \sum_{i=7}^{9} x_i \cdot x_{i+1}$$

and its star coloring are given below.

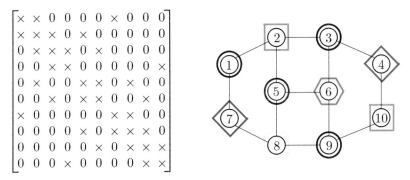

$$\begin{bmatrix} \times & \times & 0 & 0 & 0 & \times & 0 & 0 & 0 \\ \times & \times & \times & 0 & \times & 0 & 0 & 0 & 0 \\ 0 & \times & \times & \times & 0 & \times & 0 & 0 & 0 \\ 0 & 0 & \times & \times & 0 & 0 & 0 & 0 & \times \\ 0 & \times & 0 & 0 & \times & \times & 0 & \times & 0 & 0 \\ 0 & 0 & \times & 0 & \times & \times & 0 & 0 & \times & 0 \\ \times & 0 & 0 & 0 & 0 & 0 & \times & \times & 0 & 0 \\ 0 & 0 & 0 & 0 & \times & 0 & \times & \times & \times & 0 \\ 0 & 0 & 0 & 0 & 0 & \times & 0 & \times & \times & \times \\ 0 & 0 & 0 & \times & 0 & 0 & 0 & 0 & \times & \times \end{bmatrix}$$

Five colors and hence five Hessian-vector products are required to compute a compressed representation of the Hessian.

An indirect approach may again offer a lower value of p, i.e., a lower number of Hessian-vector products that have to be computed. Coleman and Cai [27] showed that the partitioning problem that has to be solved in the computation of the Hessian via an indirect model can be transformed into an *acyclic coloring* of the adjacency graph of the Hessian. Here, one has to compute a distance-1 coloring with the further restriction that every cycle in the graph uses at least three colors.

Example 13 *The sparsity pattern, the adjacency graph of the Hessian for the function given in Example 12 and its acyclic coloring are given below.*

$$
\begin{bmatrix}
\times & \times & 0 & 0 & 0 & 0 & \times & 0 & 0 & 0 \\
\times & \times & \times & 0 & \times & 0 & 0 & 0 & 0 & 0 \\
0 & \times & \times & \times & 0 & \times & 0 & 0 & 0 & 0 \\
0 & 0 & \times & \times & 0 & 0 & 0 & 0 & 0 & \times \\
0 & \times & 0 & 0 & \times & \times & 0 & \times & 0 & 0 \\
0 & 0 & \times & 0 & \times & \times & 0 & 0 & \times & 0 \\
\times & 0 & 0 & 0 & 0 & 0 & \times & \times & 0 & 0 \\
0 & 0 & 0 & 0 & \times & 0 & \times & \times & \times & 0 \\
0 & 0 & 0 & 0 & 0 & \times & 0 & \times & \times & \times \\
0 & 0 & 0 & \times & 0 & 0 & 0 & 0 & \times & \times
\end{bmatrix}
$$

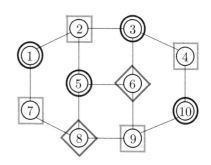

Three colors and hence only three Hessian-vector products are required to compute a compressed representation of the Hessian. However, one has to recover the entries of the sparse Hessian. Using several example problems, it was observed in [28] that the indirect method often performs better in terms of runtime than the direct method despite the need for recovery.

The direct and the indirect approaches have in common that one must employ heuristic algorithms since the computation of the minimal numbers of colors in known to be NP-hard [18, 27]. Efficient coloring algorithms have been developed [29] to provide a seed matrix such that a compressed representation of the corresponding Hessian can be evaluated. This paper contains also a detailed analysis and numerical results for different coloring techniques in the context of sparse Hessian computations. In contrast to the conditioning problem for sparse Jacobian, one can show for the indirect approach in the Hessian case that the recovery algorithms proposed in [29] are numerically stable [28].

In the case of vector-valued functions $F : \mathbb{R}^n \to \mathbb{R}^m$, the situation changes considerably. Here, the sparsity of the derivative tensor $\nabla^2 F \in \mathbb{R}^{m \times n \times n}$ is closely related to that of the Jacobian $\nabla F \in \mathbb{R}^{m \times n}$. Therefore, the seed and matrices S and W derived for the compression of ∇F can also be used for computing and reconstructing $\nabla^2 F$ as described in [2].

5.2.3   Open Problems

Several questions remain open for further studies. For the first step shown in Figure 5.1, alternative methods to detect the sparsity pattern are the subject of ongoing research. With respect to the bit vector approach, Bayesian probing [30] has been examined to some extend. Such statistical methods could and should be extended to become more established.

For the second step, the recovery of the entries of the Jacobian in an indirect approach still forms a major challenge. Alternative choices of seed matrices could provide considerable advantages. Furthermore, the determination of suitable combinations of the forward and reverse modes is still under examination.

For the graph coloring part the various graph coloring problems and their efficient solutions form also a wide field for further development.

5.3 Data Flow Reversal

The reverse mode of AD provides an efficient method to compute adjoint information. However, the memory needed by the reverse mode in its basic form is proportional to the operation count of the function evaluation. For real-world problems this fact may lead to an unacceptable memory requirement. Therefore, several checkpointing approaches have been developed.

5.3.1 Call Tree Reversals

The first checkpointing method is based on the call graph of the function evaluation. Nodes of the call graph represent actual calls of subroutines at run time, several of which may originate from the same source call, for example, within a loop. Hence, the call graph can be represented by a tree. To compute the adjoint of the complete function evaluation, there are usually numerous alternatives to reverse the forward execution of the call graph resulting in different spatial and temporal complexities.

Definition 9 *We distinguish four motions for each call of a subroutine F (each node of the call tree):*

- *The* advancing motion *of F corresponds to the usual forward evaluation mode and will be denoted by* $\boxed{\text{F}}$.

- *The* recording motion $\boxed{\text{F}}\rangle$ *of F stores in addition to the function evaluation intermediate values that are required for the adjoint computation.*

- *The* returning motion $\langle\boxed{\text{F}}$ *of F propagates adjoint values backward.*

- *The combination of* $\boxed{\text{F}}\rangle$ *and* $\langle\boxed{\text{F}}$ *into a single motion will be denoted by the symbol* $\langle\boxed{\text{F}}\rangle$.

Normally, the call $\boxed{\text{F}}$ will perform some internal calculations as well as calling some child subroutine. For preparing the adjoint calculation, each subroutine call is performed exactly once in recording motion. Here, one may use result- or argument-checkpointing to increase the overall efficiency [31]. In the case of $\langle\boxed{\text{F}}\rangle$ the storage of intermediates can be implemented locally. The four motions introduced so far provide already a wide variety of possibilities for the call tree reversal.

Example 14 *One simple example of a call tree is shown in the upper left corner of Figure. 5.2. The right upper corner of the same figure shows a complete* joint reversal *minimizing the overall memory requirement at the cost of an increased operation count assuming that the size of an argument checkpoint undercuts that of the data to be recorded significantly. The opposite strategy, i.e., a complete* split reversal, *is shown in the lower left corner. The lower right*

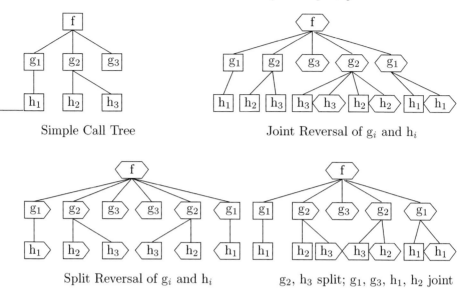

FIGURE 5.2: Different Call Tree Reversals.

corner illustrates a mixed strategy, where the overall memory requirement is again reduced at the cost of an increase in the operation count.

It is found that unless the call tree is quite deep, reversing all calls in the joint mode is not a bad idea. This strategy is frequently employed by source transformation tools. Overloading tools on the other hand reverse as default everything in split mode by saving all intermediate values during a global recording step.

An obvious task connected with a call tree reversal is the derivation of a strategy that does not exceed a given amount of memory but minimizes the overall operation count. It is shown in [31] that this combinatorial problem is NP-complete. The same is also true for the more general reversal of the computational graph of a function [32]. Therefore, the development of heuristics yielding efficient reversals of call trees and computational graphs forms an active research area, where so far only very preliminary results are available. Closely connected with this topic is the determination of suitable checkpoints that allow the restart of a function evaluation at a specific subroutine call. Appropriate data flow analyses have been proposed in [12].

5.3.2   Reversals of (Pseudo-) Timestepping Procedures

As soon as one can exploit additional information about the structure of the function evaluation, an appropriately adapted checkpointing strategy can be used. This is in particular the case if a time-stepping procedure is contained in the function evaluation allowing the usage of time-stepping oriented checkpointing. Additionally, this situation has the advantage that usually the data to be stored in one checkpoint is easily determined.

If the number of time steps is known a priori and the computational costs of the time steps are almost constant, one can compute checkpointing schedules in advance. Therefore, these checkpointing strategies are frequently called *offline checkpointing* approaches. One very popular strategy is to distribute the checkpoints equidistantly over the time interval. This so-called two-level checkpointing has been proposed several times in the literature, e.g., [33, 34], and is easy to implement. In the context of PDE-constraint optimization the equidistant checkpointing is frequently known as windowing, see, e.g., [35]. Naturally, one can apply the two-level checkpointing repeatedly for the groups of time steps that are separated by equidistant checkpoints. This approach is called *multi-level checkpointing* [7]. However, it was shown in [36] that the multi-level checkpointing is not optimal. A more advanced but still not optimal approach is the binary checkpointing used for example in [34]. In this case, the checkpoints are placed in a bisection fashion. Provably optimal checkpointing schedules can be computed in advance to achieve an optimal, i.e., minimal, run time increase for a given number of checkpoints [37, 38]. For this purpose, the special structure of the reversal process is exploited to apply dynamic programming for the determination of an optimal solution, which can then be directly derived.

Theorem 1 *Let l be the number of time steps and c the number of checkpoints. Then the minimal number of time step evaluations required for the adjoint computation equals*

$$t(c,l) = r \cdot l - \beta(c+1, r-1)$$

where r is the unique integer satisfying $\beta(c, r-1) < l \leq \beta(c,r) = \binom{c+r}{c}$.

Here, the checkpoints are placed according to certain binomial coefficients corresponding to the solution of the underlying dynamic programming problem. The scalar r is also called repetition rate because for the adjoint computation each time step is evaluated at most r times.

The binomial checkpointing strategy is integrated in the AD tool ADOL-C and used for example in [39] in the context of optimal flow control. Furthermore, the binomial checkpointing integrated in ADOL-C was used to show a very interesting effect that is worth noting: Due to the varying access speed to different memory levels it might happen that the adjoint computation with checkpointing is faster than the adjoint computation with the store-everything approach of the basic reverse mode despite the fact that several recomputations are required [40]. Recently, the binomial checkpointing approach has been extended to a multi-stage checkpointing. Here, varying access cost for different memory hierarchies are explicitly taken into account to allow the storage on disc or even the distribution on several computing nodes [41]. Hence, the offline checkpointing for more or less uniform costs of the time steps is considerably well understood.

The situation changes completely, if the computational costs of the time steps vary widely. The distribution of the step costs might be known a priori. Then one can employ a monotonicity property to derive a search algorithm for the optimal checkpointing schedule that is again based on dynamic programming and grows only quadratically with the number of time steps to be adjoined [42]. Nevertheless, often one does not know the exact distribution of the step costs a priori as assumed before. In this case, some heuristics were developed [43] which are based on combinatorial observations. However, several experiments with various applications have shown that often the exploitation of the non-uniformity was not worthwhile due to the considerably high computational cost of the heuristics. That is, the simple application of the binomial offline checkpointing yields frequently a similar or even better run time than the adapted checkpointing strategies for non-uniform step costs. With respect to this question, there might be room for further improvements.

A further complication occurs if the number of time steps to be performed is not known a priori, for example due to an adaptive time stepping scheme. This fact makes an offline checkpointing undecidable. Instead, one may still apply a straightforward checkpointing by placing a checkpoint each time when a certain number of time steps was executed. This transforms the uncertainty in the number of time steps to an uncertainty in the number of checkpoints needed. This approach is for example used by CVODES [44]. However, when the amount of memory per checkpoint is very high one certainly wants to determine the number of checkpoints required a priori. For that purpose, various online checkpointing strategies have been developed. First, a heuristic based on some simplifications of the adjoint computation was derived [45]. Recently, an algorithm yielding provably optimal online schedules was presented in [46] as long as the inequality $l \leq \beta(c, 3)$ holds for the number of checkpoints c and the unknown number of performed time steps l. In the same paper, it was shown that the algorithm yields optimal checkpointing schedules also for many l with $\beta(c, 3) < l \leq \beta(c, 4)$ or the required run time is almost optimal in that it equals $t(c, l) + 1$. For $r > 5$, checkpointing algorithms that attain provably the optimal repetition number r but not necessarily the minimal run time $t(c, l)$ are proposed in [47]. Hence, there are still open questions also for online checkpointing strategies.

On one hand, the checkpointing approaches discussed so far allow an enormous reduction of the memory required to compute the adjoint of a time-stepping procedure. On the other hand, one probably has to pay for this improvement in the form of an increase in computation time due to the recomputations. If any increase of the time needed to compute the adjoint is not acceptable, the usage of additional processors may yield a minimal temporal complexity. So far, most results for parallel reversal schedules are dedicated to the offline case, where the number of time steps is known in advance. Furthermore, we assume for simplicity from now on that the run time of one time step, the run time of one recording step and the run time of one adjoint time step is the same and normalized to one time unit. Then the minimal run time required for adjoining l time steps equals $2l$ as illustrated in the next examples. An obvious idea to achieve this minimal run time with additional processors is based on a simple bisection strategy.

Example 15 (Bisection strategy) *If $l = 8$ time steps are to be adjoined, the bisection strategy yields the parallel checkpointing schedule shown in Figure. 5.3. Here, 4 processors and 2 checkpoints are employed to perform the*

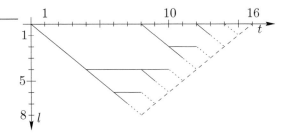

FIGURE 5.3: Parallel Reversal Schedule using Bisection Strategy.

reversal of F. The initial state is copied to the first checkpoint. Then a processor performs a forward sweep to the 4th state, i.e., the midpoint of 9 states. This state is copied to the second checkpoint. After that a forward sweep is performed to the 6th state, i.e., the midpoint of state 4 and 8. The 6th state is copied to the third checkpoint. After another time step the first recording step is performed. Subsequently the adjoint calculation starts. The number of processors needed to execute this checkpointing schedule is given by the maximal number of slanted lines crossing any vertical line. Therefore four processors are required by the parallel reversal schedule displayed in the last figure.

A slightly more general strategy, namely a uniform distribution of the checkpoint positions is proposed in [48]. Naturally, the question arises if the parallel reversal schedules determined by the bisection strategy or the recursive partition with fixed ratio as in [48] are optimal in terms of the number of processors and the number of checkpoints required. The answer is "No" because, for example, 3 processors and 2 checkpoints suffice to adjoin 8 time steps as shown in Figure 5.4. Here, a slightly different reversal technique was chosen, namely a checkpoint writing of the fifth and subsequently of the third state. Obviously, one would like to find an optimal parallel reversal schedule, i.e., a parallel reversal schedule that requires a minimal number of resources. As a first step, an upper bound $l(p, c)$ for the maximal number of time steps that can be reversed with the given number p of processors and c of checkpoints under the assumption $p \geq c$ was derived in [49]. Reversal schedules that attain this upper bound are constructed in [50]. Hence, it is shown that this upper bound is indeed sharp. Having $l(p, c)$ and any given number l of time

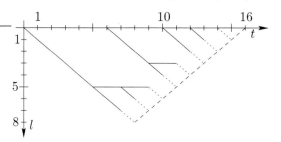

FIGURE 5.4: Improved Parallel Reversal Schedule.

steps to be adjoined one can immediately determine the minimal number of resources, i.e., the minimal number of processors and checkpoints, needed to adjoin the l time steps.

For the uniform step costs assumed here, the optimal parallel reversal schedules are determined by the Fibonacci numbers. Figure 5.5 illustrates an optimal schedule for $l = 21$ time steps including the development of the

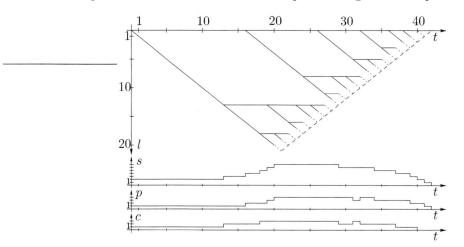

FIGURE 5.5: Parallel Reversal Schedule $l = 21$ Time Steps.

needed checkpoints c, the needed processors p and the sum s of both. There is a warm-up and a cool-down phase, where the number of processors and checkpoints required increases and decreases, respectively. Furthermore, the maximal number of processors and checkpoints, namely 7, is needed during the first recording step. This fact is true for all parallel schedules that reverse the maximal number of time steps with a given number of processors and checkpoints. Otherwise it would be possible to reverse at least one more time step. For more general cases of the run times, i.e., for non-uniform cases, the structure of the parallel reversal schedules inherits some of the properties of the uniform case but the construction gets more complicated.

There were some test implementations of the parallel reversal schedules [51]. However, currently an implementation is not available. One of the reasons might be that a satisfying theory and implementation for the distribution of the processors during the parallel reversal process is still missing. This forms one open question with respect to the parallel reversal schedules and involves a lot of combinatorial aspects. For this purpose, one has to note in addition that the communication between the processors is largely determined a priori. Moreover, assume that the given time integration is already evaluated in parallel with m processors. Let p be the number of processors that are needed by a suitable parallel reversal schedule to adjoin the l time steps. Then the number of processors required for the parallel reversal is given by the product $m \cdot p$. Further studies may be dedicated to the fact that some processors can be used to speed up the function evaluation during the warm up and cool down phase of the parallel reversal schedule instead of being idle. So far, nobody has analyzed this situation.

A further task for the future should be the development of suitable recovery strategies, if there are delays or break downs during the calculation of adjoints. For example, if there are more checkpoints than are needed by the particular parallel reversal schedule, then they can be used to manage possible delays of the processors. Additional topics for further research of the combinatorial aspects arise

With respect to online parallel reversal schedules, only one PhD thesis is available [52]. There is plenty of room for further research and development activities in this area.

5.4 Elimination Techniques

AD is based on the assumption that F decomposes into a *single assignment code* (SAC) at every point of interest as follows:

$$\text{for } j = n+1, \ldots, n+p+m$$
$$v_j = \varphi_j(v_i)_{i \prec j} \tag{5.1}$$

where $i \prec j$ denotes a direct dependence of v_j on v_i. The transitive closure of this relation is denoted by $\prec^+$. The result of each elemental function φ_j is assigned to a unique auxiliary variable v_j. The n *independent inputs* $x_i = v_i$, for $i = 1, \ldots, n$, are mapped onto m *dependent outputs* $y_j = v_{n+p+j}$, for $j = 1, \ldots, m$. Equation (5.1) involves the computation of the values of p *intermediate variables* v_k, for $k = n+1, \ldots, n+p$. The SAC induces a directed acyclic graph (DAG) $G = (V, E)$ with integer vertices $V = \{1, \ldots, n+p+m\}$

and edges $E = \{(i,j)|i \prec j\}$. The vertices are sorted topologically with respect to variable dependence, that is, $\forall i, j \in V : (i,j) \in E \Rightarrow i < j$.

The *elemental functions* φ_j are assumed to be continuously differentiable in a neighborhood of all arguments of interest. Hence the SAC can be augmented with the computation of the local partial derivatives

$$c_{j,i} = \frac{\partial \varphi_j}{\partial v_i}(v_k)_{k \prec j} \quad .$$

This process is referred to as *linearization*.

In the linearized DAG $G = (V, E)$ edges $E = \{((i,j), c_{j,i}), i, j \in V\}$ are labeled with the respective local partial derivatives and vertices $V = X \cup Z \cup Y$ are partitioned into minimal $X = \{1, \ldots, n\}$, intermediate $Z = \{n+1, \ldots, n+p\}$, and maximal $Y = \{n+p+1, \ldots, n+p+m\}$. In this section we introduce elimination methods to transform G into a subgraph G' of the complete bipartite graph $K_{n,m}$ such that the labels on the remaining edges represent exactly the nonzero entries of the Jacobian. Our objective is to minimize the computational cost of this transformation taken as the total number of floating-point operations. Generalizations to higher derivative tensors are discussed briefly in Section 5.4.5.

From the chain rule of differential calculus it follows that an entry of the Jacobian can be computed by multiplying the edge labels over all paths connecting $i \in X$ and $j \in Y$ followed by adding these products [53]. This fact can be expressed as

$$\frac{\partial v_j}{\partial v_i} = \sum_{[i \to j]} \prod_{(k,l) \in [i \to j]} c_{l,k} \quad , \tag{5.2}$$

where $[i \to j]$ denotes a path leading from i to j and $(k,l) \in E$ is an edge contained within $[i \to j]$. That is, ∇F can be accumulated by enumerating all paths connecting minimal (in X) with maximal (in Y) vertices in G.

Example 16 *Consider the DAG in Figure 5.8 (a) that will play a prominent role further below. According to Equation (5.2) the Jacobian can be computed as*

$$\begin{pmatrix} c_{5,4} \cdot c_{4,3} \cdot c_{3,1} & c_{5,4} \cdot c_{4,3} \cdot c_{3,2} \\ c_{6,4} \cdot c_{4,3} \cdot c_{3,1} & c_{6,4} \cdot c_{4,3} \cdot c_{3,2} \\ c_{7,4} \cdot c_{4,3} \cdot c_{3,1} & c_{7,4} \cdot c_{4,3} \cdot c_{3,2} \\ c_{8,4} \cdot c_{4,3} \cdot c_{3,1} + c_{8,3} \cdot c_{3,1} & c_{8,4} \cdot c_{4,3} \cdot c_{3,2} + c_{8,3} \cdot c_{3,2} \end{pmatrix} \quad .$$

The computational cost amounts to 18 multiplications and 2 additions. Obviously, this cost is far from optimal. Through exploitation of the algebraic laws of the field $(\mathbb{R}, +, \cdot)$ (associativity of multiplication and distributivity) the cost can be reduced to 11 multiplications and 1 addition as shown later in this section.

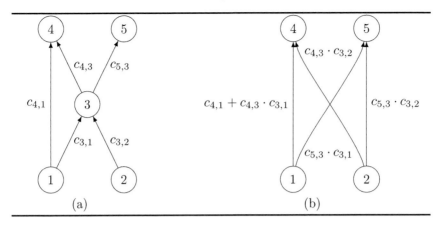

FIGURE 5.6: Vertex Elimination: Elimination of 3 in (a) yields (b) at the cost of four multiplications.

5.4.1 Vertex Elimination

The following vertex elimination rule is an immediate consequence of Equation (5.2) [54]:

A vertex $j \in V$ is eliminated by introducing *fill edges* $(i, k) \in E$ for all i with $(i, j) \in E$, k with $(j, k) \in E$, and $(i, k) \notin E$. The new edge labels are set to $c_{k,i} = c_{k,j} \cdot c_{j,i}$. For all $(i, k) \in E$ the existing edge labels are updated according to $c_{k,i} = c_{k,i} + c_{k,j} \cdot c_{j,i}$. Vertex j is removed from G together with all incident edges.

Example 17 *Figure 5.6 (b) shows the result of eliminating vertex 3 in the DAG shown in Figure 5.6 (a). Three fill edges* $(1, 5)$, $(2, 4)$, *and* $(2, 5)$ *are generated. The label of the existing edge* $(1, 4)$ *is incremented.*

The VERTEX ELIMINATION (VE) problem is to determine a vertex elimination order that minimizes the number of scalar multiplications required to accumulate the whole Jacobian.[1] The elimination cost of a single vertex i is equal to its *Markowitz degree* defined as the product $|P_i| \cdot |S_i|$, where P_i and S_i denote the sets of predecessors and successors of i, respectively.

The computational complexity of VE is unknown. Work on heuristics is based on the assumption that finding an efficient deterministic algorithm for its solution is unlikely. A greedy heuristic has been proposed in [54]. Vertices are eliminated in increasing order with respect to their Markowitz degrees. This approach turns out to perform well in many situations.

---

[1] Additions appear always in the context of fused multiply-add operations whose number is equal to that of the multiplications.

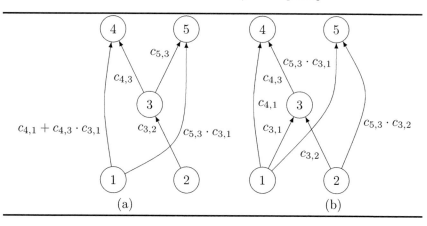

FIGURE 5.7: Edge Elimination: Front-elimination of $(1,3)$ in (a) and back-elimination of $(3,5)$ in (b).

5.4.2 Edge Elimination

Similar to vertex elimination rules for eliminating edges can be derived from Equation (5.2) [55]:

The *back-elimination* of $(i,j) \in E$ is performed by introducing new edges $(i',j) \in E$ for all i' with $(i',i) \in E$ and $(i',j) \notin E$. The new edge labels are set to $c_{j,i'} = c_{j,i} \cdot c_{i,i'}$. For all $(i',j) \in E$ the existing edge labels are updated according to $c_{j,i'} = c_{j,i'} + c_{j,i} \cdot c_{i,i'}$.

An edge $(i,j) \in E$ is *front-eliminated* by introducing new edges $(i,j') \in E$ for all j' with $(j,j') \in E$ and $(i,j') \notin E$. The new edge labels are set to $c_{j',i} = c_{j',j} \cdot c_{j,i}$. If $(i,j') \in E$, then the existing edge labels are updated according to $c_{j',i} = c_{j',i} + c_{j',j} \cdot c_{j,i}$.

In both cases (i,j) is deleted. If this deletion leads to either i or j becoming isolated (vanishing in- or outdegree), then the respective vertex is also removed from G together with all incident edges.

Example 18 *Figure 5.7 (a) shows the result of front-eliminating $(1,3)$ in the DAG shown in Figure 5.6 (a). A new fill edge $(1,5)$ is generated. The label of the existing edge $(1,4)$ is incremented.*

The back-elimination of $(3,5)$ in the DAG shown in Figure 5.6 (a) is illustrated in Figure 5.7 (b).

Similar to the VE problem the EDGE ELIMINATION (EE) problem is to determine an edge elimination order that minimizes the number of scalar multiplications required to accumulate the whole Jacobian. The cost of a front-elimination of (i,j) is equal to $|S_i|$ whereas that of a back-elimination is $|P_i|$. The computational complexity of EE is unknown.

An example graph for which edge elimination yields a lower computational cost than the optimal vertex elimination sequence is the *Lion* graph shown

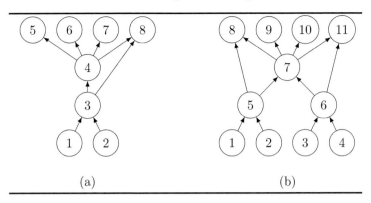

(a)   (b)

FIGURE 5.8: Linearized *Lion* (a) and *Bat* (b) Graphs.

in Figure 5.8 (a). Elimination of 3 followed by 4 results in 12 multiplications and 2 additions. The reverse vertex elimination sequence saves one addition. Back-elimination of $(4, 8)$ succeeded by a forward vertex elimination sequence accumulates the whole Jacobian at the cost of 11 multiplications and 1 addition.

5.4.3 Face Elimination

There are problems for which the optimal edge elimination sequence does not yield the minimal computational cost. An example is displayed in Figure 5.8 (b). The first and the second two columns of the (4×4)-Jacobian represented by the *Bat* graph can be computed separately at the cost of 11 multiplications and 1 addition each (see Lion graph) yielding an overall cost of 22 multiplications and 2 additions. A technical argument showing that this cost cannot be obtained by edge elimination is presented in [56]. Face elimination on the dual computational graph is proposed to overcome this discrepancy between a solution of EE and the optimum. The FACE ELIMINATION (FE) problem is formulated correspondingly. A detailed discussion of face elimination is beyond the scope of this article. So far, practical test results have failed to illustrate the superiority of face over vertex and edge eliminations on real-world application codes.

5.4.4 Computational Complexity

In this section, we discuss the main idea behind the proof for the intractability of the problem that aims to minimize the computational cost of derivative accumulation. The formal proof can be found in [57]. A polynomial reduction from ENSEMBLE COMPUTATION [19] is constructed. ENSEMBLE COMPUTATION is defined as follows:

Given a collection $C = \{C_\nu \subseteq A : \nu = 1, \ldots, |C|\}$ of subsets $C_\nu = \{c_i^\nu :$

$i = 1, \ldots, |C_\nu|\}$ of a finite set A and a positive integer Ω is there a sequence $u_i = s_i \cup t_i$ for $i = 1, \ldots, \omega$ of $\omega \le \Omega$ union operations, where each s_i and t_i is either $\{a\}$ for some $a \in A$ or u_j for some $j < i$, such that s_i and t_i are disjoint for $i = 1, \ldots, \omega$ and such that for every subset $C_\nu \in C$, $\nu = 1, \ldots, |C|$, there is some u_i, $1 \le i \le \omega$, that is identical to C_ν?

Example 19 *Let an instance of* ENSEMBLE COMPUTATION *be given by*

$$A = \{a_1, a_2, a_3, a_4\}$$
$$C = \{\{a_1, a_2\}, \{a_2, a_3, a_4\}, \{a_1, a_3, a_4\}\}$$

and $\Omega = 4$. *The answer to the decision problem is positive with a corresponding instance given by*

$$u_1 = \{a_1\} \cup \{a_2\}, \quad u_2 = \{a_3\} \cup \{a_4\}, \quad u_3 = \{a_2\} \cup u_2, \quad u_4 = \{a_1\} \cup u_2$$

where $C_1 = u_1$, $C_2 = u_3$, *and* $C_3 = u_4$.

We refer to the problem of minimizing the number of scalar multiplications and additions performed by a Jacobian accumulation code as the OPTIMAL JACOBIAN ACCUMULATION (OJA) problem. The corresponding decision version is defined as follows:

Given a linearized DAG G of a function F and a positive integer Ω is there a sequence of scalar assignments $u_k = s_k \circ t_k$, $\circ \in \{+, *\}$, $k = 1, \ldots, \omega$, where each s_k and t_k is either $c_{j,i}$ for some $(i, j) \in E$ or $u_{k'}$ for some $k' < k$ such that $\omega \le \Omega$ and for every Jacobian entry there is some identical u_k, $k \le \omega$?

Theorem 2 *OJA is NP-hard.*

The basic idea of the proof in [57] is to substitute $*$ for $\cup$ in the definition of ENSEMBLE COMPUTATION. The resulting DAGs are structurally extremely simple as the corresponding Jacobians have nonzero entries in the diagonal exclusively.

Example 20 *The instance of* OPTIMAL JACOBIAN ACCUMULATION *resulting from the reduction applied to Example 19 is a vector function* $F : \mathbb{R}^{3+4} \to \mathbb{R}^3$ *defined by the following system of equations:*

$$y_1 = x_1 * a_1 * a_2, \quad y_2 = x_2 * a_2 * a_3 * a_4, \quad y_3 = x_3 * a_1 * a_3 * a_4 \quad .$$

Its linearized DAG simply consists of three disjoint paths of (edge-)lengths two, three, and three, respectively. The Jacobian accumulation code according to Equation (5.2) is

$$f_{1,1} = a_1 * a_2, \quad f_{2,2} = a_2 * a_3 * a_4, \quad f_{3,3} = a_1 * a_3 * a_4 \quad .$$

A Jacobian accumulation code that solves the OPTIMAL JACOBIAN ACCUMULATION *problem with* $\Omega = 4$ *is given as*

$$f_{1,1} = a_1 * a_2, \quad t = a_3 * a_4, \quad f_{2,2} = a_2 * t, \quad f_{3,3} = a_1 * t \quad .$$

The key element of the proof is the presence of algebraic dependences between the local partial derivatives (equality in particular). Thus, Theorem 2 can be generalized for gradients as well as higher derivatives. The structural problem that assumes all labels to be unrelated remains open. In particular, the intractability of OJA does not imply that of VE, EE, or FE.

5.4.5 Open Problems

A number of relevant open questions need to be addressed by future research. Most problems are combinations of compiler / program analysis problems, issues regarding the stability of the underlying numerical scheme, and combinatorial questions. Here we focus on the combinatorial aspects only. However, it must be noted that a purely theoretical solution is only the first step toward usability in AD tools.

Is $O(\min(n, m))$ a sharp bound for the cost of computing a general Jacobian ∇F relative to the function $F : \mathbb{R}^n \to \mathbb{R}^m$? It remains unclear whether the known elimination techniques yield a lower computational complexity of Jacobian accumulation than classical forward or reverse modes. For scalar functions ($m = 1$) the cost of reverse vertex elimination can be undercut by a factor of less than two, which is a generalization of the corresponding statement made in [58] for *single-expression-use graphs*.

What is the worst case ratio of optimal vertex elimination to optimal edge elimination? (... optimal edge elimination to optimal face elimination?) We conjecture that the maximal discrepancy between the numbers of multiplications required by an optimal vertex elimination sequence and that of an optimal edge elimination sequence is bounded from above by two. For DAGs containing only two intermediate vertices this factor is $\frac{1}{\sqrt{2}(2-\sqrt{2})}$ [55]. Our conjecture extends to a small constant discrepancy (two?) between solutions of VE and FE.

By how much can the number of edges in a DAG be reduced using vertex and edge elimination only? This MINIMAL EDGE NUMBER (MEN) problem is motivated by the fact that, for example, in vector forward mode AD the cost of computing $\dot{Y} = \nabla F \cdot \dot{X}$, where $\dot{X} \in \mathbb{R}^{n \times l}$ is equal to $l \cdot |E|$ and $|E|$ is the number of edges in the linearized DAG of $\mathbf{y} = F(\mathbf{x})$. Preaccumulation of ∇F (reduction or, preferably, minimization of $|E|$) may lead to significant savings [56]. The exploitation of such situations for $l \gg n$ is referred to as *interface contraction* [59]. An example, illustrating the superiority of edge over vertex elimination in the context of MEN is discussed in Figure 5.9. The more general SCARSITY problem extends the elimination techniques with *rerouting* and *normalization* to potentially decrease the number of edges further [60]. So far, the superiority of these techniques over plain edge elimination could not be demonstrated on real-world problems [61].

How to adjust the cost metric of elimination techniques to account for cache / memory behavior? The impact of the memory hierarchy on the efficiency of numerical codes outweighs that of the number of floating-point operations

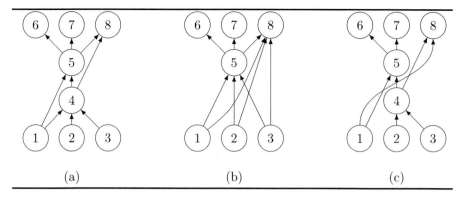

FIGURE 5.9: Superiority of edge over vertex elimination for MEN problem: The DAG in (a) contains 9 edges. There are two distinct vertex elimination sequences both of which keep the number of edges unchanged. For example, the DAG resulting from the elimination of vertex 4 has 9 edges as shown in (b). So does the final bipartite DAG. Front-elimination of $(1, 4)$ followed by back-elimination of $(5, 8)$ results in the DAG in (c) having only 8 edges.

on virtually all modern computer architectures. This fact is supported by some preliminary studies [62, 63]. It remains open whether heuristics can be developed whose applicability is not limited to a given architecture. Previous work on heuristics aiming to reduce the operations count includes [58, 64, 65].

Is it necessary to preserve symmetry in Hessian accumulation? Various ideas are being discussed regarding the accumulation of Hessian matrices using the known elimination techniques and extensions thereof, such as *vertical* $(c_{j,i} = c_{j,i}/2 + c_{j,i}/2)$ or *horizontal splitting* $(c_{j,i} = \sqrt{c_{j,i}} \cdot \sqrt{c_{j,i}})$ of edges (i, j) crossing the line of symmetry in the *Hessian DAG* [2, 66]. See Figure 5.10 for illustration. We are not aware of any heuristic that aims to exploit symmetry in the Hessian.

How to exploit identical local partial derivatives? How to exploit algebraic dependences among them? Both questions are related whereas the former represents a special case of the latter. The NP-hardness proof in [57] exploits the potential equality of local partial derivatives. Ongoing research aims to detect and exploit such knowledge for the design of appropriate heuristics. The second question is basically about the extend to which computer algebra should be included into the derivative accumulation algorithms. They would certainly not lead to a simplification of the fundamental combinatorial problems.

How to handle constant (unit) partial derivatives? Constant partial derivatives originate from linear operations in the given implementation of F. Additions and subtractions yield unit edge labels in the corresponding DAG. Their elimination at compile-time (see also [67]) appears to be desirable. The pros and cons of this approach have been discussed in [68]. It remains unclear

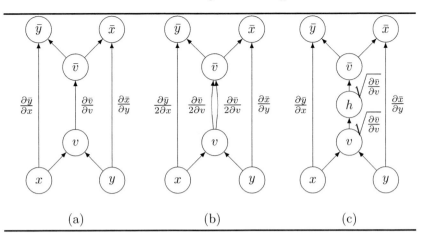

FIGURE 5.10: Should Hessian DAGs be kept symmetric? If so, then a front-[back-]elimination below the line of symmetry (los) is matched by a back-[front-]elimination above the los. The Hessian DAG in (a) represents, for example, the function $z = \sin(x \cdot y)$. It contains one eliminatable edge that crosses the los: $(v, \bar{v})$. Symmetry can be maintained by vertical splitting as shown in (b). The front-[back-]elimination of one of the resulting two parallel edges is matched by a back-[front-]elimination of the other. Newly generated edges that cross the line of symmetry are treated similarly. Alternatively, a horizontal split yields the Hessian DAG shown in (c). Note that unrestricted edge elimination combined with horizontal edge splitting is not guaranteed to terminate. For example, the front-elimination of (v, h) in (c) would be followed by a horizontal split of the resulting edge $(v, \bar{v})$ potentially yielding an infinite loop.

how to handle constant edge labels optimally with respect to the known cost metrics.

Does the "No Free Re-fill" conjecture hold? The weak version of this conjecture is about the existence of an optimal edge elimination sequence that does not generate re-fill, that is, once an edge has been eliminated it is not re-generated as fill-in by some subsequent edge elimination. A stronger formulation proposes that an elimination sequence that generates re-fill cannot be optimal. Successfully proving either version of the conjecture will have a considerable impact on the design of new heuristics for the EE problem.

What is the computational complexity of VE, EE, and FE for absorption-free DAGs? In *absorption-free DAGs* any two vertices are connected by at most one path. The resulting Jacobian accumulation codes do not perform any additions, just multiplications. Absorption-free DAGs represent an interesting special case that appears to be a promising candidate for proving the hardness of VE, EE, and FE.

5.5   Summary and Conclusion

AD has been introduced in Section 5.1 as a technique to transform numerical codes semantically such that projections of derivative tensors of arbitrary order can be computed with machine accuracy. The use of the corresponding derivative codes was investigated in the context of exploiting sparsity of the derivative tensors in Section 5.2. Trade-offs between the arithmetic complexity of adjoint codes and their memory consumption were discussed in Section 5.3. The focus of the last section of this chapter was on techniques to accumulate the whole derivative tensors with the least possible number of arithmetic operations.

The major subdomains of AD featuring extensive combinatorial problems have been covered. Deeper insight into to respective areas is provided in the cited literature. Obviously there are still plenty of unsolved (combinatorial) problems in AD. With this survey we hope to seed ideas for further research and development projects in and related to AD.

Bibliography

[1] Griewank, A., Utke, J., and Walther, A., "Evaluating Higher Derivative Tensors by Forward Propagation of Univariate Taylor Series," *Math. Comp.*, Vol. 69, 2000, pp. 1117–1130.

[2] Griewank, A. and Walther, A., *Evaluating Derivatives. Principles and Techniques of Algorithmic Differentiation (Second Edition)*, SIAM, 2008.

[3] Hascoët, L., Naumann, U., and Pascual, V., "To-be-recorded Analysis in Reverse Mode Automatic Differentiation," *Future Gen. Comp. Sys.*, Vol. 21, 2005, pp. 1401–1417.

[4] Bendtsen, C. and Stauning, O., "FADBAD, a Flexible C++ Package for Automatic Differentiation," Technical Report IMM–REP–1996–17, Department of Mathematical Modelling, Technical University of Denmark, 1996.

[5] Griewank, A., Juedes, D., and Utke, J., "Algorithm 755: ADOL-C: A Package for the Automatic Differentiation of Algorithms Written in C/C++," *ACM Trans. Math. Softw.*, Vol. 22, No. 2, 1996, pp. 131–167.

[6] Pryce, J. and Reid, J., "ADO1, a Fortran 90 code for Automatic Differentiation," Tech. Rep. RAL-TR-1998-057, Rutherford Appleton Laboratory, Chilton, Didcot, Oxfordshire, England, 1998.

[7] Giering, R. and Kaminski, T., "Recipes for Adjoint Code Construction," *ACM Trans. Math. Softw.*, Vol. 24, No. 4, 1998, pp. 437–474.

[8] Hascoët, L. and Pascual, V., "TAPENADE 2.1 User's Guide," Rapport technique 300, INRIA, Sophia Antipolis, 2004.

[9] Naumann, U. and Riehme, J., "A Differentiation-Enabled Fortran 95 Compiler," *ACM Trans. Math. Softw.*, Vol. 31, No. 4, 2005.

[10] Utke, J., Naumann, U., Wunsch, C., Hill, C., Heimbach, P., Fagan, M., Tallent, N., and Strout, M., "OpenAD/F: A modular, open-source tool for automatic differentiation of Fortran codes," *ACM Trans. Math. Softw.*, Vol. 34, No. 4, 2008.

[11] Bischof, C., Roh, L., and Mauer, A., "ADIC — An Extensible Automatic Differentiation Tool for ANSI-C," *Software–Practice and Experience*, Vol. 27, No. 12, 1997, pp. 1427–1456.

[12] Hascoët, L. and Araya-Polo, M., "The Adjoint Data-Flow Analyses: Formalization, Properties, and Applications," *[69]*, 2005, pp. 135–146.

[13] Bischof, C., Khademi, P., Bouaricha, A., and Carle, A., "Efficient Computation of Gradients and Jacobians by Transparent Exploitation of Sparsity in Automatic Differentiation," *Optimization Methods and Software*, Vol. 7, 1996, pp. 1–39.

[14] Bartholomew-Biggs, M., Bartholomew-Biggs, L., and Christianson, B., "Optimization and automatic differentiation in ADA," *Optimization Methods and Software*, Vol. 4, 1994, pp. 47–73.

[15] Tadjouddine, M., Faure, C., and Eyssette, F., "Sparse Jacobian Computation in Automatic Differentiation by Static Program Analysis," *Lecture Notes in Computer Science*, Vol. 1503, 1998, pp. 311–326.

[16] Walther, A., "Computing Sparse Hessians with Automatic Differentiation," *ACM Trans. Math. Softw.*, Vol. 34, No. 1, 2008, Article No 3.

[17] Curtis, A., Powell, M., and Reid, J., "On the Estimation of Sparse Jacobian Matrices," *J. Inst. Math. Appl.*, Vol. 13, 1974, pp. 117–119.

[18] Coleman, T. and Moré, J., "Estimation of Sparse Jacobian Matrices and Graph Coloring Problems," *SIAM J. Numer. Anal.*, Vol. 20, No. 1, 1983, pp. 187–209.

[19] Garey, M. and Johnson, D., *Computers and Intractability - A Guide to the Theory of NP-completeness*, W. H. Freeman and Company, 1979.

[20] Gebremedhin, A., Manne, F., and Pothen, A., "What Color is Your Jacobian? Graph Coloring for Computing Derivatives." *SIAM Rev.*, Vol. 47, No. 4, 2005, pp. 629–705.

[21] Newsam, G. and Ramsdell, J., "Estimation of sparse Jacobian matrices." *SIAM J. Algebraic Discrete Methods*, Vol. 4, 1983, pp. 404–417.

[22] Reichel, L. and Opfer, G., "Chebyshev-Vandermonde systems." *Math. Comput.*, Vol. 57, No. 196, 1991, pp. 703–721.

[23] Golub, G. and Van Loan, C., *Matrix computations. 3rd ed.*, The Johns Hopkins Univ. Press, 2007.

[24] Björck, A. and Pereyra, V., "Solution of Vandermonde systems of equations," *Math. Comput.*, Vol. 24, 1971, pp. 893–903.

[25] Hossain, S. and Steihaug, T., "Computing Sparse Jacobian Matrices Optimally," pp. 77–87.

[26] Coleman, T. and Moré, J., "Estimation of sparse Hessian matrices and Graph coloring problems," *Math. Program.*, Vol. 28, 1984, pp. 243–270.

[27] Coleman, T. and Cai, J., "The cyclic coloring problem and estimation of sparse Hessian matrices," *SIAM J. Alg. Disc. Meth.*, Vol. 7, No. 2, 1986, pp. 221–235.

[28] Gebremedhin, A., Pothen, A., Tarafdar, A., and Walther, A., "Efficient Computation of Sparse Hessians Using Coloring and Automatic Differentiation," *INFORMS Journal on Computing*, Vol. 21, No. 2, 2009, pp. 209–223.

[29] Gebremedhin, A., Tarafdar, A., Manne, F., and Pothen, A., "New Acyclic and Star Coloring Algorithms With Application to Computing Hessians," *SIAM J. Sci. Comput.*, Vol. 29, 2007, pp. 1042–1072.

[30] Griewank, A. and Mitev, C., "Detecting Jacobian sparsity patterns by Bayesian probing," *Math. Program. Ser. A*, Vol. 93, 2002, pp. 1–25.

[31] Naumann, U., "Call Tree Reversal is NP-complete," *[70]*, 2008, pp. 13–22.

[32] Naumann, U., "DAG Reversal is NP-complete," *Journal of Discrete Algorithms*, No. 7, 2009, pp. 402–410.

[33] Charpentier, I., "Checkpointing Schemes or Adjoint Codes: Application to the Meteorological Model Meso-NH," *SIAM J. Sci. Comput.*, Vol. 22, No. 6, 2001, pp. 2135–2151.

[34] Kubota, K., "A Fortran77 Preprocessor for Reverse Mode Automatic Differentiation with Recursive Checkpointing," *Opti. Meth. Softw.*, Vol. 10, No. 2, 1998, pp. 315–335.

[35] Berggren, M., "Numerical solution of a flow-control problem: Vorticity reduction by dynamic boundary action." *SIAM J. Sci. Comput.*, Vol. 19, No. 3, 1998, pp. 829–860.

[36] Walther, A. and Griewank, A., "Advantages of binomial checkpointing for memory-reduced adjoint calculations," *Proceedings of ENUMATH conference 2003*, edited by M. Feistauer, V. Dolejsi, P. Knobloch, and K. Najzar, Springer, Berlin, Heidelberg, New York, 2004, pp. 834–843.

[37] Griewank, A., "Achieving Logarithmic Growth of Temporal and Spatial Complexity in Reverse Automatic Differentiation," *Opti. Meth. Softw.*, Vol. 1, 1992, pp. 35–54.

[38] Griewank, A. and Walther, A., "Algorithm 799: Revolve: An Implementation of Checkpoint for the Reverse or Adjoint Mode of Computational Differentiation," *ACM Trans. Math. Softw.*, Vol. 26, No. 1, 2000, pp. 19–45.

[39] Riehme, J., Walther, A., Stiller, J., and Naumann, U., "Adjoints for Time-Dependent Optimal Control," *[70]*, 2008, pp. 175–185.

[40] Kowarz, A. and Walther, A., "Optimal Checkpointing for Time-Stepping Procedures in ADOL-C," *Computational Science – ICCS 2006*, edited by V. N. Alexandrov, G. D. van Albada, P. M. A. Sloot, and J. Dongarra, Vol. 3994 of *Lecture Notes in Computer Science*, Springer, 2006, pp. 541–549.

[41] Stumm, P. and Walther, A., "Multi-stage Approaches for Optimal Offline Checkpointing," *To appear in SIAM J. Sci. Comput.*, 2008.

[42] Walther, A., "Program reversals for evolutions with non-uniform step costs," *Acta Inf.*, Vol. 40, No. 4, 2004, pp. 235–263.

[43] Sternberg, J., "Adaptive Umkehrschemata für Schrittfolgen mit nicht-uniformen Kosten," Diploma thesis, Institute of Scientific Computing, Technical University Dresden, 2002.

[44] Serban, R. and Hindmarsh, A., "CVODES: An ODE Solver with Sensitivity Analysis Capabilities," UCRL-JP-20039, LLNL, 2003.

[45] Hinze, M. and Sternberg, J., "A-Revolve: An adaptive memory and run-time-reduced procedure for calculating adjoints; with an application to the instationary Navier-Stokes system," *Opti. Meth. Softw.*, Vol. 20, 2005, pp. 645–663.

[46] Stumm, P. and Walther, A., "New Algorithms for Optimal Online Checkpointing," *SIAM Journal on Scientific Computing*, Vol. 32, No. 2, 2010, pp. 836–854.

[47] Wang, Q., Moin, P., and Iaccarino, G., "Minimal Repetition Dynamic Checkpointing Algorithm for Unsteady Adjoint Calculation," *SIAM Journal on Scientific Computing*, Vol. 31, No. 4, 2009, pp. 2549–2567.

[48] Benary, J., "Parallelism in the reverse mode," *[71]*, 1996, pp. 137 – 147.

[49] Walther, A., "Bounding the number of processors and checkpoints needed in time-minimal parallel reversal schedules." *Computing*, Vol. 73, No. 2, 2004, pp. 135–154.

[50] Walther, A., *Program Reversal Schedules for Single- and Multi-processor Machines*, Ph.D. thesis, Institute of Scientific Computing, Technical University Dresden, Germany, 1999.

[51] Lehmann, U. and Walther, A., "The Implementation and Testing of Time-minimal and Resource-optimal Parallel Reversal Schedules," *Proceedings of ICCS conference 2002*, edited by P. M. A. Sloot, C. J. K. Tan, J. J. Dongarra, and A. G. Hoekstra, Vol. 2330 of *Lecture Notes in Computer Science*, Springer, 2002, pp. 1049–1058.

[52] Lehmann, U., *Schedules for Dynamic Bidirectional Simulations on Parallel Computers*, Ph.D. thesis, Center of High Performance Computing, Technical University Dresden, Germany, 2003.

[53] Baur, W. and Strassen, V., "The Complexity of Partial Derivatives," *Theo. Comp. Sci.*, Vol. 22, 1983, pp. 317–330.

[54] Griewank, A. and Reese, S., "On the Calculation of Jacobian Matrices by the Markovitz Rule," *[72]*, 1991, pp. 126–135.

[55] Naumann, U., *Efficient Calculation of Jacobian Matrices by Optimized Application of the Chain Rule to Computational Graphs*, Ph.D. thesis, Technical University Dresden, Feb. 1999.

[56] Naumann, U., "Optimal accumulation of Jacobian matrices by elimination methods on the dual computational graph," *Math. Prog.*, Vol. 99, 2004, pp. 399–421.

[57] Naumann, U., "Optimal Jacobian accumulation is NP-complete," *Math. Prog.*, Vol. 112, 2006, pp. 427–441.

[58] Naumann, U. and Hu, Y., "Optimal Vertex Elimination in Single-Expression-Use Graphs," *ACM Trans. Math. Softw.*, Vol. 35, No. 1, 2008.

[59] Bischof, C. and Haghighat, M., "Hierarchical Approaches to Automatic Differentiation," *[71]*, 1996, pp. 83–94.

[60] Griewank, A. and Vogel, O., "Analysis and Exploitation of Jacobian Scarcity," *Modelling, Simulation and Optimization of Complex Processes*, edited by H. Bock, E. Kostina, H. Phu, and R. Rannacher, Springer, Berlin, Heidelberg, New York, 2004, pp. 149–164.

[61] Lyons, A. and Utke, J., "On the Practical Exploitation of Scarsity," *[70]*, 2008, pp. 103–114.

[62] Forth, S., Tadjouddine, M., Pryce, J., and Reid, J., "Jacobian Code Generated by Source Transformation and Vertex Elimination can be as Efficient as Hand-Coding," *ACM Trans. Math. Softw.*, Vol. 30, No. 3, 2004, pp. 266–299.

[63] Tadjouddine, M., Forth, S., Pryce, J., and Reid, J., "Performance Issues for Vertex Elimination Methods in Computing Jacobians using Automatic Differentiation," *Proceedings of ICCS conference 2002*, edited by P. Sloot, C. Tan, J. Dongarra, and A. Hoekstra, Vol. 2330 of *Lecture Notes in Computer Science*, Springer, Berlin, Heidelberg, New York, 2002, pp. 1077–1086.

[64] Griewank, A. and Naumann, U., "Accumulating Jacobians as Chained Sparse Matrix Products," *Math. Prog.*, Vol. 95, No. 3, 2003, pp. 555–571.

[65] Naumann, U., "Cheaper Jacobians by Simulated Annealing," *SIAM J. Opt.*, Vol. 13, No. 3, 2002, pp. 660–674.

[66] Bhowmick, S. and Hovland, P., "A Polynomial-Time Algorithm for Detecting Directed Axial Symmetry in Hessian Computational Graphs," *[70]*, 2008, pp. 91–102.

[67] Utke, J., "Flattening Basic Blocks," *[69]*, 2005, pp. 121–133.

[68] Naumann, U. and Utke, J., "Optimality-Preserving Elimination of Linearities in Jacobian Accumulation," *Elec. Trans. Num. Anal. (ETNA)*, Vol. 21, 2005, pp. 134–150.

[69] Bücker, M., Corliss, G., Hovland, P., Naumann, U., and Norris, B., editors, *Automatic Differentiation: Applications, Theory, and Tools*, Vol. 50 of *Lecture Notes in Computational Science and Engineering*. Springer, Berlin, Heidelberg, New York, 2005.

[70] Bischof, C., Bücker, M., Hovland, P., Naumann, U., and Utke, J., editors, *Advances in Automatic Differentiation*, Vol. 64 of *Lecture Notes in Computational Science and Engineering*. Springer, 2008.

[71] Berz, M., Bischof, C., Corliss, G., and Griewank, A., editors, *Computational Differentiation: Techniques, Applications, and Tools*, Proceedings Series. SIAM, 1996.

[72] Corliss, G. and Griewank, A., editors, *Automatic Differentiation: Theory, Implementation, and Application*, Proceedings Series. SIAM, 1991.

Chapter 6

Combinatorial Problems in OpenAD

Jean Utke

Argonne National Laboratory

Uwe Naumann

RWTH Aachen University

6.1   Introduction

Computing derivatives of numerical models $F(\mathbf{x}) \mapsto \mathbf{y} : \mathbb{R}^n \mapsto \mathbb{R}^m$ given as a computer program P is an important but also compute-intensive task. Automatic or algorithmic[1]differentiation (AD) [1] provides the means to obtain such derivatives. OpenAD [2] implements AD as a source transformation applied to Fortran programs for both the forward and reverse modes. In this paper we describe the solutions to three combinatorial aspects of AD as implemented in OpenAD. For information regarding the general use of OpenAD please see [3]. Because OpenAD is a source transformation tool, the combinatorial problems are solved at compile time; hence, our approach **can afford costly heuristics to approximate a good solution**. Using such heuristics even with small improvements is justified when, for instance, these improvements benefit the numerical kernel of a loop and therefore the improvement accumulates over the runtime, realizing considerable savings. Our intention is

---

[1]Inertia preserves the term automatic although algorithmic may be more appropriate.

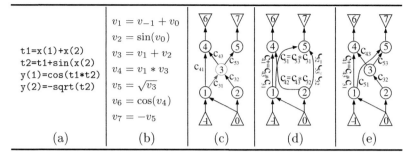

	$v_1 = v_{-1} + v_0$				
	$v_2 = \sin(v_0)$				
t1=x(1)+x(2)	$v_3 = v_1 + v_2$				
t2=t1+sin(x(2))	$v_4 = v_1 * v_3$				
y(1)=cos(t1*t2)	$v_5 = \sqrt{v_3}$				
y(2)=-sqrt(t2)	$v_6 = \cos(v_4)$				
	$v_7 = -v_5$				
(a)	(b)	(c)	(d)	(e)

FIGURE 6.1: Sequence of statements (a), corresponding sequence of elemental operations enumerating the computed values (b), the corresponding G using the value enumeration (c), elimination of vertex 3 from G (d), and front elimination of edge $(1,3)$ from G (e).

to explain the representation of the combinatorial problems in OpenAD such that the reader will find it easy to implement and investigate other solutions to these problems. The remainder of this section provides a brief overview of the concepts of AD that are relevant to the problems in question. General information on AD can be found in recent conference proceedings [4, 5], the AD community website [6], and the survey in Chapter 5.

Algorithmic differentiation relies on the chain rule applied to a sequence of elemental operations. These operations are the function calls and intrinsics built into the given programming language, such as sin, cos, or the operators + and *. Typically, right-hand-side expressions of assignment statements form single-expression-use graphs [7], and under certain conditions a sequence of single-expression-use graphs corresponding to a sequence of assignment statements in a given program can be *flattened* into a directed acyclic graph [8]. This graph is called the *computational graph* G. An example is shown in Figure 6.1. The nonminimal vertices of G represent values computed as a result of a single intrinsic function or operator call. For each such computed value v_i we can provide the local partial derivative $c_{ij} = \partial v_i / \partial v_j$ with respect to its arguments v_j. For instance, in Figure 6.1(b) we have $v_4 = v_1 * v_3$ and therefore $c_{41} = v_3$ and $c_{43} = v_1$. These partial derivatives are given as edge labels in G; see Figure 6.1(c). Following Baur's formula [9], we can compute Jacobian entries by multiplying edge labels along paths from minimal to maximal edges in G and adding parallel paths. Because of distributivity of the operations, the edge labels can be combined in many different orders yielding the same result but **differing in the number of multiplication (and addition) operations performed**; see also Section 5.4. The multiplications and additions can be expressed as graph manipulations in G. Figure 6.1(d) shows a vertex elimination where the number of multiplications is equivalent to the Markowitz degree, and Figure 6.1(e) shows a front-edge elimination where the number of multiplications is equal to the number of out-edges of the target vertex. For details and additional operations see [10, 11] and [3], Section 3.2. Each

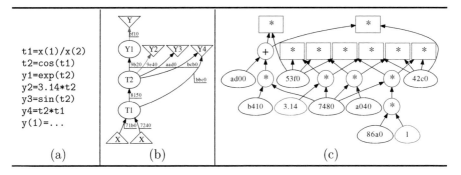

```
t1=x(1)/x(2)
t2=cos(t1)
y1=exp(t2)
y2=3.14*t2
y3=sin(t2)
y4=t2*t1
y(1)=...
```

(a)   (b)   (c)

FIGURE 6.2: Sequence of statements (a), corresponding computational graph G (b), accumulation graph resulting from edge elimination (c).

elimination step modifies G to some $G^{(k)}$. One can proceed with eliminations until the modified G has reached bipartite form G^b and the edge labels are the entries of the Jacobian $\boldsymbol{J}$ of F evaluated at some point $\mathbf{x}_0$. This description illustrates the basis of the following NP-complete [12] problem.

1. What is the minimal number of multiplications to reduce G to bipartite form?

The OpenAD representation[2] of the problem is discussed in Section 6.2.1 followed by example heuristics in Section 6.2.2. The example in Figure 6.2 is based on the so-called lion example from [11]. It can be found in the OpenAD install under **OpenAD/Examples/Lion**. The graphs in Figure 6.2(b,c) are a slightly compressed versions of what is created by the OpenAD transformation engine and can be reproduced when running **make show** in **Examples/Lion**. The edges of G in Figure 6.2(b) are decorated with addresses[3] which represent the edge labels c_{ji}. These in turn are the minimal vertices in the accumulation graph shown in Figure 6.2(c) and the correspondence between edges in (b) and minimal vertices in (c) is done via a map that for example will map edge **9b20** to vertex **86a0** and so forth. In the accumulation graph the maximal vertices represent the 8 edges of the bipartite G^b. Within the transformation engine a comparison is made between various vertex and edge elimination heuristics and the best result is returned. Note that Figure 6.2(b) shows an extra vertex connected via a unit edge at the top which is an artifact of the internal representation vs. the original lion example. This unit edge results in an additional multiplication node with 1 as operand in the accumulation graph. Still, the accumulation graph shows 12 remaining multiplications as opposed to the 11 optimal ones. The heuristic employed here happens to not find the optimal solution. Also implemented, however, is a randomized heuristic whose output

---

[2]OpenAD is actively being developed. All names of files, classes, methods, and variables mentioned here refer to versions xaifBooster: b88d9f62ae82+ , angel: ff7faed78ea6 of the respective Mercurial source code repositories (see [2]). Changes since then can be traced back by using the Mercurial web interface at http://mercurial.mcs.anl.gov/ad .

[3]We removed the invariant prefix.

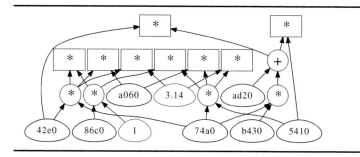

FIGURE 6.3: Accumulation graph as in Figure 6.2 but here using a randomized heuristic.

accumulation graph is shown in Figure 6.3. Again discounting the ineffective multiplication by 1 this heuristic achieves the optimal 11 multiplications. Because this is a different example the vertex addresses change and the edge `9b20` is now mapped to vertex `86c0`. The figure was produced using `make showRandom` in the same example directory.

In many practical applications one does not require the full J but only projections JS or $W^T J$ with $S \in \mathbb{R}^{n \times p}$, $W \in \mathbb{R}^{m \times p}$, where p is the number of desired directions. During the eliminations described above, the number of edges (with nonunit/nonconstant labels) at any intermediate stage $G^{(k)}$ often is smaller than that of the final G^b. Consequently, the **projection operations executed with $G^{(k)}$ are cheaper** than with G^b. The general concept is known as *scarcity* [13] and has been investigated for practical use in [14]. Figure 6.4 shows an example for a rank-1 update. The initial G has only $3n$ edges, of which $2n$ are constant. If we eliminate the intermediate vertex z (see also Figure 6.1(d)), we have n^2 nonconstant edges. Consequently, the projections JS or $W^T J$ require only $3np$ operations compared to $n^2 p$. This scenario leads to the second combinatorial problem.

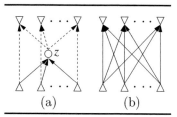

FIGURE 6.4: Computational graph (edges with constant labels are dashed) for $F(\mathbf{x}) = (\mathbf{D} + \mathbf{a}\mathbf{x}^T)\mathbf{x}$ with an intermediate $z = \mathbf{x}^T \mathbf{x}$ in (a) and state after eliminating z in (b).

2. What is a minimal representation J^* for J ?

In Section 6.2.3 we will explain the implementation of a greedy heuristic to approximate the minimal representation, with particular consideration given to the additional savings possible when one considers some edge labels to have unit value, see also Section 5.4.5. Already in the lion example used before, the effects of scarcity-aware heuristics can be observed. The bipartite G^b that is the result of the accumulations in Figure 6.2(c) has 8 edges. A scarcity aware

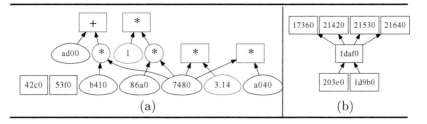

(a)   (b)

FIGURE 6.5: Accumulation graph for Figure 6.2(b) produced by a scarcity-aware heuristics (a), and the remainder graph $G^{(k)}$ corresponding to the incomplete elimination sequence (b).

heuristic will stop the elimination early when a minimal number of edges is reached which is illustrated by the smaller accumulation graph in Figure 6.5(a) that has only 5 multiplications (plus the discounted multiplication with 1) vs. the 11 multiplications of the optimal complete elimination. The corresponding remainder graph shown in Figure 6.5(b) has only 6 edges instead of the 8 of G^b and therefore the propagation through it is cheaper.

Because in OpenAD we have to build G at compile time, the limitations of an automatic code analysis[4] to identify computed values (on the left-hand side of an assignment) to their subsequent uses (in later right-hand-side-expressions) also limit the scope of any single G typically to the contents of a basic block. Thus, for an entire program P with control flow we have a sequence of s graphs G_i and, corresponding to that, a sequence of s local Jacobians J_i. The overall Jacobian $J \in \mathbb{R}^{m \times n}$ for P is therefore a chained matrix product over the J_i. Again, instead of the full Jacobian J we typically require the projections $(J_s \circ \ldots \circ (J_1 \circ S) \ldots)$, called the *forward mode*, or $(\ldots (W^T \circ J_s) \circ \ldots \circ J_1)$, called the *reverse mode*. Here we can also use the J_i^* in place of the less efficient J_i. The bracketing for the reverse mode indicates that the J_i are required in an order inverse to the execution of the original program P. Typically, because of memory limitations, one will not be able to compute and store *all* the J_i at once to then use them for the reverse sweep. Instead one will have to trade off some recomputation of the J_i from checkpoints. Determining the checkpoints and orchestrating the computation of the J_i and their use in the reverse sweep is called a *reversal scheme*. This leads to the third combinatorial problem, also shown to be NP-complete [15].

3. Which reversal scheme achieves the fewest recomputations, given constraints on the memory for checkpoints and storing the J_i?

Section 6.3.1 highlights split and joint modes as two simple cases at the respective ends of the spectrum controlled at the level of subroutine calls. Section 6.3.3 discusses the optimal solution for the special case of uniform iterations.

---

[4]This stems from *aliasing*, for example the possibility that two array elements a(i) and a(j) or two pointers p and q point to the same address.

6.2   Computational Graphs

The modular design of OpenAD separates different tasks of the source transformation. The core AD transformation engine is a module called `xaifBooster` separate from the modules for parsing, code analysis, and unparsing. An obvious candidate for the separation within the AD transformation engine is the computation of the local Jacobians J_i or J_i^*, pertaining to problems 1 and 2. We rely on the object-oriented language features of C++ to facilitate the module separation via interface classes.

6.2.1   Problem Representation

To experiment with elimination heuristics, we do not need to understand how G is constructed from code or how exactly the elimination steps are translated back into executable code. We need only an interface that describes the structure of G, the elimination steps, and (for problem 2) G^*, or *remainder graph*. The main interface is defined[5] in the following file.

`xaifBooster/algorithms/CrossCountryInterface/inc/Elimination.hpp`

The structure of G is represented by class `LinearizedComputationalGraph`, or LCG. It uses vertex and edge classes `LCGVertex` and `LCGEdge`; their definitions can be found in the same directory in header files with the respective names. This graph class (like all other graph classes in `xaifBooster`) is based on the Boost Graph Library [16]. An edge elimination step such as the front elimination of edge $(1,3)$ shown in Figure 6.1(e) yields the following two (because vertex 3 has two outedges) fused multiply-add operations: $c_{41}+=c_{43}*c_{31}$ and $c_{51}=c_{53}*c_{31}$. They are represented as instances of the class `JacobianAccumulationExpression`, or JAE. Because we need only structural information, the nonminimal vertices of a JAE represent the $*$ or $+$ operations, and the minimal vertices have references either to edges of G or to maximal vertices of other JAE that are earlier elimination results, such as the new edge $(1,5)$ in Figure 6.1(e). An instance of `JAEList` and the remainder graph G^* (which, like the input graph G, is an instance of LCG) are the results of the top-level routine `Elimination::eliminate` called from within `xaifBooster`.

Aside from the purely structural representation of G, the instances of `LCGEdge` contain a discriminator to identify whether an edge label is ± 1, constant, or generally variable. This information is needed to solve the second problem. In the following we will discuss elimination heuristics implemented by *Angel* [17, 18].

---

[5] We follow the convention used in the OpenAD manual [3], Section 2.2, by referencing all files relative to the OpenAD install directory set as the `$OPENADROOT` environment variable. We provide generated documentation of the source of all OpenAD components under the website's [2] "Documentation" link.

$\downarrow$v1 $\quad$ $\downarrow$adg $\quad$ $\downarrow$op	$\downarrow$adg $\quad$ $\downarrow\mathcal{F}$ $\quad$ (in the source code named h)
best:=$<$*some max value*$>$	cost:=0; $\quad$ v_1:=eliminatable(adg)
$\forall t \in$ v1 :	**while** $v_1 \neq \emptyset$:
$\quad$ **if** op(t,adg) $<$ best **then**	$\quad$ v_2:=$\mathcal{F}(v_1$,adg)
$\quad\quad$ v2:=$\emptyset$	$\quad$ cost+=eliminate(v_2,adg) /* adg is changed! */
$\quad$ **if** op(t,adg) $\leq$ best **then**	$\quad$ seq+=v_2
$\quad\quad$ v2:=v2$\cup\{t\}$	$\quad$ v_1:=eliminatable(adg)
$\downarrow$v2	$\downarrow$cost $\quad$ $\downarrow$seq
(a)	(b)

FIGURE 6.6: Pseudo code for `standard_heuristic_op` (a) and for `use_heuristic` (b).

6.2.2 Elimination Heuristics

The top-level driver routine for the heuristics is `Elimination::eliminate()` implemented in `angel/src/xaif_interface.cpp`. An `Elimination` instance is instantiated from `xaifBooster`, and its attributes are set via `oadDriver` command line switches a, A, m, M, and R; see [3], Section 4.1.3.4. The attributes determine which specific elimination routine is called by `eliminate`, for instance, `compute_elimination_sequence()`.[6]

Markowitz-Based Heuristics Using Markowitz (triggered by setting -M 0) as an example, we illustrate the implementation of a heuristic and **highlight the minimal set of elements to be changed for a new heuristic**. Angel internally uses plain boost graphs. The first step in `compute_elimination_sequence()` is to convert the LCG given as input via a call to `read_graph_xaif_booster`. Then we declare a stack $\mathcal{F}$ of heuristics that filters the elimination target vertices down to a single vertex. For example, the first of three such $\mathcal{F}$ (for vertex elimination) is declared in `xaif_interface.cpp`, line 1151, as

```
typedef heuristic_pair_t<lowest_markowitz_vertex_t>, reverse_mode...>
        lm_rm_t;
```
and later, on line 1154, is defined as follows.
```
lm_rm_t lm_rm_v (lowest_markowitz_vertex, reverse_mode_vertex);
```
This filter stack internally first passes vertices with the lowest Markowitz degree and then uses the reverse mode order as a tie breaker. The Markowitz filter and the reverse mode tie breaker are defined by using `standard_heuristic_op` and a function object [7] called `lmv_op_t`, defined in `angel/src/heuristics.cpp`. Figure 6.6(a) shows the pseudo code implemented in `standard_heuristic_op` taking in a vector of elimination targets v1, a graph adg, and the aforementioned function object as a formal parameter op.

---

[6]We encourage using the "search" provided in the Angel and `xaifBooster` generated source code documentation under "Documentation" at [2].

[7]C++ classes with an `operator()`.

The core of the Markowitz heuristic is encapsulated in the function object `lmv_op_t`'s `operator()`, which returns the expected product of in and out degree.

$$\texttt{in\_degree(v,cg) * out\_degree(v,cg)}$$

With this framework, the implementation of a new heuristic requires comparatively little effort, and one has the remainder of OpenAD readily available to evaluate its efficacy.

Evaluating the Cost Now that we have described how a filter stack is implemented, we return to `compute_elimination_sequence()` to look at the computation and comparison of the resulting cost. The above-mentioned three $\mathcal{F}$ together with simple forward and reverse order are passed in a call to `best_heuristic` (line 1169), which computes the elimination sequences and their respective cost individually and returns the cheapest elimination sequence. The definition is found in `heuristics_impl.hpp`. To determine the cost of each of the five $\mathcal{F}$ individually, it calls `use_heuristic`, defined in `heuristics.cpp`; see also Figure 6.6(b). It takes as input the graph G as `adg` and the filter and returns the `cost` and the elimination sequence `seq`. The filter $\mathcal{F}$ determines the next elimination target, and the cost is accumulated as incurred by the actual elimination. For the latter, one can follow the calls from `eliminate` to `vertex_elimination` to `back_edge_elimination` to see that the cost is the number of edge label multiplications. Consequently a new heuristic for this cost model does not necessitate any further changes.

Edge Eliminations Returning once more to `compute_elimination_sequence()`, we also find a variety of *edge elimination* heuristic filters declared. The logic employed for these is analogous to that applied to vertex eliminations. The cost, following the same cost model used for vertex elimination, for the best edge elimination is compared to that of the best vertex elimination. The winner is converted into an equivalent face elimination [11] sequence. Because a face elimination can be considered the elemental building block for all elimination operations, there is a common method to populate the caller-provided instance of `JAEList` which expects a sequence of face eliminations as input. An actual face elimination sequence and accompanying heuristics can be found in `compute_elimination_sequence_lsa_face`.

6.2.3 Scarcity-Preserving Heuristics

To approximate a minimal representation $\boldsymbol{J}^*$ of the Jacobian $\boldsymbol{J}$ (problem 2), OpenAD has various scarcity-preserving heuristics [14]. We describe a simple example of such a heuristic that is implemented in `compute_partial_elimination_sequence`. Its execution is triggered by the `oadDriver` command line settings `-M 1 [ -m ]`. While the heuristics discussed in Section 6.2.2 minimize the count of elimination operations, the cost here is simply the number of nonunit edge labels in the remainder graph G^*. Minimizing the elimination-induced operations count is a secondary concern. Given

↓ourLCG   (the graph G)

```
01  best:=< default instance of elimSeq_cost_t >
02  do
03     adg:=ourLCG;    curr:=< default instance of elimSeq_cost_t >
04     do
05        v₁:=eliminatable(adg) /* find eliminatable edges */
06        if v₁ ≡ ∅ then break /* this elimination is complete */
07  ‖     v₂:=reducing_edge_eliminations(v₁,adg) /* find edge count reductions */
08  ‖     if v₂ ≡ ∅ then v₂ := v₁ /* if no such target edges exist use the previous targets */
09  ‖     v₃:=refill_avoiding_edge_eliminations(v₂,adg)
10  ‖     if v₃ ≡ ∅ then v₃ := v₂
11  ‖     v₄:=lowestMarkowitzEdgeElim(v₃,adg)
12  ‖     v₅:=reverseModeEdgeElim(v₄,adg)
13        curr.elims+=v₅
14        curr.cost+=eliminate(v₅,adg) /* modifies adg */
15     if (curr < best) then best:=curr
16     if (! curr.revealedNewDependence) then break /* no new refill dependencies */
17  /* get partial elimination from 'best' until J* and populate jae_list and remainderLCG */
```

↓jae_list   ↓remainderLCG

FIGURE 6.7: Pseudo code for `compute_partial_elimination_sequence`.

the distinction of unit, constant, and variable edge labels in the `LCGEdge` class, we can disregard the multiplication of constant edge labels from the elimination operations cost because it is a compile time effort. All counters pertaining to the heuristic are kept in an instance of `elimSeq_cost_t` that is defined in `angel/include/angel_types.hpp`. The pseudo code in Figure 6.7 illustrates the core logic. Here, for simplicity, we consider only edge elimination operations. Again we use a stack of filters to narrow the eligible targets down to a single edge. However, here we do so repeatedly because in each elimination sequence we may detect a refill[8] that in a subsequent elimination sequence we attempt to avoid (line 09), thereby modifying the set v_3 and arriving at different result. After each elimination is complete, we compare it to the current best result, where `elimSeq_cost_t` holds all relevant counters, such as the minimal edge count reached along the way, and the operations incurred. If no new refill dependence is detected, we are done. The easiest entry point to **change the behavior for a new heuristic** is to modify the filter stack at any of the marked lines 07–12. To finish, we need to create G^* by replaying the best elimination sequence to the first point when the minimal edge count is reached. This also populates the instance of `JAEList`. Finally we populate the caller-provided instance of `LCG` with the contents of the Angel internal G^*. The propagation through the remainder graph is encapsulated entirely in `xaifBooster` as a code generation step. Note that the heuristic logic inside `reducing_edge_eliminations` is no longer as simple as, for instance, the Marko-

---

[8]An edge (i, j) is *refilled* if it is eliminated but subsequently recreated as a consequence of eliminating edges in an alternative path from i to j.

witz criterion because we need to **precompute the effect an elimination would have** on the edge count considering different combinations of unit/constant and variable edge labels. This precomputation is implemented in `edge_elim_effect`. Other implementations of scarcity-preserving heuristics can be found in `compute_partial_transformation_sequence` which includes logic to produce pre- and postrouting steps, and in the `_random` variants of the above, which include randomized choices and backtracking in heuristics rather than the simple greedy algorithm explained here.

6.3 Reversal Schemes

Section 6.1 introduces reversal schemes as a means to obtain the J_i in reverse order, potentially involving recomputation as a tradeoff for storage, see also Section 5.3. In OpenAD the granularity of choice for making this tradeoff is the subroutine call.

6.3.1 Simple Split and Joint Modes

For a given scope the two extreme ends of the tradeoff are called *split* mode and *joint* mode. The former stores (or "tapes") all J_i at once; the latter minimizes the storage for the J_i and the checkpoints. A split mode example is shown in Figure 6.8. The adjoint phase, that is, the propagation $(\ldots(\boldsymbol{W}^T \circ \boldsymbol{J}_s) \circ \ldots \circ \boldsymbol{J}_1)$, is colored gray.

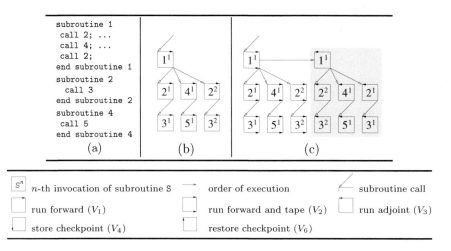

FIGURE 6.8: Example code (a), corresponding call tree (b), split mode (c), and legend.

The test code in `OpenAD/Examples/ReversalSchemes/ADMsplit` illustrates the split mode. The example source code is derived from a very simple box ocean model that is also found in `OpenAD/Examples/BoxModel` but has been augmented to gather information specific to the reversal schemes. The call graph is shown in Figure 6.9. The

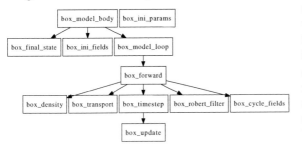

time stepping loop is in `box_model_loop` and takes 3650 iterations. The example can be built by running `make` in this directory which creates a binary `driverADMsplit`. When this driver is executed it prints the Jacobian entries, the number of checkpoints (always zero for split mode)

FIGURE 6.9: Call graph for box model (compressed from OpenAD output).

the maximal entry count in the double precision tape storage (here 135050) and the number of forward invocation of the various subroutines per column in the weight matrix W.

The joint mode is characterized by the fact that we store the J_i for each subroutine only **immediately before** the corresponding adjoint sweep. A joint mode corresponding to Figure 6.8(a) is shown in Figure 6.10 where the

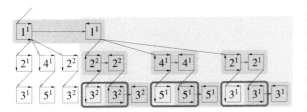

pairs of storing the J_i followed by the adjoint are colored dark gray and recomputations are framed. Both schemes exhibit a very regular structure in which each subroutine executes one of five specific variants $(V_1$-$V_4,V_6)$ generated by

FIGURE 6.10: Joint mode for Figure 6.8(a).

OpenAD.[9] The joint mode variant for the box model example can be found in `OpenAD/Examples/ReversalSchemes//ADMjoint`. The example can be built by running `make` in this directory which creates a binary `driverADMjoint`. When this driver is executed it prints the Jacobian entries and counters for the same quantities as in split mode (per column in the weight matrix W). As expected we see a dramatic reduction in the maximal double precision tape storage size accompanied by a large increase in forward computations of the subroutine for recomputations from the maximally 3657 checkpoints used at any one time during the execution. The forward computations include the taping run and illustrate the recomputation overhead being dependent on the

---

[9]Other generated variants V_5, V_7 etc. are not of relevant for this paper.

depth from the top level routine of the given subroutine within the call tree. Note that for recursive routines we have to consider a call *graph* and the recomputation overhead depends then on the runtime call stack height. In our example the overhead factor can be observed in comparison to the split mode example output and is 3 for box_forward, 4 for instance for box_density, and 5 for box_update. The large number of recomputations are the trade-off for the reduction in the tape memory requirements. In practice, instead of split and joint mode as the extremes of the spectrum, a hybrid reversal scheme is used, consisting in part of the split mode for as large a section of the model for which the corresponding J_i still fit into memory, while the higher-level parts use some checkpointing scheme and the joint mode. Because this is an application-dependent problem, we do not directly generate the entire reversal scheme in xaifBooster but rather **use a template and a postprocessing step to orchestrate the reversal**.

6.3.2   Template Mechanism

An OpenAD template is best understood as a sample subroutine with some control flow into which at predefined spots the postprocessor inserts the variants V_i. While the xaifBooster transformation generates the V_i, the control structure together with some static state information determines *which* version V_i at any invocation point in the call tree is executed. As examples, the split and joint mode templates[10] are shown in Figure 6.11 and Figure 6.12, respectively; the control structure is highlighted. The numbering of the subroutine variants (V_1-V_4,V_6) directly reflects the id number referenced in the !$PLACEHOLDER_PRAGMA$. The logic for the split mode is self-evident: the driver for the program sets the global our_rev_mode and invokes the top-level model routine

```
1   subroutine template()
2     use OAD_tape
3     use OAD_rev
4   !$TEMPLATE_PRAGMA_DECLARATIONS
5     if (our_rev_mode%plain) then
6     ! original function
7   !$PLACEHOLDER_PRAGMA$ id=1
8     end if
9     if (our_rev_mode%tape) then
10    ! taping
11  !$PLACEHOLDER_PRAGMA$ id=2
12    end if
13    if (our_rev_mode%adjoint) then
14    ! adjoint
15  !$PLACEHOLDER_PRAGMA$ id=3
16    end if
17  end subroutine template
```

FIGURE 6.11: Split mode template.

once with our_rev_mode%tape and once with our_rev_mode%adjoint set to true. The joint mode logic in the template is more complicated, but one can easily see that the execution of each subroutine variant implies which variant should be executed for its callees. The callee variant is set through calls to the respective OAD_rev* routines (highlighted in Figure 6.12); and by consulting the

---

[10]The definitions of the OAD_rev* routines can be found in $OPENADROOT/runTimeSupport/simple/OAD_rev.f90

joint scheme figure, one can easily verify the correctness of the template logic. In particular, we see that after storing the J_i for the given subroutine (line 20) the mode is set to adjoint (line 21), which then immediately follows (lines 23–27) as expected in the joint mode.

6.3.3 Reversal Scheme Using Revolve

In uniform time-stepping schemes one can assume that all checkpoints are of the same size, as are the storage requirements to produce the J_i of a single timestep. This can be used to derive an optimal reversal scheme [19] that minimizes the number of recomputations for a given total s of timesteps and permitted number p of checkpoints. Rather than a strict split or joint mode it indicates for each step, to the top-level time step routine and all its callees, which of the respective variants V_i should be executed. A Fortran version of optimal algorithm is available.[11] Figure 6.13 shows an example loop code with adjustments needed to apply OpenAD highlighted; see also [3], Section 1.3. Figure 6.14 shows the template to be applied to `loopWrapper` and used in conjunction with re-volve. The key ingredients of

```
1    subroutine template()
2      use OAD_tape
3      use OAD_rev
4      use OAD_cp
5      type(modeType) :: our_orig_mode
6      if (our_rev_mode%arg_store) then
7   !$PLACEHOLDER_PRAGMA$ id=4
8      end if
9      if (our_rev_mode%arg_restore) then
10  !$PLACEHOLDER_PRAGMA$ id=6
11     end if
12     if (our_rev_mode%plain) then
13        our_orig_mode=our_rev_mode
14        our_rev_mode%arg_store=.FALSE.
15  !$PLACEHOLDER_PRAGMA$ id=1
16        our_rev_mode=our_orig_mode
17     end if
18     if (our_rev_mode%tape) then
19        call OAD_revStorePlain
20  !$PLACEHOLDER_PRAGMA$ id=2
21        call OAD_revAdjoint
22     end if
23     if (our_rev_mode%adjoint) then
24        call OAD_revRestoreTape
25  !$PLACEHOLDER_PRAGMA$ id=3
26        call OAD_revRestoreTape
27     end if
28   end subroutine template
```

FIGURE 6.12: Joint mode template.

the template are the loop (line 16) replacing the original time-stepping loop (line 6) in `loopWrapper`. All actions are determined by calling `rvNextAction` (line 17). We distinguish storing and restoring checkpoints to file by injecting the subroutine variants V_4 (line 21) and V_6 (line 25), respectively, computing forward (lines 30–32) up to a step determined by revolve, and doing split adjoint computation (lines 34–37) for `rvFirstUTurn` and `rvUTurn`. In the latter the `loopBody` is directly injected (lines 35,37) because in the template in Figure 6.14 we explicitly replace the entire loop construct of `loopWrapper` (lines 6–8). Consequently, for the template mechanism it is important to have the time-stepping loop separated in a wrapper, as done in our example.

Referring back to our example in Section 6.3.1 we also have a variant that uses the revolve mechanism. This setup can be found in

---

[11]See http://mercurial.mcs.anl.gov/ad/RevolveF9X.

```
                                        1    !$openad XXX Template ad_revTempl.f
                                        2    subroutine loopWrapper(x,s)
                                        3      double precision :: x
    subroutine loopBody(x)              4      integer :: s
      double precision :: x             5    !$openad INDEPENDENT(x)
      x=sin(x)                          6      do i=1,s
    end subroutine                      7        call loopBody(x)
                                        8      end do
                                        9    !$openad DEPENDENT(x)
                                       10    end subroutine
```

```
program driver
  use OAD_active
  use OAD_tape
  implicit none
  external head
  type(active) :: x
  integer :: s
  call oad_tape_init()
  x%v=.5D0
  x%d=1.0D0
  write (*,fmt='(A)',advance='no') &
    'number_of_iterations_=_'
  read (*,*) s
  call loopWrapper(x,s)
  print *, 'driver_running_for_x_=',x%v
  print *, '_____yields_dy/dx_=',x%d
end program driver
```

FIGURE 6.13: Time-stepping example loop in `loopWrapper` with the loop body encapsulated in `loopBody` and a driver.

`OpenAD/Examples/ReversalSchemes/ADMrevolve`. The example can be built by running `make` in this directory which creates a binary `driverADMrevolve`. When this driver is executed it prints the Jacobian entries and counters for the same quantities as in split and joint mode (per column in the weight matrix W). The user, however, is asked to specify the maximal number of allowed checkpoints to be used at any one time for the reversal of the 3650 iterations of the time stepping loop. The reversal for all the other routines is organized such that the subroutine above `box_model_loop` are reversed in joint mode while everything below is reversed in split mode. The scheme is partially shown in Figure 6.15 and the revolve logic is marked by the $\mathcal{R}$. One can see that the joint mode portion already consumes an extra checkpoint such that when one specifies a maximum of 12 checkpoints to be used

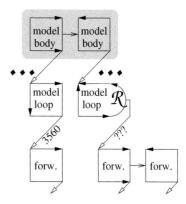

FIGURE 6.15: Box model reversal scheme with revolve.

for the revolve algorithm space for a total of 13 checkpoints will be needed. The maximal double precision tape grows only slightly to 37 entries. We see

```
 1   subroutine template ()
 2      use OAD_tape
 3      use OAD_rev
 4      use OAD_cp
 5      use revolve
 6      LOGICAL :: ini=.FALSE.
 7      TYPE(rvAction) :: rvAct
 8      CHARACTER , DIMENSION(80) :: errorMsg
 9      integer :: p, curr=0
10      write (*,fmt='(a)',advance='no') &
11         'number_of_checkpoints_=_'
12      read (*,*) p
13      ini=rvInit(s,p,errorMsg)
14      IF (.NOT.ini) WRITE(*,'(A,A)') &
15         'Error:_', errorMsg
16      do while (rvAct%actionFlag/=rvDone)
17         rvAct=rvNextAction()
18         select case (rvAct%actionFlag)
19         case (rvStore)
20            call cp_write_open (rvAct%iteration)
21   !$PLACEHOLDER_PRAGMA$ id=4
22            call cp_close
23         case (rvRestore)
24            call cp_read_open (rvAct%iteration)
25   !$PLACEHOLDER_PRAGMA$ id=6
26            curr=rvAct%iteration
27            call cp_close
28         case (rvForward)
29            call OAD_revPlain
30            do curr=curr,rvAct%iteration−1
31               call loopBody(x)
32            end do
33         case (rvFirstUTurn,rvUTurn)
34               call OAD_revTape
35               call loopBody(x)
36               call OAD_revAdjoint
37               call loopBody(x)
38         end select
39      end do
40   end subroutine template
```

FIGURE 6.14: ad_revTempl.f to be applied to loopWrapper in Figure 6.13.

the recomputation overhead factor for 12 revolve checkpoints compared to split mode to be about 6.35 but that is uniform for all routines underneath box_model_loop, that is, no longer dependent on the call tree depth. The additional level of recomputation is the tradeoff for the savings of maximally 13 checkpoints vs 3657 checkpoint in joint mode. Increasing the number of revolve checkpoints to 120 lets the recomputation factor drop to around 3.97. This however includes the first 3650 loop iterations when box_model_loop writes its checkpoint. The overhead factor that is incurred specifically by the revolve scheme therefore is only 5.35 and 2.97, respectively. It depends on the application context one balances disk space for checkpoints and recomputations.

The fact that the template in Figure 6.14 does not distinguish between rvFirstUTurn and rvUTurn is owed to the fact that we assume a joint mode reversal for the caller of loopWrapper or box_model_loop, respectively. Assuming the bulk of the work is contained within the loop this implies an inefficiency and one should consider an overall split mode approach. An example setup can be found in `OpenAD/Examples/ReversalSchemes/ADMrevolve2`. There, the slightly more complicated revolve template lets revolve place checkpoints during the taping sweep and lets it produce a tape for the last time step, i.e. when the action indicates `rvFirstUTurn` and then leaves the loop to finish the taping sweep. The internal state of revolve is retained until the adjoint sweep reaches the revolve loop and can then continue from there on. The overall split mode generally implies larger memory requirements for the tape but yields the savings of the first 3650 loop iterations shown in Figure 6.15 because they are now part of the revolve scheme.

While the above represents the solution to a very regular setup, one can use the same idea to **apply new heuristics in cases where the problem is combinatorial**, for example, when the timesteps (and therefore the associated storage requirements) are not homogeneous. We note that the template mechanism as the entry point to the overall reversal is well insulated from the remainder of the OpenAD tool chain and provides an easy access to experiment with other reversal schemes.

Acknowledgments

This work was supported by the Mathematical, Information, and Computational Sciences Division subprogram of the Office of Advanced Scientific Computing Research, Office of Science, U.S. Dept. of Energy under Contract DE-AC02-06CH11357.

Bibliography

[1] Griewank, A. and Walther, A., *Evaluating Derivatives: Principles and Techniques of Algorithmic Differentiation*, No. 105 in Other Titles in Applied Mathematics, SIAM, Philadelphia, PA, 2nd ed., 2008.

[2] "OpenAD website: downloads, manual, links,"
http://www.mcs.anl.gov/openad.

[3] Utke, J. and Naumann, U., "OpenAD/F: User Manual," Tech. rep., Argonne National Laboratory, 2009, latest version available online at http://www.mcs.anl.gov/OpenAD/openad.pdf.

[4] Bücker, H. M., Corliss, G. F., Hovland, P. D., Naumann, U., and Norris, B., editors, *Automatic Differentiation: Applications, Theory, and Implementations*, Vol. 50 of *Lecture Notes in Computational Science and Engineering*, Springer, New York, 2005.

[5] Bischof, C. H., Bücker, H. M., Hovland, P. D., Naumann, U., and Utke, J., editors, *Advances in Automatic Differentiation*, Vol. 64 of *Lecture Notes in Computational Science and Engineering*, Springer, Berlin, 2008.

[6] "AD community website: news, tools collection, bibliography," http://www.autodiff.org/.

[7] Naumann, U. and Hu, Y., "Optimal Vertex Elimination in Single-Expression-Use Graphs," *ACM Transactions on Mathematical Software*, Vol. 35, No. 1, 2008.

[8] Utke, J., "Flattening Basic Blocks," Bücker et al. [4], pp. 121–133.

[9] Baur, W. and Strassen, V., "The Complexity of Partial Derivatives," *Theoretical Computer Science*, Vol. 22, 1983, pp. 317–330.

[10] Hovland, P. D., Naumann, U., and Walther, A., "Combinatorial Problems in Automatic Differentiation," *Combinatorial Scientific Computing*, edited by U. Naumann, O. Schenk, H. Simon, and S. Toledo, Dagstuhl Seminar Proceedings, Schloss Dagstuhl - Leibniz-Zentrum fuer Informatik, Germany, Dagstuhl, Germany, 2009.

[11] Naumann, U., "Optimal Accumulation of Jacobian Matrices by Elimination Methods on the Dual Computational Graph," *Mathematical Programming, Ser. A*, Vol. 99, No. 3, 2004, pp. 399–421.

[12] Naumann, U., "Optimal Jacobian accumulation is NP-complete," *Mathematical Programming, Ser. A*, Vol. 112, 2006, pp. 427–441.

[13] Griewank, A., "A Mathematical View of Automatic Differentiation," *Acta Numerica*, Vol. 12, Cambridge University Press, 2003, pp. 321–398.

[14] Lyons, A. and Utke, J., "On the Practical Exploitation of Scarsity," Bischof et al. [5], pp. 103–114.

[15] Naumann, U., "Call Tree Reversal is NP-Complete," Bischof et al. [5], pp. 13–22.

[16] "Boost C++ Libraries website: downloads, documentation, news," http://www.autodiff.org/.

[17] "AD Nested Graph Elimination Library (angel) website: downloads, overview," http://angellib.sourceforge.net.

[18] Naumann, U. and Gottschling, P., "Simulated Annealing for Optimal Pivot Selection in Jacobian Accumulation," *Stochastic Algorithms: Foundations and Applications*, edited by A. Albrecht and K. Steinhöfel, Vol. 2827 of *Lecture Notes in Computer Science*, Springer, Berlin, Heidelberg, New York, 2003, pp. 83–97.

[19] Griewank, A. and Walther, A., "Algorithm 799: Revolve: An Implementation of Checkpoint for the Reverse or Adjoint Mode of Computational Differentiation," *ACM Transactions on Mathematical Software*, Vol. 26, No. 1, mar 2000, pp. 19–45, Also appeared as Technical University of Dresden, Technical Report IOKOMO-04-1997.

Chapter 7

Getting Started with ADOL-C

Andrea Walther

Universität Paderborn

Andreas Griewank

Humboldt Universität zu Berlin

7.1 Introduction

The C++ package ADOL-C facilitates the evaluation of first and higher derivatives of vector functions that are defined by computer programs written in C or C++ by means of algorithmic, also called automatic differentiation. This tutorial chapter describes the source code modification required for the application of ADOL-C, the most frequently used drivers to evaluate derivatives and some recent developments. ADOL-C can handle codes that may also contain classes, templates and other advanced C++-features. The resulting derivative evaluation routines may be called from C, C++, Fortran, or any other language that can be linked with C. As usual for algorithmic differentiation, see Chapter 5 or [1], the numerical values of derivatives are obtained free of truncation errors. As starting points to retrieve further information on techniques and application of algorithmic differentiation, as well as on other AD tools, we refer to the book [1]. Furthermore, the web page http://www.autodiff.org of the AD community forms a rich source of further information and pointers.

ADOL-C facilitates the simultaneous evaluation of arbitrarily high directional derivatives and the gradients of these Taylor coefficients with respect to all independent variables. Hence, ADOL-C covers the computation of standard objects required for optimization purposes as gradients, Jacobians, Hessians, Jacobian × vector products, Hessian × vector products, etc. The exploitation of sparsity is possible via a coupling with the graph coloring library ColPack [2] developed by the authors of [3] and [4]. For solution curves defined by ordinary differential equations, special routines are provided that evaluate the Taylor coefficient vectors and their Jacobians with respect to the current state vector. For explicitly or implicitly defined functions derivative tensors are obtained with a complexity that grows only quadratically in their degree. The derivative calculations involve a possibly substantial but always predictable amount of data. Most of this data is accessed strictly sequentially. Therefore, it can be automatically paged out to external files if necessary. Additionally, all derivatives are obtained at a small multiple of random access memory required by the given function evaluation program. Furthermore, ADOL-C provides a so-called tapeless forward mode, where the derivatives are directly propagated together with the function values. In this case, no additionally sequential memory or data is required.

Applications that utilize ADOL-C can be found in many fields of science and technology. This includes, e.g., fish stock assessment by the software package CASAL [5], computer-aided simulation of electronic circuits by fREEDA [6] and the numerical simulation of optimal control problems by MUSCOD-II [7]. Currently, ADOL-C is developed and maintained at the University of Paderborn by Andrea Walther and her research group. The code itself is hosted at the COIN-OR web page [8] and available under the Common Public Licence or the Gnu Public Licence.

The key ingredient of algorithmic differentiation by overloading is the concept of an *active variable*. All variables that may be considered as differentiable quantities at some time during the program execution must be of an active type. Hence, all variables that lie on the way from the input variables, i.e., the independents, to the output variables, i.e., the dependents, have to be redeclared to be of the active type. For this purpose, ADOL-C introduces the new data type adouble, whose real part is of the standard type double. In data flow terminology, the set of active variable names must contain all successors of the independent variables in the dependency graph. Variables that do not depend on the independent variables but enter the calculation, e.g., as parameters, may remain one of the *passive* types double, float, or int. There is no implicit type conversion from adouble to any passive type; thus, failure to declare variables as active when they depend on active variables will result in a compile-time error message.

The derivative calculation is based on an internal function representation, which is created during a separate so-called taping phase that starts with a call to the routine trace_on provided by ADOL-C and is finalized by calling the ADOL-C routine trace_off. All calculations involving active variables that occur between the function calls trace_on(tag,...) and trace_off(...) are recorded on a sequential data set called *pre-value tape*. Pairs of these function calls can appear anywhere in a C++ program, but they must not overlap. The nonnegative integer argument tag identifies the particular pre-value tape, i.e, the internal function representation. Once, the pre-value tape is available, the drivers provided by ADOL-C can be applied to compute the desired derivatives.

7.2 Preparing a Code Segment for Differentiation

The modifications required for the algorithmic differentiation with ADOL-C can be described by a five step procedure:

1. Include needed header-files.
   For basic ADOL-C applications the easiest way is to put
   #include "adolc.h"
   at the beginning of the file. The exploitation of sparsity and the parallel computation of derivatives require additional header files.

2. Define the region that has to be differentiated.
   That is, mark the *active section* with the two commands:

trace_on(tag,*keep*);	Start of
...	active section
trace_off(*file*);	and its end

   These two statements define the part of the program for which an internal representation is created

3. Declare all independent variables and dependent variables.
   That is define them of type adouble and mark them in the active section:

xa ≪= xp;	mark and initialize independents
...	calculations
ya ≫= yp;	mark dependents

4. Declare all active variables.
   That is define all variables on the way from the independent variables to the dependent variables of type adouble

5. Calculate derivative objects after trace_off(*file*).

Before a small example is discussed, some additional comments will be given. The optional integer argument keep of trace_on as it occurs in step 2 determines whether the numerical values of all active variables are recorded in a buffered temporary file before they will be overwritten. This option takes effect if keep $= 1$ and prepares the scene for an immediately following reverse mode differentiation as described in more detail in the sections 4 and 5 of the ADOL-C manual. By setting the optional integer argument file of trace_off to 1, the user may force a pre-value tape file to be written on disc even if it could be kept in main memory. If the argument file is omitted, it defaults to 0, so that the pre-value tape array is written onto an external file only if the length of any of the buffers exceeds BUFSIZE elements, where BUFSIZE is a user-defined size.

ADOL-C overloads the two rarely used binary shift operators $\ll=$ and $\gg=$ to identify independent variables and dependent variables, respectively, as described in step 3. For the independent variables the value of the right hand side is used as the initial value of the independent variable on the left hand side.

Choosing the set of variables that has to be declared of the adouble type in step 4 is basically up to the user. A simple strategy is the redeclaration of every floating point variable of a given source code to use the class adouble provided by ADOL-C. This can be implemented by a macro statement. Then, code changes are necessary only for a limited and usually very small part of the source files. However, due to the resulting taping effort, this simple approach may result in an unnecessary higher run time of both the function evaluation and the derivative calculations. Alternatively, one may originally only retype the independents, the dependents, and all other variables that one is certain to lie on a dependency path between the independents and dependents. A compilation attempt will then reveal by error diagnostics all statements where passive values are computed from active ones. Each one of these left hand sides may then either be activated, i.e., the corresponding variable is retyped, or in some cases the value of the right hand side may be deliberately deactivated using the member function value() of the adouble-class. This process must be repeated until no more compile-time errors arise. Compared to the global change strategy described above, a reduced set of augmented variables is created resulting in a smaller internal function representation generated during the taping step.

To illustrate the required source code modification, we consider the following lighthouse example from [1], where the light beam of a light house hits the quay-wall at a point $y = (y_1, y_2)$. Introducing a coordinate system and applying planar geometry yields that the coordinates of y are given by

$$y_1 = \frac{\nu * \tan(\omega * t)}{\gamma - \tan(\omega * t)} \quad \text{and} \quad y_2 = \frac{\gamma * \nu * \tan(\omega * t)}{\gamma - \tan(\omega * t)}.$$

The two symbolic expressions might be evaluated by the simple program shown below in Figure 7.1. Here, the distance $x_1 = \nu$, the slope $x_2 = \gamma$, the angular velocity $x_3 = \omega$, and the time $x_4 = t$ form the independent variables. Subsequently, six statements are evaluated using arithmetic operations and

```
...
int main()
{ double x1, x2, x3, x4;          /* inputs */
  double v1, v2, v3;              /* intermediates */
  double y1, y2;                  /* outputs */

  x1 = 3.7; x2 = 0.7;            /* some input values */
  x3 = 0.5; x4 = 0.5;

  v1 = x3*x4;                     /* function evaluation */
  v2 = tan(v1);
  v1 = x2-v2;
  v3 = x1*v2;
  v2 = v3/v1;
  v3 = v2*x2;

  y1 = v2; y2 = v3; }            /* output values */
```

FIGURE 7.1: Source Code for Lighthouse Example (double Version).

elementary functions. Finally, the last two intermediate values are assigned to the dependent variables y_1 and y_2.

For the computation of derivatives with ADOL-C, one has to perform the changes of the source code as described in the previous section. This yields the code segment given on the left hand side of Figure 7.2, where modified lines are marked with /* ! */. Note that the function evaluation itself is completely unchanged.

If this adouble version of the program is executed, ADOL-C generates an internal function representation contained in the pre-value tape. The pre-value tape for the lighthouse example is sketched on the right hand side of Figure 7.2. Once the internal representation is generated, drivers provided by ADOL-C can be used to compute the desired derivatives.

Note that the internal representation of the function evaluation that is contained in the tape is a special format for the computational graph. So far, this fact has not been exploited. However, one may applied the elimination techniques discussed in Chapter 5 or 13 also to the representation of the computational graph given by the tapes. This could yield a considerably reduced runtime for the whole Jacobian or Jacobian-matrix products. Also new approaches may be derived from this silghtly different setting.

```
/* ! */   #include "adolc.h"
           ...
           int main()
/* ! */   { adouble x1, x2, x3, x4;
/* ! */       adouble v1, v2, v3;
/* ! */       adouble y1, y2;

/* ! */       trace_on(1);
/* ! */       x1 <<= 3.7; x2 <<= 0.7;
/* ! */       x3 <<= 0.5; x4 <<= 0.5;

              v1 = x3*x4;
              v2 = tan(v1);
              v1 = x2-v2;
              v3 = x1*v2;
              v2 = v3/v1;
              v3 = v2*x2;

/* ! */       y1 >>= v2; y2 >>= v3;
/* ! */       trace_off();   }
```

ADOL–C tape
trace_on, tag
<<=, x1, 3.7
<<=, x2, 0.7
<<=, x3, 0.5
<<=, x4, 0.5
*, x3, x4, v1
tan, v1, v2
−, x2, v2, v1
*, x1, v2, v3
/, v1, v3, v2
*, v2, x2, v3
>>=, v2, y1
>>=, v3, y2

FIGURE 7.2: Lighthouse example (adouble version) and pre-value tape.

7.3   Easy-to-Use Drivers

For the convenience of the user, ADOL-C provides several easy-to-use drivers that compute the most frequently required derivative objects. Throughout, it is assumed that after the execution of an active section, the corresponding pre-value tape with the identifier tag contains a detailed record of the computational process by which the final values y of the dependent variables were obtained from the values x of the independent variables. This functional relation between the input variables x and the output variables y is given by

$$F : \mathbb{R}^n \mapsto \mathbb{R}^m, \qquad x \to F(x) \equiv y.$$

The presented drivers are all C functions and, therefore, can be used within C and C++ programs. Fortran-callable companions can be found in the appropriate header files.

Drivers for Optimization Purposes

For the calculation of whole derivative vectors and matrices up to order 2, ADOL-C provides the following procedures:

```
int gradient(tag,n,x,g)
short int tag;              // pre-value tape identification
int n;                      // number of independents n and m = 1
double x[n];                // independent vector x
double g[n];                // resulting gradient ∇F(x)


int jacobian(tag,m,n,x,J)
short int tag;              // pre-value tape identification
int m;                      // number of dependent variables m
int n;                      // number of independent variables n
double x[n];                // independent vector x
double J[m][n];             // resulting Jacobian F'(x)


int hessian(tag,n,x,H)
short int tag;              // pre-value tape identification
int n;                      // number of independents n and m = 1
double x[n];                // independent vector x
double H[n][n];             // resulting Hessian matrix ∇²F(x)
```

The driver routine hessian computes only the lower half of $\nabla^2 f(x)$ so that all values H[i][j] with $j > i$ of H allocated as a square array remain untouched during the call of hessian. Hence only $i+1$ doubles need to be allocated starting at the position H[i].

To use the full capability of algorithmic differentiation when the product of derivatives with certain weight vectors or directions are needed, ADOL-C offers the following three drivers:

```
int jac_vec(tag,m,n,x,v,z)          // result z = F'(x)v
int vec_jac(tag,m,n,repeat,x,u,z)   // result z = uᵀF'(x)
int hess_vec(tag,n,x,v,z)           // result z = ∇²F(x)v
```

where repeat=0 signals the first call at the argument x and repeat=1 allows the reusage of intermediate results during a repeated call at the same argument x.

A detailed description of the interface of these drivers can be found in the ADOL-C documentation. Furthermore, ADOL-C provides several drivers for special cases of derivative calculation. For solution curves defined by ordinary differential equations, special routines are provided that evaluate the Taylor coefficient vectors and their Jacobians with respect to the current state vector.

In addition to the routines for derivative evaluation, ADOL-C provides functions for an appropriate memory allocation. Using these facilities, one may compute derivatives of the lighthouse example presented in the last section by the code segment given in Figure 7.3.

Drivers for Sparse Derivative Matrices

Quite often, the Jacobians and Hessians that have to be computed are sparse

```
...                              /* as above */
trace_off();

double *xp, *yp;                 /* passive inputs and outputs */
xp = myalloc(4); yp = myalloc(2);
xp[0] = 3.7; xp[1] = 0.7; xp[2] = 0.5; xp[3] = 0.5;

double** J;
double *u, *v, *Jv, *uJ;

J = myalloc(2,4);
u = myalloc(2); v = myalloc(4); Jv = myalloc(2); uJ = myalloc(4);
jacobian(1, 2, 4, xp, J);        /* Calculate F' */

xp[0] = 2.0; xp[1] = 1.0;        /* change independents */
v[0] = 1.0; v[1] = 0.0; v[2] = 0.0; v[3] = 0.0;
jac_vec(1, 2, 4, xp, v, Jv);     /* Calculate Jv = F'(xp)*v */
u[0] = 1.0; u[1] = 0.0;
vev_jac(1, 2, 4, 0, xp, u, uJ);  /* Calculate uJ = u^T*F'(xp) */
...                              /* do something with derivatives */
```

FIGURE 7.3: Derivative Calculation with ADOL-C.

matrices. Therefore, ADOL-C provides additionally drivers that allow the exploitation of sparsity. The exploitation of sparsity is frequently based on *graph coloring* methods, discussed for example in [3] and [4], and involves deepened insights in combinatorial Scientific Computing. Here, it is quite interesting that two different field of this research area can be efficiently combined for the calculation of sparse derivative matrices.

To compute the entries of sparse Jacobians and sparse Hessians, respectively, in coordinate format with ADOL-C one may use the drivers:

```
int sparse_jac(tag,m,n,repeat,x,&nnz,&rind,&cind,&values,&options)
int sparse_hess(tag,n,repeat,x,&nnz,&rind,&cind,&values,&options)
```

Once more, a detailed description of the calling structure can be found in the documentation of ADOL-C.

Computation of Higher Derivatives

Frequently, applications in scientific computing need second- and higher-order derivatives. Often, one does not require full derivative tensors but only the derivatives in certain directions $s_i \in \mathbb{R}^n$. Suppose a collection of p directions $s_i \in \mathbb{R}^n$ is given, which form a matrix

$$S = [s_1, s_2, \ldots, s_p] \in \mathbb{R}^{n \times p}.$$

One possible choice is $S = I_n$ with $p = n$, which leads to full tensors being evaluated. ADOL-C provides the function tensor_eval to calculate the derivative tensors

$$\nabla_S^k F(x_0) = \frac{\partial^k}{\partial z^k} F(x_0 + Sz)\Big|_{z=0} \in \mathbb{R}^{p^k} \quad \text{for} \quad k = 0, \ldots, d \qquad (7.1)$$

simultaneously. The function tensor_eval has the following calling sequence and parameters:

```
void tensor_eval(tag,m,n,d,p,x,tensor,S)
short int tag;              // pre-value tape identification
int m;                      // number of dependent variables m
int n;                      // number of independent variables n
int d;                      // highest derivative degree d
int p;                      // number of directions p
double x[n];                // values of independent variables x₀
double tensor[m][size];     // (7.1) in compressed form
double S[n][p];             // seed matrix S
```

Using the symmetry of the tensors defined by (7.1), the memory requirement can be reduced enormously. The collection of tensors up to order d comprises $\binom{p+d}{d}$ distinct elements. Hence, the second dimension of tensor must be greater or equal to $\binom{p+d}{d}$. To compute the derivatives, tensor_eval propagates internally univariate Taylor series along $\binom{n+d-1}{d}$ directions. Using this derivative information, the desired tensor values are computed as described in [9].

The access of individual entries in symmetric tensors of higher order is a little tricky. We always store the derivative values in the two dimensional array tensor and provide two different ways of accessing them.

For example, suppose some active section involving $m \geq 5$ dependents and $n \geq 2$ independents has been executed and taped. We may select $p = 2$, $d = 3$ and initialize the $n \times 2$ seed matrix S with two columns s_1 and s_2. Then we are able to execute the code segment

```
double**** tensorentry = (double****) tensorsetup(m,p,d,tensor);
tensor_eval(tag,m,n,d,p,x,tensor,S);
```

This way, we evaluated all tensors defined in (7.1) up to degree 3 in both directions s_1 and s_2 at some argument x. The full manual of the ADOL-C package contains a detailed description of these drivers and alternatives to access the tensor and some examples illustrating their usage.

Derivatives of Implicit and Inverse Functions

Sometimes, one needs derivatives of variables $y \in \mathbb{R}^m$ that are implicitly defined as functions of some variables $x \in \mathbb{R}^{n-m}$ by an algebraic system of equations

$$G(z) = 0 \in \mathbb{R}^m \quad \text{with} \quad z = (y, x) \in \mathbb{R}^n.$$

Naturally, the n arguments of G need not be partitioned in this regular fashion, and we wish to provide flexibility for a convenient selection of the $n - m$ *truly* independent variables. Let $P \in \mathbb{R}^{(n-m)\times n}$ be a $0 - 1$ matrix that picks out these variables so that it is a column permutation of the matrix $[0, I_{n-m}] \in \mathbb{R}^{(n-m)\times n}$. Then the nonlinear system $G(z) = 0, Pz = x$ has a regular Jacobian, wherever the implicit function theorem yields y as a function of x. Hence, we may also write

$$F(z) = \left(\begin{array}{c} G(z) \\ Pz \end{array} \right) \equiv \left(\begin{array}{c} 0 \\ Pz \end{array} \right) \equiv S x, \qquad (7.2)$$

where $S = [0, I_p]^T \in \mathbb{R}^{n\times p}$ with $p = n-m$. Now, we have rewritten the original implicit functional relation between x and y as an inverse relation $F(z) = Sx$. In practice, we may implement the projection P simply by marking $n - m$ of the independents also dependent.

Given any $F : \mathbb{R}^n \mapsto \mathbb{R}^n$ that is locally invertible and an arbitrary seed matrix $S \in \mathbb{R}^{n\times p}$ we may evaluate all derivatives of $z \in \mathbb{R}^n$ with respect to $x \in \mathbb{R}^p$ by calling the following routine:

```
void inverse_tensor_eval(tag,n,d,p,z,tensor,S)
short int tag;              // pre-value tape identification
int n;                      // number of variables n
int d;                      // highest derivative degree d
int p;                      // number of directions p
double z[n];                // values of independent variables z
double tensor[n][size];     // partials of z with respect to x
double S[n][p];             // seed matrix S
```

The results obtained in tensor are exactly the same as if we had called tensor_eval with tag pointing to a pre-value tape for the evaluation of the inverse function $z = F^{-1}(y)$ for which naturally $n = m$. Note that the columns of S belong to the domain of that function.

As an example consider the following two nonlinear expressions

$$G_1(z_1, z_2, z_3, z_4) = z_1^2 + z_2^2 - z_3^2, \qquad G_2(z_1, z_2, z_3, z_4) = \cos(z_4) - z_1/z_3 \ .$$

The equations $G(z) = 0$ describe the relation between the Cartesian coordinates (z_1, z_2) and the polar coordinates (z_3, z_4) in the plane. Now, suppose we are interested in the derivatives of the second Cartesian $y_1 = z_2$ and the second (angular) polar coordinate $y_2 = z_4$ with respect to the other two variables $x_1 = z_1$ and $x_2 = z_3$. Then the active section could look simply like

```
for (j=1; j < 5; j++)    z[j] <<= zp[j];
g[1] = z[1]*z[1]+z[2]*z[2]-z[3]*z[3];
g[2] = cos(z[4]) - z[1]/z[3];
g[1] >>= gp[1];          g[2] >>= gp[2];
z[1] >>= zp[1];          z[3] >>= zp[2];
```

where the double variable gp stores the actual function value of $G(z)$ and zp[1] and zp[2] are dummy arguments. In the last line the two independent variables z[1] and z[3] are made simultaneously dependent thus generating a square system that can be inverted (at most arguments). The corresponding projection and seed matrix are

$$P = \begin{pmatrix} 1 & 0 & 0 & 0 \\ 0 & 0 & 1 & 0 \end{pmatrix} \quad \text{and} \quad S^T = \begin{pmatrix} 0 & 0 & 1 & 0 \\ 0 & 0 & 0 & 1 \end{pmatrix}.$$

Provided the vector zp is consistent in that its Cartesian and polar components describe the same point in the plane the resulting tuple gp must vanish. The call to inverse_tensor_eval with $n = 4$, $p = 2$ and d as desired will yield the implicit derivatives, provided tensor has been allocated appropriately of course and S has the value given above. The example is untypical in that the implicit function could also be obtained explicitly by symbolic manipulations. It is typical in that the subset of z components that are to be considered as truly independent can be selected and altered with next to no effort at all.

7.4 Reusing the Pre-Value Tape for Arbitrary Input Values

In some situations it may be desirable to calculate the value and derivatives of a function at arbitrary arguments by using a pre-value tape of the function evaluation at one argument and reevaluating the function and its derivatives using the given ADOL-C routines. This approach can significantly reduce run times, and it also allows to port problem functions, in the form of the corresponding pre-value tape files, into a computing environment that does not support C++ but does support C or Fortran. Therefore, the routines provided by ADOL-C for the evaluation of derivatives can be used at arguments x other than the point at which the pre-value tape was generated, provided all comparisons involving adoubles yield the same result. The last condition implies that the control flow is unaltered by the change of the independent variable values. Therefore, this sufficient condition is tested by ADOL-C and if it is not met the ADOL-C routine called for derivative calculations indicates this contingency through its return value. Currently, there are six return values, see Table 7.1.

In Figure 7.4 these return values are illustrated. If the user finds the return value of an ADOL-C routine to be negative the taping process simply has to be repeated by executing the active section again. The crux of the problem lies in the fact that the pre-value tape records only the operations that are executed during one particular evaluation of the function. If there are branches conditioned on adouble comparisons one may hope that re-taping becomes

TABLE 7.1: Description of return values

+3	The function is locally analytic.
+2	The function is locally analytic but the sparsity structure (compared to the situation at the taping point) may have changed, e.g., while at taping arguments fmax(a,b) returned a we get b at the argument currently used.
+1	At least one of the functions fmin, fmax or fabs is evaluated at a tie or zero, respectively. Hence, the function to be differentiated is Lipschitz-continuous but possibly non-differentiable.
0	Some arithmetic comparison involving adoubles yields a tie. Hence, the function to be differentiated may be discontinuous.
-1	An adouble comparison yields different results from the evaluation point at which the pre-value tape was generated.
-2	The argument of a user-defined quadrature has changed from the evaluation point at which the pre-value tape was generated.

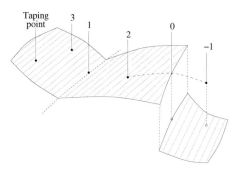

FIGURE 7.4: Return values around the taping point.

unnecessary when the points settle down in some small neighborhood, as one would expect for example in an iterative equation solver.

7.5   Suggestions for Improved Efficiency

Since the type adouble has a nontrivial constructor, the mere declaration of large adouble arrays may take up considerable run time. The user should be warned against the usual Fortran practice of declaring fixed-size arrays that can accommodate the largest possible case of an evaluation program with variable dimensions. If such programs are converted to or written in C, the overloading in combination with ADOL-C will lead to very large run time increases for comparatively small values of the problem dimension, because the actual computation is completely dominated by the construction of the large adouble arrays. The user is advised to create dynamic arrays of adoubles by using the C++ operator new and to destroy them using delete. For storage efficiency it is desirable that dynamic objects are created and destroyed in a last-in-first-out fashion.

Whenever an adouble is declared, the constructor for the type adouble assigns it a nominal address, which we will refer to as its *location*. The location is of the type locint defined in the header file <usrparms.h>. Active vectors occupy a range of contiguous locations. As long as the program execution never involves more than 65,536 active variables, the type locint may be defined as unsigned short. Otherwise, the range may be extended by defining locint as (unsigned) int or (unsigned) long, which may nearly double the overall mass storage requirement. Sometimes one can avoid exceeding the accessible range of unsigned shorts by using more local variables and deleting adoubles created by the new operator in a last-in-first-out fashion. When memory for adoubles is requested through a call to malloc() or other related C memory-allocating functions, the storage for these adoubles is allocated; however, the C++ adouble constructor is never called. The newly defined adoubles are never assigned a location and are not counted in the stack of live variables. Thus, any results depending upon these pseudo-adoubles will be incorrect. For these reasons **DO NOT use malloc() and related C memory-allocating functions when declaring adoubles.**

To avoid the storage and manipulation of structurally trivial derivative values, one should pay careful attention to the naming of variables. Ideally, the intermediate values generated during the evaluation of a vector function should be assigned to program variables that are consistently either active or passive, in that all their values either are or are not dependent on the independent variables in a nontrivial way. For example, this rule is violated if a temporary variable is successively used to accumulate inner products involving first only passive and later active arrays. Then the first inner product and all its successors in the data dependency graph become artificially active and the derivative evaluation routines will waste time allocating and propagating trivial or useless derivatives. Sometimes even values that do depend on the independent variables may be of only transitory importance and may not affect the dependent variables. For example, this is true for multipliers that are used to scale linear equations, but whose values do not influence the dependent variables in a mathematical sense. Such dead-end variables can be deactivated by the use of the value function, which converts adoubles to doubles. The deleterious effects of unnecessary activity are partly alleviated by run time activity flags in the derivative routine hov_reverse described in detail in the full manual of ADOL-C.

7.6   Advance Algorithmic Differentiation in ADOL-C

External Differentiated Functions

Ideally, AD is applied to a given function as a whole. In practice, however, sophisticated projects usually evolve over a long period of time. Within this process, a heterogeneous code base for the project develops, which may include the incorporation of external solutions, changes in programming paradigms or even of programming languages. Equally heterogeneous, the computation of derivative values appears. Hence, different AD-tools may be combined with hand-derived codes based on the same or different programming languages. ADOL-C supports such settings by the concept of external differentiated functions. Hence, an external differentiated function itself is not differentiated by ADOL-C. The required derivative information have to be provided by the user.

For this purpose, it is assumed that the external differentiated function has the signature

int ext_func(int n, double *yin, int m, double *yout);

where the function names can be chosen by the user as long as the names are unique. This double version of the external differentiated function has to be *registered* using the ADOL-C function

edf = reg_ext_fct(ext_func);.

This function starts to initialize the structure edf. The full manual of the ADOL-C package provides a detailed description of the complete initialization process. Subsequently, the call to the external differentiated function in the function evaluation can be substituted by the call of

int call_ext_fct(edf, n, xp, x, m, yp, y);

such that the externally provided derivative information is taken into account during the computation of the overall derivatives. The ADOL-C package provides an example that shows the application of this functionality.

Advance Algorithmic Differentiation of Time Integration Processes

For many time-dependent applications, the corresponding simulations are based on ordinary or partial differential equations. Furthermore, frequently there are quantities that influence the result of the simulation and can be seen as control of the systems. If there are only a few parameters to be optimized the forward mode of AD is usually the right method for the computation of the required derivative. However, in the case of a distributed control, i.e., numerous independent variables the number of which might even scale with the discretization the reverse mode has to be applied.

However, the corresponding derivative calculation may become extremely tedious if possible at all because of the sheer size of the intermediate data that has to be stored for the derivative calculation. To overcome this difficulty many checkpointing techniques have been developed. Here, only a few intermediate states are stored as checkpoints. Subsequently, the required forward information is recomputed piece by piece from the checkpoints according to the requests of the adjoint calculation. Obviously, the question arises where one should place the checkpoints.

If the number of time steps for integrating the differential equation describing the state is known a priori, one very popular checkpointing strategy is to distribute the checkpoints equidistantly over the time interval. However, it was shown in [36] using combinatorial arguments that this approach is not optimal. Obviously, the technique of dynamic programming may yield optimal checkpointing strategies. However, the resulting cost is unnecessary high. Exploiting the situation one faces for the adjoint computation one can derive optimal so-called binomial checkpointing schedules in advance to achieve for a given number of checkpoints an optimal, i.e., minimal, run time increase [10]. This procedure is referred to as offline checkpointing and implemented in the package revolve [10].

However, in the context of flow control, the partial differential equations to be solved are usually stiff, and the solution process relies therefore on some adaptive time stepping procedure. Since in these cases the number of time steps performed is known only after the complete integration an offline checkpointing strategy is not applicable. Instead, one may apply a straightforward checkpointing by placing a checkpoint each time a certain number of time steps has been executed. This transforms the uncertainty in the number of time steps to a uncertainty in the number of checkpoints needed. However, when the amount of memory per checkpoint is very high one certainly wants to determine the number of checkpoints required a priori. For that purpose, also several procedures for online checkpointing exist that distribute a given number of checkpoints on the fly during the integration procedure. The algorithms proposed in [11] yields for a given number of checkpoints an almost time-optimal adjoint computation for a wide range of applications. If the number of time steps is really high or only very few checkpoints are available the results of [11] are not applicable. For these cases, checkpointing schedules that are optimal with respect to the repetition number, i.e., the maximal time a certain time step is evaluated, are proposed in [12]. However, provable time-optimal checkpointing strategies for the general case are still missing and form an open research field of combinatorial Scientific Computing.

For a memory-reduced adjoint calculation for a time interval $[0, T]$ within ADOL-C, one may apply an appropriate integration scheme given by

some initializations yielding x_0
for $i = 0, \ldots, N - 1$

$$x_{i+1} = F(x_i, u_i, t_i)$$
evaluation of the target function

where $x_i \in \mathbb{R}^n$ denotes the state and $u_i \in \mathbb{R}^m$ the control at time t_i for a given time grid $t_0, \dots, t_N$ with $t_0 = 0$ and $t_N = T$. The operator $F : \mathbb{R}^n \times \mathbb{R}^m \times \mathbb{R} \mapsto \mathbb{R}^n$ defines the time step to compute the state at time t_i. Note that we do not assume a uniform time step size.

When computing derivatives of the target function with respect to the control, the consequences for the tape generation using the "basic" taping approach as implemented in ADOL-C so far are shown in the left part of Figure 7.5. As can be seen, the iterative process is completely unrolled due to

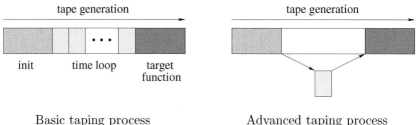

Basic taping process   Advanced taping process

FIGURE 7.5: Different taping approaches.

the taping proces, since the pre-value tape contains an internal representation of each time step. Hence, the overall tape comprises a significant amount of redundant information as illustrated by the light grey rectangles in Figure 7.5.

To overcome the repeated storage of essentially the same information, a *nested taping* mechanism has been incorporated into ADOL-C as illustrated on the right part of Figure 7.5. This new capability allows the encapsulation of the time-stepping procedure such that only the last time step $x_N = F(x_{N-1}, u_{N-1}, t_{N-1})$ is taped as one representative of the time steps in addition to a function pointer to the evaluation procedure F of the time steps. The function pointer has to be stored for a possibly necessary retaping during the derivative calculation as explained below.

Instead of storing the complete pre-value tape of the iterative process, only a very limited number of intermediate states are kept in memory. They serve as checkpoints, such that the required information for the backward integration is generated piecewise during the adjoint calculation. For this modified adjoint computation the optimal checkpointing schedules provided by revolve [10] are employed. An adapted version of the software package revolve is part of ADOL-C and automatically integrated in the ADOL-C library. Based on the optimal checkpointing strategy, c checkpoints are distributed such that computational effort is minimized for the given number of checkpoints and time steps N. It is important to note that the overall tape size is drastically reduced due to the advanced taping strategy. For the implementation of this nested taping we introduced a so-called "differentiating context" that enables ADOL-C to handle different internal function representations during the taping procedure and the derivative calculation. This approach allows the generation of a new tape inside the overall tape, where the coupling of the different tapes is based on the *external differentiated function* described above.

Written under the objective of minimal user effort, the checkpointing routines of ADOL-C need only very limited information. The user must provide two routines as implementation of the time-stepping function F, i.e.,

int time_step_function(int n, adouble *u);
int time_step_function(int n, double *u);

where the function names can be chosen by the user as long as the names are unique. It is possible that the result vector of one time step iteration overwrites the argument vector of the same time step. Then, no copy operations are required to prepare the next time step.

At first, the adouble version of the time step function has to be *registered* using the ADOL-C function

CP_Context cpc(time_step_function);.

This function starts to initialize the structure cpc. The full manual provided by the ADOL-C package provides a detailed description of the complete initialization process. Subsequently, the time loop in the function evaluation can be substituted by a call of the function

int cpc.checkpointing();

Then, ADOL-C computes derivative information using the optimal checkpointing strategy provided by revolve internally, i.e., completely hidden from the user. The source of the ADOL-C package contains an example illustrating the use of this advance AD technique.

Advance Algorithmic Differentiation of Fixed Point Iterations

Quite often, the state of the considered system denoted by $x \in \mathbb{R}^n$ depends on some design parameters denoted by $u \in \mathbb{R}^m$. One example for this setting forms the flow around an aircraft wing. Here, the shape of the wing that is defined by the design vector u determines the flow field x. The desired quasi-steady state x_* fulfills the fixed point equation

$$x_* = F(x_*, u) \tag{7.3}$$

for a given continuously differentiable function $F : \mathbb{R}^n \times \mathbb{R}^m \mapsto \mathbb{R}^n$. A fixed point property of this kind is also exploited by many other applications.

Assume that one can apply the iteration

$$x_{k+1} = F(x_k, u) \tag{7.4}$$

to obtain a linear converging sequence $\{x_k\}$ generated for any given control $u \in \mathbb{R}^n$. Then the limit point $x_* \in \mathbb{R}^n$ fulfils the fixed point equation (7.3). Moreover, suppose that $\|\frac{dF}{dx}(x_*, u)\| < 1$ holds for any pair (x_*, u) satisfying equation (7.3). Hence, there exists a differentiable function $\phi : \mathbb{R}^m \mapsto \mathbb{R}^n$, such that $\phi(u) = F(\phi(u), u)$, where the state $\phi(u)$ is a fixed point of F according to a control u. To optimize the system described by the state vector $x = \phi(u)$ with respect to the design vector u, derivatives of ϕ with respect to u are of particular interest.

To exploit the advanced algorithmic differentiation of such fixed point iterations ADOL-C provides a special function fp_iteration(...) that uses a subtape for the fixed point iteration. For efficiency, the user has to provide pointers to functions, that compute for x and u a single iteration step $y = F(x, u)$ in a double and a adouble version. Furthermore, pointers to functions computing the norm of a vector are required. These functions together with values for the variables eps, eps_deriv, N_max, and N_max_deriv control the iterations. Thus the following loops are performed:

$$
\begin{array}{ll}
\textbf{do} & \textbf{do} \\
\quad k = k + 1 & \quad k = k + 1 \\
\quad x = y & \quad \zeta = \xi \\
\quad y = F(x, u) & \quad (\xi^T, \bar{u}^T) = \zeta^T F'(x_*, u) + (\bar{x}^T, 0^T) \\
\textbf{while } \|y - x\| \geq \varepsilon \text{ and } k \leq N_{max} & \textbf{while } \|\xi - \zeta\|_{deriv} \geq \varepsilon_{deriv} \\
& \quad \text{and } k \leq N_{max,deriv}
\end{array}
$$

The full manual of the ADOL-C package provides a detailed description of the handling of this specialized function for the handling of fixed point iterations. Additionally, the ADOL-C package includes an example illustrating the application of this feature.

Advance Algorithmic Differentiation of OpenMP Parallel Programs

ADOL-C allows to compute derivatives in parallel for functions containing OpenMP parallel loops. This implies that an explicit loop-handling approach is applied. A typical situation is shown in Figure 7.6, where the OpenMP-parallel loop is preceded by a serial startup calculation and followed by a serial finalization phase.

Initialization of the OpenMP-parallel regions for ADOL-C is only a matter of adding a macro to the outermost OpenMP statement. Two macros are available that only differ in the way the global tape information is handled. Using ADOLC_OPENMP, this information, including the values of the augmented variables, is always transferred from the serial to the parallel region using *firstprivate* directives for initialization. For the special case of iterative codes where parallel regions, working on the same data structures, are called repeatedly the ADOLC_OPENMP_NC macro can be used. Then, the information transfer is performed only once within the iterative process upon encounter of the first

function eval.   derivative calcul.

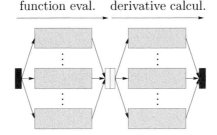

FIGURE 7.6: Basic layout of OpenMP-based derivation process.

parallel region through use of the *threadprivate* feature of OpenMP that makes use of thread-local storage, i.e., global memory local to a thread. Due to the inserted macro, the OpenMP statement has the following structure:

#pragma omp ... ADOLC_OPENMP   or
#pragma omp ... ADOLC_OPENMP_NC

Inside the parallel region, separate tapes may be created. Each single thread works in its own dedicated AD-environment. All serial facilities of ADOL-C are applicable as usual. The global derivatives can be computed using the tapes created in the serial and parallel parts of the function evaluation.

For the usage of the parallel facilities, the manual of the ADOL-C package describes the required compilation and source modifications in more detail. Furthermore, corresponding examples are provided.

7.7   Tapeless Forward Differentiation

Up to version 1.9.0, the development of the ADOL-C software package was based on the decision to store all data necessary for derivative computations on tapes, since these tapes enable ADOL-C to offer a very broad functionality. However, really large-scale applications may require the tapes to be written out to corresponding files. In almost all cases this means a considerable drawback in terms of run time due to the excessive memory accesses. Nevertheless, there are several cases, where derivative computations do not require a tape.

Starting with version 1.10.0, ADOL-C now features a tapeless forward mode for computing first order derivatives in scalar mode, i.e., $\dot{y} = F'(x)\dot{x}$, and in vector mode, i.e., $\dot{Y} = F'(x)\dot{X}$. These tapeless variants coexist with the more universal tape based mode in the package. Because of the different implementation strategy, also the required modification of the source code are different as illustrated in Figure 7.7 for the scalar mode. After defining the variables as adoubles only two things are left to do. First one needs to initialize

```
/* ! */   #ADOLC_TAPELESS
/* ! */   #include "adolc.h"
/* ! */   typedef adtl::adouble adouble;
          ...
          int main()
/* ! */   { adouble x1, x2, x3, x4;
/* ! */       adouble v1, v2, v3;
/* ! */       adouble y1, y2;

              x1 = 3.7; x2 = 0.7; x3 = 0.5; x4 = 0.5; /* initialization of x */

/* ! */       x1.setADValue(1); x2.setADValue(0);     /* initialization of ẋ */
/* ! */       x3.setADValue(0); x4.setADValue(0);

              v1 = x3*x4; v2 = tan(v1); v1 = x2-v2;  /* same as before */
              v3 = x1*v2; v2 = v3/v1; v3 = v2*x2;
              y1 = v2; y2 = v3;

/* ! */       cout << y1.getADValue() << " " << y2.getADValue() << endl;}
```

FIGURE 7.7: Source Code for Lighthouse Example (Tapeless Version).

the values of the independent variables for the function evaluation. This can be done by assigning the variables a double value. Then, the corresponding derivative value is set to zero. Alternatively, ADOL-C offers a function named setValue for setting the value of a variable without changing the derivative part. To set the derivative components of the independent variables, ADOL-C provides two possibilities:

- Using the constructor

  adouble x1(2,1), x2(4,0), y;

  This would create the three variables x_1, x_2 and y. Obviously, the latter remains uninitialized. The variable x_1 holds the value 2, x_2 the value 4 whereas the derivative values are initialized to $\dot{x}_1 = 1$ and $\dot{x}_2 = 0$.

- Setting point values directly

  adouble x1=2, x2=4, y;
  ...
  x1.setADValue(1);
  x2.setADValue(0);

  The same example as above but now using setADValue-method for initializing the derivative values.

The derivatives can be obtained at any time during the evaluation process by calling the getADValue-method

> adouble y;
>
> ...
>
> cout ≪ y.getADValue();

Similar to the tapeless scalar forward mode, the tapeless vector forward mode can be applied by defining ADOLC_TAPELESS and an additional pre-processor macro named NUMBER_DIRECTIONS. This macro takes the max-imal number of directions to be used within the resulting vector mode. Just as ADOLC_TAPELESS the new macro must be defined before including the adolc.h header file since it is ignored otherwise. A more detailed description of the tapeless forward mode can be found in the manual of ADOL-C.

7.8 Conclusions and Further Developments

Advanced differentiation techniques have been integrated recently in the ADOL-C tool. This comprises for example the optimal checkpointing for time integrations when the number of time steps is known in advance. For this purpose, ADOL-C employs the routine revolve for a binomial checkpointing [10] to achieve an enormous reduction of the memory required for calculation the adjoint of the time-dependent process. Furthermore, ADOL-C allows now the exploitation of fixed point iterations by providing drivers for the reverse accumulation [13] for the memory reduced computation of adjoints. Recently, ADOL-C has been coupled with the IPOPT package [14] for a convienient derivative provision in large-scale optimization.

First drivers for the differentiation of OpenMP parallel programs are included in the current version of ADOL-C. The differentiation of MPI-parallel programs with ADOL-C is the subject of current research. It is planed to integrate corresponding routines into ADOL-C as soon as possible. Finally, it is planed to extend the tapeless forward mode also for the computation of higher order derivatives.

Bibliography

[1] Griewank, A. and Walther, A., *Evaluating Derivatives. Principles and Techniques of Algorithmic Differentiation (Second Edition)*, SIAM, Philadelphia, 2008.

[2] ColPack, http://www.cscapes.org/coloringpage/software.htm.

[3] Gebremedhin, A., Manne, F., and Pothen, A., "What Color is Your Jacobian? Graph Coloring for Computing Derivatives." *SIAM Rev.*, Vol. 47, No. 4, 2005, pp. 629–705.

[4] Gebremedhin, A., Tarafdar, A., Manne, F., and Pothen, A., "New Acyclic and Star Coloring Algorithms With Application to Computing Hessians," *SIAM Journal of Scientific Computing*, Vol. 29, 2007, pp. 1042–1072.

[5] Bull, B., Francis, R. I. C. C., Dunn, A., McKenzie, A., Gilbert, D. J., and Smith, M. H., "CASAL (C++ algorithmic stock assessment laboratory) – User Manual," Tech. Rep. 127, NIWA, Private Bag 14901, Kilbirnie, Wellington, New Zealand, 2005.

[6] Hart, F. P., Kriplani, N., Luniya, S. R., Christoffersen, C. E., and Steer, M. B., "Streamlined Circuit Device Model Development with fREEDA® and ADOL-C," *Automatic Differentiation: Applications, Theory, and Implementations*, edited by H. M. Bücker, G. Corliss, P. Hovland, U. Naumann, and B. Norris, Lecture Notes in Computational Science and Engineering, Springer, New York, 2005, pp. 295–307.

[7] Bock, H. G., Leineweber, D. B., Schafer, A., and Schloder, J. P., "An Efficient Multiple Shooting Based Reduced SQP Strategy for Large-Scale Dynamic Process Optimization – Part II: Software Aspects and Applications," *Computers and Chemical Engineering*, Vol. 27, No. 2, 2003, pp. 167–174.

[8] ADOL-C, http://www.coin-or.org/projects/ADOL-C.xml.

[9] A. Griewank, J. U. and Walther, A., "Evaluating higher derivative tensors by forward propagation of univariate Taylor series," *Mathematics of Computation*, Vol. 69, 2000, pp. 1117–1130.

[10] Griewank, A. and Walther, A., "Algorithm 799: Revolve: An Implementation of Checkpoint for the Reverse or Adjoint Mode of Computational Differentiation," *ACM Trans. Math. Softw.*, Vol. 26, No. 1, 2000, pp. 19–45.

[11] Stumm, P. and Walther, A., "New Algorithms for Optimal Online Checkpointing," *SIAM Journal on Scientific Computing*, Vol. 32, No. 2, 2010, pp. 836–854.

[12] Wang, Q., Moin, P., and Iaccarino, G., "Minimal Repetition Dynamic Checkpointing Algorithm for Unsteady Adjoint Calculation," *SIAM Journal on Scientific Computing*, Vol. 31, No. 4, 2009, pp. 2549–2567.

[13] Christianson, B., "Reverse accumulation and attractive fixed points," *Optimization Methods and Software*, Vol. 3, 1994, pp. 311–326.

[14] Ipopt, http://www.coin-or.org/projects/Ipopt.xml.

Chapter 8

Algorithmic Differentiation and Nonlinear Optimization for an Inverse Medium Problem

Johannes Huber, Olaf Schenk

University of Basel

Uwe Naumann, Ebadollah Varnik

RWTH Aachen University

Andreas Wächter

IBM Research

8.1 Introduction

Numerical optimization, e.g., for the solution of inverse problems for systems described by partial-differential equations (PDEs), is an important aspect of scientific computing. Using the example of an inverse medium problem (Section 8.2), we demonstrate how combinatorial techniques play a significant role for the efficient solution of large-scale optimization problems. In particular, the fast computation of sparse derivative matrices using algorithmic differentiation (Section 8.5) is made possible by coloring algorithms, and the efficient solution of sparse linear systems (Section 8.6) requires weighted graph matching algorithms.

8.2　The Inverse Medium Problem

Many engineering and science problems—in such diverse areas as wave propagation in ultrasound tomography, wireless communication, geophysical seismic imaging, and other areas such as atmospheric sciences, image registration, medicine, structural-fluid interactions, and chemical process industry — can be expressed in the form of a PDE-constrained optimization problem. In this chapter we are mainly interested in inverse problems that arise in computational wave propagation and consider the problem of investigating the interior of a solid body, e.g., a human body, steel, or a part of the Earth's crust. A distinct feature of waves propagating through a homogeneous medium is their ability to travel over long distances while retaining much of their shape and initial energy. Thus waves are ubiquitous for remote-sensing of well-defined bodies (e.g., micro-cracks, land mines) or more general inhomogeneities (e.g., tumor cells in medical imaging, or oil deposits in seismic imaging). A typical common feature in these biological and geophysical application areas is that non-destructive testing methods must be applied and, very often, it is not possible to investigate the objects by a direct probe.

The prediction of the scattered fields from known incident waves and given material parameter functions is called *simulation* or *forward problem*. The forward problem in computational wave propagation is typically modeled by the Helmholtz equation. In contrast, estimating the parameter function (e.g., detecting the tumor cells) from measured scattered fields and known incident waves is called the *inverse problem* and, in wave propagation, the *inverse medium problem* [1]. In this chapter we use algorithmic differentiation techniques and large-scale nonlinear optimization algorithms in order to solve two-dimensional inverse medium problems that arise in computational wave propagation. The optimization problem is formulated as a PDE-constrained optimization problem, in which the Helmholtz equation represents the underlying PDE.

In Section 8.2.1, we discuss the forward problem, its boundary conditions, and we discretize the PDE using finite-difference discretizations. Our main objective is the inverse problem, which is analyzed in Section 8.2.2. By using a discretize-then-optimize approach, we can formulate the inverse problem as a finite dimensional optimization problem of the form

$$\min_{x \in \mathbb{R}^n} F(x)$$
$$\text{s.t. } c_{\mathcal{E}}(x) = 0 \qquad (8.1)$$
$$c_{\mathcal{I}}(x) \geq 0,$$

where the objective $F : \mathbb{R}^n \to \mathbb{R}$ and the constraints $c_{\mathcal{E}} : \mathbb{R}^n \to \mathbb{R}^p$ and $c_{\mathcal{I}} : \mathbb{R}^n \to \mathbb{R}^q$ are continuously differentiable. Here, the objective function F quantifies the misfit of the predicted and the measured wave fields, and the constraint functions $c_{\mathcal{E}}$ represents the discretized Helmholtz equation. We model previous knowledge on the material parameter functions by using inequality constraints $c_{\mathcal{I}}(x)$, and the number of (in-)equality constraints may be large (say in the millions).

There are several algorithms [2, 3] available to approach these challenging problems, such as *full space* methods and variants of the *reduced space* method. We are following the full space approach, where optimality and feasibility are reached simultaneously. The infinite-dimensional differential equations are discretized, and only linearized state equations have to be solved in every optimization step, which can improve the overall performance tremendously [4, 5, 6]. Reduced space algorithms have strengths in terms of solving large-scale problems, but also suffer from some disadvantages when compared to full-space techniques. For example, a potential pitfall is the expense of solving the nonlinear discretized PDE operator exactly multiple times during each iteration to calculate the exact reduced gradient in order to achieve superlinear local convergence. In addition, these approaches do not allow for general inequality constraints or bound constraints on dependent (state) variables.

Drawbacks of the full space are a significantly increased number of variables and the necessity to handle the equality constraints explicitly. However, the first and second derivatives of the discretized constraints and objective are typically readily available and sparse. Currently, the most promising optimization methods for large-scale nonconvex optimization problems are sequential quadratic programming (SQP) and interior-point methods (IPM). Since we want the method to take advantage of existing additional information, a large number of inequality constraints may also appear. Due to the combinatorial complexity of finding the set of active constraints, prevailing SQP methods are usually not efficient for this class of problems. In this chapter, we use a primal-dual interior-point algorithm that has been implemented in the open-source optimization library IPOPT. All efficient methods for solving large-scale instances (with the number of variables and constraints ranging up into the millions) rely on the use of first and possibly second derivatives of the functions $F(x)$, $c_{\mathcal{E}}(x)$, and $c_{\mathcal{I}}(x)$. Hence, efficient means to compute derivatives are essential, and algorithmic differentiation (AD) tools (see Section 8.5) have become indispensable for users of optimization algorithms. For example, AD techniques are used in optimization modeling language environments (such as AMPL [7] and GAMS [8]), where the user writes the formulae defining an optimization problem in an intuitive text format, and their derivatives are efficiently computed behind the scenes. In many cases, however, the function values for the optimization problem are evaluated by computer code (e.g., when modeling environments are unavailable or too inefficient, or when legacy models of chemical plant equipment are used), and here AD tools working with this source code are employed increasingly more often [9].

Another ingredient shared by most large-scale optimization algorithms is the necessity to solve large linear systems during the computation of a new trial iterate. In many cases, the efficient solution of these linear systems relies on the exploitation of the sparsity structure of the matrix, particularly when the linear system is solved by a direct factorization approach. All state-of-the-art algorithms for solving sparse linear systems make heavy use of combinatorial scientific computing techniques, such as weighted graph matching [10, 11, 12] (see Section 8.6). Furthermore, the linear system involves derivative matrices of the optimization problem functions. Therefore, the AD tools must provide features to efficiently determine the sparsity structure of the derivatives, and must also be able to exploit this sparsity structure during the evaluation of the derivative values. Here, further combinatorial scientific computing techniques such as coloring [13] play a crucial role (see Section 8.5).

8.2.1   Computational Wave Propagation

For the numerical simulation of waves in inhomogeneous media, numerical methods based on the discretization of the underlying partial differential equations remain the method of choice. While standard finite difference or finite element methods are well understood and established, their use leads to new computational challenges. The discretization of the wave equation in the frequency domain (Helmholtz equation) leads to very large sparse linear systems. Furthermore, the inverse medium problem is not only highly nonlinear and ill-posed, but also nonconvex.

In the frequency domain, the wave equation reduces to the Helmholtz equation

$$\Delta u^t(x) + k(x)^2 u^t(x) = f(x), \qquad x \in \mathbb{R}^n, n = 2, 3, \tag{8.2}$$

where Δ denotes the Laplace Operator $\Delta = \sum_{i=1}^d \frac{\partial^2}{\partial x_i}$, $k(x)$ the position dependent wave number, and u^t the complex valued amplitude of the total field. For simplicity, we shall assume that the area of interest is contained in a bounded region $\Omega \subset \mathbb{R}^n$, which is itself embedded in a homogeneous unbounded medium; that is,

$$k(x) = k_0 \in \mathbb{R} \qquad \forall x \notin \Omega.$$

Now, we consider the typical situation with $f = 0$, where an incident wave impinges on the local scattering region Ω. Hence, the total field $u^t = u^i + u$ can be separated into the given incident field u^i and the scattered field u. The

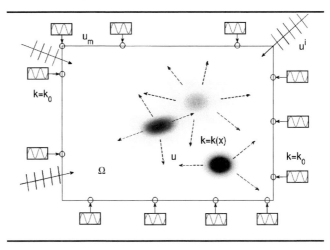

FIGURE 8.1: (See color insert.) To detect bodies or anomalies in a solid body, it is irradiated from several directions. The waves are scattered at inhomogeneities and the scattered waves are detected at, for example, the computational domain boundary. By minimizing the misfit of simulation and measurement, the bodies and inhomogeneities can be recovered.

incident field satisfies the homogeneous Helmholtz equation

$$\Delta u^i(x) + k_0^2 u^i(x) = 0, \qquad x \in \mathbb{R}^n,$$

and generates a purely outgoing scattered field u. Using (8.2), we immediately find

$$\Delta u + k^2 u = -\Delta u^i - k^2 u^i = (k_0^2 - k^2)u^i. \tag{8.3}$$

Since $k(x) = k_0$ outside Ω, the scattered wave is purely outgoing in the exterior, and we shall use absorbing boundary conditions (ABC), such as the Sommerfeld boundary conditions, higher order ABC [14], or perfectly matched layer (PML), in order to prevent waves from reflecting back into the computational domain. For simplicity, we restrict ourselves in the following discussion of the inverse medium problem to the Sommerfeld boundary condition

$$\frac{\partial u}{\partial n} = i\,k\,u.$$

Next, we discretize (8.3). For simplicity reasons, we only demonstrate the discretization for two-dimensional domains and second-order finite differences. Here, h_1 and h_2 represent the spatial discretization stepsize in the x and y directions. We denote the discrete approximation with $\mathbf{u}$ and $\mathbf{k}$. A regular grid Ω_h is introduced, with N_x times N_y grid points, and by using a lexicographical numbering $\nu : \mathbb{N}_0^2 \to \mathbb{N}, \nu(s,t) = s \cdot N_x + t + 1$, we can derive the discretized

PDE

$$\mathbf{c}_{\nu(s,t)} = -\left(\frac{2}{h_1^2} + \frac{2}{h_2^2} + \mathbf{k}_{\nu(s,t)}^2\right)\mathbf{u}_{\nu(s,t)}$$
$$+\frac{1}{h_1^2}\mathbf{u}_{\nu(s-1,t)} + \frac{1}{h_2^2}\mathbf{u}_{\nu(s,t-1)} + \frac{1}{h_1^2}\mathbf{u}_{\nu(s+1,t)} + \frac{1}{h_2^2}\mathbf{u}_{\nu(s,t+1)} \quad (8.4)$$
$$-\left(\mathbf{k}_{\nu(s,t)}^2 - k_0^2\right)\mathbf{u^i}_{\nu(s,t)}$$

for interior grid points (s,t). At the boundary, ghost point elimination is used, which in our case reduces, e.g., at the upper bound, to

$$\mathbf{c}_{\nu(s,t)} = -\left(\frac{2}{h_1^2} + \frac{2}{h_2^2} + \mathbf{k}_{\nu(s,t)}^2 - \frac{2i}{h_2}\mathbf{k}_{\nu(s,t)}\right)\mathbf{u}_{\nu(s,t)}$$
$$+\frac{1}{h_1^2}\mathbf{u}_{\nu(s-1,t)} + \frac{2}{h_2^2}\mathbf{u}_{\nu(s,t-1)} + \frac{1}{h_1^2}\mathbf{u}_{\nu(s+1,t)} \quad (8.5)$$
$$-\left(\mathbf{k}_{\nu(s,t)}^2 - k_0^2\right)\mathbf{u^i}_{\nu(s,t)}.$$

This discretization leads to the sparse linear system

$$\mathbf{c}(k,\mathbf{u}) = \mathbf{A}(k)\mathbf{u} - \mathbf{b}(k,u^i) = \mathbf{0}, \quad (8.6)$$

where $\mathbf{A}$ is a large-scale, sparse, complex valued, indefinite and, in general, nonsymmetric matrix.

8.2.2   Inverse Medium Problem Formulation

We consider a setup of N_e incident waves from different directions. The waves are measured at N_m boundary positions $x_m, m = 1, \ldots, N_m$, resulting in the discrete measured value $u_{e,m}$ for the incident wave $e \in \{1, \ldots, N_e\}$. Following a discretize-then-optimize approach, we discretize the functions and operators, including the fields $u_e \mapsto \mathbf{u}_e$ ($\mathbf{u}_e$ is a vector, not a component) and the control variables $k \mapsto \mathbf{k}$ by collocation; i.e., $\mathbf{k}_{\nu(s,t)} = k(x_{\nu(s,t)})$. The optimization variables are thus $\mathbf{x} = (\mathbf{u}_1^\top, \ldots, \mathbf{u}_{N_e}^\top, \mathbf{k}^\top)^\top$. The measurements should be reproduced as correctly as possible while the corresponding fields u_e must fulfill the Helmholtz equation in (8.3). To tackle ill-posedness, a regularization term is added to the objective function. Thus, the optimization problem can be formulated as

$$\min_{\mathbf{x}} \quad F(\mathbf{x}) = \tfrac{1}{2}\sum_{e=1}^{N_e}\sum_{m=1}^{N_m}|\mathbf{I}_m\mathbf{u}_e - u_{e,m}|^2 + \tfrac{\alpha}{2}R(\mathbf{k})$$

$$s.t. \quad \mathbf{c}_{\mathcal{E}}(\mathbf{x}) = \begin{pmatrix}\mathbf{A}(k) & & \\ & \ddots & \\ & & \mathbf{A}(k)\end{pmatrix}\begin{pmatrix}\mathbf{u_1} \\ \vdots \\ \mathbf{u_{N_e}}\end{pmatrix} - \begin{pmatrix}\mathbf{b}(k, u_1^i) \\ \vdots \\ \mathbf{b}(k, u_{N_e}^i)\end{pmatrix} = \mathbf{0},$$

$$\mathbf{c}_{\mathcal{I}}(\mathbf{x}) = \begin{pmatrix}k^+ - k \\ k - k^-\end{pmatrix} \geq \mathbf{0},$$

$$(8.7)$$

where $\mathbf{I}_m$ denotes a matrix for interpolation at x_m and $R(\mathbf{k})$ is a discretization of the regularization $R(\mathbf{k}) = \|\nabla k\|_2^2$, see e.g., [15]. The inequality constraints $\mathbf{c}_{\mathcal{I}}(\mathbf{k}) \geq \mathbf{0}$ represent additional knowledge such as box constraints, which are available for many practical cases.

Depending on the discretization and the number of incoming waves, the resulting optimization problem is computationally challenging. As an example, consider a simulation domain in two dimensions with 400 grid points in each dimension; i.e., $N = 1.6 \times 10^5$ mesh points. We assume to measure three incoming waves $(N_e = 3)$, for which the corresponding complex values are optimization variables. Including the control variables for $\mathbf{k}$, the discretized optimization problem has all in total 1.12×10^6 optimization variables. However, due to the mixed nonlinear product of the optimization variables $\mathbf{k}$ and $\mathbf{u}$ in the equality constraints representing the discretized PDE (8.6), the optimization problem is nonlinear and nonconvex. As a result, a number of local, but non-global minima for the inverse problem may exist. By taking previous knowledge into account, e.g., by using the formerly mentioned box constraints $\mathbf{c}_{\mathcal{I}}$, it is possible to gear the optimization algorithm to a good solution.

8.3 Large-Scale Nonlinear Optimization and `IPOPT`

In this section, we aim to demonstrate the effectiveness of combinatorial scientific computing techniques in nonlinear optimization, where we are concerned with finding local solutions of optimization problems of the form

$$\min_{x \in \mathbb{R}^n} F(\mathbf{x}) \quad \text{subject to } \mathbf{c}(\mathbf{x}) = \mathbf{0}, \quad \mathbf{x} \geq \mathbf{0}, \tag{8.8}$$

where the objective function $F : \mathbb{R}^n \longrightarrow \mathbb{R}$ and the constraint functions $\mathbf{c} : \mathbb{R}^n \longrightarrow \mathbb{R}^m$ are assumed to be sufficiently smooth (usually C^2). Note that problems with general nonlinear inequality constraints like (8.1) or (8.7) can be reformulated into the above form by means of slack variables.

In the remainder of this chapter, we briefly describe the particular optimization algorithm implemented in the open-source software package `IPOPT` [16], used for the numerical experiments in Section 8.7, in conjunction with the AD tool `dcc` (Section 8.5) and the parallel sparse linear solver package `PARDISO` (Section 8.6).

`IPOPT` implements an IPM [17] for nonlinear optimization. IPMs avoid the high combinatorial complexity introduced by inequality constraints by replacing them with a barrier term which is added to the objective function. One then obtains the barrier problem

$$\min_{\mathbf{x} \in \mathbb{R}^n} \varphi_\mu(\mathbf{x}) = F(\mathbf{x}) - \mu \sum_{i=0}^{n} \ln(\mathbf{x}_i) \quad \text{subject to} \quad \mathbf{c}(\mathbf{x}) = \mathbf{0}, \tag{8.9}$$

where $\mu > 0$ is a barrier parameter, and $\mathbf{x}_i$ denotes the ith component of the vector $\mathbf{x}$. Note that the logarithmic term in the objective function pushes the variable components away from their zero boundary, and that the degree of influence of this barrier term depends on the size of the barrier parameter. Under standard assumptions it can be shown that locally optimal solutions $\mathbf{x}_*(\mu)$ of (8.9) converges to local optima $\mathbf{x}_*$ of the original problem (8.8) as $\mu \to 0$. Therefore, the algorithm solves a sequence of barrier problems with fixed μ_l that is eventually driven to zero. Each barrier problem is solved to a convergence tolerance $O(\mu_l)$.

To describe the solution process of a particular barrier problem, consider its first-order optimality conditions

$$\begin{aligned} \nabla \varphi_{\mu_l}(\mathbf{x}) + \nabla \mathbf{c}(\mathbf{x})\lambda &= \mathbf{0}, \\ \mathbf{c}(\mathbf{x}) &= \mathbf{0}. \end{aligned} \tag{8.10}$$

Under standard assumptions, these conditions are satisfied at a local solution $\mathbf{x}_*$ of (8.9) for some suitable Lagrangian multipliers $\lambda_* \in \mathbb{R}^m$. Therefore, the algorithm consists of applying a variant of Newton's method to this nonlinear system of equations; i.e., search directions $(\Delta\mathbf{x}^k, \Delta\lambda^k)$ are obtained from the linearization of (8.10),

$$\begin{bmatrix} \mathbf{H}^k + \Sigma^k & \nabla \mathbf{c}(\mathbf{x}^k) \\ \nabla \mathbf{c}(\mathbf{x}^k)^\top & 0 \end{bmatrix} \begin{pmatrix} \Delta\mathbf{x}^k \\ \Delta\lambda^k \end{pmatrix} = - \begin{pmatrix} \nabla \varphi_\mu(\mathbf{x}^k) + \nabla \mathbf{c}(\mathbf{x}^k)\lambda^k \\ \mathbf{c}(\mathbf{x}^k) \end{pmatrix}. \tag{8.11}$$

Here,

$$\mathbf{H}^k = \nabla^2 F(\mathbf{x}^k) + \sum_{j=1}^m \lambda_j^k \nabla^2 \mathbf{c}_j(\mathbf{x}^k),$$

is the Hessian of the Lagrangian function for (8.8), and Σ^k is a diagonal matrix approximating the Hessian of the logarithmic barrier term in (8.9). Once a search direction $\Delta\mathbf{x}^k$ has been determined, a back-tracking line search is performed to find a step size $\alpha^k \in (0,1]$ that provides an "acceptable" trial point that makes sufficient progress towards a solution of the problem. Finally, the new iterates are obtained from $\mathbf{x}^{k+1} = \mathbf{x}^k + \alpha^k \Delta\mathbf{x}^k$ and $\lambda^{x+1} = \lambda^k + \alpha^k \Delta\lambda^k$. Details of the optimization algorithm in IPOPT can be found in [18].

8.4   Closed Form of Derivatives

The inverse medium problem can be stated as a PDE-constrained optimization problem, which induces the nonlinear optimization problem (8.7) after discretization. This finite dimensional nonconvex optimization problem with equality and inequality constraints is solved by an IPM, resulting in the linear system (8.11), which is solved by PARDISO. To setup the actual matrix, the appearing derivatives are evaluated in code generated by the algorithmic differentiation techniques implemented in dcc. In this section, we derive the matrix blocks analytically by hand. In doing so, we find evaluation functions and sparsity patterns that are recovered by the algorithmic differentiation tool, whose relative ease of use is illustrated. While hand-coding turns out to be a feasible alternative in the given situation, it is likely to become much harder for more complex problems, for example, if adaptive nonregular grids are used in 3D. The benefits of using tools for algorithmic differentiation will become even more obvious in such cases. Nevertheless, we consider the side-by-side discussion of both approaches enlightening in the context of this tutorial-style article.

For problem (8.7) the objective function is

$$F(\mathbf{x}) = \frac{1}{2}\sum_{e=1}^{N_e}\sum_{m=1}^{N_m}|\mathbf{I}_m\mathbf{u_e} - u_{e,m}|^2 + \frac{\alpha}{2}R(\mathbf{k}) \tag{8.12}$$

$$= \frac{1}{2}\sum_{e=1}^{N_e}\sum_{m=1}^{N_m}\mathfrak{Re}(\mathbf{I}_m\mathbf{u_e} - u_{e,m})^2 + \mathfrak{Im}(\mathbf{I}_m\mathbf{u_e} - u_{e,m})^2 + \frac{\alpha}{2}R(\mathbf{k}).$$

Due to the Sommerfeld boundary conditions, the discretized fields $\mathbf{u_e}$ are complex valued; F in (8.12) has to be differentiated with respect to the real part of each scattered wave $\mathfrak{Re}(\mathbf{u_e})$, its imaginary part $\mathfrak{Im}(\mathbf{u_e})$, and the control variable $\mathbf{k}$:

$$\nabla_{\mathfrak{Re}(\mathbf{u_e})}F = \sum_{m=1}^{N_m}\mathbf{I}_m^{\top}\mathfrak{Re}(\mathbf{I}_m\mathbf{u_e} - u_{e,m}),$$

$$\nabla_{\mathfrak{Im}(\mathbf{u_e})}F = \sum_{m=1}^{N_m}\mathbf{I}_m^{\top}\mathfrak{Im}(\mathbf{I}_m\mathbf{u_e} - u_{e,m}),$$

$$\nabla^2_{\mathfrak{Re}(\mathbf{u_e})}F = \nabla^2_{\mathfrak{Im}(\mathbf{u_e})}F = \sum_{m=1}^{N_m}\mathbf{I}_m^{\top}\mathbf{I}_m.$$

As regularization term, we chose the squared L^2-norm of the gradient $R(\mathbf{k}) = \|\nabla k\|_2^2$ exemplarily. Because $\Omega = (0,1)^2$, this is discretized as follows:

$$\|\nabla k\|^2 = \int_{\Omega}\nabla k\cdot\nabla k = \int_0^1\int_0^1\partial_1 k(x,y)^2 dy dx + \int_0^1\int_0^1\partial_2 k(x,y)^2 dx dy$$

$$= h_1\sum_{s=1}^{N_x}\int_0^1\partial_1 k\left((s - \text{\textonehalf})\,h_1, y\right)^2 dy + \mathcal{O}(h_1^2)$$

$$+ h_2\sum_{s=1}^{N_y}\int_0^1\partial_2 k\left(x, (s - \text{\textonehalf})\,h_2\right)^2 dx + \mathcal{O}(h_2^2)$$

$$= h_1 h_2 \left(\sum_{s,t} \partial_1 k \left((s - 1/2) h_1, t h_2 \right)^2 + \sum_{s,t} \partial_2 k \left(t h_1, (s - 1/2) h_2 \right)^2 \right)$$
$$+ \mathcal{O}(h_1^2 + h_2^2)$$

$$= h_1 h_2 \left(\sum_{s,t} \left(\frac{k(s h_1, t h_2) - k((s-1) h_1, t h_2)}{h_1} \right)^2 \right.$$
$$+ \left. \sum_{s,t} \left(\frac{k(t h_1, s h_2) - k(t h_1, (s-1) h_2)}{h_2} \right)^2 \right)$$
$$+ \mathcal{O}(h_1^2 + h_2^2)$$

$$= \frac{h_2}{h_1} \sum_{s,t} \left(\mathbf{k}_{\nu(s,t)} - \mathbf{k}_{\nu(s-1,t)} \right)^2 + \frac{h_1}{h_2} \sum_{s,t} \left(\mathbf{k}_{\nu(t,s)} - \mathbf{k}_{\nu(t,s-1)} \right)^2$$
$$+ \mathcal{O}(h_1^2 + h_2^2)$$

$$= \frac{h_2}{h_1} (\mathbf{D}_{\mathbf{o},1} \, \mathbf{k})^\top \mathbf{D}_{\mathbf{o},1} \, \mathbf{k} + \frac{h_1}{h_2} (\mathbf{D}_{\mathbf{o},2} \, \mathbf{k})^\top \mathbf{D}_{\mathbf{o},2} \, \mathbf{k} + \mathcal{O}(h_1^2 + h_2^2),$$

where in midpoint quadrature rule, trapezium quadrature rule, and central difference quotient, and the fact $k(x_{\nu(s,t)}) = k_0 = \mathbf{k}_{\nu(s,t)}$ on $\partial \Omega$ was used. $\mathbf{D}_{\mathbf{o},1}, \mathbf{D}_{\mathbf{o},2}$ denote one-sided difference matrices. Thus, the discretization of the regularization results in

$$R(\mathbf{k}) = \frac{h_2}{h_1} \mathbf{k}^\top \mathbf{D}_{\mathbf{o},1}^\top \mathbf{D}_{\mathbf{o},1} \, \mathbf{k} + \frac{h_1}{h_2} \mathbf{k}^\top \mathbf{D}_{\mathbf{o},2}^\top \mathbf{D}_{\mathbf{o},2} \, \mathbf{k},$$

and the differentiation terms read as

$$\nabla_{\mathbf{k}} R(\mathbf{k}) = 2 \left(\frac{h_2}{h_1} \mathbf{D}_{\mathbf{o},1}^\top \mathbf{D}_{\mathbf{o},1} \, \mathbf{k} + \frac{h_1}{h_2} \mathbf{D}_{\mathbf{o},2}^\top \mathbf{D}_{\mathbf{o},2} \, \mathbf{k} \right),$$

$$\nabla_{\mathbf{k}}^2 R(\mathbf{k}) = 2 \left(\frac{h_2}{h_1} \mathbf{D}_{\mathbf{o},1}^\top \mathbf{D}_{\mathbf{o},1} + \frac{h_1}{h_2} \mathbf{D}_{\mathbf{o},2}^\top \mathbf{D}_{\mathbf{o},2} \right),$$

Therefore, the Hessian matrix of the objective function is

$$\nabla_{\mathbf{x}}^2 F(\mathbf{x}) = \begin{pmatrix} \sum_m \mathbf{I}_m^\top \mathbf{I}_m & & & \\ & \ddots & & \\ & & \sum_m \mathbf{I}_m^\top \mathbf{I}_m & \\ & & & 2 \left(\frac{h_2}{h_1} \mathbf{D}_{\mathbf{o},1}^\top \mathbf{D}_{\mathbf{o},1} + \frac{h_1}{h_2} \mathbf{D}_{\mathbf{o},2}^\top \mathbf{D}_{\mathbf{o},2} \right) \end{pmatrix},$$
$$\tag{8.13}$$

which is a block-diagonal matrix, of which the blocks are sparse, symmetric and positive semidefinite, implying convexity of the objective function.

Taking a look at the constraints, we see that the inequality constraint terms are

$$
\mathbf{c}_{\mathcal{I}}(\mathbf{k}) = \begin{pmatrix} \mathbf{k}^{+} - \mathbf{k} \\ \mathbf{k} - \mathbf{k}^{-} \end{pmatrix},
$$

$$
\nabla_{\mathbf{k}} \mathbf{c}_{\mathcal{I}}(\mathbf{k}) = \begin{pmatrix} -\mathbf{I} \\ \mathbf{I} \end{pmatrix},
$$

$$
\nabla_{\mathbf{k}}^{2} \mathbf{c}_{\mathcal{I}}(\mathbf{k}) = \mathbf{0}.
$$

The structure of the equality constraint terms depends on the discretization scheme. For the general case we find from (8.7) that

$$
\nabla_{\mathbf{x}} \mathbf{c}_{\mathcal{E}}(\mathbf{x}) = \begin{pmatrix} \mathbf{A}(\mathbf{k}) & & & \mathbf{A}'(\mathbf{k}, \mathbf{u}_{1}) - \mathbf{b}'(\mathbf{k}, \mathbf{u}_{1}^{i}) \\ & \mathbf{A}(\mathbf{k}) & & \mathbf{A}'(\mathbf{k}, \mathbf{u}_{2}) - \mathbf{b}'(\mathbf{k}, \mathbf{u}_{2}^{i}) \\ & & \ddots & \vdots \\ & & \mathbf{A}(\mathbf{k}) & \mathbf{A}'(\mathbf{k}, \mathbf{u}_{N_{e}}) - \mathbf{b}'(\mathbf{k}, \mathbf{u}_{N_{e}}^{i}) \end{pmatrix},
$$

where $\mathbf{A}(\mathbf{k})$ is the discretized Helmholtz operator and

$$
\mathbf{A}'(\mathbf{k}, \mathbf{u}_{2}) - \mathbf{b}'(\mathbf{k}, \mathbf{u}_{2}^{i}) = \nabla_{\mathbf{k}}(\mathbf{A}(\mathbf{k})\mathbf{u}_{e}) - \mathbf{b}(\mathbf{k}, \mathbf{u}_{e}^{i})).
$$

The finite difference discretization of (8.4) and (8.5) results in

$$
\frac{\partial c_{\nu(s,t)}}{\partial k_{\nu(s,t)}} = -2\left(k_{\nu(s,t)} u_{\nu(s,t)} + k_{\nu(s,t)} u_{e}^{i}(x_{\nu(s,t)})\right)
$$

and

$$
\frac{\partial c_{\nu(s,t)}}{\partial k_{\nu(s,t)}} = -2\left(-\frac{i}{h_{2}} u_{\nu(s,t)} + k_{\nu(s,t)} u_{e}^{i}(x_{\nu(s,t)})\right) \tag{8.14}
$$

for an interior and upper boundary grid point (s, t), respectively, and zero for differentiation with respect to other components of $\mathbf{k}$. We denote this by

$$
\mathbf{A}'(\mathbf{k}, \mathbf{u}_{e}) - \mathbf{b}'(\mathbf{k}, \mathbf{u}_{e}^{i}) = -2(\tilde{\mathbf{K}}\mathbf{U}_{e} - \mathbf{K}\mathbf{U}_{e}^{i}), \tag{8.15}
$$

where $\mathbf{U}_{e} = \text{diag}(\mathbf{u}_{e})$, $\mathbf{K} = \text{diag}(\mathbf{k})$, and $\tilde{\mathbf{K}}$ is a diagonal matrix with entries according to (8.14). Note that with this discretization (8.15) is a (potentially singular) diagonal matrix. Since all second-order derivatives of (8.4) and (8.5) are zero except for $\frac{\partial^{2} c_{\nu}}{\partial k_{\nu}^{2}} = -2(\mathbf{u}_{\nu} + u_{e}^{i}(x_{\nu}))$ and $-2u_{e}^{i}(x_{\nu})$, and $\nabla_{\mathbf{k}, \mathbf{u}_{e}}^{2} c_{\mathcal{E}} = \tilde{\mathbf{K}}$, the Hessian block part at the upper left in (8.11) looks like

$$
\sum_{s \in \mathcal{E}} \lambda_{s} \nabla^{2} \mathbf{c}_{s}(\mathbf{x}) = \begin{pmatrix} & & & \mathbf{L}_{1}\tilde{\mathbf{K}} \\ & \mathbf{0} & & \vdots \\ & & & \mathbf{L}_{N_{e}}\tilde{\mathbf{K}} \\ \mathbf{L}_{1}\tilde{\mathbf{K}} & \cdots & \mathbf{L}_{N_{e}}\tilde{\mathbf{K}} & -2\sum_{e}\mathbf{L}_{e}(\tilde{\mathbf{U}}_{e} + \mathbf{U}_{e}^{i}) \end{pmatrix},
$$

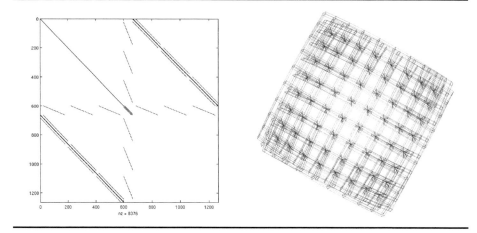

FIGURE 8.2: Sparsity pattern of the saddle-point linear system and the visualization of the graph with Graphviz.

where $\mathbf{L_e}$ denotes a diagonal matrix with entries corresponding to the Lagrangian multipliers of the **e**th discretized PDE.

Figure 8.2 shows the block structure of the step computation matrix (8.11). The submatrices of the Jacobian matrix consists of several discretized PDEs, in which each PDE is represented by a block-tridiagonal five-point finite-difference stencil matrix. On the right hand side a Graphviz visualization [19] of this matrix is presented. Note that the discretization mesh is clearly visible and the layers represent the real and imaginary part of the Helmholtz operator which are connected by the boundary conditions. On each interior grid point the layers are connected to an additional **k**-layer. In the next section, we describe algorithmic differentiation techniques that efficiently and automatically compute the Jacobian and Hessian matrices.

8.5 Algorithmic Differentiation

We use algorithmic differentiation (AD) [1] to compute the first and second derivatives required by IPOPT to solve (8.7). Our derivative code compiler dcc implements both forward (or tangent-linear) and reverse (or adjoint) modes by source code transformation for a restricted but sufficiently large subset of C++. Higher derivative codes can be generated by reapplication of dcc to its own output. Some user intervention may be required to implement the data-flow reversal in adjoint versions of large-scale numerical simulation codes.

For many nonlinear programming problems the generation of first and higher derivative codes using dcc is entirely automatic.

The user implements both the constraint functions $\mathbf{c}(\mathbf{x}, p) \in \mathbb{R}^m$ and the objective function $F(\mathbf{x}, p) \in \mathbb{R}$. Here, $\mathbf{x} \in \mathbb{R}^n$ are the optimization variables, and $p \in \mathbb{R}^{n_p}$ are input parameters (e.g., data) that are kept constant during the optimization. The IPOPT-dcc interface expects generic signatures

```
void constraints(
   int* ip,   // (in)   integer parameters
   double* x, // (in)   active state
   double* p, // (in)   passive double parameters
   double* c  // (out)  active residual  );
```

and

```
void objective(
   int* ip,     // (in)   integer parameters
   double* x,   // (in)   active state
   double* p,   // (in)   passive double parameters
   double& obj  // (out)  active objective )
```

with the indicated input/output-structure ((in)/(out)). All array sizes are expected to be static. They are provided by the user through corresponding preprocessor macros; for example,

```
#define N_IP 2  // number of integer parameters
#define N_X 42  // state size (n)
#define N_P 45  // number of passive double parameters
#define N_C 24  // number of constraints (m)
```

While most nonlinear programming problems can be implemented in the subset of C++ currently accepted by dcc, ongoing development aims to relax the current restrictions on the accepted syntax and semantics. The focus, however, is clearly on the generation of efficient first- and higher-order tangent-linear and adjoint code rather than on language coverage.

For given implementations of $\mathbf{c}(\mathbf{x}, p)$ and $F(\mathbf{x}, p)$, dcc generates derivative codes for the evaluation of the tangent-linear models

$$\mathbf{c}^{(1)} = \nabla_x \mathbf{c}(\mathbf{x}, p) \cdot \mathbf{x}^{(1)} \in \mathbb{R}^m,$$
$$F^{(1)} = \nabla_x F(\mathbf{x}, p) \cdot \mathbf{x}^{(1)} \in \mathbb{R},$$

and the adjoint models

$$\mathbf{x}_{(1)}^T = \nabla_x \mathbf{c}(\mathbf{x}, p)^T \cdot \mathbf{c}_{(1)} \in \mathbb{R}^m,$$
$$\mathbf{x}_{(1)}^T = \nabla_x F(\mathbf{x}, p)^T \cdot F_{(1)} \in \mathbb{R}^n.$$

Superscripts/subscripts of variables denote the order of differentiation in tangent-linear/adjoint mode. Reapplication of dcc to its own output in tangent-linear mode yields the second-order tangent-linear models

$$\mathbf{c}^{(1,2)} = \left\langle \nabla_{xx}\mathbf{c}(\mathbf{x},p), \mathbf{x}^{(1)}, \mathbf{x}^{(2)} \right\rangle \in \mathbb{R}^m,$$

$$F^{(1,2)} = \left\langle \nabla_{xx}F(\mathbf{x},p), \mathbf{x}^{(1)}, \mathbf{x}^{(2)} \right\rangle \in \mathbb{R},$$

and the second-order adjoint models

$$\mathbf{x}_{(1)}^{(2)} = \left\langle \mathbf{c}_{(1)}, \nabla_{xx}\mathbf{c}(\mathbf{x},p), \mathbf{x}^{(2)} \right\rangle \in \mathbb{R}^n, \tag{8.16}$$

$$\mathbf{x}_{(1)}^{(2)} = \left\langle F_{(1)}, \nabla_{xx}F(\mathbf{x},p), \mathbf{x}^{(2)} \right\rangle \in \mathbb{R}. \tag{8.17}$$

We use $\langle \mathbf{T}, \mathbf{x}, \mathbf{y} \rangle$ and $\langle \mathbf{z}, \mathbf{T}, \mathbf{x} \rangle$ to denote projections of the symmetric (Hessian) tensor $\mathbf{T} = (t_{i,j,k}) \in \mathbb{R}^{m \times n \times n}$ onto $\mathbb{R}^m$ and $\mathbb{R}^n$, respectively. In particular, $\langle \nabla_{xx}F(\mathbf{x},p), \mathbf{x}^{(1)}, \mathbf{x}^{(2)} \rangle = (\mathbf{x}^{(1)})^T \cdot \nabla_{xx}F(\mathbf{x},p) \cdot \mathbf{x}^{(2)}$ and $\langle f_{(1)}, \nabla_{xx}F(\mathbf{x},p), \mathbf{x}^{(2)} \rangle = f_{(1)} \cdot \nabla_{xx}F(\mathbf{x},p) \cdot \mathbf{x}^{(2)}$. This notation is well-defined as $t_{i,j,k} = t_{i,k,j}$ for $i = 1, \ldots, m$ and $j, k = 1, \ldots, n$.

While dcc is able to generate second-order adjoint code in adjoint-over-tangent-linear and adjoint-over-adjoint modes, the resulting codes typically lag behind the result of tangent-linear-over-adjoint mode in terms of computational efficiency. Hence, the latter turns out to be most suitable for the computation of the derivatives required by IPOPT that are discussed in the following section.

8.5.1 Derivative Codes by dcc

IPOPT requires three types of derivatives. The Jacobian of the constraints $\nabla_x \mathbf{c}(\mathbf{x},p)^\top$ can be computed by the tangent-linear or adjoint versions of $\mathbf{c}(\mathbf{x},p)$ generated by dcc. Adjoint mode should be favored whenever $K \cdot m < n$, where K denotes the constant runtime ratio of a single execution of the adjoint versus a single execution of the tangent-linear code. This factor varies depending on the AD tool,[1] its underlying algorithmic strategy, e.g., operator overloading (see, for example, [5]) or source transformation, and the compilers' program analysis and code generation capabilities. A large fraction of the ongoing research in the field of AD is aimed towards the minimization of K. Typically, its value ranges between 3 and 50. For code generated by dcc we observe factors between 3 and 10, depending on the properties of the input code.

The gradient of the objective $\nabla_x F(\mathbf{x},p)$ should be computed by the adjoint code if $K < n$. Pragmatically, one might choose to relax this condition to $P \cdot K < n$, where $P \geq 1$. Remember that the generation of adjoint code may not be entirely automatic if, for example, the memory requirement of a standard adjoint code exceeds the available resources. Some restructuring of the original code may be required in order to exploit dcc's *joint* call graph

---

[1] Refer to www.autodiff.org for a list of AD tools available.

reversal scheme [1] for a trade-off between memory requirement and arithmetic complexity. Feasible *subroutine argument checkpoints* need to be defined by the user, implying the need for considerable insight into the logic of the adjoint code generation algorithm. In some situations the user may be willing to accept reduced efficiency as the price for a fully automatically generated derivative code. In large-scale nonlinear optimization the computation of the gradient of the objective in tangent-linear mode is infeasible.

The Hessian of the (weighted) Lagrangian

$$\nabla_{xx}\mathcal{L}(\mathbf{x}, p, v, \lambda) \equiv v \cdot \nabla_{xx}F(x, p) + \langle \lambda, \nabla_{xx}\mathbf{c}(\mathbf{x}, p) \rangle$$

can be computed by a second-order adjoint code obtained by applying dcc in tangent-linear mode to the previously generated adjoint version of the Lagrangian. In fact, the same method is applied to compute a scaled Hessian of the objective and a projection onto $\mathbb{R}^{n \times n}$ of the Hessian of the constraints. For this purpose we set $\mathbf{c}_{(1)} = \lambda$ and $F_{(1)} = v$ in (8.16) and (8.17), respectively.

First Derivative Codes

Let implementations of the constraints and the objective be given in a file named f.cpp. We run

$(PATH\_TO\_DCC\_EXE)/dcc\ f.cpp\ t1.conf$

to build the tangent-linear code in t1_f.cpp. Mode and order of differentiation are specified through respective entries in the configuration file t1.conf. A prefix t1_ is attached to names generated in first-order tangent-linear mode. For example, the signature of the tangent-linear constraints becomes

```
void t1_constraints (
  int* ip,
  double* x,  double* t1_x,
  double* p,  double* t1_p,
  double* c,  double* t1_c
)
```

The tangent-linear code can be used to compute directional derivatives of c with respect to both x and p. We set t1_p=0 for use with IPOPT.

The Jacobian of the constraints can be computed in tangent-linear mode by letting t1_x range over the Cartesian basis vectors in $\mathbb{R}^n$. The columns of the Jacobian $Jac\_c \equiv \nabla_x\mathbf{c}(\mathbf{x}, p)^\top$ are returned through t1_c as illustrated by the following code fragment.

```
for (int i=0;i<N_X;i++) {
  t1_x[i]=1;
  t1_constraints(ip,x,t1_x,p,t1_p,c,t1_c);
  for (int j=0;j<N_C;j++) Jac_c[j][i]=t1_c[j];
  t1_x[i]=0;
}
```

Adjoint constraints should be used for the accumulation of $\nabla_x \mathbf{c}(\mathbf{x}, p)$ if the value of N_X is considerably larger than N_C, as discussed at the beginning of this section. Detection and exploitation of sparsity is crucial for large problems and will be discussed in Section 8.5.2.

With f.cpp given as before we run

$(PATH_TO_DCC_EXE)/dcc f.cpp a1.conf

to build the adjoint code a1_f.cpp. The configuration file a1.conf contains information on variable names and static sizes of the stacks used for data-flow reversal in addition to mode and order of differentiation. The prefix a1_ is attached to names generated in first-order adjoint mode. For example, the signature of the adjoint objective becomes

```
void a1_objective(
  int a1_mode,
  int* ip,
  double* x, double* a1_x,
  double* p, double* a1_p,
  double& obj, double& a1_obj
)
```

An integer parameter (a1_mode) selects between various modes that become relevant only in the context of interprocedural adjoint code. Passing a1_mode =1 runs the adjoint code. The gradient of the objective is computed by a single run of the adjoint code with a1_obj set to one.

```
a1_obj=1;
a1_objective(1,ip,x,a1_x,p,a1_p,obj,a1_obj);
```

The gradient is returned in a1_x.

Second Derivative Codes

With a1_f.cpp containing implementations of the adjoint constraints and the adjoint objective, we run

$(PATH_TO_DCC_EXE)/dcc a1_f.cpp t2_a1.conf

to build the second-order adjoint code t2_a1_f.cpp containing second-order adjoint versions of the constraints

```
void t2_a1_constraints(
  int a1_mode,
  int* ip,
  double* x, double* t2_x, double* a1_x, double* t2_a1_x,
  double* p, double* t2_p, double* a1_p, double* t2_a1_p,
  double* c, double* t2_c, double* a1_c, double* t2_a1_c
)
```

and of the objective

```
void t2_a1_objective (
  int al_mode,
  int* ip,
  double* x, double* t2_x, double* a1_x, double* t2_a1_x,
  double* p, double* t2_p, double* a1_p, double* t2_a1_p,
  double* obj, double* t2_obj, double* a1_obj, double*
      t2_a1_obj
)
```

The Hessian of the Lagrangian $v \cdot \nabla_{xx} F(\mathbf{x}, p) + \langle \lambda, \nabla_{xx} \mathbf{c}(\mathbf{x}, p) \rangle$ is computed in two steps. For $\langle \lambda, \nabla_{xx} \mathbf{c}(\mathbf{x}, p) \rangle$ we use the second-order adjoint constraints with a1_c = λ.

```
for (int j=0;j<N_X;j++) {
  for (int i=0;i<N_X;i++) a1_x[i]=t2_x[i]=t2_a1_x[i]=0;
  t2_x[j]=1;
  for (int i=0;i<N_C;i++) a1_c[i]=lambda[i];
  t2_a1_constraints (1,ip,
                     x,t2_x,a1_x,t2_a1_x,
                     p,t2_p,a1_p,t2_a1_p,
                     c,t2_c,a1_c,t2_a1_c);
  for (int i=0;i<N_X;i++) hessL[j][i]=t2_a1_x[i];
}
```

The resulting projection of the Hessian of the constraints into $\mathbb{R}^{n \times n}$ is added to the v-scaled Hessian of the objective. The latter is computed using the second-order adjoint objective with a1_obj = v.

```
for (int j=0;j<N_X;j++) {
  for (int i=0;i<N_X;i++) a1_x[i]=t2_x[i]=t2_a1_x[i]=0;
  for (int i=0;i<N_P;i++) t2_p[i]=a1_p[i]=t2_a1_p[i]=0;
  obj=t2_obj=t2_a1_obj=0;
  t2_x[j]=1;
  v=1;
  t2_a1_objective (1,ip,
                   x,t2_x,a1_x,t2_a1_x,
                   p,t2_p,a1_p,t2_a1_p,
                   obj,t2_obj,v,t2_a1_obj);
  for (int i=0;i<N_X;i++)
    hessL[j][i]=hessL[j][i]+t2_a1_x[i];
}
```

Again, the detection and exploitation of sparsity in the Hessian of the objective and the constraints is crucial when dealing with large-scale nonlinear programming problems.

8.5.2 Detection and Exploitation of Sparsity

The theoretical and algorithmic challenges behind the detection and exploitation of sparsity in first and second derivative tensors are discussed in

Chapter 5. Three types of derivatives need to be considered in the context of nonlinear continuous optimization with IPOPT: the gradient $\nabla_x F(\mathbf{x}, p)$ of the objective, the Jacobian $\nabla_x \mathbf{c}(\mathbf{x}, p)^\top$ of the constraints $\mathbf{c}$, and the Hessian $\nabla_{xx} \mathcal{L}(\mathbf{x}, p, v, \lambda)$ of the Lagrangian $\mathcal{L}$. The gradient $\nabla_x F(\mathbf{x}, p) \in \mathbb{R}^n$ of the objective can be expected to be dense. We take a closer look at the handling of the remaining two derivative matrices.

Jacobian $\nabla_x \mathbf{c}(\mathbf{x}, p)^\top$ of the Constraints

IPOPT requires the Jacobian of the constraints in the standard sparse (row index, column index, value) coordinate format. We use ColPack [13] for the exploitation of sparsity in $\nabla_x \mathbf{c}^\top$. ColPack requires the sparsity pattern of $\nabla_x \mathbf{c}^\top$ as input. Therefore, the runtime library of dcc provides the special data type jsp together with overloaded versions of the arithmetic operations and intrinsic functions. dcc is used to change the types of all floating-point variables to jsp. A single execution of the overloaded implementation of the constraints computes dependencies of the outputs c on the active inputs x. The data type jsp augments the original variables with sets of appropriate variable references. While technical details of the implementation are beyond the scope of this chapter, we note that the efficiency of this propagation depends to a large extend on the chosen container types and the speed of the union operation. Our present version of jsp exhibits satisfactory performance as documented in Section 8.7.

ColPack expects the sparsity pattern as an integer matrix SP $\in \{\mathbb{N} \cup 0\}^{m \times \hat{n}}$, where $0 \le \hat{n} \le n$ denotes the maximal number of nonzero entries per row in $\nabla_x \mathbf{c}^\top$. It lists for each constraint $\mathbf{c}_j$, $j = 0, \ldots, m - 1$, the indexes i of those $\mathbf{x}_i$ that the value of c_j depends on. Internally, the bipartite graph of $\nabla_x \mathbf{c}^\top$ is built and colored using a sequential partial distance-2 coloring algorithm with an appropriate heuristic to order the vertices, for example, *smallest-last* [13]. The number of columns k in the resulting seed matrix S $\in \{0, 1\}^{n \times k}$ is equal to the number of colors in ColPack's approximate solution to the NP-hard coloring problem.

The compressed Jacobian of the constraints CJac_c $\in \mathbb{R}^{m \times k}$ is obtained by executing the tangent-linear version of the constraints generated by dcc k times computing directional derivatives of the contraints with respect to x in the directions given by the columns of S.

```
double** CJac_c=new double *[N_C];
for(int i=0; i<N_C; i++) CJac_c[i] = new double[k];
for (int j=0; j<k; j++) {
    for (int i=0; i<N_X; i++) t1_x[i] = S[i][j];
    t1_constraints(ip,x,t1_x,p,t1_p,c,t1_c);
    for (int i=0; i<N_C; i++) CJac_c[i][j] = t1_c[i];
}
```

ColPack's recovery routine extracts $\nabla_x \mathbf{c}^\top$ in coordinate format. It returns three vectors of equal lengths containing the nonzero entries of $\nabla_x \mathbf{c}^\top$ as well as the corresponding row and column indexes.

Hessian $\nabla_{xx}\mathcal{L}(\mathbf{x}, p, v, \lambda)$ of the Lagrangian

Detection and exploitation of sparsity in the Hessian $\nabla_{xx}\mathcal{L}(\mathbf{x}, p, v, \lambda) = v \cdot \nabla_{xx}F(\mathbf{x}, p) + \langle \lambda, \nabla_{xx}\mathbf{c}(\mathbf{x}, p) \rangle$ of the Lagrangian is performed separately for the two terms $v \cdot \nabla_{xx}F$ and $\langle \lambda, \nabla_{xx}\mathbf{c} \rangle$ followed by adding them to get $\nabla_{xx}\mathcal{L}$ in sparse coordinate format.

The sparsity patterns of the Hessians $v \cdot \nabla_{xx}F \in \mathbb{R}^{n \times n}$ and $\langle \lambda, \nabla_{xx}\mathbf{c} \rangle \in \mathbb{R}^{n \times n}$ are computed using the special data type hsp provided by the runtime library of dcc. Similar to the first-order dependencies in jsp, overloaded versions of the arithmetic operations and intrinsic functions enable the propagation of second-order dependencies of variables of type hsp. A careful approach to the implementation of the hsp library is crucial to ensure satisfactory runtime performance. Here we use a conservative overestimation of the Hessian sparsity pattern based on partial separability [20]. Ongoing research and development aims to push the efficiency limits of available implementations. Refer to Section 8.7 for evidence on the good quality of the current version of hsp.

ColPack applies its star coloring algorithm to the adjacency graphs of $v \cdot \nabla_{xx}F$ and $\langle \lambda, \nabla_{xx}\mathbf{c} \rangle$ followed by the generation of both seed matrices S_f and S_c. The two second-order adjoint codes

```
t2_a1_objective (1 , ip ,
                 x , t2_x , a1_x , t2_a1_x ,
                 p , t2_p , a1_p , t2_a1_p ,
                 obj , t2_obj , a1_obj , t2_a1_obj );
```

and

```
t2_a1_constraints (1 , ip ,
                   x , t2_x , a1_x , t2_a1_x ,
                   p , t2_p , a1_p , t2_a1_p ,
                   c , t2_c , a1_c , t2_a1_c );
```

generated by dcc are seeded individually. To get $v \cdot \nabla_{xx}F$ we set a1_obj $= v$ and we let t2_x range over the columns of S_f. The columns of the compressed scaled Hessian CHess_f are returned through t2_a1_x. Similarly, for $\lambda^T \nabla_{xx}\mathbf{c}$ we set a1_c $= \lambda$ and we let t2_x range over the columns of S_c. Again, the columns of the compressed projected Hessian tensor CHess_c are returned through t2_a1_x.

Representations of the two Hessians in coordinate format are retrieved by ColPack's corresponding recovery routine. Pairwise merges of the three vectors containing the nonzero entries as well as the corresponding row and column indexes, respectively, yield the Hessian of the Lagrangian in coordinate format.

8.6   Sparse Linear Algebra and PARDISO

The efficient solution of the saddle-point matrix in (8.11) is of utmost importance for the nonlinear optimization process. Due to the discretization of the PDE-constrained optimization problem, linear symmetric indefinite systems with millions to hundreds of millions of unknowns have to be solved. To do this within reasonable time requires efficient use of powerful parallel computers and/or advanced combinatorial and numerical algorithms based on direct or approximate direct factorizations [21]. To date, only very limited software for such large systems is generally available and the PARDISO software addresses this issue.[2]

In this section, some important combinatorial aspects and main algorithmic features for solving sparse systems will be reviewed. The algorithmic improvements of the past twenty years have reduced the time required to factor sparse symmetric matrices by almost three orders of magnitude. Combined with significant advances in the performance to cost ratio of computing hardware during this period, current sparse solver technology makes it possible to solve quickly and easily problems that, until recently, might have been considered far too large. This section discusses the basic and the latest developments for sparse direct solution methods that have led to modern LDL^T decomposition techniques for the Jacobian and Hessian matrices arising in large-scale nonlinear optimization [10, 23, 24].

It is the purpose of this chapter to describe the main combinatorial and numerical algorithms used in the parallel linear solver PARDISO.

As described in Section 8.3, the solution of symmetric indefinite saddle-point matrices represents the main computational kernel in nonlinear programming. In this section, a pivoting method is described that aims at both robustness and computational performance. We will use a Level-3 BLAS left-looking factorization as described in [25, 11, 26] based on a supernode pivoting approach. An interchange among the rows and columns of a supernode of diagonal size n_s, referred to as supernode Bunch–Kaufman pivoting, has no effect on the overall fill-in, and this is the mechanism for finding a suitable pivot in the LDL^T factorization method used in PARDISO. However, there is no guarantee that the numerical factorization algorithm would always succeed in finding a suitable pivot within a diagonal block related to a supernode. When the algorithm reaches a point where it cannot factor the supernode based on a 1×1 and 2×2 pivoting approach, it uses a pivot perturbation strategy similar to that described in [27]. The magnitude of the potential pivot is tested against a constant threshold of $\epsilon \cdot ||A||_1$, where ϵ is a half-machine precision perturbation. Therefore, any tiny pivots encountered during elimination are set to $\text{sign}(a_{ii}) \cdot \epsilon \cdot ||A||_1$ — this trades off some numerical stability for the abil-

---

[2]The interested reader can find a comparative study for symmetric indefinite matrices in [22].

$\gamma_1 := |a_{r1}| = \max_{k=2,\ldots,n_s} |a_{k1}|$
    with diagonal block of size n_s
$\gamma_r \geq \gamma_1$ is the magnitude of the largest
    off-diagonal in the r-row of the block
if $\max(|a_{11}|, \gamma_1) \leq \epsilon \cdot ||A||_1$:
    use pivot perturbation:
        $\tilde{a}_{11} = sign(a_{11}) \cdot \epsilon \cdot ||A||_1$
    use perturbed $\tilde{a}_{11}$ as a 1×1 pivot.
else if $|a_{11}| \geq \alpha\gamma_1$:
    use a_{11} as a 1×1 pivot.
else if $|a_{11}| \cdot \gamma_r \geq \alpha\gamma_1^2$:
    use $|a_{11}|$ as a 1×1 pivot.
else if $|a_{rr}| \geq \alpha\gamma_r$:
    use $|a_{rr}|$ as a 1×1 pivot.
else

    use $\begin{pmatrix} a_{11} & a_{r1} \\ a_{r1} & a_{rr} \end{pmatrix}$ as a 2×2 pivot

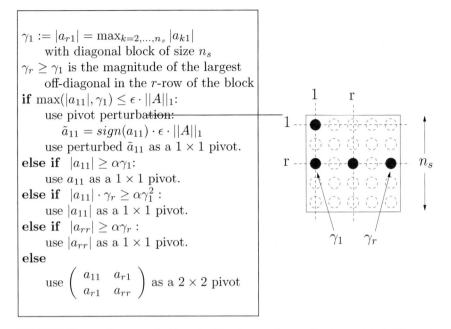

FIGURE 8.3: Supernode Bunch–Kaufman pivot selection with half-machine precision perturbation.

ity to keep pivots from getting too small. The result of this pivoting approach is that the factorization is, in general, not accurate, and iterative refinement is needed.

Figure 8.3 describes the usual 1×1 and 2×2 Bunch–Kaufman pivoting strategy within the diagonal block corresponding to a supernode of size n_s. The pivoting strategy is supplemented with half-machine precision perturbation techniques. The Bunch–Kaufman pivoting method computes two scalars γ_1 and γ_r. The scalar γ_1 is the largest off-diagonal element, e.g., $|a_{r1}|$, in the first column of the diagonal block corresponding to the supernode of size n_s. γ_r is the largest off-diagonal element in the corresponding row r. The scalar $\alpha = (\sqrt{17} + 1)/8$ is chosen to minimize the element growth. With this choice, the element growth in the diagonal block after k steps is bounded by the factor $(2.57)^{k-1}$.

The algorithm then selects either 1×1 and 2×2 pivots for the factorization. If both $|a_{11}|$ and $|\gamma_1|$ are too small, e.g., smaller than $\epsilon \cdot ||A||_1$, we apply the pivot perturbation technique as described above.

8.6.1 Graph-Based Pivoting Methods

This subsection introduces graph algorithms used to preprocess sparse linear systems of equations. After an introduction of the basic concepts, the graph matching algorithms are briefly reviewed. In the second part of the subsection, motivations for symmetric algorithms are given, and the approaches used in PARDISO are introduced. When looking back to classical partial pivoting approaches [28], the corresponding approach to the nonsymmetric a priori pivoting would be an a priori symmetric 2×2 block pivoting. The first to propose such an approach were Duff and Pralet, and Schenk and Gärtner [29, 11]. Further elaborations on the topic were given by Hagemann and Schenk [12] in the context of iterative methods.

Numerical stability in the decomposition is typically maintained through partial pivoting, which can have a significant impact on the factorization speed. Row interchanges due to partial pivoting can unpredictably affect the nonzero structure of the factor, thus making it impossible to statically allocate data structures on high-performance architectures. Using nonsymmetric row or column permutations to ensure a *nonzero diagonal* or to maximize the product of the absolute diagonal values are among the techniques often used as preprocessing steps for LU factorizations in order to reduce the number of dynamic pivoting steps. The original idea, on which these nonsymmetric permutations are based, is to find a *maximum weighted matching* of a *bipartite graph*. Finding a maximum weighted matching is a well-known assignment problem in operation research and combinatorial analysis, and it will now be applied to challenging problems in numerical linear algebra and nonlinear optimization.

As discussed above, dynamic pivoting has been a central tool by which nonsymmetric sparse linear solvers gain stability. Therefore, improvements in speeding up direct factorization methods were limited to the uncertainties that have arisen from using pivoting. Certain techniques, like the column elimination tree [30], have been useful for predicting the sparsity pattern despite pivoting. However, in the symmetric case, the situation becomes more complicated since only symmetric reorderings, applied to both columns and rows, are required, and no a priori choice of pivots is given. This makes it almost impossible to predict the elimination tree in a sensible manner, and the use of cache-oriented level-3 BLAS is impossible.

With the introduction of symmetric maximum weighted matchings [11] as an alternative to complete pivoting [31], it is now possible to treat symmetric indefinite systems similarly to symmetric positive definite systems. This allows us to predict fill using the elimination tree, and thus allows us to set up the data structures that are required to predict dense submatrices (also known as supernodes). This in turn means that one is able to exploit level-3 BLAS applied to the supernodes. Consequently, the classical Bunch–Kaufman pivoting approach needs to be performed only inside the supernodes. This approach has recently been successfully implemented in symmetric indefinite version of PARDISO [11]. As a major consequence of this novel approach, the sparse indefinite solver has been improved to become almost as efficient as its symmetric positive definite analogy.

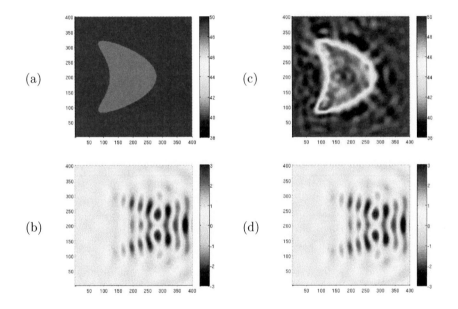

FIGURE 8.4: (See color insert.) Reconstruction: True control function (a), one incident wave (b), and reconstructed control function (c) and wave profile according to the incident wave (d).

TABLE 8.1: Size of the optimization problem for four different discretizations.

Mesh Size	# Optimization Variables	# Equality Constraints	# Inequality Constraints	# nnz in Newton System
100^2	69'604	60'000	59'208	1'044'876
200^2	279'204	240'000	78'408	4'208'676
300^2	628'804	540'000	177'608	9'492'476
400^2	1'118'404	960'000	316'808	16'896'276

TABLE 8.2: Runtime in seconds of the different phases (sparsity pattern evaluation, sparse matrix coloring, Hessian and Jacobian evaluation, linear solver, remainder IPM algorithm) and number of Newton-type iterations to solve the inverse medium optimization problem.

Mesh Size	Sparsity Pat. (s)	Matrix Col. (s)	Function Eval. (s)	Linear Solver (s)	Rem. IPM Alg. (s)	# Iter.
100^2	0.8	0.3	9	359	8	22
200^2	3.2	1.4	53	3'083	56	29
300^2	6.6	3.7	115	13'088	194	28
400^2	11.7	6.9	210	29'257	440	29

8.7 Numerical Experiments

We applied the nonlinear optimization and algorithmic differentiation techniques to recover the k-parameter function in computational wave propagation. Here, a kite-shaped object with $k(x) = 48$, $x \in \Omega_{\text{Kite}}$, is hidden in a background medium of $k(x) = 40$ $x \in \Omega/\Omega_{\text{Kite}}$ (see Figure 8.4 (a)). Since the background medium is known and covers the hidden object completely, we fixed the control parameter k on the boundary to the true value. In order to obtain the boundary values $u_{e,m}$, we solved the forward problem for three different scattered waves ($e = (1, 2, 3)$), and linearly interpolated them at 160 uniformly distributed measurement points at the boundary. An example for a simulated scattered wave is shown in Figure 8.4 (b), where the 2D object is irradiated from the left. We then applied the NLP and AD techniques to the inverse medium problem (8.7), where the Hessian and Jacobian matrices were automatically generated using the differentiation tool dcc and solved by the sparse linear algebra package PARDISO. Our fast and automatic inversion approach is able to reconstruct the hidden object in a highly scattering medium, and the reconstructed k-profile and the corresponding wave is shown in Figures 8.4 (c) and 8.4 (d).

The problem was solved on several 2D grids of size 100^2 up to 400^2 grid

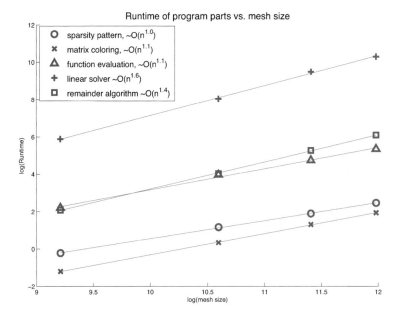

FIGURE 8.5: Complexity of the different phases (coloring, Hessian and Jacobian evaluation, linear solver) in the inverse medium optimization problem.

points resulting in large-scale nonlinear optimization problems up to $1.1 \cdot 10^6$ optimization variables. Table 8.1 shows a few details of the inverse medium problems, namely the number of optimization variables, the number of equality and inequality constraints, and the number of nonzero entries in the Hessian and Jacobian. Due to the fixed control parameter on the boundary, the number of variables does not scale exactly with the mesh size. The source code presented to dcc contains 258 lines of code (loc) and yields a tangent-linear code with 1134 loc and first- and second-order adjoint codes with 1758 loc and 8341 loc, respectively.

The interior-point optimization method converged in 22 up to 29 iterations. The overall runtime can be subdivided into the initialization, which mainly consists of sparsity pattern evaluation and sparse structure coloring; the generation and the evaluation of the derivative functions to set up the Hessian and Jacobian; the solution of the saddle-point matrices; and the remainder of the interior point algorithm. These data are shown in Table 8.2. It is interesting to observe that the generation of the Jacobian and Hessian matrices can be done completely automatically in a very small amount of runtime. The conservative algorithm for computing the Hessian sparsity pattern outperforms exact algorithms such as the one described in [32] by several orders of magnitude. For a problem of size 400^2 less than 10 seconds are needed as opposed to 3400 seconds when computing the exact pattern. Note that similar compression rates can be achieved. Smallest-Last ordering for sequential coloring yields a total of six evaluations of the second-order adjoint code. Natural ordering reduces this number to five. In the given case study, the gap between the number of colors used for the compression of the Jacobians and Hessians and the achievable minimum is typically very small and it vanishes for appropriate choices of the ordering heuristic. Otherwise, indirect compression may turn out superior to direct methods. See also Chapter 5.

Given the forward simulation code, it is today possible to solve highly non-linear 2D inverse medium problems in a resonable amount of time. Figure 8.5 fits the data from Table 8.2 to a monomial in order to find the total experimental complexity. From the theory it is well known that the complexity for the factorization of the saddle-point matrices is $O(n^{1.5})$, where n respresents the matrix size, that is, the number of optimization variables and constraints. This can also be observed in Figure 8.5.

The algorithms and the software that have been described in the previous sections can be very beneficial for PDE-constrained optimization applications. Overall, it is shown that by using NLP and AD techniques, highly nonlinear inverse medium problems with over one million inversion parameters, including state and control variables, and millions of nonzeros in the KKT system can be automatically inverted in less than 9 hours on a desktop computer.

Bibliography

[1] Colton, D. and Kress, R., *Inverse acoustic and electromagnetic scattering theory*, Springer Verlag, 1998.

[2] Biegler, L., Ghattas, O., Heinkenschloss, M., and van Bloemen Waanders, B., *Large-Scale PDE Constrained Optimization*, Vol. 30, Lecture Notes in Computational Science and Engineering, Springer-Verlag, 2003.

[3] Nocedal, J. and Wright, S., *Numerical Optimization*, Springer, New York, USA, 2006.

[4] Biros, G. and Ghattas, O., "Parallel Lagrange-Newton–Krylov-Schur methods for PDE-constrained optimization. II: The Lagrange-Newton solver and its application to optimal control of steady viscous flows." *SIAM J. Scientific Computing*, Vol. 27, No. 2, 2005, pp. 714–739.

[5] Biros, G. and Ghattas, O., "Parallel Lagrange-Newton-Krylov-Schur methods for PDE-constrained optimization. I: The Krylov-Schur solver." *SIAM J. Scientific Computing*, Vol. 27, No. 2, 2005, pp. 687–713.

[6] Haber, E. and Ascher, U., "Preconditioned all-at-once methods for large, sparse parameter estimation problems," *Inverse Problems*, Vol. 17, 2001, pp. 1847–1864.

[7] Fourer, R., Gay, D. M., and Kernighan, B. W., *AMPL: A Modeling Language for Mathematical Programming*, 2nd ed., 2002.

[8] Brook, A., Kendrick, D., and Meeraus, A., "GAMS, a User's Guide," *ACM SIGNUM Newsletter*, Vol. 23, No. 3-4, 1988, pp. 10–11.

[9] Griewank, A. and Walther, A., *Evaluating derivatives: principles and techniques of algorithmic differentiation*, Society for Industrial and Applied Mathematics (SIAM), 2008.

[10] Schenk, O., Wächter, A., and Hagemann, M., "Matching-based Preprocessing Algorithms to the Solution of Saddle-Point Problems in Large-Scale Nonconvex Interior-Point Optimization," Vol. 36, No. 2-3, April 2007, pp. 321–341.

[11] Schenk, O. and Gärtner, K., "On fast factorization pivoting methods for symmetric indefinite systems," *Elec. Trans. Numer. Anal.*, Vol. 23, No. 1, 2006, pp. 158–179.

[12] Hagemann, M. and Schenk, O., "Weighted Matchings for Preconditioning Symmetric Indefinite Linear Systems," *SIAM J. Scientific Computing*, Vol. 28, 2006, pp. 403–420.

[13] Gebremedhin, A., Manne, F., and Pothen, A., "What color is your Jacobian? Graph coloring for computing derivatives," *SIAM review*, Vol. 47, No. 4, 2005, pp. 629.

[14] Bayliss, A., Gunzburger, M., and Turkel, E., "Boundary Conditions for the Numerical Solution of Elliptic Equations in Exterior Regions," *SIAM Journal on Applied Mathematics*, Vol. 42, No. 2, 1982, pp. 430–451.

[15] Akcelik, V., Biros, G., and Ghattas, O., "Parallel multiscale Gauss-Newton-Krylov methods for inverse wave propagation," *Proceedings of the 2002 ACM/IEEE conference on Supercomputing*, Supercomputing '02, IEEE Computer Society Press, Los Alamitos, CA, USA, 2002, pp. 1–15.

[16] Wächter, A. and Laird, C., "IPOPT — Software package for large-scale nonlinear optimization," 2010, Available at `https://projects.coin-or.org/Ipopt`.

[17] Forsgren, A., Gill, P. E., and Wright, M. H., "Interior Methods for Nonlinear Optimization," *SIAM Review*, Vol. 44, No. 4, 2002, pp. 525–597.

[18] Wächter, A. and Biegler, L. T., "On the Implementation of a Primal-Dual Interior Point Filter Line Search Algorithm for Large-Scale Nonlinear Programming," Vol. 106, No. 1, 2006, pp. 25–57.

[19] "Graphviz—Graph Visualization Software," 2010, Available at http:// www.graphviz.org.

[20] Griewank, A. and Toint, P. L., "On the Unconstrained Optimization of Partially Separable Functions," *Nonlinear Optimization 1981*, edited by M. J. D. Powell, Academic Press, New York, 1982, pp. 301–312.

[21] Schenk, O., Bollhöfer, M., and Römer, R. A., "On Large Scale Diagonalization Techniques For The Anderson Model Of Localization," *SIAM Review*, Vol. 50, No. 1, 2008, pp. 91–112, SIGEST Paper.

[22] Gould, N., Hu, Y., and Scott, J., "A numerical evaluation of sparse direct solvers for the solution of large sparse, symmetric linear systems of equations," *ACM Transactions on Mathematical Software (TOMS)*, Vol. 33, No. 2, 2007.

[23] Schenk, O., Wächter, A., and Weiser, M., "Inertia Revealing Preconditioning For Large-Scale Nonconvex Constrained Optimization," *SIAM J. Scientific Computing*, Vol. 31, No. 2, 2008, pp. 939–960.

[24] Curtis, F. E., Schenk, O., and Wächter, A., "An Interior-Point Algorithm for Large-Scale NonlinearOptimization with Inexact Step Computations," *SIAM J. Scientific Computing*, Vol. 32, No. 6, 2010, pp. 3447–3475.

[25] Schenk, O. and Gärtner, K., "Solving Unsymmetric Sparse Systems of Linear Equations with PARDISO," *Journal of Future Generation Computer Systems*, Vol. 20, No. 3, 2004, pp. 475–487.

[26] Schenk, O., Gärtner, K., and Fichtner, W., "Efficient sparse LU factorization with left-right looking strategy on shared memory multiprocessors," *BIT*, Vol. 40, No. 1, 2000, pp. 158–176.

[27] Li, X. S. and Demmel, J. W., "SuperLU_DIST: A Scalable Distributed-Memory Sparse Direct Solver for Unsymmetric Linear Systems," *ACM Transactions on Mathematical Software*, Vol. 29, No. 2, 2003, pp. 110–140.

[28] Golub, G. H. and Van Loan, C. F., *Matrix Computations*, 3rd ed., 1996.

[29] Duff, I. S. and Pralet, S., "Strategies for scaling and pivoting for sparse symmetric indefinite problems," Vol. 27, No. 2, 2005, pp. 313–340.

[30] Demmel, J. W., Eisenstat, S. C., Gilbert, J. R., Li, X. S., and Liu, J. W. H., "A Supernodal Approach to Sparse Partial Pivoting," Vol. 20, No. 3, 1999, pp. 720–755.

[31] Duff, I. S. and Koster, J., "The design and use of algorithms for permuting large entries to the diagonal of sparse matrices," Vol. 20, No. 4, 1999, pp. 889–901.

[32] Gebremedhin, A. H., Tarafdar, A., Pothen, A., and Walther, A., "Efficient Computation of Sparse Hessians Using Coloring and Automatic Differentiation," *INFORMS J. on Computing*, Vol. 21, No. 2, 2009, pp. 209–223.

Chapter 9

Combinatorial Aspects/Algorithms in Computational Fluid Dynamics

Rainald Löhner

George Mason University

9.1   System of Conservation Laws

Most of the equations used to describe the behaviour of continua are of the form:

$$\mathbf{u}_{,t} + \nabla \cdot (\mathbf{F}^a - \mathbf{F}^v) = \mathbf{S}(\mathbf{u}) \quad , \tag{9.1}$$

where $\mathbf{u}, \mathbf{F}^a, \mathbf{F}^v, \mathbf{S}(\mathbf{u})$ denote the vector of unknowns, advective and diffusive flux tensors, and source-terms respectively. In the case of compressible gases, we have

$$\mathbf{u} = (\rho \; ; \; \rho v_i \; ; \; \rho e) \quad ,$$

$$\mathbf{F}^a_j = (\rho v_j \; ; \; \rho v_i v_j + p\delta_{ij} \; ; \; v_j(\rho e + p)) \quad ,$$

$$\mathbf{F}^v_j = (0 \; ; \; \sigma_{ij} \; ; \; v_l \sigma_{lj} + kT_{,j}) \quad . \tag{9.2}$$

Here ρ, p, e, T, k, v_i denote the density, pressure, specific total energy, temperature, conductivity and fluid velocity in direction x_i respectively. This set of equations is closed by providing an equation of state, e.g., for a polytropic gas:

$$p = (\gamma - 1)\rho[e - \frac{1}{2}v_j v_j] \quad , \quad T = c_v[e - \frac{1}{2}v_j v_j] \quad , \qquad (9.3\ a,b)$$

where γ, c_v are the ratio of specific heats and the specific heat at constant volume respectively. Furthermore, the relationship between the stress tensor σ_{ij} and the deformation rate must be supplied. For water and almost all gases, Newton's hypothesis

$$\sigma_{ij} = \mu(\frac{\partial v_i}{\partial x_j} + \frac{\partial v_j}{\partial x_i}) + \lambda \frac{\partial v_k}{\partial x_k}\delta_{ij} \qquad (9.4)$$

complemented with Stokes' hypothesis

$$\lambda = -\frac{2\mu}{3} \qquad (9.5)$$

is an excellent approximation. The compressible Euler equations are obtained by neglecting the viscous fluxes, i.e., setting $\mathbf{F}^v = 0$. The incompressible Euler or Navier-Stokes equations are obtained by assuming that the density is constant and that pressure does not depend on temperature. The Maxwell equations of electromagnetics, the heat conduction equations of solids, and the equations describing elastic solids undergoing small deformation can readily be written in the form given by Equation (9.1).

The spatial approximation of Equation (9.1) is obtained by describing the unknowns in terms of known functions N^i and unknown coefficients $\hat{u}_i$:

$$\mathbf{u} \approx \mathbf{u}^h = N^i \hat{u}_i \quad , \qquad (9.6)$$

where the Einstein summation convention over repeated indices has been assumed (as it will throughout the chapter). The idea is that the functions N^i are linearly independent, and as the number of functions N^i increases, the difference between $\mathbf{u}$ and $\mathbf{u}^h$ should diminish. A system of ODEs is then obtained by projecting the so-called residual (i.e., the PDE with $\mathbf{u}^h$ instead of $\mathbf{u}$) into weighting functions W^j and integrating over space [55, 56]:

$$\int_\Omega W^i(\mathbf{u}^h_{,t} + \nabla \cdot \mathbf{F}(\mathbf{u}^h) - \mathbf{S}(\mathbf{u}^h))d\Omega = 0 \quad , \qquad (9.7)$$

or

$$\int_\Omega W^i \left[N^j(\hat{u}_j)_{,t} + \nabla \cdot \mathbf{F}(N^j \hat{u}_j) - \mathbf{S}(N^j \hat{u}_j) \right] d\Omega = 0 \quad . \qquad (9.8)$$

As before, as the number of functions N^j, W^i increases, the residual should vanish. In order to minimize integration costs, and in order to conform to geometry and apply boundary conditions in a natural way, the functions N^i, W^i

are chosen to be non-zero only in a small region of space. Figure 9.1 shows typical linear (triangular elements) and bilinear (quadrilateral elements) functions in 2-D. Local functions in other dimensions and/or of different order are obtained in a similar way [56].

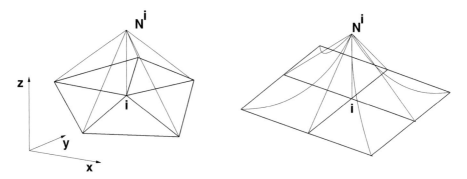

FIGURE 9.1: Linear and Bilinear Functions in 2-D.

As Equation (9.8) is composed of integrals (i.e., sums), these regions, called elements, can then be joined together into what is called a mesh. The automatic construction of meshes has been an active area of research for the last two decades. See [14, 15, 50, 29, 16] for some overviews. All integrals are evaluated using the **element subdomain paradigm**

$$\int_{\Omega} .. = \sum_{el} \int_{\Omega_{el}} ..\tag{9.9}$$

Note that the only data required to perform the evaluation of the integrals are

- The points that belong to an element

- The coordinates of the points forming the mesh

- The unknowns of the elements or points forming the mesh

As the functions in each element are known and simple, in many cases the evaluations of the integrals can be performed in closed form. If not (higher order elements, curved faces/edges), the integrals are obtained via numerical quadrature (Gauss, Lobatto, etc. [56]). Readers with a pure finite difference background may not associate Equation (9.9) with so-called "stencils." However, stencils are obtained from Equation (9.9) by setting N^i polynomial and weighing them with Dirac-deltas [31]. Depending on the choice of N^i, W^i different schemes are obtained:

- Simple Finite Difference Schemes: N^i polynomial, W^i Dirac-deltas;

- Finite Volume Schemes: N^i polynomial, W^i locally constant functions;

- Galerkin Finite Elements: $N^i = W^i$, polynomial;

- Spectral Elements: N^i trigonometric/special polynomials, W^i Dirac-deltas or trigonometric/special polynomials.

9.2 Grid Size Estimates

Let us consider the meshing (and solver) requirements for typical aerodynamic and hydrodynamic problems purely from the approximation theory standpoint. Defining the Reynolds-number (i.e., the ratio of inertial/inviscid to viscous forces for the problem at hand) as

$$Re = \frac{\rho |\mathbf{v}_\infty| l}{\mu} \ , \tag{9.10}$$

where $\rho, \mathbf{v}, \mu$ and l denote the density, free stream velocity, and viscosity of the fluid, as well as a characteristic object length, we have the following estimates for the boundary layer thickness and velocity gradient at the wall for flat plates from boundary layer theory [41]:

a) <u>Laminar Flow</u>:

$$\frac{\delta(x)}{x} = 5.5 \cdot Re_x^{-1/2} \ , \quad \left.\frac{\partial v}{\partial y}\right|_{y=0} = 0.332 \cdot Re_x^{1/2} \tag{9.11}$$

b) <u>Turbulent Flow</u>:

$$\frac{\delta(x)}{x} = 5.5 \cdot Re_x^{-1/5} \ , \quad \left.\frac{\partial v}{\partial y}\right|_{y=0} = 0.0288 \cdot Re_x^{4/5} \tag{9.12}$$

This implies that the minimum element size required to capture the main vortices of the boundary layer (via a large-eddy simulation (LES)) will be $h \approx Re^{-1/2}$ and $h \approx Re^{-1/5}$ for the laminar and turbulent cases. In order to capture the laminar sublayer (i.e., the wall gradient, and hence the friction) properly, the first point off the wall must have a (resolved) velocity that is only a fraction ϵ of the free-stream velocity v_∞:

$$\left.\frac{\partial v}{\partial y}\right|_{y=0} h_w = \epsilon v_\infty \ , \tag{9.13}$$

implying that the element size close to the wall must be inversely proportional to the gradient. This leads to element size requirements of $h \approx Re^{-1/2}$ and $h \approx Re^{-4/5}$, i.e., considerably higher for the turbulent case. Let us consider in some more detail a wing of aspect ratio Λ. As is usual in aerodynamics, we will base the Reynolds-number on the root chord length of the wing. We suppose that the element size required near the wall of the wing will be of the form (see above)

$$h \approx \frac{1}{\alpha Re^q} \quad , \qquad (9.14)$$

and that, at a minimum, β layers of this element size will be required in the direction normal to the wall. If we assume, conservatively, that most of the points will be in the (adaptively/ optimally gridded) near-wall region, the total number of points will be given by

$$n_p = \Lambda \beta \alpha^2 Re^{2q} \quad . \qquad (9.15)$$

One could argue that the mesh in the laminar sublayer does not have to be isotropic. For this reason the following estimates are conservative. Assuming we desire an accurate description of the vortices in the flowfield, the (significant) advective timescales will have to be resolved with an explicit time-marching scheme. The number of timesteps required will then be at least proportional to the number of points in the chord/ streamwise direction, i.e., of the form:

$$n_t = \frac{\gamma}{h} = \alpha \gamma Re^q \quad . \qquad (9.16)$$

Consider now the best case scenario: $\alpha = \beta = \gamma = \Lambda = 10$. In the sequel, we will label this case "Very Large Eddy Simulation" (VLES). A more realistic set of numbers for typical LES simulations would be: $\alpha = 100$, $\beta = \gamma = \Lambda = 10$. While the wing is typical for aerodynamics calculations, similar estimates can be obtained for cars, trucks, trains, ships and submarines. The number of points required for simulations based on these estimates summarized in Tables 9.1 and 9.2. Recall that the Reynolds-number for for cars and trucks lies in the range $Re = 10^6 - 10^7$, for aeroplanes $Re = 10^7 - 10^8$, and for naval vessels $Re = 10^8 - 10^9$. This implies that at present any direct simulation of Navier-Stokes (DNS) is out of the question for the Reynolds-numbers encountered in aerodynamic and hydrodynamic engineering applications.

TABLE 9.1: Estimate of grid and timestep requirements

Simulation Type	npoin	ntime
Laminar	$10^4 Re$	$10^2 Re^{1/2}$
VLES	$10^4 Re^{2/5}$	$10^2 Re^{1/5}$
LES	$10^6 Re^{2/5}$	$10^3 Re^{1/5}$
DNS	$10^4 Re^{8/5}$	$10^2 Re^{4/5}$

TABLE 9.2: Estimate of grid and timestep requirements

Re	n_p VLES	n_t VLES	n_p LES	n_t LES	n_p DNS	n_t DNS
10^6	$10^{6.4}$	$10^{3.2}$	$10^{8.4}$	$10^{4.2}$	$10^{13.6}$	$10^{6.8}$
10^7	$10^{6.8}$	$10^{3.4}$	$10^{8.8}$	$10^{4.4}$	$10^{15.2}$	$10^{7.6}$
10^8	$10^{7.2}$	$10^{3.6}$	$10^{9.2}$	$10^{4.6}$	$10^{16.8}$	$10^{8.4}$
10^9	$10^{7.6}$	$10^{3.8}$	$10^{9.6}$	$10^{4.8}$	$10^{18.4}$	$10^{9.2}$

9.3 Work Estimates for Different Shape-Functions

With the widespread current interest in high-order (finite difference, finite volume, finite element, discontinuous Galerkin, spectral, etc.) methods [5, 25, 3, 2, 6, 23, 35, 39], a recurring question is when these schemes pay off. In order to obtain a qualitative answer, let us assume that the present mesh already allows for a uniform (optimal) distribution of the error. One example in fluid dynamics would be an isotropic mesh of uniform element size for the calculation of homogeneous turbulence. Another would be an adapted mesh. Let us suppose further that we have a way to solve for the unknown coefficients $\hat{u}$ of the resulting discrete system in optimal time complexity, i.e., it takes $O(N \log_{2^d}(n_{el}))$, where N denotes the number of matrix entries, n_{el} the number of elements and d the dimension of the problem. This would imply a fixed number of multigrid cycles and is certainly a lower bound, seldomly achieved in practice [18, 45, 52]. Without loss of generality, let us also assume that we have finite elements with Lagrange polynomials in tensor form as show in 2-D in Figure 9.2. (similar, if not worse, estimates will ensue for discontinuous Galerkin schemes).

Denoting by h the element size and p the order of the polynomial, the degrees of freedom per element will be

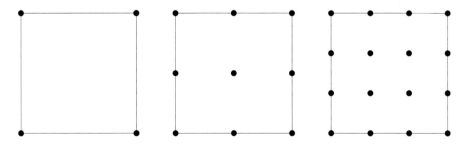

FIGURE 9.2: Degrees of Freedom for Lagrange Elements in 2-D.

$$DOF_{el} = (1+p)^d \quad , \tag{9.17}$$

the number of matrix entries per element

$$n_{mat} = (1+p)^{2d} \quad , \tag{9.18}$$

and the number of elements

$$n_{el} = O(h^{-d}) \quad , \tag{9.19}$$

yielding a work estimate of

$$W_{d,p} = c_1 h^{-d}(1+p)^{2d} \log_{2^d}(h^{-d}) \quad . \tag{9.20}$$

On the other hand, the error is given by

$$\epsilon_u = \|u - u^h\| = c_2 h^{p+1}|u|_{p+1} \quad . \tag{9.21}$$

In order to see what happens for a solution with a certain degree of spatial frequency content, let us set

$$u = u_0 e^{i\omega x} \quad . \tag{9.22}$$

In this case, $u_{p+1} = \omega^{p+1} u_0 e^{i\omega x}$ and

$$\epsilon_u = c_2 h^{p+1} \omega^{p+1}|u|_0 = c_2 \left(\frac{2\pi h}{\lambda}\right)^{p+1} |u|_0 \quad , \tag{9.23}$$

where λ denotes the spatial wavelength corresponding to ω. The relative error (which is what is usually sought) therefore becomes:

$$\epsilon = \frac{\epsilon_u}{|u|_0} = c_2 \left(\frac{2\pi h}{\lambda}\right)^{p+1} \quad , \tag{9.24}$$

implying

$$h = \frac{\lambda}{2\pi} \left(\frac{\epsilon}{c_2}\right)^{\frac{1}{p+1}} \quad . \tag{9.25}$$

Thus, the work changes with the required relative error ϵ according to

$$W_{d,p} = c_1 c_2^{\frac{d}{p+1}} \left(\frac{2\pi}{\lambda}\right)^d \epsilon^{-\frac{d}{p+1}}(1+p)^{2d} \log_{2^d}\left(\epsilon^{-\frac{d}{p+1}}\right) \quad . \tag{9.26}$$

or

$$W_{d,p} = c_1 c_2^{\frac{d}{p+1}} \left(\frac{2\pi}{\lambda}\right)^d d\epsilon^{-\frac{d}{p+1}}(1+p)^{2d-1} \log_{2^d}\left(\epsilon^{-1}\right) \quad . \tag{9.27}$$

Note that the constants c_1, c_2 depend on d, p, and are not always easy to estimate. In the sequel, we will assume (and believe there is some empirical evidence that this is correct) that the main effects of d, p are in the remaining terms, i.e., the desired relative error ϵ and d, p.

9.3.1 Work Estimates for Linear Problems

Let us consider the ratios of work for linear ($p = 1$) and cubic ($p = 3$) elements in 3D ($d = 3$). We have

$$r_{1,3}^{3D} = \frac{W_{3,1}}{W_{3,3}} \approx \frac{\epsilon^{-\frac{3}{2}} 2^5}{\epsilon^{-\frac{3}{4}} 4^5} = \epsilon^{-\frac{3}{4}} 2^{-5} \quad . \tag{9.28}$$

One can see from this expression that the ratio depends on the relative error desired. For $\epsilon = 10^{-2}$ (typical of engineering interest), we obtain $r_{1,3}^{3D} \approx 10^{\frac{3}{2}} 2^{-5} \approx 0.99$,
i.e., the linear and cubic elements are similar in speed. For $\epsilon = 10^{-4}$ (typical of high precision simulations), we obtain $r_{1,3}^{3D} \approx 10^3 2^{-5} \approx 31$,
i.e., the linear element should be more than an order of magnitude slower (!). Table 9.3 lists the ratio of work as compared to linear elements for different relative error levels. Because we assumed that the work was proportional to the number of elements (as well as a log-factor) and the number of matrix entries per element, these same ratios apply to the storage requirements of the different schemes. One can see that higher order elements only offer significant advantages for relative errors that are below $\epsilon = 10^{-3}$. Given typical uncertainties in engineering runs (physical model, model constants, boundary and initial conditions, etc.), this relative error is seldomly achieved in practice. It would thus appear that higher order elements do not offer an advantage for these applications.

TABLE 9.3: Estimated work ratio of linear versus higher order elements in 3D (linear PDE, $N \log N$ estimate)

Element	$\epsilon = 10^{-2}$	$\epsilon = 10^{-3}$	$\epsilon = 10^{-4}$	$\epsilon = 10^{-5}$
p=2	1.32	4.16	13.17	41.64
p=3	0.99	5.56	31.25	175.73
p=4	0.65	5.13	40.77	323.82
p=5	0.41	4.12	41.15	411.52

9.3.2 Work Estimates for Nonlinear Problems

The situation is even more in favour of low-order elements for nonlinear cases in which, for each iteration, we require a new matrix build. The more complex shape-functions, as well as the appearance of Gaussian integration points, will increase the work estimate. Neglecting the effect of the more complex shape-functions, and assuming that the number of Gauss-points increases as the square-root of the number of degrees of freedom (again, a lower bound), we have:

$$W_{d,p} = c_1 c_2^{\frac{d}{p+1}} \left(\frac{2\pi}{\lambda}\right)^d d\epsilon^{-\frac{d}{p+1}} (1+p)^{\frac{5d}{2-1}} \log_{2^d}\left(\epsilon^{-1}\right) \quad . \tag{9.29}$$

The corresponding ratios have been compiled in Table 4.

TABLE 9.4: Estimated work ratio of linear versus higher order elements in 3D (nonlinear PDE, $N \log N$ estimate)

Element	$\epsilon = 10^{-2}$	$\epsilon = 10^{-3}$	$\epsilon = 10^{-4}$	$\epsilon = 10^{-5}$
p=2	0.72	2.27	7.17	22.67
p=3	0.35	1.96	11.05	62.13
p=4	0.16	1.30	10.31	81.92
p=5	0.08	0.79	7.92	79.20

Note that as before, for every level of relative accuracy desired a particular higher order of element is optimal. Moreover, even for $\epsilon = 10^{-3}$, the linear element is at most slower than a factor of 2, a factor that can easily be absorbed by careful coding. Only from $\epsilon = 10^{-4}$ onwards, do the higher order elements yield a considerable advantage.

9.3.3 Possible Objections

A number of possible objections may be raised for the estimates given. We list the most common ones:

a) Cache and Local Memory
The most obvious objection to the estimates given is that the re-use of cache and local memory may not be the same for different element types. After all, higher order elements should lend themselves to a higher degree of cache and local memory re-use per multiplication if coded properly. This may be true, but it depends on coding. And if experience is any guide, properly coded low-order elements can achieve a very high degree of cache and local memory re-use as well.

b) Derivatives
Another objection to the estimates given is that for many engineering applications one is interested in derivative information. A typical example may be shear stress on walls. Over the years, derivative "recovery" methods have been designed [11, 24, 53, 36, 43, 32], and superconvergence estimates have been obtained for both low and high order elements. From these, it appears that the quality of derivatives that can be obtained is similar to that of the unknowns. This would imply that the estimates derived above still hold.

The comparison of error and work estimates shows that for relative accuracy in the 1% range, which is typical of engineering interest, it may prove very difficult to improve on linear elements. It is interesting to note that the empirical evidence in this respect corroborates the estimates derived.

As expected, the estimates also show that the optimal order of element in terms of work and storage demands depends on the desired relative accuracy.

9.4 Basic Data Structures and Loops

Given that most of the production CFD codes are used for problems where the physics (inflow conditions, material parameters, equations of state, turbulent viscosity, reaction rates for chemically reacting flows, ...) and geometry may not be known to better than 1% relative accuracy, it is not surprising that they are usually based on low order elements. As stated before, the integration of the weighted residual statement given by Equation (11) is performed using the **element subdomain paradigm**

$$\int_{\Omega} .. = \sum_{el} \int_{\Omega_{el}} ..$$ (9.33)

For low-order elements, the unknowns, as well as the fluxes, can be approximated using the same functions without noticeable loss of accuracy. This, then, leads for the discrete unknowns $\hat{u}$ to a system of nonlinear ODEs of the form

$$\mathbf{M}_c \cdot \hat{\mathbf{u}}_{,t} = \mathbf{C} \cdot \hat{\mathbf{F}} + \mathbf{S} \ .$$ (9.34)

For viscous fluxes, as well as source-terms, the straightforward Galerkin approximation will yield optimal order of convergence. Much more severe difficulties are encountered for the advective fluxes (which also encompass divergence-constraints in the incompressible limit). Not only are these terms highly nonlinear, but the Galerkin fluxes can be shown to be unstable in the Euler limit. The recourse taken is to replace the simple Galerkin fluxes by so-called "numerically consistent" or "stabilized" fluxes. For the limit of linear advection, this can be shown to correspond to the addition of some form of artificial viscosity [21, 1, 37, 48, 22, 12, 44, 46, 17, 4, 33, 51, 10]. Over the last two decades, the accuracy and sophistication of these so-called consistent numerical fluxes (or approximate Riemann solvers) has increased steadily [44], to the point that at present they consume the bulk of the CPU requirements. Fluxes may be evaluated at different locations depending on where the variables are placed. For finite volume codes where the conserved variables are placed at the nodes, fluxes are computed at edges (see Figure 9.3).

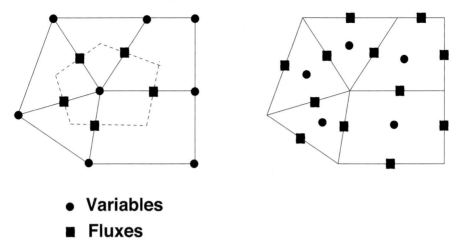

• **Variables**

■ **Fluxes**

FIGURE 9.3: Placement of Variables and Fluxes Evaluation.

For finite element codes based on linear elements, it is convenient to switch from element-based loops to edge-based loops. Both the operation count and the number of indirect address fetches and stores are reduced significantly. For codes where variables are placed in the center of elements, fluxes are computed at faces.

9.4.1 Basic Loop

Consider then the evaluation of an edge right-hand-side (RHS), typical of explicit timestepping or the residual evaluation of iterative solvers, given by the following loop (the same basic principle applies to element-based or face-based solvers):

<u>Loop 1</u>

```
      do 1600 iedge=1,nedge
      ipoi1=lnoed(1,iedge)
      ipoi2=lnoed(2,iedge)
      redge=geoed( iedge)*(unkno(ipoi2)-unkno(ipoi1))
      rhspo(ipoi1)=rhspo(ipoi1)+redge
      rhspo(ipoi2)=rhspo(ipoi2)-redge
 1600 continue
```

The operations to be performed can be grouped into the following three major steps:

a) Gather point information into the edge;

b) Perform the required mathematical operations at edge level (for the loop shown—a Laplacian—this is is a simple multiplication and subtraction; for an Euler solver the approximate Riemann solver and the equation of state would be substituted in);

c) Scatter-Add the edge RHS to the assembled point RHS.

From this simple loop one can already see that the amount of computation is small as compared to the amount of data transfered in and out of memory. This implies that on none of the modern microprocessors peak performances even remotely close to the theoretical peak will be achieved. Typical numbers lie in the range of 10–15% of theoretical peak performance.

The transfer of information to and from memory required in steps a),c) is proportional to the number of nodes in the edge (element, face) and the number of unknowns per node. If the nodes within each edge (element, face) are widely spaced in memory, cache-misses are likely to occur. If, on the other hand, all the points within an element are 'close' in memory, cache-misses are minimized. From these considerations, it becomes clear that cache-misses are directly linked to the bandwidth of the equivalent matrix system (or graph). Point renumbering to reduce bandwidths has been an important theme for many years in traditional finite element applications [56]. The aim was to reduce the cost of the matrix inversion, which was considered to be the most expensive part of any finite element simulation. The optimal renumbering of points in such a way that spatial (or near-neighbour) locality is mirrored in memory is a problem of formidable algorithmic complexity. Fortunately, most of the benefits of renumbering points are already obtained from near-optimal heuristic renumbering techniques. Among the most commonly used we mention wavefront, reverse Cuthill-McKee [7], bins and octrees as well as various space-filling curves [31]. Figure 9.4 shows a few of these for a small 2-D mesh. All of these are of complexity $O(N)$ or at most $O(N \log N)$, and are well woth the effort. When possible cache-misses in each edge have been minimized, the next step is to minimize cache-misses as different edges are processed. In an ideal setting, as the loop over the edges progresses the data accessed from and sent to point arrays should increase as monotonically as possible. This would preserve data locality optimally. As before, a number of sorting techniques (heaps, linked-lists, etc.) can be employed that achieve this aim and are of complexity $O(N)$ [31].

9.4.2   Vector Machines

The possibility of a memory overlap that would occur if two of the nodes within the same group of edges being processed by the vector registers are the same inhibits vectorization. If the edges can be reordered into groups such that within each group none of the nodes are accessed more than once, vec-torization can be enforced using compiler directives. Denoting the grouping

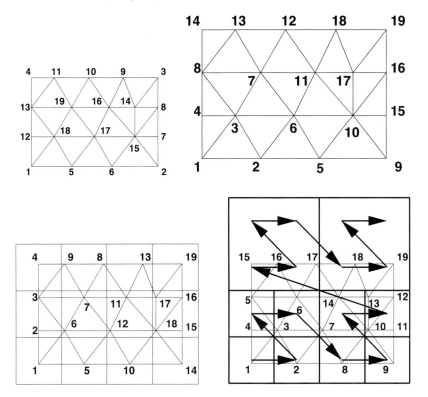

FIGURE 9.4: Point Renumbering Techniques (Original, Adv. Front, Bin, Space-Filling Curve).

array by edpas, the vectorized RHS assembly would look as follows:

<u>Loop 2</u>

```
      do 1400 ipass=1,npass
      nedg0=edpas(ipass)+1
      nedg1=edpas(ipass+1)
!dir$ ivdep                          ! Ignore Vector DEPendencies
        do 1600 iedge=nedg0,nedg1
        ipoi1=lnoed(1,iedge)
        ipoi2=lnoed(2,iedge)
        redge=geoed(~~iedge)*(unkno(ipoi2)-unkno(ipoi1))
        rhspo(ipoi1)=rhspo(ipoi1)+redge
        rhspo(ipoi2)=rhspo(ipoi2)-redge
 1600  continue
 1400 continue
```

Algorithms that group the edges into nonconflicting groups fall into the category of coloring schemes [31].

9.4.3　Shared Memory Multicore Machines

In order to obtain optimal performance on shared multicore processors via OpenMP, care has to be taken not to increase cache misses while allowing for parallelization and vectorization. The compromise typically chosen is similar to that of domain decomposition: assign to each processor groups of edges that work on the same group of points [30]. In order to avoid the explicit declaration of local and shared variables, is it convenient to use the sub-subroutine technique [31]. For Loop 2, this translates into

Master OMP Loop 2

```
      do 1000 imacg=1,npasg,nproc
      imac0=          imacg
      imac1=min(npasg,imac0+nproc-1)
c$omp parallel do private(ipasg)    ! Parallelization directive
         do 1200 ipasg=imac0,imac1
         call loop2p(ipasg)
 1200   continue
 1000 continue
```

Loop2p becomes a subroutine of the form:

```
      subroutine loop2p(ipasg)
      npas0=edpag(ipasg)+1
      npas1=edpag(ipasg+1)
      do 1400 ipass=npas0,npas1
      nedg0=edpas(ipass)+1
      nedg1=edpas(ipass+1)
!dir$ ivdep                         ! Ignore Vector DEPendencies
         do 1600 iedge=nedg0,nedg1
         ipoi1=lnoed(1,iedge)
         ipoi2=lnoed(2,iedge)
         redge=geoed(  iedge)*(unkno(ipoi2)-unkno(ipoi1))
         rhspo(ipoi1)=rhspo(ipoi1)+redge
         rhspo(ipoi2)=rhspo(ipoi2)-redge
 1600   continue
 1400 continue
```

For relatively simple Riemann solvers and simple equations of state, this type of loop structure works well. However, once more esoteric Riemann solvers and equations of state are required, one is forced to split the inner loop even further. In particular, for equations of state one may have a table look-up with many if-tests, and even (oh woeful news!) parts that do not vectorize well.

The usual recourse is to split the inner edge-loop even further:

```
c       inner loop, processed in groups
        do 1600 iedge=nedg0,nedg1,medgl
        iedg0=iedge-1
        iedg1=min(iedg0+medgl,nedg1)
        nedgl=iedg1-iedg0
c
c       transcribe variables for local edge-group
        call tranvariab(...)
c
c       obtain fluxes for local edge-group
        call apprxriem(....)
c
c       obtain RHS for local edge-group
        call rhsvisc(...)
c
!dir$ ivdep                       ! Ignore Vector DEPendencies
        do 1800 iedgl=1,nedgl
        ipoi1=lnoel(1,iedgl)
        ipoi2=lnoel(2,iedgl)
        rhspo(ipoi1)=rhspo(ipoi1)+redgl(iedgl)
        rhspo(ipoi2)=rhspo(ipoi2)-redgl(iedgl)
 1800   continue
 1600 continue
```

Note that the transcription of variables, the fluxes and the RHS obtained at the local edge-group level may involve many braches (if-tests), but that care has been taken that these innermost loops (which reside in separate subroutines) are, if at all possible, vectorized.

9.4.4 Distributed Memory Machines

For field solvers based on grids, parallel execution with distributed memory machines is best achieved by domain splitting. The elements are grouped together into subdomains, and are allocated to a processor. Each subdomain can then advance the solution independently, enabling parallel execution. At the boundaries of these subdomains, information needs to be transferred as the computation proceeds. Because of this association between processors and subdomains, these terms will be used interchangeably within the same context.

The computational work required by field solvers is proportional to the number of elements. This proportionality is linear for explicit time-marching or iterative schemes, of the $N \log(N)$ type for multigrid solvers, and of the $N^p, p > 1$ type for implicit time-marching or direct solvers. The factor p

depends on the bandwidth of the domain, which in turn not only depends on the number of elements, but also the shape of the domain, the ratio of the number of boundary points to the domain points, the local numbering of the points and elements, etc. This implies that the optimal load balance is achieved if each processor is allocated the same number of elements, and all processors have subdomains with approximately the same shape. Besides the CPU time required in each processor to advance the solution to the next level (an iteration, a timestep), the communication overhead between processors has to be taken into account. This overhead is directly proportional to the amount of information that needs to be transferred between processors. For field solvers based on grids, this amount of information is proportional to the number of subdomain boundary points. Therefore, an optimal load and communication balance is achieved by allocating to each processor the same number of elements while minimizing the number of boundary points in each subdomain.

In many situations, the computational work required in an element or subdomain of elements may change by orders of magnitude during a run. For grids that are continuously refined/derefined every 5-10 timesteps (e.g., strongly unsteady flows with traveling shock waves [27]), the number of elements in a subdomain may increase or decrease rapidly. With only 2 levels of h-refinement, the number of elements increases by a factor of 64 in 3-D. For flows that are chemically reacting, the CPU requirements can vary by factors in excess of 100 depending on whether the element is in the cold, unreacted region, the flame, or the hot, reacted region [38]. Both of these cases illustrate the need for fast, dynamic load balancing algorithms. A number of load balancing algorithms have been devised in recent years. They can be grouped into three families: simulated annealing [9, 13, 54], recursive bisection [54, 19, 42, 47, 49], and diffusion [8, 40, 28]. These classic algorithms have been described in other chapters of this book, so there is no need to repeat them here. In order to be able to change the unknowns at subdomain interfaces, an exchange of information between processors has to be allowed. Two ways of accomplishing this transfer are possible (see Figure 9.5):

a) <u>Right-Hand Side Information</u>:
Given by the following set of operations:

- Assemble all RHS contributions in each processor

- Exchange RHS contributions between processors, adding

- Update the unknowns at the interfaces.

A similar sequence is required for min/max comparisons among nearest neighbours (e.g., for limiting purposes). The advantage of this approach is that the total amount of information (elements, points) required for the parallel version is the same as for the serial version. However, the original serial code has to be modified extensively. These low level modifications are also reflected in

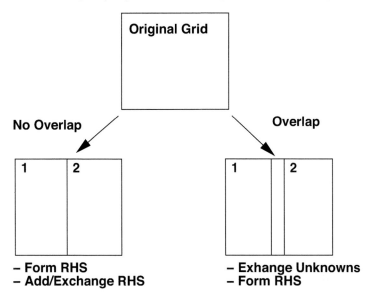

FIGURE 9.5: RHS assembly for parallel flow solvers.

a relatively large number of transfer operations among processors if limiting or other involved nearest neighbour operations are required when building a RHS.

b) <u>Unknown Information</u>:
If an additional layer of elements is added to each subdomain, the build-up of a RHS can be accomplished as follows:

- Update each subdomain as for the serial, 1-domain case;

- Exchange the updated unknown information between processors.

In this approach, the total amount of information (elements, points) required for the parallel version is larger than that required for the serial version. In order to see how big a penalty is incurred, let us consider a cube of regular cartesian cells. The number of cells along a dimension n_c will be $n_c = n_p^1/3$, where n_p denotes the total number of points. Assuming both work and storage grow linearly with the number of points, the ratio between the split and unsplit domains will be: $r_w = (1 + 2/n_c)^3$. Furthermore, the ratio of communication required for the overlapped and non-overlapped splitting will be $r_c = (1 + 2/n_c)(1 + 1/n_c)$. The ratio of points exchanging information over the total number of points is given by: $c_r = 6/n_c$. Table 9.5 summarizes these ratios for different mesh sizes, assuming (as is typical for unstructured grids) that the number of elements n_e is approximately $n_e = 5.5n_p$.

TABLE 9.5: Extra work and communication estimates for split domains

n_e	n_p	n_c	r_w	r_c	c_r
10,000	1,818	12	1.58	1.259	0.492
20,000	3,636	15	1.44	1.204	0.390
50,000	9,090	20	1.32	1.148	0.287
100,000	18,181	26	1.25	1.117	0.228
200,000	36,363	33	1.19	1.092	0.181
500,000	90,090	44	1.14	1.068	0.133
1,000,000	181,818	56	1.11	1.054	0.106

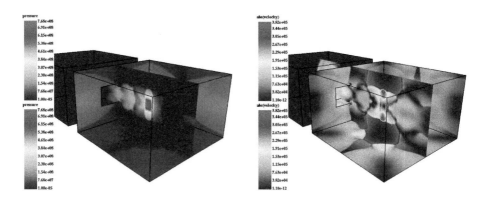

FIGURE 9.6: (See color insert.) Blast in Room: pressures and velocities.

Note that for reasonable domain sizes the extra amount of information (storage) and work is not large. On the other hand, the code employed in each subdomain is exactly the same as for the serial case. This advantage outweighs all possible disadvantages and has made this second possibility the method of choice for most codes used in practice.

9.5   Example: Blast in Room

This example is included to demonstrate the use of the techniques described in the chapter. It considers the compressible flow resulting from a blast in a room. The geometry, together with the solution, can be discerned from Figure 9.6.

The case was chosen as it demonstrates the parallel performace for what is perhaps the worst-case scenario for any kind of machine: extremely low-cost explicit solvers, i.e., cases where the ratio of communication vs. computation is high. The compressible Euler equations are solved using an edge-based FEM-FCT technique [26, 34, 31]. The initialization was performed by interpolating the results of a very detailed 1-D (spherically symmetric) run. The machine used was an SGI ICE with 1,920 nodes (or blades), where each node was composed of 2 Quad-core Xeon processors Nehalem E5440 series running at 2.8 Ghz, giving a total of 15,360 cores. The maching had a theoretical peak performance of 172 Teraflops. The interconection network was composed of two InfinibBand interconnect (10/20 Gb/s), with one interconnect dedicated to I/O and the other to MPI traffic. The operating system was Suse Linux Enterprise 10 tuned for SGI. The code was compiled using the Intel compiler, version 10.1.017. The following compiler options were used:

```
mpiifort -mtune=pentium4 -132 -i4 -r8 -W1 -w95 -Vaxlib -zero -fpp
-fpe0 -O3
```

The timings obtained for different meshes and numbers of processors are summarized in Table 9.6.

TABLE 9.6: Blast in room (ERDC)

nelem	nproc	nprol	CPU [sec]	S	S(Ideal)	nelem/ncore	ncore*time/ elem/step
40M	64	1	402	1.00	1.00	0.625M	1.29e-6
40M	64	4	127	3.16	4.00	0.156M	1.63e-6
40M	64	8	88	4.56	8.00	0.078M	2.25e-6
40M	256	1	124	3.24	4.00	0.156M	1.59e-6
233M	512	1	317	1.00	1.00	0.455M	1.39e-6
233M	512	4	109	2.91	4.00	0.113M	1.92e-6
233M	512	8	86	3.68	8.00	0.056M	3.02e-6
233M	1024	1	207	1.53	1.00	0.227M	1.82e-6
233M	1024	4	83	3.81	4.00	0.056M	2.91e-6

Note that the processor-independent "figure of merit": `ncroc*time/nelem/nstep` shows a drastic decay in performance once the number of elements falls below 300Kels per processor. This should come as no surprise if one recalls Table 9.5. Note also that the penalty for using shared vs. distributed memory is insignificant.

9.6 Conclusions and Outlook

This chapter has attempted to summarize the modus operandi of most field solvers for systems of conservation laws. Although the emphasis was on computational fluid dynamics, the same procedures apply to computational electromagnetics, computational heat transfer and computational structural dynamics. Starting from general approximation functions, the discrete coefficients were obtained via the weighted residual methodology. The concept and rationale for local approximation functions was described. Grid size as well as work estimates for higher order elements were derived. The main 'right-hand side' loop that is at the core of most field solvers was optimized for low cache misses, avoidance of memory contention and avoidance of dirty cache lines on shared memory machines using a variety of renumbering techniques. Domain decomposition algorithms for distributed memory machines were also touched upon. Finally, a typical example with timings was given.

Bibliography

[1] Anderson J.D.A., *Computational Fluid Dynamics: The Basics with Applications*; McGraw Hill, New York, (1995).

[2] Atkins H. L. and C. W. Shu, Quadrature Free Implementation of Discontinuous Galerkin Method for Hyperbolic Equations; *AIAA J.* 36, 5 (1998).

[3] Bey K.S., J.T. Oden, and A. Patra, A Parallel hp-Adaptive Discontinuous Galerkin Method for Hyperbolic Conservation Laws; *Applied Numerical Mathematics* 20, 321-336 (1996).

[4] Blazek J., *Computational Fluid Dynamics: Principles and Applications*; Elsevier Science, Blazek, Amsterdam, (2001).

[5] Cockburn B., S. Hou, and C. W. Shu, TVD Runge-Kutta Local Projection Discontinuous Galerkin Finite Element Method for Conservation Laws IV: The Multidimensional Case; *Mathematics of Computation* 55, 545-581, Berlin, (1990).

[6] Cockburn B., G.E. Karniadakis, and C. W. Shu (eds.), *Discontinuous Galerkin Methods, Theory, Computation, and Applications*; Springer Lecture Notes in Computational Science and Engineering 11, Springer-Verlag (2000).

[7] Cuthill E., and J. McKee, Reducing the Bandwidth of Sparse Symmetric Matrices; *Proc. ACM Nat. Conf.*, New York 1969, 157-172 (1969).

[8] Cybenko G., Dynamic Load Balancing of Distributed Memory Multiprocessors; *J. Parallel and Distr. Comp.* 7, 279-301 (1989).

[9] Dahl E.D., Mapping and Compiled Communication on the Connection Machine; *Proc. Distributed Memory Computer Conf.*, Charleston, S.C., April (1990).

[10] Donea J., and A. Huerta, *Finite Element Methods for Flow Problems*; John Wiley & Sons, New York, (2002).

[11] Douglas J., and T. Dupont, Galerkin Approximation for the Two-Point Boundary-Value Problem Using Continuous Piecewise Polynomial Spaces; *Num. Math.* 22, 99-109 (1974).

[12] Ferziger J.H., and M. Peric, *Computational Methods for Fluid Dynamics*; Springer Verlag, New York, (1999).

[13] Flower J., S. Otto, and M. Salama, Optimal Mapping of Irregular Finite Element Domains to Parallel Processors; 239-250. In Parallel Computations and Their Impact on Computational Mechanics, AMD-Vol 86, ASME, New York (1987).

[14] George P.L., F. Hecht, and E. Saltel, Fully Automatic Mesh Generation for 3D Domains of any Shape; *Impact of Computing in Science and Engineering* 2, 3, 187-218 (1990).

[15] George P.L., *Automatic Mesh Generation*; J. Wiley & Sons, New York, (1991).

[16] George P.L., and H. Borouchaki, *Delaunay Triangulation and Meshing*; Editions Hermes, Paris (1998).

[17] Gresho P.M., and R.L. Sani, *Incompressible Flow and the Finite Element Method; Volume 1: Advection-Diffusion, Volume 2: Isothermal Laminar Flow*; John Wiley & Sons, New York, (2000)

[18] Hackbusch W., and U. Trottenberg (eds.) - Multigrid Methods; *Lecture Notes in Mathematics* 960, Springer Verlag (1982).

[19] von Hanxleden R., and L.R. Scott, Load Balancing on Message Passing Architectures; *J. Parallel and Distr. Comp.* 13, 312-324 (1991).

[20] Helenbrook B.T., D. Mavriplis, and H. L. Atkins, Analysis of p-Multigrid for Continuous and Discontinuous Finite Element Discretizations; *AIAA*-03-3989 (2003).

[21] Hirsch C., *Numerical Computation of Internal and External Flow*; J. Wiley & Sons (1991).

[22] Hoffmann K.A., and S.T. Chiang - *Computational Fluid Dynamics*; Engineering Education System (1998).

[23] Kroll N., ADIGMA - A European Project on the Development of Adaptive Higher Order Variational Methods for Aerospace Applications; in *Proc. ECCOMAS CFD 2006* (P. Wesseling, E. Oñate, and J. Priaux eds.) TU Delft, Egmond and Zee, September 5–8, 2006, The Netherlands (2006).

[24] Levine N.D., Superconvergent Recovery of the Gradient from Piecewise Linear Finite-Element Approximations; *IMA J. Num. Anal.* 5, 407-427 (1985).

[25] Lin Y., and Y. S. Chin, Discontinuous Galerkin Finite Element Method for Euler and Navier-Stokes Equations; *AIAA J.* 31, 2016-2023 (1993).

[26] Löhner R., K. Morgan, J. Peraire, and M. Vahdati, Finite Element Flux-Corrected Transport (FEM-FCT) for the Euler and Navier-Stokes Equations; ICASE Rep. 87-4, *Int. J. Num. Meth. Fluids* 7, 1093-1109 (1987).

[27] Löhner R., and J.D. Baum, Adaptive H-Refinement on 3-D Unstructured Grids for Transient Problems; *Int. J. Num. Meth. Fluids* 14, 1407-1419 (1992).

[28] Löhner R., and R. Ramamurti A Load Balancing Algorithm for Unstructured Grids; *Comp. Fluid Dyn.* 5, 39-58 (1995).

[29] Löhner R., Automatic Unstructured Grid Generators; *Finite Elements in Analysis and Design* 25, 111-134 (1997).

[30] Löhner R., Renumbering Strategies for Unstructured-Grid Solvers Operating on Shared-Memory, Cache-Based Parallel Machines; *Comp. Meth. Appl. Mech. Eng.* 163, 95-109 (1998).

[31] Löhner R., *Applied CFD Techniques, 2nd Edition*; J. Wiley & Sons (2008).

[32] Löhner R., S. Appanaboyina and J. R. Cebral, Parabolic Recovery of Boundary Gradients; *Comm. Num. Meth. Eng.* 24, 1611-1615 (2008).

[33] Lomax H., T.H. Pulliam, and D.W. Zingg, *Fundamentals of Computational Fluid Dynamics*; Springer Verlag, Berlin (2001).

[34] Luo H., J.D. Baum, and R. Löhner, Edge-Based Finite Element Scheme for the Euler Equations; *AIAA J.* 32, 6, 1183-1190 (1994).

[35] Luo H., J.D. Baum, and R. Löhner, On the Computation of Steady-State Compressible Flows Using a Discontinuous Galerkin Method; *Int. J. Num. Meth. Eng.* 73, 597-623 (2008).

[36] Mackinnon R.J., and G.F. Carey, Superconvergent Derivatives: A Taylor Series Analysis; *Int. J. Num. Meth. Eng.* 28, 489-509 (1989).

[37] Oran E., and J.P. Boris, *Numerical Simulation of Reactive Flow*; Elsevier, New York (1987).

[38] Patnaik G., K. Kailasanath, and E.S. Oran, Effect of Gravity on Flame Instabilities in Premixed Gases; *AIAA J.* 29, 12, 2141-2147 (1991).

[39] Persson P.-O., J. Bonet, and J. Peraire, Discontinuous Galerkin Solution of the NavierStokes Equations on Deformable Domains; *Comp. Meth. Appl. Mech. Eng.* 198, 15851595 (2009). The Netherlands (2006).

[40] Pruhs K.R., T.F. Znati, and R.G. Melhem, Dynamic Mapping of Adaptive Computations onto Linear Arrays; pp.285-300 in *Unstructured Scientific Computation on Scalable Multiprocessors* (P. Mehrota, J. Saltz and R. Voight eds.), MIT Press (1992).

[41] Schlichting H., *Boundary Layer Theory*; McGraw-Hill, New York (1979).

[42] Simon H., Partitioning of Unstructured Problems for Parallel Processing; *NASA Ames Tech. Rep.* RNR-91-008 (1991).

[43] Thomée V., J. Xu, and N. Y. Zhang Superconvergence of the Gradient in Piecewise Linear Finite-Element Approximation to a Parabolic Problem; *SIAM J. Num. Anal.* 26, 3, 553573 (1989).

[44] Toro E.F., *Riemann Solvers and Numerical Methods for Fluid Dynamics*; Springer Verlag, Berlin (1999).

[45] Trottenberg U., C.W. Oosterlee, and A. Schuller, *Multigrid*; Academic Press, New York (2001).

[46] Turek S., *Efficient Solvers for Incompressible Flow Problems: An Algorithmic and Computational Approach*; Springer Verlag, Berlin (1999).

[47] Venkatakrishnan V., H.D. Simon, and T.J. Barth, A MIMD Implementation of a Parallel Euler Solver for Unstructured Grids; *NASA Ames Tech. Rep.* RNR-91-024 (1991).

[48] Versteeg H.K., and W. Malalasekera, *An Introduction to Computational Fluid Dynamics: The Finite Volume Method*; Addison-Wesley, New York (1996).

[49] Vidwans A., Y. Kallinderis, and V. Venkatakrishnan, A Parallel Load Balancing Algorithm for 3-D Adaptive Unstructured Grids; *AIAA*-93-3313-CP (1993).

[50] Weatherill N.P., and O. Hassan, Efficient Three-Dimensional Delaunay Triangulation with Automatic Point Creation and Imposed Boundary Constraints; *Int. J. Num. Meth. Eng.* 37, 2005-2039 (1994).

[51] Wesseling P., *Principles of Computational Fluid Dynamics*; Springer Verlag (2001).

[52] Wesseling P., *An Introduction to Multigrid Methods*; Edwards, Heidelberg (2004).

[53] Wheeler M.F., and J. Whiteman, Superconvergent Recovery of Gradients on Subdomain from Piecewise Linear Finite Element Approximations; *Num. Meth. P.D.E.'s* 3, 357-374 (1987).

[54] Williams D., Performance of Dynamic Load Balancing Algorithms for Unstructured Grid Calculations; *CalTech Rep.* C3P913 (1990).

[55] Zienkiewicz O.C., and K. Morgan, *Finite Elements and Approximation*; J. Wiley & Sons, New York (1983).

[56] Zienkiewicz O.C., and R. Taylor, *The Finite Element Method*; McGraw Hill, New York (1988).

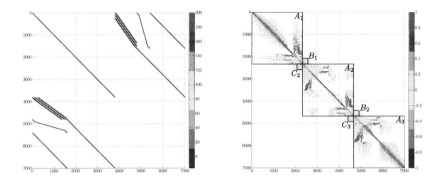

FIGURE 4.2: Reordering of a saddle-point matrix (left) to the `PSPIKE` structure (right) by applying the scaling and reordering techniques to the matrix.

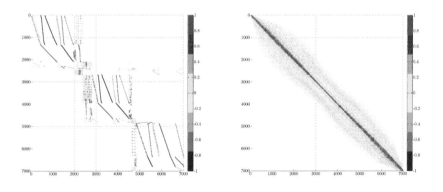

FIGURE 4.3: Partitioning obtained by the partitioner Mondriaan (left) and
the ordering obtained by the spectral method HSL MC73 (right) for the previous matrix. The second matrix would be partitioned with horizontal and
vertical lines while respecting the permutation.

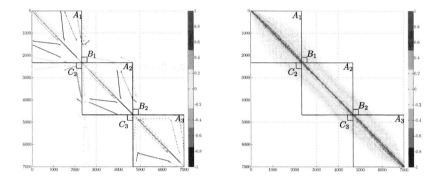

FIGURE 4.4: Permutation of the weighted graph matching implementation HSL MC64 applied to the partitioned matrix by `Mondriaan` (left) versus applied to the spectral method reordered matrix (right).

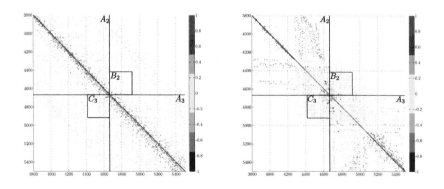

FIGURE 4.5: Reordered matrix by applying a heuristic to improve the weight of the coupling blocks without reordering (left) and with reordering (right).

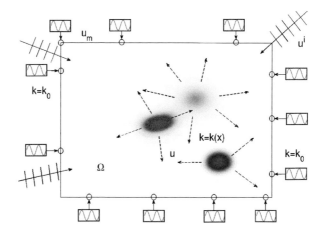

FIGURE 8.1: To detect bodies or anomalies in a solid body, it is irradiated from several directions. The waves are scattered at inhomogeneities and the scattered waves are detected at, for example, the computational domain boundary. By minimizing the misfit of simulation and measurement, the bodies and inhomogeneities can be recovered.

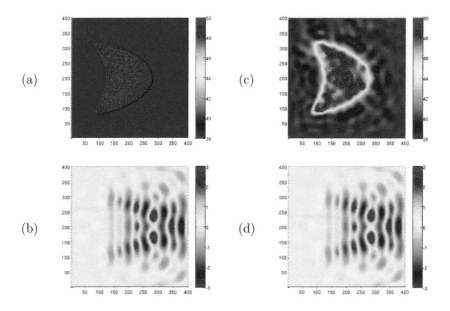

FIGURE 8.4: Reconstruction: True control function (a), one incident wave (b), and reconstructed control function (c) and wave profile according to the incident wave (d).

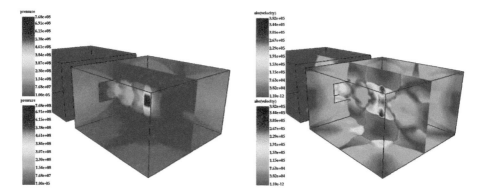

FIGURE 9.6: Blast in Room: pressures and velocities.

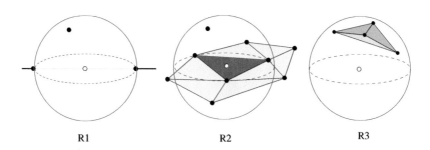

FIGURE 11.6: Point generating rules. The new points (empty dots) are chosen at the circumcenters of an encroached subsegment (left), an encroached subface (middle), and a bad-quality tetrahedron (right).

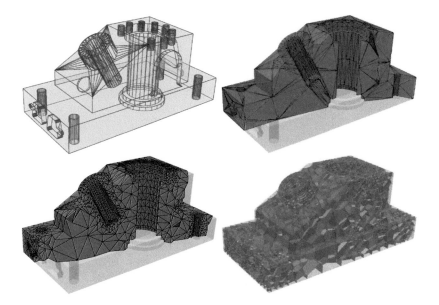

FIGURE 11.11: Example meshes produced with `TetGen`. Top left: The input PLC. The minimum facet-facet angle is about 76°. Top right: The initial coarse mesh; i.e., the input to DELAUNAYREFINEMENT. Bottom left: The refined Delaunay mesh by DELAUNAYREFINEMENT. Bottom right: The dual Voronoi partition (Voronoi faces are randomly colored for visualization).

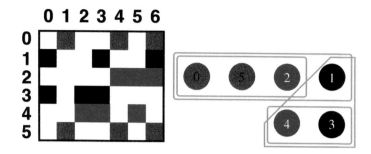

FIGURE 12.2: Sparse matrix from Figure 12.1 partitioned into 3 parts by a 1D row partitioning. Rows 0 and 5 are assigned to processor 0 (red), rows 1 and 3 to processor 1 (black), and rows 2 and 4 to processor 2 (blue). Left: the matrix, with nonzeros painted by the color of their processor. Right: the column-net hypergraph created from the matrix, with only the cut nets shown (in green). The vertices are painted in the same color as their corresponding row. Column 6 has $\lambda_6 = 3$ processors, causing $\lambda_6 - 1 = 2$ communications during sparse matrix–vector multiplication, and the corresponding net $\{0, 1, 2, 5\}$ has three parts. This means that vector component v_6 has to become known to all three processors, and hence it has to be sent by the source processor to the other two processors. The other cut columns are columns 2, 3, 4; they each have two parts and cause one communication. Columns 0, 1, and 5 do not cause communication. The load balance happens to be perfect. The communication volume as defined by Equation (12.3) equals $V = 2+1+1+1 = 5$, which is the sum of the number of communications for the separate columns.

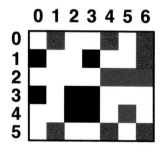

FIGURE 12.3: Sparse matrix from Figure 12.1 partitioned into 3 parts by a 2D partitioning, using the Mondriaan package with the hybrid approach of this chapter. Rows 1 and 4, and columns 4 and 6 each cause one communication. The load imbalance is one nonzero, i.e., $\epsilon = 1/6$. The total communication volume is 4.

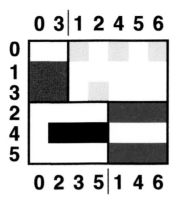

FIGURE 12.4: Sparse matrix from Figure 12.1 partitioned into 4 parts by a Mondriaan partitioning. The first split is in the row direction, whereas the second and third splits are in the column direction. The matrix is shown in the local view, displaying the local submatrices of the processors. The column indices are different for the top and the bottom part. The submatrix $I_0 \times J_0$ in the upper left corner contains the nonzeros assigned to processor 0 (red), where $I_0 = \{0, 1, 3\}$ and $J_0 = \{0, 3\}$.

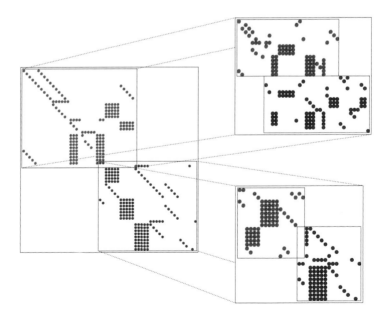

FIGURE 12.6: Split of the 59×59 matrix `impcol b` with 312 nonzeros into four parts. The first split (left) is by the fine-grain method; the second split (top right) is by rows; and the third split (bottom right) is again by the fine-grain method. The matrix is permuted correspondingly by moving, after each split, the newly cut rows and columns to the middle and the uncut ones to the left or right, or top or bottom. The matrix structure seen on the left is the (doubly) separated block-diagonal (SBD) structure, see [3].

FIGURE 15.7: Left: Geometric partitioning on 16 processors. Right: K-way multilevel partitioning on 16 processors. Both partitionings have been computed using ParMETIS.

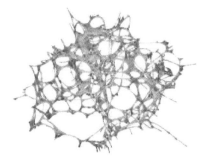

boneS10. $|V| = 914898, |E| = 27276762$ cvxbqp1. $|V| = 40000, |E| = 120000.$

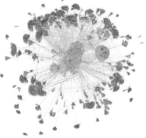

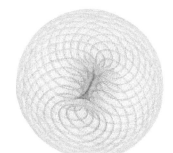

connectus. $|V| = 392366, |E| = 1124842$ aircraft. $|V| = 11271, |E| = 20267.$

FIGURE 19.9: Drawing of some graphs from the University of Florida Sparse Matrix Collection [43]. More drawings can be found at [26].

Chapter 10

Unstructured Mesh Generation

Jonathan Richard Shewchuk

University of California, Berkeley

10.1 Introduction

One of the central tools of scientific computing is the fifty-year old *finite element method*—a numerical method for approximating solutions to partial differential equations. The finite element method and its cousins, the finite volume method and the boundary element method, simulate physical phenomena including fluid flow, heat transfer, mechanical deformation, and electromagnetic wave propagation. They are applied heavily in industry and science for marvelously diverse purposes—evaluating pumping strategies for petroleum extraction, modeling the fabrication and operation of transistors and integrated circuits, optimizing the aerodynamics of aircraft and car bodies, and studying phenomena from quantum mechanics to earthquakes to black holes.

The aerospace engineer Joe F. Thompson, who commanded a multi-institutional mesh generation effort called the National Grid Project [1], wrote in 1992 that

> An essential element of the numerical solution of partial differential equations (PDEs) on general regions is the construction of a grid (mesh) on which to represent the equations in finite form. ... [A]t present it can take orders of magnitude more man-hours to construct the grid than it does to perform and analyze the PDE solution on the grid. This is especially true now that PDE codes of wide applicability are becoming available, and grid generation has been cited repeatedly as being a major pacing item. The PDE codes now available typically require much less esoteric expertise of the knowledgeable user than do the grid generation codes.

Two decades later, meshes are still a recurring bottleneck. The *automatic mesh generation problem* is to divide a physical domain with a complicated geometry—say, an automobile engine, a human's blood vessels, or the air around an airplane—into small, simple pieces called *elements*, such as triangles or rectangles (for two-dimensional geometries) or tetrahedra or rectangular prisms (for three-dimensional geometries). Millions or billions of elements may be needed.

A mesh must satisfy nearly contradictory requirements: it must conform to the shape of the object or simulation domain; its elements may be neither too large nor too numerous; it may have to grade from small to large elements over a relatively short distance; and it must be composed of elements that are of the right shapes and sizes. "The right shapes" typically include elements that are nearly equilateral and equiangular, and typically exclude elements that are long and thin, e.g., shaped like a needle or a kite. However, some applications require

good   bad

anisotropic elements that are long and thin, albeit with specified orientations and eccentricities, to interpolate fields with anisotropic second derivatives or to model anisotropic physical phenomena such as laminar air flow over an airplane wing. Some problems in mesh generation—for example, dividing an arbitrary three-dimensional domain into a modest number of nicely shaped hexahedra—are suspected to have no general solution.

By my reckoning, the history of mesh generation falls into three periods, conveniently divided by decade. The pioneering work was done by researchers from several branches of engineering, especially mechanics and fluid dynamics, during the 1980s (though as we shall see, the earliest work dates back to at least 1970). This period brought forth most of the techniques used today: the octree, Delaunay, and advancing front methods for mesh generation, and mesh "clean-up" methods such as smoothing and topological transformations. Unfortunately, nearly all the algorithms developed during this period are fragile, and produce unsatisfying meshes when confronted by complex domain geometries (e.g., domain faces meeting at small angles or in topologically complicated ways), stringent demands on element shape, or numerically imprecise CAD models.

Around 1988, these problems attracted the interest of researchers in computational geometry, a branch of theoretical computer science. Whereas most engineers were satisfied with mesh generators that usually work for their chosen domains, computational geometers set a loftier goal: *provably good mesh generation*, the design of algorithms that are mathematically guaranteed to produce a satisfying mesh, even for domain geometries unimagined by the algorithm designer. This work flourished during the 1990s.

During the first decade of the 2000s, mesh generation became bigger than the finite element methods that gave birth to it. Computer animation uses triangulated surface models extensively, and the most novel new ideas for using, processing, and generating meshes often debut at computer graphics conferences. In economic terms, the videogame and motion picture industries probably now exceed the finite element industries as users of meshes.

Meshes are used for multivariate interpolation in hundreds of other applications. For example, an airborne laser scanning technology called LIDAR (Light Detection and Ranging) enables an airplane to measure the altitude of the ground at millions of locations in a few hours. A cartographer can generate a mesh of triangles whose corners lie at those locations, estimate the altitude at any other point by interpolation, and generate a contour map of the terrain. Similarly, meshes find heavy use in image processing, geographic information systems, radio propagation analysis, shape matching, and population sampling. Mesh generation has become a truly interdisciplinary topic.

10.2 Meshes

Meshes are categorized according to their dimensionality and choice of elements. *Triangular meshes*, *tetrahedral meshes*, *quadrilateral meshes*, and *hexahedral meshes* are named according to the shapes of their elements. The two-dimensional elements (triangles and quadrilaterals) serve both in modeling two-dimensional domains and in *surface meshes* embedded in three dimensions, which are prevalent in computer graphics, boundary element methods, and simulations of thin plates and shells.

Quadrilateral elements are four-sided polygons; their sides need not be parallel. Hexahedral elements are brick-like polyhedra, but their faces need not be parallel or even planar. Simplicial meshes (triangular and tetrahedral) are much easier to generate, but for some applications, quadrilateral and hexahedral meshes offer more accurate interpolation and approximation. Non-simplicial elements sometimes make life easier for the numerical analyst; simplicial elements nearly always make life easier for the mesh generator. For topological reasons, hexahedral meshes can be extraordinarily difficult to generate for geometrically complicated domains.

As a compromise, some applications can use *hex-dominant* meshes (and occasionally their two-dimensional analog, *quad-dominant* meshes) in which most of the elements are hexahedra, but other element shapes—tetrahedra, pyramids, wedges—fill the regions where the meshing algorithm cannot produce good hexahedra.

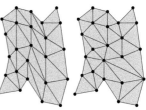

Meshes are also categorized as structured or unstructured. A *structured mesh* (near right), such as a regular cubical grid, has the property that its vertices can be numbered so that simple arithmetic suffices to determine which vertices share an element with a selected vertex. This chapter discusses only *unstructured meshes* (far right), which entail explicitly storing each vertex's neighboring vertices or elements. Structured meshes have been studied extensively [2]; they are suitable primarily for domains that have tractable geometries and do not require a strongly graded mesh.

The elements comprising a mesh must entirely cover the domain without overlapping each other. For most applications their faces must intersect "nicely," so that if two elements intersect, their intersection is a vertex or edge or entire face of both. (Formally, a mesh must be a *cell complex*.) The mesh generation problem becomes superficially easier if we permit *nonconforming elements* like the triangle at right that shares one edge with two other triangles each having half of that edge. But nonconforming elements rarely alleviate the underlying numerical problems, so they are rarely used in unstructured meshes.

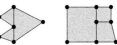

The goal of mesh generation is to create elements that *conform* to the shape of the geometric domain and meet constraints on their sizes and shapes. The next two sections discuss domain conformity and element quality.

10.2.1 Domain Conformity

Mesh generation algorithms vary immensely in what domains they can mesh and how those domains are specified. The input to a mesh generator—particularly one in the theory literature—might be a simple polygon or polyhedron. Meshing becomes more difficult if the domain can have *internal boundaries* that no element is permitted to cross, such as a boundary between two materials in a heat transfer simulation. Meshing is substantially more difficult for domains that have curved edges and surfaces, called *ridges* and *patches*, which are typically represented as splines or subdivision surfaces. Consider these geometries in turn.

Simply cutting a polyhedron into tetrahedra without producing unreasonably many is thorny, even if the element sizes and shapes are unconstrained. Illustrated at right is a polyhedron of Chazelle [3] that has n vertices and $\mathcal{O}(n)$ edges,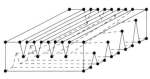
but cannot be divided into fewer than $\Theta(n^2)$ convex bodies. Chazelle and Palios [4] show that the worst-case complexity of subdividing a polyhedron is related to its number of *reflex edges*: edges at which the interior dihedral angle exceeds 180°. (A *dihedral angle* is the angle at which two faces meet in three dimensions. A simple polyhedron is convex if and only if none of its interior dihedral angles exceeds 180°.) Specifically, some polyhedra with r reflex edges cannot be divided into fewer than $\Omega(n + r^2)$ convex bodies. Conversely, Chazelle and Palios give an algorithm that divides any polyhedron into $\mathcal{O}(n + r^2)$ tetrahedra.

Internal boundaries exist to help apply boundary conditions for partial differential equations and to support discontinuities in physical properties, like differences in heat conductivity in a multi-material simulation. A boundary, whether internal or external, must be represented by a union of edges or faces of the mesh. Elements cannot cross boundaries, and where two materials meet, their meshes must have matching edges and faces. This requirement may seem innocuous, but it makes meshing much harder if the domain has small angles.

Curved domains introduce more difficulties. Some applications require elements that curve to match a domain. Others approximate a domain with a piecewise linear mesh at the cost of introducing inaccuracies in shape, finite element solutions, and surface normal vectors (which are important for computer graphics). In finite element methods, curved domains are sometimes approximated with elements whose faces are described by parametrized quadratic, cubic, bilinear, or trilinear patches. Regardless of the choice, it is tricky to mesh a *cusp* where two curved patches meet at a dihedral angle of zero. If the two patches are broken into linear triangles with insufficient care, they can easily intersect each other where they should not.

Curved domains are often represented ambiguously. In many CAD models, two patches that are supposed to meet at a shared ridge do not actually meet, or they extend too far and pass through each other. These defects are usually tiny and do no harm when the model is visualized or used for manufacturing, but they can be fatal to a mesh generator that tries to conform to the defects. The topology of the ridges and patches (i.e., the ways that they meet) is sometimes ambiguous, and so is the question of what is inside the domain and what is outside. The problem of *CAD clean-up*—the preparation of solid models so they are suitable as input to mesh generators—has received substantial attention in industry, albeit little from academia. A notable virtue of the Finite Octree mesh generator of Shephard and Georges [5] is that it queries CAD modeling programs directly and copes with their inconsistencies.

10.2.2   What Is a Good Element?

Most applications of meshes place constraints on both the shapes and sizes of the elements. These constraints come from several sources. First, large angles (near $180°$) can cause large interpolation errors. In the finite element method, these errors induce a large *discretization error*—the difference between the computed approximation and the true solution of the PDE. Second, small angles (near $0°$) can cause the stiffness matrices associated with the finite element method to be ill-conditioned. (Small angles do not harm interpolation accuracy, and many applications can tolerate them.) Third, smaller elements offer more accuracy, but cost more computationally. Fourth, small or skinny elements can induce instability in the explicit time integration methods employed by many time-dependent physical simulations. Consider these four constraints in turn.

The first constraint forbids large angles. Most applications of triangulations use them to interpolate a multivariate function whose true value might or might not be known. The most popular interpolation methods are the simplest ones: piecewise linear interpolation over triangles and tetrahedra (though quadratic and higher-order elements are not rare), piecewise bilinear interpolation over quadrilaterals, and piecewise trilinear interpolation over hexahedra. There are two types of interpolation error that matter for most applications: the difference between the interpolated function and the true function, and the difference between the *gradient* of the interpolated function and the gradient of the true function. Element shape is largely irrelevant for the first of these—the way to reduce interpolation error is to use smaller elements.

However, the error in the gradient depends on both the shapes and the sizes: it can grow arbitrarily large as an element's largest angle approaches $180°$ [6, 7, 8, 9]. (In three dimensions, it is the element's largest dihedral angle that usually matters most [10, 9].) This error is the reason why large angles are forbidden. In the finite element method, this error contributes to the discretization error [7]. In surface meshes for computer graphics, it causes triangles to have normal vectors that poorly approximate the normal to the true surface, and these can create visual artifacts in rendering. Nonconvex quadrilateral and hexahedral elements, with angles exceeding $180°$, sabotage interpolation and the finite element method.

The second constraint forbids small angles, albeit not for all applications. If your application is the finite element method, then the eigenvalues of the stiffness matrix associated with the method ideally should be clustered as close together as possible. Matrices with poor eigenvalue spectra affect linear equation solvers by slowing down iterative methods and introducing large roundoff errors into direct methods. The relationship between element shape and matrix conditioning depends on the PDE being solved and the basis functions and test functions used to discretize it, but as a rule of thumb, it is the small angles that are deleterious: the largest eigenvalue approaches infinity as an element's smallest angle approaches zero [11, 9]. Fortunately, most linear equation solvers cope well with a few bad eigenvalues.

Although large and sometimes small angles are bad as they approach the limits of 180° and 0°, there are two circumstances in which the ideal element is not equilateral but anisotropic. First, if you wish to interpolate a function as accurately as possible with a fixed number of elements, you should tailor their anisotropy according to the function. This is a vague statement, because it depends on the order of the interpolation method; to be more concrete, if you use linear interpolation over triangular or tetrahedral elements, their anisotropy should be determined by the anisotropy of the second derivatives of the function being interpolated, with the greatest density of elements along the directions maximizing the second directional derivatives. That is why aerodynamics simulations use extremely thin wedge-shaped elements to model laminar air flow at the surface of an aerofoil: the velocities and pressures vary little along the surface of the aerofoil, but vary rapidly as you move away. These elements make simulations possible that would be computationally intractable with isotropic elements, because they would require too many of them. Second, if you are solving an anisotropic PDE—for instance, modeling a material with anisotropic material properties, such as wood with its grain—you can achieve better matrix conditioning with elements having matching anisotropy. It is possible, though not common, for these two drivers of anisotropy (interpolation error and matrix conditioning) to conflict sharply in their preferences.

The third constraint governs element size. Many mesh generation algorithms take as input not just the domain geometry, but also a space-varying *size field* that specifies the ideal size, and sometimes anisotropy, of an element as a function of its position in the domain. (The size field is often implemented by interpolation over a *background mesh*.) A large number of *fine* (small) elements may be required in one region where they are needed to attain good accuracy—often where the physics is most interesting, as amid turbulence in a fluid flow simulation—whereas other regions might be better served by *coarse* (large) elements, to keep their number small and avoid imposing an overwhelming computational burden on the application. Ideally, a mesh generator should be able to *grade* from very small to very large elements over a short distance. However, overly aggressive grading introduces skinny elements in the transition region. Hexahedral meshes are especially difficult to grade without including bad elements. Hence, the size field alone does not determine element size: mesh generators often create elements smaller than specified to maintain good element quality in a graded mesh, and to conform to small geometric features of a domain.

There is a chicken-and-egg problem: element size and anisotropy should be tailored to the function being interpolated, but in the finite element method, you need a mesh to discover what that function is. This motivates *adaptive refinement*: use a coarse mesh to generate a rough initial solution of a PDE, then generate a better, finer mesh by using the coarse mesh and the approximate solution to determine the size field and perhaps the anisotropy. Solve the PDE again, more accurately. Repeat as desired.

The fourth constraint forbids unnecessarily small or skinny elements for time-dependent PDEs solved with explicit time integration methods. The stability of explicit time integration is typically governed by the Courant–Friedrichs–Lewy condition [12], which implies that the computational time step must be small enough that a wave or other time-dependent signal cannot cross more than one element per time step. Therefore, elements with short edges or short altitudes may force a simulation to take unnecessarily small time steps, at great computational cost, or risk introducing a large dose of spurious energy that causes the simulation to "explode."

The constraints on domain conformity and element quality are more difficult to satisfy together than separately. The simplest algorithms for triangulating a polygon [13] or polyhedron [4] usually produce very skinny triangles or tetrahedra. Conversely, we know how to tile space or slabs with acute tetrahedra [14] or perfectly cubical hexahedra, but the same is not true for interestingly shaped domains.

Some meshing problems are impossible. An obvious example is a polygonal domain that has a corner with a 1° angle, and a demand to mesh it with triangles having all their angles greater than 30°. The problem becomes possible again if we relax it, and demand angles greater than 30° everywhere *except* in the 1° corner. Polygonal domains with small angles can be meshed by cutting off the sharp corners before applying a standard meshing algorithm to the remainder [15]. But for domains with internal boundaries, the problem is impossible again—the illustration at right de- 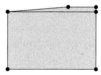 picts a domain composed of two polygons glued together that, surprisingly, provably has no mesh whose new angles are all over 30° [16]. Simple polyhedra in three dimensions inherit this hurdle—even without internal boundaries. Three-dimensional domains with small angles and internal boundaries, wherein an arbitrary number of ridges and patches can meet at a single vertex, are one of the biggest challenges in mesh generation.

10.3 Methods of Mesh Generation

This section compares three classes of mesh generation algorithms: advancing front methods, wherein elements crystallize one by one, coalescing from the boundary of a domain to its center; Delaunay refinement algorithms, which construct a Delaunay triangulation and insert additional vertices to ensure that its elements are good; and grid, quadtree, and octree algorithms, which overlay a structured background grid and use it as a guide to subdivide a domain. A fourth subsection describes mesh improvement algorithms, which take an existing mesh and make it better through local optimization. The few fully unstructured mesh generation algorithms that do not fall into one of these four categories are not yet in widespread use.

I give short shrift to many worthy topics. Schneiders [17] surveys quadrilateral and hexahedral meshing far more deeply. Bern and Eppstein [13] survey triangulation algorithms that optimize many criteria not discussed here. I ignore rich literatures on surface meshing [18] and parallelism.

Before diving in, I want to highlight a startling 1970 article by Frederick, Wong, and Edge [19] that describes, to the best of my knowledge, the first Delaunay mesh generation algorithm, the first advancing front method, and the first algorithm for Delaunay triangulations in the plane besides slow exhaustive search—all one and the same. The irony of this distinction is that the authors appear to have been unaware that the triangulations they create are Delaunay. Moreover, a careful reading of their paper reveals that their meshes are *constrained* Delaunay triangulations. Perhaps this paper is still so little known because it was two decades ahead of its time.

10.3.1 Advancing Front Mesh Generation

Advancing front methods construct elements one by one, starting from the domain boundary and advancing inward—or occasionally outward, as when meshing the air around an airplane. The frontier where elements meet unmeshed domain is called the *front*, which ventures forward until the domain is paved with elements and the front vanishes.

Advancing front methods are characterized by exceptionally high quality elements at the domain boundary. The worst elements appear where the front collides with itself, and assuring their quality is difficult, especially in three dimensions; there is no literature on provably good advancing front algorithms. Advancing front methods have been 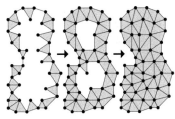 particularly successful in fluid mechanics, because it is easy to place extremely anisotropic elements or specialized elements at the boundary, where they are needed to model phenomena such as laminar air flow.

Most early methods created vertices then triangulated them in two separate stages [19, 20, 21]. For instance, Frederick, Wong, and Edge [19] use "a magnetic pen to record node point data and a computer program to generate element data." The simple but crucial next insight—arguably, the "true" advancing front technique—was to interleave vertex creation with element creation, so the front can guide the placement of vertices. Alan George [22] took this step in 1971, but it was forgotten and reinvented in 1980 by Sadek [23] and again in 1987 by Peraire, Vahdati, Morgan, and Zienkiewicz [24], who also introduced support for anisotropic triangles. Immediately thereafter, methods of this design appeared for tetrahedral meshing [25, 26].

10.3.1.1 A Generic Advancing Front Method

Advancing front algorithms mesh the domain features in order of increasing dimension: first, the domain edges and ridges are subdivided into mesh edges whose lengths are specified by the size field; then, the domain faces and patches are subdivided into triangles or quadrilaterals; then, the domain volumes are subdivided into tetrahedra or hexahedra. Each such subdivision is usually done with an advancing front subroutine. (There are some odd exceptions. Marcum and Weatherill [27] use a Delaunay mesh generator to create a complete tetrahedralization, then throw it all away except for the surface mesh, which they use to seed their advancing front algorithm.)

For concreteness, consider meshing a volume with tetrahedra. The algorithm maintains a *front* composed of *active* triangles. A triangle is active if we desire to construct a tetrahedron upon it: either it is a domain boundary triangle with no tetrahedron yet adjoining it, or it is a face of one and only one tetrahedron and should be shared by a second adjoining tetrahedron.

The core of the algorithm is a loop that selects an active triangle and constructs a new tetrahedron upon it. Often, the new tetrahedron adjoins more than one active triangle. To detect this circumstance, the algorithm maintains the active triangles in a dictionary (e.g., a hash table) that maps a triple of vertex indices to the triangle having those vertices, if that triangle is active. Each triangle is assigned an *orientation* that specifies which of its two sides is active. Consider the following generic advancing front algorithm:

ADVANCINGFRONT(B)
{ B is a set of boundary faces, with orientations. }
1 $D \leftarrow$ a dictionary containing the oriented faces in B
2 **while** D is not empty
    { Loop invariant: at this point, D contains all the active faces. }
3   Remove a face f from D
4   Choose a vertex v (new or preexisting) so that $t = \mathrm{conv}(f \cup v)$ is
         nicely shaped and nicely intersects the front
5   Output the element t
6   **for** each face g of t except f
7   **if** g is in the dictionary
8   Remove g from the dictionary
9   **else** Enter g in the dictionary, oriented facing away from t

Implementations of this generic algorithm can vary in many ways.
- How do we choose which face to advance (Line 3)?
- How do we choose a new or preexisting vertex (Line 4)?
- How do we *quickly* find the preexisting vertices that are good candidates?
- How do we determine if a new element intersects the front nicely (i.e., forms a cell complex)?

The first two questions crucially affect not only the quality of the mesh, but also whether the algorithm will complete its task at all. The next two are questions solely of speed: they can be done by exhaustive checks, at great computational cost, or they can be accelerated by spatial search algorithms that identify candidates close to the active face f.

10.3.1.2 Some Specific Algorithms

Let us see how different researchers address these four questions. Consider the choice of face to advance. Peraire, Peiró, and Morgan [28] find that for triangular meshes, advancing the shortest edge gives the best results, because it makes it easier to generate good triangles where fronts collide. For tetrahedral meshes, they choose the active face with the shortest altitude. Marcum [29],

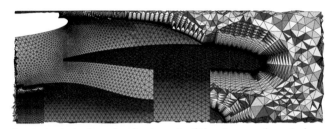

who specializes in meshes for aerodynamics, prefers to lay down a whole row of elements before starting another, so that the elements near the boundary are as structured as possible. Sadek [23] pri-

A tetrahedral advancing front mesh of the air around an airplane nacelle. Courtesy of David Marcum.

oritizes elements within a row by noting corners in the front where the angle is not close to 180° and filling them in early. He constructs all the elements adjoining a corner simultaneously to find the best compromise among them.

Second, consider the choice of a vertex (Line 4). For a triangular mesh and an active edge f, Peraire et al. compute a preexisting candidate vertex v' and a new candidate vertex v''. The preexisting candidate is chosen by the Delaunay empty circle criterion (see Section 10.3.2.1), subject to the constraint that the candidate triangle $\text{conv}(f \cup v')$ must nicely intersect the front. (It appears that their two-dimensional algorithm nearly always produces a constrained Delaunay triangulation.) The new candidate v'' is chosen on f's bisector, so the candidate triangle $\text{conv}(f \cup v'')$ is isosceles. The lengths of this candidate triangle's two new edges are dictated by the size field, subject to the constraint that the candidate triangle not be too skinny.

Peraire et al. choose the new candidate v'' over the preexisting candidate v' if and only if the circle centered at v'' and passing through the endpoints of f encloses no valid preexisting candidate, as illustrated at right. In other words, the angle of the triangle $\text{conv}(f \cup v'')$ at v'' is at least twice the angle of the triangle $\text{conv}(f \cup v')$ at v', or there is no valid candidate v'.

Thus, they never create a new vertex too close to a preexisting one.

Unfortunately, sometimes every candidate vertex—v'' and every preexisting vertex—yields an invalid new element. This circumstance occurs where two fronts collide, especially when a front with small elements meets a front with much larger elements, and the latter front lacks good candidate vertices for the former. A new candidate vertex must be generated close enough to f to create a valid element. Ideally, the size field prevents such disparities between colliding fronts; but sometimes small geometric features of the domain boundary or the inherent difficulty of filling a polyhedral cavity with tetrahedra interfere.

For tetrahedral meshes, the process of generating a new candidate vertex and comparing it with preexisting vertices is more complicated. I omit details, but Peraire et al. do not use the Delaunay criterion, which is less effective in three dimensions (see Section 10.4).

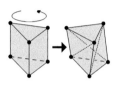

Colliding fronts in tetrahedral meshes can be more difficult to resolve than you might guess. The difficulty arises because there are polyhedra that cannot be subdivided into tetrahedra without creating new vertices. At right is Schönhardt's untetrahedralizable polyhedron [30], which has three reflex edges. If an advancing front algorithm produces a cavity of this shape, it must introduce a new vertex inside the cavity to complete the mesh. One vertex at the center of Schönhardt's polyhedron will do, but if the mesher is not careful about where it places the vertex, it will need to introduce more. In general, it is NP-hard to determine how many new vertices are required to tetrahedralize a cavity [31]. I am not aware of any tetrahedral advancing front algorithm that guarantees it can fill any polyhedral cavity.

Advancing front methods generate anisotropic meshes easily. The user specifies the locally desired anisotropy as a space-varying tensor field akin to the size field. The mesher simply applies an affine transformation to the mesh when it performs a front-advancing step; the candidate vertices v' and v'' are computed and compared in the transformed space, so the algorithm constructs an anisotropic element whose orientation and eccentricity are approximately as specified.

Marcum [29] notes that for meshes in aerodynamics, especially anisotropic ones, it is sometimes better to create new candidate vertices with an *advancing-point* or *advancing-normal* placement, which means that vertices near important surfaces are aligned along lines orthogonal to those surfaces.

Our third question is, how do we screen out preexisting vertices as candidates without checking every one of them? Peraire et al. store vertices in a spatial search data structure called an *alternating digital tree*, which is similar enough to a quadtree or octree that the differences are unimportant. With this data structure, one can query all the front vertices in a ball around f, and the query is usually efficient in practice (though there is no guarantee).

For triangular meshes, Mavriplis [32] maintains a constrained Delaunay triangulation of the unmeshed part of the domain, and uses it as a search structure. It has the nice property that the best preexisting candidate vertex for an edge f is already connected to f by a triangle. Other preexisting candidate vertices (e.g., for anisotropic meshing) can be found efficiently by a depth-first search through adjoining triangles. For tetrahedral meshes, Marcum uses the same strategy of maintaining a triangulation of the unmeshed region, but the triangulation is generally not Delaunay. Each newly constructed tetrahedron is introduced into this triangulation (if it is not already there) by a sequence of tetrahedral transformations: tetrahedron subdivisions to introduce new vertices, and topological flips (see Section 10.3.4) to force the edges and faces of the new tetrahedron into the triangulation. There is no guarantee that these flips will always work!

Our fourth question is, how do we determine if a new element t intersects the front illegally? For triangular meshes, it suffices to check whether t's new edges intersect any active edges (and if so, *how* they intersect them). For tetrahedral meshes, we must check t's new faces for edge-face intersections with the active faces. Löhner and Parikh [25] describe how to perform these tests for any pair of triangular faces. But how do we avoid the computational cost of checking *every* active face?

One simple solution is to use a quadtree or octree, and store each active face in every leaf octant it intersects. (These octants are identified by a simple depth-first search.) For each prospective new element, determine which leaf octants its new faces intersect, look up which active faces intersect those octants, and check only those faces—in practice, a small fraction of the total—against the new faces. Unfortunately, this data structure can store a face in many leaves. Peraire et al. avoid this cost by casting the problem as a six-dimensional *orthogonal range search* query: enclose each face in an axis-aligned bounding box, represent each box as a point in six-dimensional space (listing the three-dimensional coordinates of the minimum and maximum corners of the box), and store these points in a six-dimensional octree. A query asks which stored bounding boxes intersect another axis-aligned bounding box; this query is answered with a simple depth-first search of the octree.

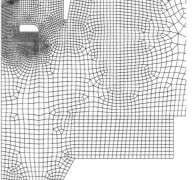

A quadrilateral mesh generated by Paving [33]. Courtesy of Ted Blacker.

Mavriplis and Marcum use an entirely different method: they exploit the fact that they have a triangulation of the unmeshed region by tracing the new edges/faces through the triangles/tetrahedra of that triangulation.

In practice, these search structures improve the speed substantially, and advancing front meshers typically generate an n-vertex mesh in $\mathcal{O}(n \log n)$ time. There are pathological circumstances in which the running time is asymptotically greater, but they are rare in practice.

Perhaps the most famous quadrilateral mesh generator is an advancing front method, Blacker and Stephenson's *Paving* algorithm [33]. It is substantially more intricate than even the tetrahedral methods. Consider the following difficulty. Because every quadrilateral has four edges, and every edge in the interior of the domain is shared by exactly two quadrilaterals, a quadrilateral mesh must have an even number of edges on its boundary. As the fronts advance into a domain, their collisions sometimes subdivide  a cavity into two smaller cavities. If each smaller cavity has an odd number of edges, the mesher is doomed. To prevent this peril, Paving sometimes deforms a row of elements so that two colliding fronts meet with the correct parity.

Hexahedral meshing is fundamentally harder, and no reliable unstructured hexahedral meshing algorithm exists yet. Early efforts at advancing front methods where thwarted by cavities that could not be finished. A fascinating recent idea called *Unconstrained Plastering* [34] advances the front one sheet at a time without immediately deciding how to subdivide the sheet into hexahedra. An element is formed where three such sheets intersect. It seems to be crucial *not* to pre-mesh the boundary with quadrilaterals.

10.3.2   Delaunay Mesh Generation

The Delaunay triangulation is a geometric structure introduced by Boris Nikolaevich Delaunay [35] in 1934. In two dimensions, it has a striking advantage: among all possible triangulations of a fixed set of vertices, the Delaunay triangulation maximizes the minimum angle [36]. Sadly, this property does not generalize to tetrahedra, but Delaunay triangulations of any dimensionality optimize several geometric criteria related to interpolation accuracy [37, 38, 39].

Delaunay refinement algorithms construct a Delaunay triangulation and *refine* it by inserting new vertices, chosen to eliminate skinny or oversized elements, while always maintaining the Delaunay property of the mesh. The triangulation guides the algorithm to place new vertices far from existing ones, so that short edges and skinny elements are not created unnecessarily.

Most Delaunay methods, unlike advancing front methods, create their worst elements near the domain boundary and their best elements in the interior, which suggests the wisdom of combining the two methods and using Delaunay refinement in place of nasty advancing front collisions; see Section 10.3.2.4. The advantage of Delaunay methods, besides their optimality properties, is that they can be designed to have mathematical guarantees: that they will always construct a valid mesh and, at least in two dimensions, that they will never produce skinny elements.

10.3.2.1 Delaunay and Constrained Delaunay Triangulations

Let V be a set of vertices in the plane, not all collinear. The *circumscribing circle*, or *circumcircle*, of a triangle is the unique circle that passes through all three of its vertices. A triangle whose vertices are in V is *Delaunay* if its circumcircle encloses no vertex in V. Note that any number of vertices might lie *on* the triangle's circumcircle; so long as the circle is *empty*—encloses no vertex—the triangle is Delaunay. A *Delaunay triangulation* of V is a subdivision of V's convex hull into Delaunay triangles with disjoint interiors.

It is not obvious that a triangulation of V exists whose triangles are all Delaunay. The easiest way to see this is through a *lifting map* that transforms V's Delaunay triangles into faces of a convex polyhedron in three dimensions. The first lifting map, by Brown [40], projects V onto a sphere, but the *parabolic lifting map* of Seidel [41] is numerically better behaved. It maps each vertex $(x, y) \in V$ to a lifted point $(x, y, x^2 + y^2)$ in E^3. Consider the convex hull of the lifted points $\{(x, y, x^2 + y^2) : (x, y) \in V\}$, a polyhedron illustrated above. Taking the underside of this convex hull and projecting it down to the x-y plane yields the Delaunay triangulation of V.

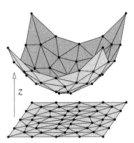

This relationship between Delaunay triangulations and convex hulls has two consequences. First, from the fact that every finite point set has a convex hull, it follows that every finite point set has a Delaunay triangulation. Second, it brings to meshing the power of a huge literature on polytope theory and algorithms: every convex hull algorithm is a Delaunay triangulation algorithm!

The Delaunay triangulation of V is unique if and only if no four vertices in V lie on a common empty circle. (In that case, the Delaunay triangulation is the planar dual of the well-known *Voronoi diagram*.) If it is not unique, the lifting map produces a convex hull that has faces with more than three sides; any arbitrary triangulation of these faces yields a Delaunay triangulation.

Delaunay triangulations extend to any dimensionality. The *circumsphere* of a tetrahedron is the unique sphere that passes through all four of its vertices. A tetrahedron is *Delaunay* if its circumsphere encloses no vertex. A set of vertices in three dimensions, not all coplanar, has a Delaunay tetrahedralization, which is unique if and only if no five vertices lie on a common empty sphere. The lifting map relates the triangulation to a convex, four-dimensional polyhedron.

As Delaunay triangulations in the plane maximize the minimum angle, do they solve the problem of triangular mesh generation? No, for two reasons illustrated at right. First, skinny triangles might appear anyway. Second, the Delaunay triangulation of a domain's vertices might not respect the domain's boundary. Both these problems can be solved by inserting additional vertices, as illustrated. Section 10.3.2.3 explains where to place them.

An alternative solution to the second problem is to maintain a *constrained Delaunay triangulation* (CDT), illustrated below. A CDT is defined with respect to a set of vertices and *segments* that demarcate the domain boundary (including internal boundaries). Every segment is required to appear as an edge of the CDT. The triangles of a CDT are not required to be Delaunay; instead, they must be *constrained Delaunay*. The definition, which I omit, partly relaxes the empty circumcircle criterion to compensate for the demand to include every segment.

One merit of a CDT is that it can respect arbitrary segments without requiring the insertion of any additional vertices besides the endpoints of the segments. CDTs were

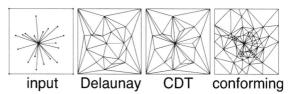

input   Delaunay   CDT   conforming

mathematically formalized by Lee and Lin [42] in 1986, though algorithms that unwittingly construct CDTs appeared much sooner [19, 43]. Lee and Lin show that the CDT inherits the Delaunay triangulation's optimality: among all possible triangulations of the vertex set *that include all the segments*, the CDT maximizes the minimum angle.

CDTs generalize to three dimensions [44]. However, there exist simple polyhedra that do not have CDTs—recall Schönhardt's and Chazelle's polyhedra [30, 3], which cannot be subdivided into tetrahedra without new vertices that are not vertices of the polyhedra. To create a CDT that respects the boundary of a polyhedral domain, you sometimes must add new vertices on the domain's segments. Once the (unconstrained) Delaunay triangulation of the vertices respects the segments, the augmented domain has a CDT; no additional vertices are needed on the domain faces.

10.3.2.2   Algorithms for Constructing Delaunay Triangulations

The first published Delaunay triangulation algorithm I know of tests every possible d-simplex—every combination of $d + 1$ vertices, where d is the dimension—to see if its circumsphere is empty, taking $\mathcal{O}(n^{d+2})$ time. This brute force algorithm (in its dual, Voronoi form) and a three-dimensional implementation appear in a 1967 paper by John Bernal, a father of structural biology who discovered the structure of graphite and was awarded a Stalin Peace Prize, and John Finney, his last Ph.D. student [45].

The next-simplest Delaunay triangulation algorithms are *gift-wrapping algorithms*, which are advancing front methods minus the creation of new vertices. This class includes the pioneering 1970 algorithm of Frederick, Wong, and Edge [19]. The first gift-wrapping algorithm for three or more dimensions appeared in 1979 when Brown [40] published his lifting map and observed that the 1970 gift-wrapping algorithm of Chand and Kapur [46] for computing general-dimensional convex hulls can construct Voronoi diagrams. The two- and three-dimensional algorithms have been rediscovered many times since.

In two dimensions, the empty circumcircle property implies a simple gift-wrapping rule: to advance an active edge (Line 4 of ADVANCINGFRONT), choose the candidate vertex v that maximizes the angle at v in the new triangle. (For numerical reasons, an implementation should not actually compute this angle; it should choose among the vertices with a polynomial test described by Guibas and Stolfi [47]—which, unlike the angle test, also works in three dimensions.)

A useful observation from computational geometry is that we can exploit the uniqueness of the Delaunay triangulation (enforced by the careful use of symbolic weight perturbations [48, Section 5.4]) to remove the tests for invalid element intersections from Line 4 of ADVANCINGFRONT, thereby saving a lot of time. However, keeping those tests intact enables gift-wrapping to compute a *constrained* Delaunay triangulation, and that is what Frederick et al. did, and what Nguyen [43] did for tetrahedra, long before the mathematics were developed to justify the rightness of what they were doing.

Gift-wrapping has the disadvantage that the running time of a naïve implementation is quadratic in the number of vertices—sometimes cubic in three dimensions. For uniform vertex distributions, clever bucketing strategies can improve Delaunay gift-wrapping to expected linear time, even in three or more dimensions [49]. Nevertheless, this leaves us in want of a fast algorithm for graded vertex distributions and for CDTs.

In the decision-tree model of computation, sets of n vertices in the plane sometimes require $\Omega(n \log n)$ time to triangulate. The first algorithm to run in optimal $\mathcal{O}(n \log n)$ time was the *divide-and-conquer algorithm* of Shamos and Hoey [50], subsequently simplified by Lee and Schachter [51] and Guibas and Stolfi [47] and sped up by Dwyer [52]. The algorithm partitions a set of vertices into two halves separated by a line, recursively computes the Delaunay triangulation of each subset, and merges the two triangulations into one. This algorithm remains the fastest planar Delaunay triangulator in practice [16], but the divide-and-conquer strategy is not fast in three dimensions.

Incremental insertion algorithms insert vertices into a Delaunay triangulation one at a time, always restoring the Delaunay property to the triangulation before inserting another vertex. Incremental insertion algorithms are the fastest three-dimensional Delaunay triangulators in practice. The difference between a Delaunay triangulation algorithm and a modern Delaunay mesh generator is that the former is given all the vertices at the outset, whereas the latter uses the triangulation to decide where to insert additional vertices, making incremental insertion obligatory.

Lawson [36] invented an algorithm for incremental insertion, but it works only in two dimensions. A faster algorithm that works in any dimensionality was discovered independently by Bowyer [53], Hermeline [54, 55], and Watson [56]. Bowyer and Watson simultaneously submitted it to *Computer Journal* and found their articles published side by side.

The idea of vertex insertion is simple. Suppose you are inserting a new vertex v into a triangulation. If a triangle's circumcircle (or a tetrahedron's circumsphere) encloses v, that element is no 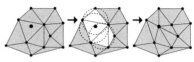 longer Delaunay, so it must be deleted. This suggests the *Bowyer–Watson algorithm*.

- Find one element whose circumsphere encloses v.
- Find all the others by a depth-first search in the triangulation.
- Delete them all, creating a polyhedral cavity; see the figure above.
- For each edge/face of this cavity, create a new element joining it with v.

It is not difficult to prove that the new elements are Delaunay.

The first step is called *point location*. Most Delaunay mesh generation algorithms generate new vertices inside the circumspheres of badly shaped or oversized elements, in which case point location is free. However, this is not true for the input vertices—vertices of the domain and other vertices a user wishes to incorporate. Locating these points in the triangulation is sometimes the most costly and complicated part of the algorithm. Incremental insertion algorithms are differentiated mainly by their point location methods.

In one of the great papers of computational geometry, Clarkson and Shor [57] show (among many other wonders) that a point location method based on a *conflict graph* makes incremental insertion run in expected $\mathcal{O}(n \log n)$ time in the plane and expected $\mathcal{O}(n^2)$ time in three dimensions. Both times are worst-case optimal. Crucial to this performance is the idea that the vertices are inserted in random order, which is only possible for vertices that are known from the start. The three-dimensional algorithm often runs in $\mathcal{O}(n \log n)$ time in practice [58], but certainly not always—it is possible for a Delaunay triangulation to have $\Theta(n^2)$ tetrahedra! If you wish to implement a three-dimensional Delaunay triangulator, I recommend an improved variant of the Clarkson–Shor algorithm by Amenta, Choi, and Rote [58].

All three types of algorithm extend to CDTs. Gift-wrapping algorithms are the simplest, but the collision tests make them slow. A divide-and-conquer algorithm of Chew [59] constructs planar CDTs in $\mathcal{O}(n \log n)$ time, but it is complicated. The algorithm most commonly used in practice begins with a Delaunay triangulation of the domain vertices, then incrementally inserts the domain edges into the CDT [60]. The running time could be as slow as $\Theta(n^3)$ but is usually in $\mathcal{O}(n \log n)$ in practice, and the algorithm's simplicity suggests that the theoretically optimal algorithms would rarely be competitive. Three-dimensional CDTs can be constructed by incremental face insertion in $\mathcal{O}(n^2 \log n)$ time [61], but this algorithm also is likely to be fast in practice.

10.3.2.3   A Generic Delaunay Refinement Algorithm

The early Delaunay mesh generators, like the early advancing front methods, created vertices and triangulated them in two separate stages [19, 62, 63]; and as with advancing front methods, the era of modern meshing began in 1987 with the insight, care of Frey [64], to use the triangulation as a search structure to decide where to place the vertices. *Delaunay refinement* is the notion of maintaining a Delaunay triangulation while inserting vertices in locations dictated by the triangulation itself.

Unlike advancing front algorithms, most Delaunay algorithms do not triangulate the domain boundary before creating tetrahedra, but there are exceptions. The following generic Delaunay refinement algorithm inserts vertices for two reasons: to enforce boundary conformity, and to eliminate bad elements.

DELAUNAYREFINEMENT(V, R, P)
{ V, R, and P are sets of boundary vertices, ridges, and patches. }
{ Each patch is bounded by ridges in R; each ridge's endpoints are in V. }
1    $T \leftarrow$ the Delaunay triangulation of V
2   **while** T contains a skinny or oversized element **or**
                        T does not respect all the ridges in R and patches in P
3   **if** T fails to respect some ridge $r \in R$
4            $v \leftarrow$ a point on the disrespected portion of r
5   **else if** T fails to respect some patch $p \in P$
6            $v \leftarrow$ a point on p, on or near the disrespected portion
7   **else if** T contains a skinny or oversized element t
8            $v \leftarrow$ a point inside t's circumsphere
9   Insert v into T, updating T so it is still Delaunay
10   Delete from T the elements that lie outside the domain

Implementations of this generic algorithm vary primarily in two ways:

- How do we choose a new vertex to eliminate a bad element (Line 8)?
- How do we choose vertices to enforce domain conformity (Lines 3–6)?

10.3.2.4   Some Specific Algorithms

Let us see how different researchers address these two questions. Where should we position a new vertex to remove an element t (Line 8)? Any point inside t's circumsphere will suffice. Frey [64] chooses t's *circumcenter*—the center of its circumsphere, illustrated at right. Weatherill [65] chooses t's centroid. But circumcenters are usually a better choice. Why? If a new vertex is too close to a prior vertex, the two vertices generate a short edge, which gives rise to skinny elements. The centroid of an element with a large angle can be very close to the vertex at the large angle. A Delaunay circumcenter, being the

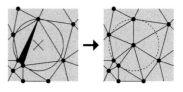

center of an empty circumsphere, cannot. It is therefore an excellent place
to put a new vertex. This is a powerful advantage of the Delaunay property,
essential to proving that Delaunay algorithms are good (Section 10.4). The
guiding principle is "create no short edges."

If a skinny triangle's circumcircle is substantially larger than its shortest
edge, it is often better to place the new vertex closer to the short edge, so
they form an acceptable new triangle. The effect is to make Delaunay refinement behave like an advancing front method—at least until the fronts collide,
whereupon Delaunay refinement performs better. Frey [64] and Üngör [66]
report that these "off-centers" give an excellent compromise between quality
and number of triangles. A more aggressive algorithm of Erten and Üngör [67]
optimizes the placement of a new vertex in a circumcircle; it generates triangular meshes that often have no angle smaller than $41°$.

Line 8 carries a potential hazard: the new vertex v might lie too close to
the domain boundary, begetting an undesirably small or skinny element in
the gap. Sometimes v lies outside the domain, and cannot be inserted.

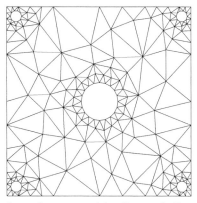

Most algorithms prevent vertices from
lying too close to the boundary by one of
three alternatives. First, if an algorithm
begins by spacing vertices on the boundary closely enough (relative to the size
field and the sizes of the domain's geometric features), no element's circumcenter or
centroid will be too close to the boundary [64, 69, 70]. Second, an algorithm can
explicitly check if a new vertex is too close
to, or beyond, the boundary (a condition
called *encroachment*); if so, move the vertex onto the boundary before inserting it,
applying the vertex placement rules discussed below for disrespected boundary

A mesh generated by Ruppert's Delaunay refinement algorithm [68].

ridges and patches [68, 71, 72]. Third, an algorithm can fix a new vertex
after insertion by smoothing it (see Section 10.3.4).

The biggest differences among Delaunay refinement algorithms are in how
they enforce domain conformity (Lines 3–6). The earliest algorithms assure
boundary integrity simply by refining the boundary enough before generating
a mesh [64, 69, 70]. But it is difficult to know how much is enough, especially
if the mesh should be graded.

We want each ridge to be approximated by a
sequence of edges of the Delaunay mesh. If we keep
track of which vertices of the mesh lie on a ridge, as
well as their order along the ridge, we know what
that sequence of edges should be, and we can check
if those edges are actually present. Where an edge
is missing, Line 4 generates a new vertex on the

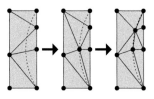

ridge, and Line 9 inserts it into the mesh, as illustrated at right. Ideally, two new edges will appear in the Delaunay triangulation to bridge the gap. If not, the missing edges are refined in turn. Schroeder and Shephard [73] call this process *stitching*.

Schroeder and Shephard place the new vertex where the ridge intersects an edge or face of the mesh. By contrast, provably good algorithms usually place the new vertex as far as possible from other vertices on the ridge, to avoid creating short edges. For polyhedral geometries, the new vertex goes at the midpoint of the missing edge [65, 68, 71].

Stitching works for patches much as it does for ridges: we insert vertices on a patch until the mesh conforms to it. But what is a patch's analog to a missing edge? Each patch should be approximated by triangular faces of the Delaunay mesh. Assume that the ridges that bound each patch have been recovered—Line 3 gives ridges priority over patches. We look at the union of the mesh faces that have all three vertices on the patch, and check for holes.

A hole implies that some edge of the mesh must pass through the patch. Schroeder and Shephard stitch vertices where mesh edges intersect the patch. This strategy carries some risk of inadvertently creating short edges. For poly-hedral domains (i.e., all the patches are flat), a provably good algorithm of my own design [71] maintains a two-dimensional Delaunay triangulation of each polygonal patch. If a patch triangulation includes a triangular face that is missing from the tetrahedral mesh, the algorithm generates a new vertex at the circumcenter of the missing triangle, and inserts it into the tetrahe-dral mesh *and* the patch triangulation. Weatherill [65] stitches new vertices at centroids of missing triangles.

If a domain has ridges or patches sepa-rated by small angles, stitching is difficult, because the vertices stitched into one ridge  or patch can knock edges or triangles out of another, as illustrated above. A *conforming Delaunay triangulation* is a Delaunay mesh such that every do-main edge is a union of mesh edges, and every domain face is a union of mesh faces. Computational geometers have studied this problem and concluded that a conforming Delaunay triangulation often entails the addition of many, many new vertices. In two dimensions, Bishop [74] shows that every set of n vertices and segments has a conforming Delaunay triangulation with $\mathcal{O}(n^{2.5})$ addi-tional vertices. It might be possible to improve this bound, but segment sets are known for which $\Theta(n^2)$ additional vertices are needed. Closing the gap between the $\mathcal{O}(n^{2.5})$ and $\Omega(n^2)$ bounds remains an open problem.

In three dimensions, no polynomial bound is known! Algorithms do exist; Murphy, Mount, and Gable [75] gave the first conforming Delaunay tetrahe-dralization algorithm, and better subsequent algorithms have been coupled with Delaunay refinement [76]. But if you insist on a truly Delaunay mesh, you will probably always need to tolerate an unnecessarily large number of vertices near small domain angles.

CDTs obviate these problems in two dimensions, without the need for any new vertices at all. In three dimensions, CDTs starkly reduce the number of new vertices that must be added, but a polynomial bound is still not known.

The advantage of CDTs over Delaunay triangulations is less pronounced when coupled with stringent demands for element quality. When a bad element's circumcenter strays too close to the domain boundary, Delaunay refinement algorithms must refine the boundary, CDT or not. Nevertheless, algorithms that employ CDTs refine less and deliver meshes with fewer elements. Examples include the triangular meshers of Chew [69, 77] and the tetrahedral mesher of Si [78]. Boivin and Ollivier-Gooch [72] show that CDTs can also aid the recovery of curved ridges in triangular meshes.

Another advantage of CDTs is that the simplices outside the domain can be deleted (Line 10) before the refinement loop (Lines 2–9), saving time and unnecessary refinement. CDTs are invaluable when a user has subdivided the boundary of a domain, and demands that it not be subdivided further.

Many Delaunay tetrahedral meshers, including algorithms by George, Hecht, and Saltel [79], Weatherill [80], and Hazlewood [81], choose to sacrifice the Delaunay (and constrained Delaunay) property so that ridges and patches can be recovered with fewer new vertices, and so that stitching cannot knock them back out. They use bistellar flips (see Section 10.3.4) to recover ridges and patches when possible, resorting to vertex insertion when necessary. The final meshes are "almost Delaunay" and are usually good in practice, but their quality cannot, so far as we know, be mathematically guaranteed, because circumcenters can be arbitrarily close to prior vertices.

Delaunay mesh generators, like advancing front methods, adapt easily to anisotropy. George and Borouchaki [82] modify the Bowyer–Watson algorithm [53, 56] to use circumellipses instead of circumspheres. However, if the anisotropy field varies over space, the newly created elements might not have empty circumellipses, and the quality of the mesh cannot be guaranteed.

10.3.3 Grid, Quadtree, and Octree Mesh Generation

Another class of mesh generators is those that overlay a domain with a background grid whose resolution is small enough that each of its cells overlaps a very simple, easily triangulated portion of the domain. A variable-resolution grid, usually a quadtree or octree, yields a graded mesh. (Readers not familiar with quadtrees may consult Samet [83].) Element quality is usually assured by warping the grid so that no short edges appear when the cells are triangulated, or by improving the mesh afterward.

Grid meshers place excellent elements in the domain interior, but the elements near the domain boundary are worse than with other methods. Other disadvantages are the tendency for most mesh edges to be aligned in a few preferred directions, which may influence subsequent finite element solutions, and the difficulty of creating anisotropic elements that are not aligned with the grid. Their advantages are their speed, their ease of parallelism, the fact that some of them have mathematical guarantees, and most notably, their robustness for meshing imprecisely specified geometry and dirty CAD data.

10.3.3.1   A Generic Octree Mesher

Most quadtree and octree meshers perform roughly the following steps:

OCTREEMESH(B)
{ B is the set of boundary vertices, ridges, and patches. }
1   Create an octree, refined as dictated by B and the size field
2   Balance the octree
3   Warp the octree by moving its vertices as dictated by B
4   **for** each leaf octant O that intersects the domain
5   Triangulate the intersection of O with the domain

The rules for refining the octree (Line 1) differ radically between algorithms, and are usually complicated. Most algorithms refine any octant containing more than one vertex of B, so every vertex lies in its own leaf octant. Most algorithms also refine any octant that intersects two patches, ridges, or vertices that do not intersect each other. A mesher that supports curved surfaces will subdivide an octant if its intersection with a ridge or patch is not well approximated by a single edge or polygon.

Most algorithms require the octree to satisfy some *balance constraint* (Line 2). The most common balance constraint is to require that any two leaf octants that share an edge differ in width by no more than a factor of two. (Octants that do not intersect the domain are exempted.) Wherever this constraint is violated, the larger octant is refined.

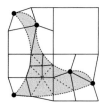

The purpose of warping the octree in Line 3 is to ensure that the elements created by Line 5 have good quality—in particular, that they have no short edges. Warping means to move octree vertices, usually onto the domain boundary. Typically, for each domain vertex in B, the nearest octree vertex is warped (moved) onto the domain vertex, as illustrated at right. Next, for each domain ridge that passes through an octant face, either an octree vertex is warped onto the ridge, or a new vertex is created at the intersection point. The same happens where a domain patch intersects an octant edge.

Finally, the mesher produces a triangulation for each leaf octant, covering the intersection of the (possibly warped) octant with the domain (Lines 4 and 5). The vertices of this triangulation include octant vertices, domain vertices, and possibly new vertices introduced wherever a ridge intersects an octant face or a patch intersect an octant edge. In some algorithms, there are only a small number of cases governing how the domain can intersect a warped octant, and for each case there is a fixed *template* or *stencil* that determines how to triangulate the octant [84, 85, 86]. Meshers using templates can be very fast. Other algorithms use a Delaunay or advancing front triangulation to fill each octant with elements [15, 5]. Yet others triangulate simply by adding a new vertex in the interior of each octant [87].

10.3.3.2   Some Specific Algorithms

The first quadtree mesher, published in 1983 by Yerry and Shephard [84], set the pattern for the generic OCTREEMESH algorithm. Yerry's warping step follows the triangulation step, but this difference is cosmetic. The intersection between a leaf quadrant and the domain is approximated by a square clipped by a single line, with the clipping line passing through two of sixteen points on the quadrant's boundary. Therefore, there is a finite set of possible leaf quadrants, and these are triangulated by templates. There are templates for both triangular and quad-dominant meshes. The approximation implies that the mesh usually does not have the same shape as the domain, so a postprocessing step warps the mesh boundary vertices onto the domain boundary. Finally, the element quality is improved by smoothing (see Section 10.3.4). Best results are obtained if the user aligns the domain vertices with quadtree vertices. The following year, Yerry and Shephard [85] generalized their method to three dimensions.

Bern, Eppstein, and Gilbert [15] designed a provably good triangular mesher for polygons. To guarantee high quality, it refines the quadtree severely—for instance, each input vertex must be in the center square of a 5×5 grid of equal-sized quadrants, empty except for that one vertex. Quadtree balancing and warping follow the generic OCTREEMESH algorithm closely, with an interesting twist: sometimes, the intersection of an octant with the domain comprises two or more connected components. These are meshed independently—the algorithm creates one copy of the quadrant for each connected component, and each copy is warped and perhaps even refined independently. After warping, each leaf quadrant's component has at most eight vertices, and is subdivided into optimal (e.g., Delaunay) triangles.

Probably the most famous octree mesher is the Finite Octree method of Shephard and Georges [5]. Rather than warping the octree before triangulating the octants, Shephard and Georges create a vertex wherever a patch intersects an octant edge or a ridge intersects an octant face, and triangulate each octant with an advancing front or Delaunay method; then they use edge contractions

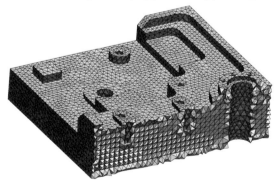

A tetrahedral mesh generated by a Finite Octree successor in the software *Simmetrix*. A cut shows the interior octree structure. Courtesy of Mark Beall.

to remove short edges. Finally, they rely on mesh improvement methods (Section 10.3.4) to ensure the quality of the tetrahedra.

A noteworthy feature of the Finite Octree method is its ability to interface with a variety of CAD packages, and to work robustly with dirty CAD data. Recognizing that it is expensive to compute intersections of patches with faces of octants, the algorithm tries to infer as much domain geometry as possible from patch-edge and ridge-face intersections, which are single points. If the intersection of a patch with an octant is a topological disk, these points often (but not always) suffice to infer how the octant should be triangulated.

The provably good *isosurface stuffing* algorithm of Labelle and Shewchuk [86] generates tetrahedral meshes of domains that have entirely smooth surfaces (i.e., no vertices or ridges), like biological tissues. The sole interface required between the algorithm and the domain geometry is the ability to query whether a specified point is inside or outside the domain; this lends the algorithm some robustness to imprecise geometry. Isosurface stuffing uses an octree to generate a tetrahedral background mesh, and it warps that mesh instead of the octree itself. Wherever the surface intersects an edge of the background mesh, either a new vertex is created (if neither edge endpoint is close enough) or an endpoint of the edge is warped onto the intersection. Background tetrahedra are subdivided into mesh tetrahedra according to a set of specified stencils. The parameters that determine whether a background mesh vertex warps or not were chosen by the authors to optimize the worst possible mesh tetrahedron.

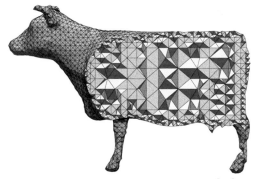

A tetrahedral mesh by isosurface stuffing [86]. A cut shows the graded interior tetrahedra.

Some quadrilateral and hexahedral meshers fill the deep interior of a domain with a regular grid, then bridge the gap by projecting or pulling the vertices on the grid boundary to the domain boundary. See Schneiders [17] for details.

10.3.4 Mesh Improvement

From nearly the beginning of the field, most mesh generation systems have included a mesh "clean-up" component that improves the quality of a finished mesh. Today, simplicial mesh improvement heuristics offer by far the highest quality of all the methods, and excellent control of anisotropy. Their disadvantages are the requirement for an initial mesh and a lack of mathematical guarantees. (They can guarantee they will not make the mesh worse.)

The ingredients of a mesh improvement method are a set of local transformations, which replace small groups of tetrahedra with other tetrahedra of better quality, and a schedule that searches for opportunities to apply them. *Smoothing* is the act of moving a vertex to improve the quality of the elements adjoining it. Smoothing does not change the topology (connectivity) of the mesh. *Topological transformations* are operations that change the mesh topology by removing elements from a mesh and replacing them with a different configuration of elements occupying the same space.

Smoothing is commonly applied to each interior vertex of the mesh in turn, perhaps for several passes over the mesh. Domain boundary vertices can be smoothed too, subject to constraints that keep them on the boundary. The simplest and most famous way to smooth an interior vertex is to move it to the centroid of the vertices that adjoin it. This method is called *Laplacian smoothing* because of its interpretation as a Laplacian finite difference operator. It dates back at least to a 1970 paper by Kamel and Eisenstein [88]. It usually works well for triangular meshes, but it is unreliable for tetrahedra, quadrilaterals, and hexahedra.

A better smoothing method begins by recalling that the Delaunay triangulation of a fixed vertex set is the triangulation that minimizes the volume bounded between a triangulation lifted to the parabolic lifting map and the paraboloid itself. What if the domain's interior vertices are not fixed? Chen and Xu [89] suggest globally smoothing the interior vertices by minimizing the same volume. Alliez, Cohen-Steiner, Yvinec, and Desbrun [90] implemented an iterative tetrahedral mesh improvement method that alternates between this global smoothing step and recomputing the Delaunay triangulation—in other words, it alternates between numerical and combinatorial optimization of the triangulation with respect to the same objective function.

A slower but even better smoother is the nonsmooth optimization algorithm of Freitag, Jones, and Plassmann [91], which can optimize the worst element in a group—for instance, maximizing the minimum dihedral angle among the tetrahedra that share a specified vertex. The figure at right shows smoothing to maximize the minimum plane angle.

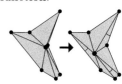

The simplest topological transformation is the *edge flip* in a triangular mesh, which replaces two triangles with two different triangles. Also illustrated at right are several analogous transformations for tetrahedra. These are examples of what mathematicians call *bistellar flips*. Similar flips exist for quadrilaterals and hexahedra; see Bern, Eppstein, and Erickson [92] for a list.

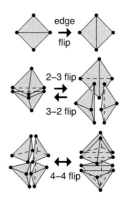

Some topological transformations are more complicated than the basic flips. *Edge contraction* removes an edge from the mesh by merging its two endpoints into a single vertex. The triangles or tetrahedra that share the contracted edge are deleted. It is wise to use smoothing to position the new vertex. Edge contraction is useful both to remove elements that have poor quality because an edge is too short, and to coarsen a mesh where its tetrahedra are unnecessarily small.

Edge removal, proposed by Brière de l'Isle and George [93], is a transformation that deletes all the tetrahedra that share a specified edge, then retriangulates the cavity with other tetrahedra chosen to maximize the quality of the worst new tetrahedron. This choice can be efficiently made by a dynamic programming algorithm of Klincsek [94]. The 3-2 and 4-4 flips are the simplest examples of edge removal, which replaces m tetrahedra with $2m - 4$.

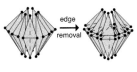

The job of a schedule is to search the mesh for tetrahedra that can be improved by local transformations, ideally as quickly as possible. Most schedules use simple hill-climbing optimization: they apply a transformation only if it improves the mesh. This usually means that the quality of the worst element created by the transformation is better than the quality of the worst element deleted; but in some implementations it means that the average element quality improves. There is a large literature on how to evaluate the quality of an element; see Field [95] for a survey.

Influential mesh improvement schedules include those by Joe [96] and Freitag and Ollivier-Gooch [97]. The latter authors present a schedule that uses flips, edge removal, and optimization-based smoothing to eliminate most poorly shaped tetrahedra, and they offer empirical recommendations about what makes some schedules better than others. This work was extended by Klingner and Shewchuk [98], whose most notable addition is a local transformation that inserts new vertices and thereby eliminates skinny tetrahedra that other transformations cannot touch. Their additions make mesh improvement more reliable: instead of removing most skinny tetrahedra, you can generally remove them all. They report that in their improved test meshes, no dihedral angle is smaller than 31° or larger than 149°.

Mesh improvement is a surprisingly powerful way to introduce anisotropy into a mesh. Löhner and Morgan [99] used flips and vertex insertion to make triangular meshes anisotropic as early as 1986, and the same has subsequently been done for tetrahedral meshes [100, 101]. An exciting new application of mesh improvement is *dynamic meshing*, the modeling of materials undergoing radical changes in shape, such as flow or fracture [102, 103, 101], that a fixed mesh cannot model.

Effective mesh improvement methods also exist for quadrilateral meshes [104]. There does not seem to have been much work on applying hexahedral flips.

Mesh improvement algorithms can be surprisingly slow—often slower than mesh generators—because they make multiple passes over the mesh, and computing element quality is expensive. When simulation accuracy is crucial or a single mesh will be used for many simulations, the cost is worth it.

10.4   Guaranteed-Quality Mesh Generation

The story of guaranteed-quality meshing is an interplay of ideas between Delaunay methods and methods based on grids, quadtrees, and octrees.

Provably good mesh generation began in 1988, when Baker, Grosse, and Rafferty [105] gave an algorithm to triangulate a polygon so that all the new angles in the mesh are between 14° and 90°. They overlay the polygon with a fine square grid, create new vertices at some grid points and at some intersections between grid lines and the polygon boundary, and triangulate them with a complicated case analysis.

The following year, Paul Chew [69] gave a more practical algorithm, based on Delaunay refinement, that guarantees angles between 30° and 120°. (The input polygon may not have an angle less than 30°.) He begins by subdividing the polygon's edges so that all the edge lengths are in a range $[h, \sqrt{3}h]$, where h is chosen small enough that such a subdivision exists with no two edge endpoints closer to each other than h. He computes the CDT of the subdivision, then repeatedly inserts a vertex at the circumcenter of a skinny triangle.

At the core of Chew's paper, we find a powerful idea that has appeared in some form in every guaranteed-quality Delaunay refinement algorithm. A skinny element can always be refined away, so there are only two possible outcomes: eventually the algorithm will delete the last skinny element and succeed, or it will run forever, creating skinny elements as fast as it deletes them. If we prove it does not run forever, we prove it succeeds. Can we do that?

We call an element "skinny" only if its circumradius—the radius of its circumcircle—is longer than its shortest edge. When we insert a new vertex at the center of that empty circle, all the newly created edges must be at least as long as the circumradius, and therefore at least as long as the shortest prior edge. It follows that *the Delaunay refinement algorithm never creates an edge shorter than the shortest edge already existing.* This fact places a perpetual lower bound on the distance between two vertices. The algorithm cannot run forever, because first it will run out of places to put new vertices! To complete the proof for Chew's algorithm, we just need to verify that no circumcenter ever falls outside the domain. (Hint: no edge is ever shorter than h; no circumcenter ever comes closer to the boundary than $h/2$.)

But are those really the skinny elements? In two dimensions, yes. A triangle's circumradius is longer than its shortest edge if and only if its smallest angle is less than 30°. What about tetrahedra?

Dey, Bajaj, and Sugihara [70] generalize Chew's algorithm to generate tetrahedral meshes of convex polyhedral domains. They refine every tetrahedron whose circumradius is more than twice as long as its shortest edge. (The factor of two is lost to ensure that no new vertex is placed too close to the boundary.) Their algorithm is guaranteed to terminate by the same reasoning as for Chew's. But a few bad tetrahedra slip through. Although most thin tetrahedra have large circumradii, there are exceptions: a type of tetrahedron called a *sliver* or *kite*.

The canonical sliver is formed by arranging four vertices around the equator of a sphere, equally spaced, then perturbing one of the vertices slightly off the equator. A sliver can have a circumradius shorter than its shortest edge, yet have dihedral angles arbitrarily close to 0° and 180°. Slivers have no two-dimensional analog, but they occur frequently in three-dimensional Delaunay triangulations. In practice, slivers are not difficult to remove [71]—they die when you insert vertices at their circumcenters, just like any other element. But you cannot *prove* that the algorithm will not run forever. Also, slivers removed by refinement sometimes leave undesirably small elements in their wake. Provably good sliver removal is one of the most difficult theoretical problems in mesh generation, although mesh improvement algorithms beat slivers consistently in practice.

None of these algorithms produce graded meshes. The quadtree algorithm of Bern, Eppstein, and Gilbert [15], discussed in Section 10.3.3.2, meshes a polygon so no new angle is less than 18.4°. It has been influential in part because the meshes it produces are not only graded, but *size-optimal*: the number of triangles in a mesh is at most a constant factor times the number in the smallest possible mesh (measured by triangle count) having no angle less than 18.4°. Ironically, the algorithm produces too many triangles to be practical—but only by a constant factor. Neugebauer and Diekmann [106] improve the algorithm by replacing square quadrants with rhomboids. They produce triangles with angles between 30° and 90°, many of them equilateral.

Mitchell and Vavasis [87] developed an octree algorithm that offers guarantees on dihedral angles (i.e., the very worst slivers are ruled out), grading, and size-optimality. The bounds are not strong enough to be satisfying in practice, and are not explicitly stated. Nevertheless, these papers broadened the ambitions of provably good meshing tremendously.

A seminal triangular meshing paper by Jim Ruppert [68] brought guaranteed good grading and size-optimality to Delaunay refinement algorithms. Ruppert's was the first guaranteed-quality mesh generation algorithm to be truly satisfying in practice [16]. It accepts nonconvex domains with internal boundaries and produces graded meshes that are not overrefined in practice. It began to grapple with the difficulty of domains with small angles, and it established the theoretical strength of the DELAUNAYREFINEMENT template.

I generalized Ruppert's algorithm to three dimensions [71]. This tetrahedral algorithm has much in common with Dey's, but like Ruppert's, it accepts nonconvex domains with internal boundaries and offers guaranteed good grading. However, it is not guaranteed to eliminate slivers, which implies (for technical reasons) that it cannot guarantee size-optimality.

Subsequently, provably good Delaunay refinement algorithms have been extended to curved surface meshing [77, 108, 109], anisotropy [110, 109], and domains with curved boundaries [72, 111, 107, 112]. Researchers have also studied how to guarantee a subquadratic running time even if the Delaunay triangulation of the input has a quadratic number of tetrahedra [113]. Most of these algorithms are practical as well as provably good.

Only a few results are known on guaranteed quality quadrilateral meshing, and none for hexahedra. Bern and Eppstein [114] use circle packings to quadrilateralize polygons, producing no angle greater than 120°. Atalay, Ramaswami, and Xu [115] use quadtrees to quadrilateralize polygons, producing no angle less than 18.4°.

Today, the two research issues presenting the most difficulty in tetrahedral meshing are sliver removal and domains with small angles.

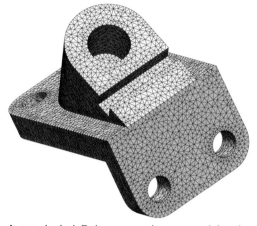

A tetrahedral Delaunay mesh generated by the software *DelPSC* [107]. Courtesy of Josh Levine.

A method of Chew [116] that randomly perturbs a circumcenter and the *sliver exudation* algorithm of Cheng, Dey, Edelsbrunner, Facello, and Teng [117, 118] both remove the very worst slivers from a Delaunay mesh, but the quality bounds are very weak and are not explicitly stated. Significant provable bounds on dihedral angles (1° or over) are virtually unheard of outside of space-filling or slab-filling tetrahedralizations. There are two exceptions. An algorithm by Labelle [119] is a true hybrid of the Delaunay and octree methods: it refines a Delaunay triangulation by inserting vertices at lattice points near the circumcenters of skinny tetrahedra. The lattice helps to guarantee that deep in the interior of the mesh, all the dihedral angles are between 30° and 135°. These bounds do not hold near the boundary—another argument for combining the advancing front and Delaunay methods.

The *isosurface stuffing* algorithm [86], discussed in Section 10.3.3.2, uses these insights about lattices to mesh volumes bounded by smooth surfaces. Smooth boundaries are just easy enough to permit strong quality guarantees: the dihedral angles of the tetrahedra are all between 10.7° and 164.8°. These bounds are derived by a computer-assisted proof that checks many cases. Unfortunately, it appears to be excruciatingly difficult to extend guarantees like this to domains with sharp corners or sharp ridges.

The problem of domains with small angles is solved for triangular meshes by a remarkably elegant tweak by Miller, Pav, and Walkington [120] that perfects Ruppert's algorithm. In three dimensions, Cheng, Dey, Ramos, and Ray [121] have a satisfying algorithm for meshing polyhedra, but the problem remains difficult for domains with internal boundaries. Existing algorithms

[76] sometimes refine much more than necessary because they must maintain the Delaunay property.

Another research direction deserving investigation is guaranteed-quality mesh improvement. Although sliver exudation can be seen as a provably good method that uses bistellar flips to locally improve a Delaunay mesh, there are no provably good local mesh improvement methods for meshes that are not Delaunay. There seem to be deep questions about how "tangled" a tetrahedralization can be, and whether it can always be fixed by local transformations. The same questions loom for quadrilateral and hexahedral meshes.

Acknowledgments

This chapter was supported in part by the National Science Foundation under Awards CCF-0635381 and IIS-0915462 and in part by the University of California Lab Fees Research Program under Grant 09-LR-01-118889-OBRJ.

I thank Mark Beall, Ted Blacker, Josh Levine, David Marcum, Heather Myers, and Mark Shephard for providing meshes. I thank Anand Kulkarni and Martin Wicke for suggesting improvements to drafts of this chapter.

Bibliography

[1] Thompson, J. F., "The National Grid Project," *Computer Systems in Engineering*, Vol. 3, No. 1–4, 1992, pp. 393–399.

[2] Thompson, J. F., Soni, B. K., and Weatherill, N. P., editors, *Handbook of Grid Generation*, CRC Press, Boca Raton, FL, 1999.

[3] Chazelle, B., "Convex Partitions of Polyhedra: A Lower Bound and Worst-Case Optimal Algorithm," *SIAM Journal on Computing*, Vol. 13, No. 3, Aug. 1984, pp. 488–507.

[4] Chazelle, B. and Palios, L., "Triangulating a Nonconvex Polytope," *Discrete & Computational Geometry*, Vol. 5, No. 1, Dec. 1990, pp. 505–526.

[5] Shephard, M. S. and Georges, M. K., "Automatic Three-Dimensional Mesh Generation by the Finite Octree Technique," *International Journal for Numerical Methods in Engineering*, Vol. 32, No. 4, Sept. 1991, pp. 709–749.

[6] Synge, J. L., *The Hypercircle in Mathematical Physics*, Cambridge University Press, New York, 1957.

[7] Babuška, I. and Aziz, A. K., "On the Angle Condition in the Finite Element Method," *SIAM Journal on Numerical Analysis*, Vol. 13, No. 2, April 1976, pp. 214–226.

[8] Jamet, P., "Estimations d'Erreur pour des Éléments Finis Droits Presque Dégénérés," *RAIRO Analyse Numérique*, Vol. 10, No. 1, 1976, pp. 43–60.

[9] Shewchuk, J. R., "What Is a Good Linear Element? Interpolation, Conditioning, and Quality Measures," *Eleventh International Meshing Roundtable*, Ithaca, New York, Sept. 2002, pp. 115–126.

[10] Křížek, M., "On the Maximum Angle Condition for Linear Tetrahedral Elements," *SIAM Journal on Numerical Analysis*, Vol. 29, No. 2, April 1992, pp. 513–520.

[11] Bank, R. E. and Scott, L. R., "On the Conditioning of Finite Element Equations with Highly Refined Meshes," *SIAM Journal on Numerical Analysis*, Vol. 26, No. 6, Dec. 1989, pp. 1383–1394.

[12] Courant, R., Friedrichs, K., and Lewy, H., "Über die Partiellen Differenzengleichungen der Mathematischen Physik," *Mathematische Annalen*, Vol. 100, Aug. 1928, pp. 32–74.

[13] Bern, M. and Eppstein, D., "Mesh Generation and Optimal Triangulation," *Computing in Euclidean Geometry*, edited by D.-Z. Du and F. Hwang, World Scientific, Singapore, 1992, pp. 23–90.

[14] Eppstein, D., Sullivan, J. M., and Üngör, A., "Tiling Space and Slabs with Acute Tetrahedra," *Computational Geometry: Theory and Applications*, Vol. 27, No. 3, March 2004, pp. 237–255.

[15] Bern, M., Eppstein, D., and Gilbert, J. R., "Provably Good Mesh Generation," *31st Annual Symposium on Foundations of Computer Science*, 1990, pp. 231–241.

[16] Shewchuk, J. R., "Triangle: Engineering a 2D Quality Mesh Generator and Delaunay Triangulator," *Applied Computational Geometry: Towards Geometric Engineering*, Vol. 1148 of *Lecture Notes in Computer Science*, Springer, Berlin, May 1996, pp. 203–222, From the First ACM Workshop on Applied Computational Geometry.

[17] Schneiders, R., "Quadrilateral and Hexahedral Element Meshes," *Handbook of Grid Generation*, chap. 21, CRC Press, Boca Raton, FL, 1999.

[18] Boissonnat, J.-D., Cohen-Steiner, D., Mourrain, B., Rote, G., and Vegter, G., "Meshing of Surfaces," *Effective Computational Geometry for Curves and Surfaces*, edited by J.-D. Boissonnat and M. Teillaud, Mathematics and Visualization, chap. 5, Springer, 2006, pp. 181–229.

[19] Frederick, C. O., Wong, Y. C., and Edge, F. W., "Two-Dimensional Automatic Mesh Generation for Structural Analysis," *International Journal for Numerical Methods in Engineering*, Vol. 2, 1970, pp. 133–144.

[20] Cavendish, J. C., "Automatic Triangulation of Arbitrary Planar Domains for the Finite Element Method," *International Journal for Numerical Methods in Engineering*, Vol. 8, No. 4, 1974, pp. 679–696.

[21] Lo, S. H., "A New Mesh Generation Scheme for Arbitrary Planar Domains," *International Journal for Numerical Methods in Engineering*, Vol. 21, No. 8, Aug. 1985, pp. 1403–1426.

[22] George, J. A., *Computer Implementation of the Finite Element Method*, Ph.D. thesis, Stanford University, Stanford, CA, March 1971, Technical report STAN-CS-71-208.

[23] Sadek, E. A., "A Scheme for the Automatic Generation of Triangular Finite Elements," *International Journal for Numerical Methods in Engineering*, Vol. 15, No. 12, Dec. 1980, pp. 1813–1822.

[24] Peraire, J., Vahdati, M., Morgan, K., and Zienkiewicz, O. C., "Adaptive Remeshing for Compressible Flow Computations," *Journal of Computational Physics*, Vol. 72, No. 2, Oct. 1987, pp. 449–466.

[25] Löhner, R. and Parikh, P., "Generation of Three-Dimensional Unstructured Grids by the Advancing-Front Method," *International Journal for Numerical Methods in Fluids*, Vol. 8, 1988, pp. 1135–1149.

[26] Peraire, J., Peiró, J., Formaggia, L., Morgan, K., and Zienkiewicz, O. C., "Finite Element Euler Computations in Three Dimensions," *International Journal for Numerical Methods in Engineering*, Vol. 26, No. 10, Oct. 1988, pp. 2135–2159.

[27] Marcum, D. L. and Weatherill, N. P., "Unstructured Grid Generation Using Iterative Point Insertion and Local Reconnection," *Twelfth AIAA Applied Aerodynamics Conference*, June 1994, AIAA paper 94-1926.

[28] Peraire, J., Peiró, J., and Morgan, K., "Advancing Front Grid Generation," *Handbook of Grid Generation*, edited by J. F. Thompson, B. K. Soni, and N. P. Weatherill, chap. 17, CRC Press, Boca Raton, FL, 1999.

[29] Marcum, D. L., "Unstructured Grid Generation Using Automatic Point Insertion and Local Reconnection," *Handbook of Grid Generation*, chap. 18, CRC Press, 1999.

[30] Schönhardt, E., "Über die Zerlegung von Dreieckspolyedern in Tetraeder," *Mathematische Annalen*, Vol. 98, 1928, pp. 309–312.

[31] Ruppert, J. and Seidel, R., "On the Difficulty of Triangulating Three-Dimensional Nonconvex Polyhedra," *Discrete & Computational Geometry*, Vol. 7, No. 3, 1992, pp. 227–253.

[32] Mavriplis, D. J., "An Advancing Front Delaunay Triangulation Algorithm Designed for Robustness," *Journal of Computational Physics*, Vol. 117, 1995, pp. 90–101.

[33] Blacker, T. D. and Stephenson, M. B., "Paving: A New Approach to Automated Quadrilateral Mesh Generation," *International Journal for Numerical Methods in Engineering*, Vol. 32, No. 4, Sept. 1991, pp. 811–847.

[34] Staten, M. L., Owen, S. J., and Blacker, T. D., "Paving & Plastering: Progress Update," *Proceedings of the 15th International Meshing Roundtable*, Birmingham, AL, Sept. 2006, pp. 469–486.

[35] Delaunay, B. N., "Sur la Sphère Vide," *Izvestia Akademia Nauk SSSR, VII Seria, Otdelenie Matematicheskii i Estestvennyka Nauk*, Vol. 7, 1934, pp. 793–800.

[36] Lawson, C. L., "Software for C^1 Surface Interpolation," *Mathematical Software III*, Academic Press, New York, 1977, pp. 161–194.

[37] Rajan, V. T., "Optimality of the Delaunay Triangulation in $\mathbb{R}^d$," *Proceedings of the Seventh Annual Symposium on Computational Geometry*, North Conway, NH, June 1991, pp. 357–363.

[38] Melissaratos, E. A., "L_p Optimal d Dimensional Triangulations for Piecewise Linear Interpolation: A New Result on Data Dependent Triangulations," Tech. Rep. RUU-CS-93-13, Department of Computer Science, Utrecht University, Utrecht, The Netherlands, April 1993.

[39] Waldron, S., "The Error in Linear Interpolation at the Vertices of a Simplex," *SIAM Journal on Numerical Analysis*, Vol. 35, No. 3, 1998, pp. 1191–1200.

[40] Brown, K. Q., "Voronoi Diagrams from Convex Hulls," *Information Processing Letters*, Vol. 9, No. 5, Dec. 1979, pp. 223–228.

[41] Seidel, R., "Voronoi Diagrams in Higher Dimensions," Diplomarbeit, Institut für Informationsverarbeitung, Technische Universität Graz.

[42] Lee, D.-T. and Lin, A. K., "Generalized Delaunay Triangulations for Planar Graphs," *Discrete & Computational Geometry*, Vol. 1, 1986, pp. 201–217.

[43] Nguyen, V. P., "Automatic Mesh Generation with Tetrahedral Elements," *International Journal for Numerical Methods in Engineering*, Vol. 18, 1982, pp. 273–289.

[44] Shewchuk, J. R., "General-Dimensional Constrained Delaunay Triangulations and Constrained Regular Triangulations I: Combinatorial Properties," *Discrete & Computational Geometry*, Vol. 39, No. 1–3, March 2008, pp. 580–637.

[45] Bernal, J. D. and Finney, J. L., "Random Close-Packed Hard-Sphere Model. II. Geometry of Random Packing of Hard Spheres," *Discussions of the Faraday Society*, Vol. 43, 1967, pp. 62–69.

[46] Chand, D. R. and Kapur, S. S., "An Algorithm for Convex Polytopes," *Journal of the ACM*, Vol. 17, No. 1, Jan. 1970, pp. 78–86.

[47] Guibas, L. J. and Stolfi, J., "Primitives for the Manipulation of General Subdivisions and the Computation of Voronoi Diagrams," *ACM Transactions on Graphics*, Vol. 4, No. 2, April 1985, pp. 74–123.

[48] Edelsbrunner, H. and Mücke, E. P., "Simulation of Simplicity: A Technique to Cope with Degenerate Cases in Geometric Algorithms," *ACM Transactions on Graphics*, Vol. 9, No. 1, 1990, pp. 66–104.

[49] Dwyer, R. A., "Higher-Dimensional Voronoi Diagrams in Linear Expected Time," *Discrete & Computational Geometry*, Vol. 6, No. 4, 1991, pp. 343–367.

[50] Shamos, M. I. and Hoey, D., "Closest-Point Problems," *16th Annual Symposium on Foundations of Computer Science*, Berkeley, CA, Oct. 1975, pp. 151–162.

[51] Lee, D.-T. and Schachter, B. J., "Two Algorithms for Constructing a Delaunay Triangulation," *International Journal of Computer and Information Sciences*, Vol. 9, No. 3, 1980, pp. 219–242.

[52] Dwyer, R. A., "A Faster Divide-and-Conquer Algorithm for Constructing Delaunay Triangulations," *Algorithmica*, Vol. 2, No. 2, 1987, pp. 137–151.

[53] Bowyer, A., "Computing Dirichlet Tessellations," *Computer Journal*, Vol. 24, No. 2, 1981, pp. 162–166.

[54] Hermeline, F., *Une Methode Automatique de Maillage en Dimension n*, Ph.D. thesis, Université Pierre et Marie Curie, 1980.

[55] Hermeline, F., "Triangulation Automatique d'un Polyèdre en Dimension N," *RAIRO Analyse Numérique*, Vol. 16, No. 3, 1982, pp. 211–242.

[56] Watson, D. F., "Computing the *n*-dimensional Delaunay Tessellation with Application to Voronoi Polytopes," *Computer Journal*, Vol. 24, No. 2, 1981, pp. 167–172.

[57] Clarkson, K. L. and Shor, P. W., "Applications of Random Sampling in Computational Geometry, II," *Discrete & Computational Geometry*, Vol. 4, No. 1, Dec. 1989, pp. 387–421.

[58] Amenta, N., Choi, S., and Rote, G., "Incremental Constructions con BRIO," *Proceedings of the Nineteenth Annual Symposium on Computational Geometry*, June 2003, pp. 211–219.

[59] Chew, L. P., "Constrained Delaunay Triangulations," *Algorithmica*, Vol. 4, No. 1, 1989, pp. 97–108.

[60] Anglada, M. V., "An Improved Incremental Algorithm for Constructing Restricted Delaunay Triangulations," *Computers and Graphics*, Vol. 21, No. 2, March 1997, pp. 215–223.

[61] Shewchuk, J. R., "Updating and Constructing Constrained Delaunay and Constrained Regular Triangulations by Flips," *Proceedings of the Nineteenth Annual Symposium on Computational Geometry*, San Diego, CA, June 2003, pp. 181–190.

[62] Cavendish, J. C., Field, D. A., and Frey, W. H., "An Approach to Automatic Three-Dimensional Finite Element Mesh Generation," *International Journal for Numerical Methods in Engineering*, Vol. 21, No. 2, Feb. 1985, pp. 329–347.

[63] Jameson, A., Baker, T. J., and Weatherill, N. P., "Calculation of Inviscid Transonic Flow over a Complete Aircraft," *Proceedings of the 24th AIAA Aerospace Sciences Meeting*, Reno, NV, Jan. 1986, AIAA paper 86-0103.

[64] Frey, W. H., "Selective Refinement: A New Strategy for Automatic Node Placement in Graded Triangular Meshes," *International Journal for Numerical Methods in Engineering*, Vol. 24, No. 11, Nov. 1987, pp. 2183–2200.

[65] Weatherill, N. P., "Delaunay Triangulation in Computational Fluid Dynamics," *Computers and Mathematics with Applications*, Vol. 24, No. 5/6, Sept. 1992, pp. 129–150.

[66] Üngör, A., "Off-Centers: A New Type of Steiner Points for Computing Size-Optimal Quality-Guaranteed Delaunay Triangulations," *Computational Geometry: Theory and Applications*, Vol. 42, No. 2, Feb. 2009, pp. 109–118.

[67] Erten, H. and Üngör, A., "Triangulations with Locally Optimal Steiner Points," *Symposium on Geometry Processing 2007*, July 2007, pp. 143–152.

[68] Ruppert, J., "A Delaunay Refinement Algorithm for Quality 2-Dimensional Mesh Generation," *Journal of Algorithms*, Vol. 18, No. 3, May 1995, pp. 548–585.

[69] Chew, L. P., "Guaranteed-Quality Triangular Meshes," Tech. Rep. TR-89-983, Department of Computer Science, Cornell University, 1989.

[70] Dey, T. K., Bajaj, C. L., and Sugihara, K., "On Good Triangulations in Three Dimensions," *International Journal of Computational Geometry and Applications*, Vol. 2, No. 1, 1992, pp. 75–95.

[71] Shewchuk, J. R., "Tetrahedral Mesh Generation by Delaunay Refinement," *Proceedings of the Fourteenth Annual Symposium on Computational Geometry*, Minneapolis, Minnesota, June 1998, pp. 86–95.

[72] Boivin, C. and Ollivier-Gooch, C., "Guaranteed-Quality Triangular Mesh Generation for Domains with Curved Boundaries," *International Journal for Numerical Methods in Engineering*, Vol. 55, No. 10, 20 Aug. 2002, pp. 1185–1213.

[73] Schroeder, W. J. and Shephard, M. S., "Geometry-Based Fully Automatical Mesh Generation and the Delaunay Triangulation," *International Journal for Numerical Methods in Engineering*, Vol. 26, No. 11, Nov. 1988, pp. 2503–2515.

[74] Bishop, C. J., "Nonobtuse Triangulation of PSLGs," Unpublished manuscript.

[75] Murphy, M., Mount, D. M., and Gable, C. W., "A Point-Placement Strategy for Conforming Delaunay Tetrahedralization," *Proceedings of the Eleventh Annual Symposium on Discrete Algorithms*, Jan. 2000, pp. 67–74.

[76] Rand, A. and Walkington, N., "Collars and Intestines: Practical Conformal Delaunay Refinement," *Proceedings of the 18th International Meshing Roundtable*, Salt Lake City, UT, Oct. 2009, pp. 481–497.

[77] Chew, L. P., "Guaranteed-Quality Mesh Generation for Curved Surfaces," *Proceedings of the Ninth Annual Symposium on Computational Geometry*, San Diego, California, May 1993, pp. 274–280.

[78] Si, H., "Constrained Delaunay Tetrahedral Mesh Generation and Refinement," *Finite Elements in Analysis and Design*, Vol. 46, 2010, pp. 33–46.

[79] George, P.-L., Hecht, F., and Saltel, E., "Automatic Mesh Generator with Specified Boundary," *Computer Methods in Applied Mechanics and Engineering*, Vol. 92, No. 3, Nov. 1991, pp. 269–288.

[80] Weatherill, N. P. and Hassan, O., "Efficient Three-Dimensional Grid Generation Using the Delaunay Triangulation," *Proceedings of the First European Computational Fluid Dynamics Conference*, Brussels, Belgium, Sept. 1992, pp. 961–968.

[81] Hazlewood, C., "Approximating Constrained Tetrahedrizations," *Computer Aided Geometric Design*, Vol. 10, 1993, pp. 67–87.

[82] George, P.-L. and Borouchaki, H., *Delaunay Triangulation and Meshing: Application to Finite Elements*, Hermès, Paris, 1998.

[83] Samet, H., "The Quadtree and Related Hierarchical Data Structures," *Computing Surveys*, Vol. 16, 1984, pp. 188–260.

[84] Yerry, M. A. and Shephard, M. S., "A Modified Quadtree Approach to Finite Element Mesh Generation," *IEEE Computer Graphics and Applications*, Vol. 3, January/February 1983, pp. 39–46.

[85] Yerry, M. A. and Shephard, M. S., "Automatic Three-Dimensional Mesh Generation by the Modified-Octree Technique," *International Journal for Numerical Methods in Engineering*, Vol. 20, No. 11, Nov. 1984, pp. 1965–1990.

[86] Labelle, F. and Shewchuk, J. R., "Isosurface Stuffing: Fast Tetrahedral Meshes with Good Dihedral Angles," *ACM Transactions on Graphics*, Vol. 26, No. 3, July 2007, pp. 57.1–57.10, Special issue on Proceedings of SIGGRAPH 2007.

[87] Mitchell, S. A. and Vavasis, S. A., "Quality Mesh Generation in Three Dimensions," *Proceedings of the Eighth Annual Symposium on Computational Geometry*, 1992, pp. 212–221.

[88] Kamel, H. A. and Eisenstein, K., "Automatic Mesh Generation in Two- and Three-Dimensional Inter-Connected Domains," *Symposium on High Speed Computing of Elastic Structures*, Les Congreès et Colloques de l'Université de Liège, Liège, Belgium, 1970.

[89] Chen, L. and Xu, J., "Optimal Delaunay Triangulations," *Journal of Computational Mathematics*, Vol. 22, No. 2, 2004, pp. 299–308.

[90] Alliez, P., Cohen-Steiner, D., Yvinec, M., and Desbrun, M., "Variational Tetrahedral Meshing," *ACM Transactions on Graphics*, Vol. 24, 2005, pp. 617–625, Special issue on Proceedings of SIGGRAPH 2005.

[91] Freitag, L. A., Jones, M., and Plassmann, P. E., "An Efficient Parallel Algorithm for Mesh Smoothing," *Fourth International Meshing Roundtable*, Albuquerque, NM, Oct. 1995, pp. 47–58.

[92] Bern, M., Eppstein, D., and Erickson, J., "Flipping Cubical Meshes," *Engineering with Computers*, Vol. 18, No. 3, Oct. 2002, pp. 173–187.

[93] Brière de l'Isle, E. and George, P.-L., "Optimization of Tetrahedral Meshes," *Modeling, Mesh Generation, and Adaptive Numerical Methods for Partial Differential Equations*, Vol. 75 of *IMA Volumes in Mathematics and its Applications*, Springer-Verlag, 1995, pp. 97–127.

[94] Klincsek, G. T., "Minimal Triangulations of Polygonal Domains," *Annals of Discrete Mathematics*, Vol. 9, 1980, pp. 121–123.

[95] Field, D. A., "Qualitative Measures for Initial Meshes," *International Journal for Numerical Methods in Engineering*, Vol. 47, 2000, pp. 887–906.

[96] Joe, B., "Construction of Three-Dimensional Improved-Quality Triangulations Using Local Transformations," *SIAM Journal on Scientific Computing*, Vol. 16, No. 6, Nov. 1995, pp. 1292–1307.

[97] Freitag, L. A. and Ollivier-Gooch, C., "Tetrahedral Mesh Improvement Using Swapping and Smoothing," *International Journal for Numerical Methods in Engineering*, Vol. 40, No. 21, Nov. 1997, pp. 3979–4002.

[98] Klingner, B. M. and Shewchuk, J. R., "Aggressive Tetrahedral Mesh Improvement," *Proceedings of the 16th International Meshing Roundtable*, Seattle, Washington, Oct. 2007, pp. 3–23.

[99] Löhner, R. and Morgan, K., "Improved Adaptive Refinement Strategies for Finite Element Aerodynamic Computations," *AIAA 24th Aerospace Sciences Meeting*, 1986, AIAA paper 86-0499.

[100] Pain, C. C., Umpleby, A. P., de Oliveira, C. R. E., and Goddard, A. J. H., "Tetrahedral Mesh Optimisation and Adaptivity for Steady-State and Transient Finite Element Calculations," *Computer Methods in Applied Mechanics and Engineering*, Vol. 190, 2001, pp. 3771–3796.

[101] Wicke, M., Ritchie, D., Klingner, B. M., Burke, S., Shewchuk, J. R., and O'Brien, J. F., "Dynamic Local Remeshing for Elastoplastic Simulation," *ACM Transactions on Graphics*, Vol. 29, No. 3, July 2010, Special issue on Proceedings of SIGGRAPH 2010.

[102] Cardoze, D., Cunha, A., Miller, G. L., Phillips, T., and Walkington, N. J., "A Bézier-Based Approach to Unstructured Moving Meshes," *Proceedings of the Twentieth Annual Symposium on Computational Geometry*, Brooklyn, New York, June 2004, pp. 310–319.

[103] Mauch, S., Noels, L., Zhao, Z., and Radovitzky, R. A., "Lagrangian Simulation of Penetration Environments via Mesh Healing and Adaptive Optimization," *Proceedings of the 25th Army Science Conference*, Nov. 2006.

[104] Kinney, P., "CleanUp: Improving Quadrilateral Finite Element Meshes," *Sixth International Meshing Roundtable*, Park City, UT, Oct. 1997, pp. 449–461.

[105] Baker, B. S., Grosse, E., and Rafferty, C. S., "Nonobtuse Triangulation of Polygons," *Discrete & Computational Geometry*, Vol. 3, No. 2, 1988, pp. 147–168.

[106] Neugebauer, F. and Diekmann, R., "Improved Mesh Generation: Not Simple but Good," *Fifth International Meshing Roundtable*, Pittsburgh, PA, Oct. 1996, pp. 257–270.

[107] Cheng, S.-W., Dey, T. K., and Levine, J. A., "A Practical Delaunay Meshing Algorithm for a Large Class of Domains," *Proceedings of the 16th International Meshing Roundtable*, Oct. 2007, pp. 477–494.

[108] Boissonnat, J.-D. and Oudot, S., "Provably Good Sampling and Meshing of Surfaces," *Graphical Models*, Vol. 67, No. 5, Sept. 2005, pp. 405–451.

[109] Cheng, S.-W., Dey, T. K., Ramos, E. A., and Wenger, R., "Anisotropic Surface Meshing," *Proceedings of the Seventeenth Annual Symposium on Discrete Algorithms*, Miami, FL, Jan. 2006, pp. 202–211.

[110] Labelle, F. and Shewchuk, J. R., "Anisotropic Voronoi Diagrams and Guaranteed-Quality Anisotropic Mesh Generation," *Proceedings of the Nineteenth Annual Symposium on Computational Geometry*, San Diego, CA, June 2003, pp. 191–200.

[111] Pav, S. E. and Walkington, N. J., "Delaunay Refinement by Corner Lopping," *Proceedings of the 14th International Meshing Roundtable*, San Diego, CA, Sept. 2005, pp. 165–181.

[112] Rineau, L. and Yvinec, M., "Meshing 3D Domains Bounded by Piecewise Smooth Surfaces," *Proceedings of the 16th International Meshing Roundtable*, Seattle, WA, Oct. 2007, pp. 443–460.

[113] Hudson, B., Miller, G. L., and Phillips, T., "Sparse Voronoi Refinement," *Proceedings of the 15th International Meshing Roundtable*, Springer, Birmingham, AL, Sept. 2006, pp. 339–358.

[114] Bern, M. and Eppstein, D., "Quadrilateral Meshing by Circle Packing," *Sixth International Meshing Roundtable*, Oct. 1997, pp. 7–19.

[115] Atalay, F. B., Ramaswami, S., and Xu, D., "Quadrilateral Meshes with Bounded Minimum Angle," *Proceedings of the 17th International Meshing Roundtable*, Oct. 2008, pp. 73–91.

[116] Chew, L. P., "Guaranteed-Quality Delaunay Meshing in 3D," *Proceedings of the Thirteenth Annual Symposium on Computational Geometry*, Nice, France, June 1997, pp. 391–393.

[117] Cheng, S.-W., Dey, T. K., Edelsbrunner, H., Facello, M. A., and Teng, S.-H., "Sliver Exudation," *Journal of the ACM*, Vol. 47, No. 5, Sept. 2000, pp. 883–904.

[118] Edelsbrunner, H. and Guoy, D., "An Experimental Study of Sliver Exudation," *Tenth International Meshing Roundtable*, Newport Beach, CA, Oct. 2001, pp. 307–316.

[119] Labelle, F., "Sliver Removal by Lattice Refinement," *International Journal of Computational Geometry and Applications*, Vol. 18, No. 6, Dec. 2008, pp. 509–532.

[120] Miller, G. L., Pav, S. E., and Walkington, N. J., "When and Why Ruppert's Algorithm Works," *International Journal of Computational Geometry and Applications*, Vol. 15, No. 1, Feb. 2005, pp. 25–54.

[121] Cheng, S.-W., Dey, T. K., Ramos, E. A., and Ray, T., "Quality Meshing for Polyhedra with Small Angles," *Proceedings of the Twentieth Annual Symposium on Computational Geometry*, Brooklyn, NY, June 2004, pp. 290–299.

Chapter 11

3D Delaunay Mesh Generation

Klaus Gärtner, Hang Si

WIAS Berlin

Alexander Rand

University of Texas-Austin

Noel Walkington

Carnegie Mellon University

11.1 Introduction

Given a geometric domain, the mesh generation problem is to partition it into a set of simple (non-overlapping) cells. Various properties on the partition, such as the size, shape, or the number of cells, may be required. Algorithms are needed which generate such meshes from arbitrary domains in optimal time. This chapter considers the development of meshing algorithms which partition the underlying domain into simplices.

A *Delaunay triangulation* of a domain in $\mathbb{R}^d$ is a simplicial partition with the property that the circumball of every simplex does not contain any vertex of the partition in its interior; see Figure 11.1. These partitions have many desirable structural and combinatorial properties, and have numerous applications [1, 2, 3]. For example, the edges of Delaunay meshes are perpendicular to, and bisected by, the faces of the dual Voronoi diagram. This property

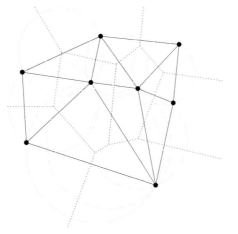

FIGURE 11.1: A 2D Delaunay triangulation and Voronoi diagram.

is exploited by "finite volume" algorithms for the numerical solution of partial differential equations. Especially in cases with nonlinearities and limited smoothness, requirements including positivity, dissipativity, and other conservation properties often make finite volume schemes a good choice (see [4] for a general description).

Additional properties of boundary-conforming Delaunay meshes [5] can be used prove existence of discrete solutions on finite size grids fulfilling qualitative properties of their analytic counterparts. Techniques similar to the analytic ones (test functions, discrete weak maximum principle, discrete versions of Sobolev embedding theorems, see [6, 7, 8]) can be used to prove these properties and to evaluate functionals with higher precision, for instance. These possibilities together with further challenges like the construction of the "best," in the sense of the evaluation of functionals defined by applications, boundary-conforming Delaunay meshes with a restricted number of vertices are one motivation to push the development and research of proven and fast algorithms for Delaunay mesh generation.

Delaunay refinement is one of the classical techniques to solve the meshing problem. Introduced by Chew [9] and Ruppert [10] for meshing 2D domains, this algorithm produces a *Delaunay mesh*, that is, a Delaunay triangulation which solves the meshing problem; see Figure 11.2 for an example. The basic idea is to incrementally update a Delaunay triangulation of the domain by judiciously adding vertices to (i) to produce a triangulation of the domain (boundary conformity), and (ii) eliminate badly shaped elements. The original algorithms in [9, 10] only required the addition of vertices at the circumcenters of triangles and edges. A triangle is determined to be "badly shaped" if it has any small angles, which is a sufficient (but not necessary [11]) criteria for many finite element applications. For the class of non-acute domains, Ruppert

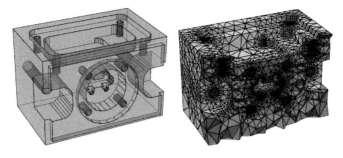

FIGURE 11.2: A 3D domain (left) and a Delaunay mesh (right).

proved that his algorithm will create a Delaunay mesh consisting of well-shaped elements and the number of elements is size optimal. The key step of the analysis was to introduce a geometric sizing function (Ruppert's local feature size) determined solely by the (input) domain. Given this function, the proof of termination reduces to a short, simple induction argument. With this conceptual breakthrough, meshing has developed into a rigorous science.

Ruppert initially dismissed the idea of using his algorithm to generate 3D meshes since it is not possible to eliminate certain poorly shaped tetrahedra (slivers) by circumcenter addition: only tetrahedra with large circumradius-to-shortest-edge, or radius-edge, ratio are eliminated [12, 13]. With this measure of tetrahedron quality, Ruppert's algorithm and proof of termination extend directly for non-acute domains. Experiments in 3D show that this algorithm achieves good mesh quality (modulo slivers) and size. To date, the only mesh generation algorithms guaranteed to terminate for arbitrary 3D domains have been extensions of Ruppert's algorithm [14, 15]. "Sliver removal" is typically also attempted either during refinement or as a post-process. If slivers can be removed, the result is a bounded aspect ratio mesh which is suitable for most finite element methods. This is a reasonable target for a general-purpose mesh generator: in 3D, weaker geometric conditions on the mesh are closely tied technical data-regularity requirements [16, 17] which may or may not be satisfied in specific applications.

Acute input angles pose a fundamental challenge to Delaunay refinement algorithms; without special consideration, termination cannot be guaranteed. Specifically, in order to attempt to guarantee that input segments appear in the mesh as a union of edges in the triangulation, the algorithm may insert vertices in acutely adjacent segments (i.e., adjacent segment that meet at an acute angle) indefinitely; see Figure 11.3. In two dimensions acute input angles which prevent termination are well understood [18] and by slightly relaxing the output criteria, failure of the algorithm can be prevented. Acutely adjacent input segments must be split at equal length and certain skinny triangles near these input angles are allowed to remain. This very simple modification to Ruppert's algorithm accommodates general input in a way that is both

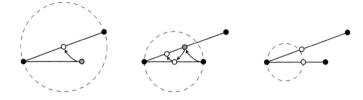

FIGURE 11.3: A segment encroached by a vertex on an adjacent input segment (left). Inserting segment midpoints leads to an infinite encroachment sequence (middle). No encroachment occurs with adjacent segments of equal length (right).

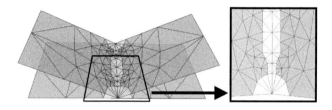

FIGURE 11.4: Acutely adjacent faces with consistent triangulation (the white triangles) along the shared segment (left) and a closer view (right).

theoretically sound and effective in practice [19]. In three dimensions there are many different types of acute input angles (between pairs of segments and/or faces) leading to much more complicated extensions of Ruppert's algorithm. Acutely adjacent faces each need a consistent triangulation along their common segment; see Figure 11.4 for an example.

Theoretical guarantees make Delaunay refinement a particularly attractive mesh generation technique, but in three dimensions this approach faces additional challenges not seen in two-dimensional variants. We describe a Delaunay refinement algorithm which admits a reasonably general class of inputs followed by a presentation of the current status, including limitations, of the analysis.

Outline. Section 11.2 contains a description of the 3D variant of Ruppert's algorithm. This is followed in Section 11.3 by an analysis of the termination requirements and output size guarantees which are consistent with the common 2D results. For inputs which do not meet the minimum input angle requirements, a procedure for protecting acute input angles before performing Delaunay refinement is described and analyzed in Section 11.4. Finally Section 11.5 discusses a few details important to the implementation of the algorithms given and contains several examples of meshes produced by these algorithms.

11.2   Delaunay Refinement

This section describes a basic Delaunay refinement algorithm to construct a 3D Delaunay mesh with well-shaped elements.

Definitions. A *polyhedron* is the union of convex polyhedra, $P = \bigcup_{i \in I} \bigcap H_i$, where I is a finite index set and each H_i is a finite set of closed half-spaces [3]. The *dimension* of P is the largest dimension of a convex polyhedron in P. Note that P is a topological subspace of $\mathbb{R}^d$. The *boundary* of P, bd(P), is the set of closure points of P. A *face* of P is any set of the form $F \subseteq \mathrm{bd}(P) \cap H_i$. F is again a polyhedron of lower dimension.

Input to the Delaunay refinement algorithm is a *piecewise linear complex* (PLC) $\mathcal{X}$, which is a collection of polyhedra such that: *(i)* $P \in \mathcal{X}$ implies that all faces of P are in $\mathcal{X}$, and *(ii)* the intersection of any two polyhedra of $\mathcal{X}$ is either empty or a face of both of them. The *dimension* of a PLC is the largest dimension of its polyhedra. For an example, a 3D polyhedron and all its faces form a PLC. The *underlying space* $|\mathcal{X}|$ of $\mathcal{X}$ is the union of all polyhedra of $\mathcal{X}$, i.e., $|\mathcal{X}| = \bigcup_{P \in \mathcal{X}} P$. $|\mathcal{X}|$ is a topological subspace of $\mathbb{R}^d$. The collection $\mathcal{X}$ induces a topology on its underlying space $|\mathcal{X}|$ [5].

The *radius-edge ratio*, ρ, of a tetrahedron is the ratio between the radius of its circumscribed ball and the length of its shortest edge. The regular tetrahedron has the minimum value $\sqrt{6}/4 \approx 0.612$. Most of the badly shaped tetrahedra will have a large radius-edge ratio except *slivers*, which are nearly degenerate tetrahedra whose radius-edge ratio may be as small as $\sqrt{2}/2 \approx 0.707$. Strictly speaking ρ is not a shape measure; nevertheless, it is useful in the analysis of this algorithm.

Algorithm. Let $\mathcal{X}$ be a 3D PLC. We call 1- and 2-faces of $\mathcal{X}$ *segments* and *facets*. Each segment and facet will be represented by a subcomplex of a tetrahedral mesh $\mathcal{T}$ of $\mathcal{X}$. We call 1- and 2-simplices of these subcomplexes *subsegments* and *subfaces* to distinguish them from other simplices of $\mathcal{T}$. A subsegment (or a subface) σ is said to be *encroached* if the interior of its smallest circumscribed ball contains vertices of $\mathcal{T}$.

The algorithm described in Figure 11.5 is from Shewchuk [14] which generalizes Ruppert's 2D algorithm [10]. After the initialization, the algorithm runs in a loop (lines $2-6$). During each iteration, a new point $\mathbf{v}$ is found by one of the three *point generating rules*; see Figure 11.6.

DELAUNAYREFINEMENT $(\mathcal{X}, \rho_0)$
$\mathcal{X}$ is a 3D PLC; ρ_0 is a radius-edge ratio bound.

---

1.   initialize a DT $\mathcal{D}$ of the vertex set of $\mathcal{X}$;
2.   **while** ($\exists$ encroached subsegment or subface)
3.   **or** ($\exists \, \tau \in \mathcal{D}$ such that $\rho(\tau) > \rho_0$), **do**
4.   create a new point $\mathbf{v}$ by rule i, $i \in \{1, 2, 3\}$;
5.   update $\mathcal{D}$ to be the DT of vert$(\mathcal{D}) \cup \{\mathbf{v}\}$;
6.   **endwhile**
7. $\mathcal{T} := \mathcal{D} \setminus \{\tau \mid \tau \in \mathcal{D} \text{ and } \tau \nsubseteq |\mathcal{X}|\}$
8.   **return** $\mathcal{T}$;

FIGURE 11.5: The Delaunay refinement algorithm [10, 14]. It generates a conforming Delaunay mesh $\mathcal{T}$ of a 3D PLC $\mathcal{X}$ such that no tetrahedra of $\mathcal{T}$ has radius-edge ratio larger than ρ_0.

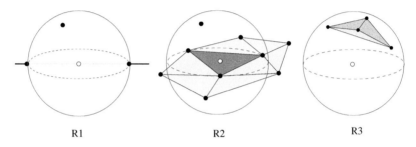

      R1   R2   R3

FIGURE 11.6: (See color insert.) Point generating rules. The new points (empty dots) are chosen at the circumcenters of an encroached subsegment (left), an encroached subface (middle), and a bad-quality tetrahedron (right).

$R1$: If a subsegment is encroached upon, $\mathbf{v}$ is its midpoint. $\mathbf{v}$ is an $R1$-*vertex*.
$R2$: If a subface is encroached upon, $\mathbf{v}$ is the circumcenter of its diametric ball. $\mathbf{v}$ is an $R2$-*vertex*. If $\mathbf{v}$ encroaches upon a subsegment use $R1$ to return a vertex $\mathbf{v}$; otherwise, return $\mathbf{v}$.
$R3$: If a tetrahedron τ has radius-edge ratio $\rho(\tau) > \rho_0$, let $\mathbf{v}$ be its circumcenter. $\mathbf{v}$ is an $R3$-*vertex*. If $\mathbf{v}$ encroaches upon a subsegment or subface, then use $R1$ or $R2$ to return a vertex $\mathbf{v}$; otherwise, return $\mathbf{v}$.

    Rule $R1$ has the highest priority, and $R3$ has the lowest. Priorities of the rules are important; for example, they ensure that the new point $\mathbf{v}$ lies either inside the mesh domain or on its boundary.

11.3   Termination and Output Size

Although Ruppert's analysis provides much insight into the behavior of this algorithm, many theoretical questions in the 3D algorithm remain unsolved. We address the termination and output size of this algorithm.

11.3.1   Proof of Termination

Termination of the algorithm DELAUNAYREFINEMENT (Figure 11.5) can be proved by showing new vertices are not inserted "close" to any existing vertex. For this purpose, it is crucial to find the relationship between a new edge and an existing edge.

Definitions. Among the points responsible for the addition of a vertex $\mathbf{v}$, we define a *parent* $p(\mathbf{v})$ as follows. If $\mathbf{v}$ is an $R1$- or $R2$-vertex, $p(\mathbf{v})$ is the point encroaching the subsegment or subface; $p(\mathbf{v})$ may be a Delaunay vertex or a rejected circumcenter. If there are several encroaching points then $p(\mathbf{v})$ is the one closest to $\mathbf{v}$. If two encroaching points are at the same distance to $\mathbf{v}$, choose an input vertex if possible; otherwise choose the one which was added earliest. If $\mathbf{v}$ is an $R3$-vertex, let τ be the tetrahedron $\mathbf{v}$ splits. Then $p(\mathbf{v})$ is the endpoint of the shortest edge of τ which is an input vertex if possible, or the one added earlier otherwise. If the input points are ordered, the parent function is uniquely defined by selecting the first input point whenever there is a choice. By convention an input point is not assigned a parent.

For each new point $\mathbf{v}$ its *insertion radius* is $r(\mathbf{v}) = \|\mathbf{v} - p(\mathbf{v})\|$, where $\|\cdot\|$ denotes the Euclidean distance function. For an input point $\mathbf{u} \in \mathcal{X}$ we define $r(\mathbf{u}) = \|\mathbf{u} - \mathbf{w}\|$, where $\mathbf{w} \in \mathcal{X}$ is an input vertex nearest to $\mathbf{u}$.

There are three types of *input angles* in $\mathcal{X}$. If two segments intersect, the angle between them is a *segment-segment angle*; if a facet intersects a non-coplanar segment, any line inside the facet and the segment form an angle and the smallest such angle is a *segment-facet angle*; if two facets intersect, their dihedral angle (i.e., the angle between their normals) is a *facet-facet angle*.

Let $\mathbf{v}$ be a new vertex, and $\mathbf{p} = p(\mathbf{v})$ be its parent. If the input PLC $\mathcal{X}$ contains no acute input angle, the relations between $r(\mathbf{v})$ and $r(\mathbf{p})$ established by Ruppert in [10] is

$$r(\mathbf{v}) \geq \frac{\rho_0}{C} r(\mathbf{p}), \qquad (11.1)$$

where $C = \sqrt{2}$ if $\mathbf{v}$ is an $R1$- or $R2$-vertex, and $C = 1$ if $\mathbf{v}$ is an $R3$-vertex. Note that if $\mathbf{p}$ is a vertex in the mesh then the Delaunay triangulation contained an edge of length $r(\mathbf{p})$ prior to the insertion of $r(\mathbf{v})$. If $\mathbf{p}$ is a rejected $R3$-vertex, we can re-apply (11.1) to an existing vertex $\mathbf{q} = p(\mathbf{p})$ to obtain $r(\mathbf{v}) \geq$

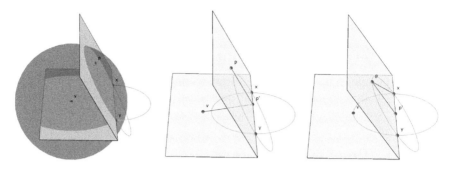

FIGURE 11.7: Proof of termination when two incident facets form an acute facet-facet angle.

$\frac{\rho_0}{C} r(\mathbf{p}) \geq \frac{\rho_0}{C^2} r(\mathbf{q})$. It follows that there is always an existing edge shorter than $r(\mathbf{v})$ provided $\rho_0 \geq C^2$. This establishes termination of the algorithm when no acute angles are present in the input and $\rho_0 \geq 2$.

It is possible to relax the acute input angle assumptions slightly with a more detailed analysis. Consider the situation where the insertion of $\mathbf{v}$ arises due to the existence of an acute angle θ in $\mathcal{X}$. There are three cases:

(1) θ is a segment-segment angle.
(2) θ is a segment-facet angle.
(3) θ is a facet-facet angle.

The shortest edge lengths at $\mathbf{v}$ and $\mathbf{p}$ for cases *(1)* and *(2)* have been established by Shewchuk [14], i.e., if $\theta \geq 45°$, then:

$$r(\mathbf{v}) \geq \frac{1}{2\cos\theta}\, r(\mathbf{p}). \tag{11.2}$$

It immediately follows that $r(\mathbf{v}) \geq r(\mathbf{p})$ when $\theta \geq \arccos\frac{1}{2} = 60°$. Hence, in cases *(1)* and *(2)*, an acute input angle no smaller than $60°$ does not affect the termination. However, the inequality (11.2) does not hold in case *(3)*. This case is excluded by the requirement: $\theta \geq 90°$ in the original analysis [14].

Let two facets F_1 and F_2 in $\mathcal{X}$ form an acute facet-facet angle θ, $\mathbf{v} \in F_1$, and $\mathbf{p} \in F_2$; see Figure 11.7 left. Note that $\mathbf{v}$ must not encroach the segment $\mathbf{xy}$ on $F_1 \cap F_2$, and $\mathbf{p}$ must encroach a subface whose circumcenter is $\mathbf{v}$.

In order to find the relation between $r(\mathbf{v})$ and $r(\mathbf{p})$, we introduce an auxiliary point $\mathbf{p}'$ which is the projection of $\mathbf{p}$ onto $\mathbf{xy}$; see Figure 11.7 middle. Note that $\angle\mathbf{vp'p} \geq \theta$. If $\theta \geq \arctan\sqrt{2} \approx 54.74°$ within the triangle $\mathbf{vpp'}$ [20] proved that the following relation holds

$$\|\mathbf{v} - \mathbf{p}\| \geq \frac{2}{3\sqrt{2}\cos\theta}\|\mathbf{p} - \mathbf{p}'\|. \tag{11.3}$$

Although $\|\mathbf{p} - \mathbf{p}'\|$ is not the length of an existing edge, it is easy to see that

$$\|\mathbf{p} - \mathbf{p}'\| \geq \frac{1}{\sqrt{2}} \|\mathbf{p} - \mathbf{y}\|. \qquad (11.4)$$

The case when the equality holds in (11.4) is shown in Figure 11.7 right.

Since $r(\mathbf{v}) = \|\mathbf{v} - \mathbf{p}\|$ and $\|\mathbf{p} - \mathbf{y}\| \geq r(\mathbf{p})$, combining (11.3) and (11.4) we get the relation: if $\theta \geq \arctan \sqrt{2} \approx 54.74°$, then

$$r(\mathbf{v}) \geq \frac{1}{3 \cos \theta} \, r(\mathbf{p}). \qquad (11.5)$$

Equation (11.5) guarantees the fact that an acute input facet-facet angle no smaller than $\arccos \frac{1}{3}$ will not cause a newly created edge shorter than an existing edge. Hence the algorithm must terminate.

Theorem 11.3.1 ([20]) *Let θ_m be the smallest input angle in $\mathcal{X}$ and $\rho_0 \geq 2$. The algorithm* DELAUNAYREFINEMENT *terminates if $\theta_m \geq \arccos \frac{1}{3} \approx 70.53°$.*

11.3.2 Output Size

It is observed that the algorithm DELAUNAYREFINEMENT (Figure 11.5) in general runs linearly with respect to the output size. Thus bounding the output size is important for bounding the runtime. Define lfs($\mathbf{x}$), the *local feature size* at a point $\mathbf{x}$ in $|\mathcal{X}|$ (see Section 11.4). The following estimation on m is the standard generalization of Ruppert's results to 3D [10].

$$m = \Theta \left(\int_{|\mathcal{X}|} \frac{1}{\text{lfs}(\mathbf{x})^3} d\mathbf{x} \right). \qquad (11.6)$$

In practice, one would prefer an estimation of m which is bounded by the input size n. It is not explicitly given in (11.6). Note that in general the output size of this algorithm may not be solely determined by input size. The topology and geometry of the input PLC $\mathcal{X}$ must also be taken into account.

This problem has recently been studied by Miller et al. [21] and Hudson et al. [22]. By simplifying the input to a point set, they show that a worst case upper bound on the integral in (11.6) is $O(n \log \Delta)$, where Δ is the spread of the point set (explained below). However, this result is based on a distribution assumption on the input point set. They commented that this assumption is not necessary [22].

In the following, we use a combinatorial approach to get an upper bound of the total number of output points without any assumption on the input.

Let S be a *periodic point set* [23], i.e., $S \subseteq [0, 1)^3$ is a finite set of points, and duplicated within each integer unit cube: $S' = S + \mathbb{Z}^3$, where $\mathbb{Z}^3$ is the three-dimensional integer grid. Let L denote the diameter of S, while s denotes the smallest pairwise distance among input features. Then the ratio $\Delta = L/s$ is referred as the *spread* of S [24]. The algorithm DELAUNAYREFINEMENT

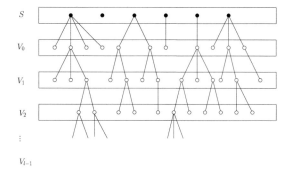

FIGURE 11.8: Sort vertices into trees by the parent relation.

(Figure 11.5) uses only $R3$ for refining a periodic point set. We would like to estimate its output size.

The parent relation associates each new vertex $\mathbf{v}$ uniquely to another vertex $\mathbf{p} = p(\mathbf{v})$ in the tetrahedralization. Moreover, we can form a *parent sequence* of vertices, $\{\mathbf{v}_i\}_{i=0}^{j+1}$, such that $\mathbf{v}_0$ (the first vertex) is a new vertex, for $i = 1, 2, ..., j$, $\mathbf{v}_{i+1} = p(\mathbf{v}_i)$, and $\mathbf{v}_{j+1}$ (the last vertex) is a vertex in S.

We first sort all new vertices produced by the algorithm into a collection of subsets of V by the following approach: Let $\mathbf{v}$ be a new vertex and $\{\mathbf{v}_i\}_{i=0}^{j+1}$ be its parent sequence. Then it is called a *rank-j* vertex, i.e., there are j $R3$-vertices between $\mathbf{v}$ and $\mathbf{v}_{j+1}$. Denote the set of all rank-j vertices by V_j. Hence $V = \{V_0, V_1, ..., V_{l-1}\}$; see Figure 11.8. Obviously, $V_i \cap V_j = \emptyset$ for any $V_i, V_j \in V$. Hence we have

$$|V| = |S| + |V_0| + |V_1| + \cdots + |V_{l-1}|. \tag{11.7}$$

Next we show that the collection V has a finite size. Note that an edge incident to rank-m vertex $\mathbf{v}$ has a length at least $r(\mathbf{v})$. Based upon (11.1) for $R3$-vertices, $r(\mathbf{v}) \geq \rho_0^{m+1} r(\mathbf{p})$, where $\mathbf{p} = g(\mathbf{v})$. Since here $\mathbf{p}$ is an input vertex, $r(\mathbf{p}) \geq \text{lfs}(\mathbf{p}) \geq s$. Also note that $r(\mathbf{v}) \leq L$. Hence $L \geq r(\mathbf{v}) \geq \rho_0^{m+1} \geq s \implies \rho_0^{m+1} \leq L/s \implies m + 1 \leq \log_{\rho_0}(L/s)$. The largest number of subsets in V is:

$$l = \lceil \log_{\rho_0} \Delta \rceil - 1. \tag{11.8}$$

Next we show a key fact: the set V_0 has the "capability" to have the largest cardinality among all sets in V. Note it does not mean that the cardinality of V_0 must be the largest in V. It means that no set in V will have a cardinality larger than the maximal possible cardinality of V_0.

Let $\mathcal{D}(S)$ be the Delaunay tetrahedralization for S, and let $\mathcal{V}(S)$ be the dual Voronoi diagram of $\mathcal{D}(S)$. $\mathcal{V}(S)$ forms a space partition of the convex hull, $\text{conv}(S)$ of S. All vertices in $\mathcal{V}(S)$ are rank-0 candidates to be added by this algorithm. Let S_0 be the set formed by all vertices of S and only rank-0 vertices (no rank-1 vertices are added yet). $\mathcal{V}(S_0)$ is a new partition of $\text{conv}(S)$. A rank-1 vertex candidate is a Voronoi vertex in $\mathcal{V}(S_0)$ which is the circumcenter of a bad quality tetrahedron which must have a short edge formed by two rank-0 vertices. Obviously, the candidates for rank-1 vertices are locally restricted in $\mathcal{V}(S_0)$ since they depend on the locations of rank-0 vertices. This shows that the space containing rank-0 vertices that can potentially be added is much larger than the corresponding space for rank-1 vertices. The same holds for vertices having higher ranks, since a new vertex is always added at the "locally sparsest" location. The updated Voronoi diagram becomes sparser after new vertices are added. Hence V_0 has the capability to have the largest cardinality among other sets in $\mathcal{V}$.

It remains to determine the largest possible number of rank-0 vertices. Let $\mathbf{v}$ be an input vertex. Let e, an edge incident to $\mathbf{v}$, be the shortest edge of a bad quality tetrahedron t_0 in a Delaunay mesh $\mathcal{T}$. The worst radius-edge ratio of t_0, $\rho(t_0) \leq L/s = \Delta$. The Delaunay refinement algorithm will remove t_0 by adding its circumcenter $\mathbf{c}_0$. Let t_1 be the tetrahedron $t_1 \ni e$, created by the insertion of $\mathbf{c}_0$. It must have $\rho(t_1) \leq \frac{1}{2}\Delta$. If $\rho(t_1) > \rho_0$, then t_1 is again removed by adding its circumcenter $\mathbf{c}_1$. This will create a new tetrahedron $t_2 \ni e$, with $\rho(t_2) \leq \frac{1}{4}\Delta$. This process will repeat until a tetrahedron $t_k \ni e$ with $\rho(t_k) \leq (\frac{1}{2})^k \Delta \leq \rho_0$ is created. Hence a short edge e could produce at most $k \leq \lceil \log_2 \Delta \rceil$ new vertices.

After adding at most $\lceil \log_2 \Delta \rceil$ rank-0 vertices, $\mathbf{v}$ gets an edge which does not belong to any bad quality tetrahedron. The degree of a vertex in a quality radius-edge triangulation must be bounded [25, 20]; let δ_0 denotes the degree bound associated with a radius-edge bound of ρ_0. There are at most δ_0 edges connected at $\mathbf{v}$, thus, after inserting at most $\delta_0 \lceil \log_2 \Delta \rceil$ vertices, all edges incident to $\mathbf{v}$ do not belong to any bad quality tetrahedron. Since there are n input vertices, thus, the total number of rank-0 vertices is

$$|V_0| \leq (\delta_0 \lceil \log_2 \Delta \rceil)n. \tag{11.9}$$

Note that $\delta_0 \lceil \log_2 \Delta \rceil$ is a constant which does not depend on n. The total size of V can be estimated by substituting (11.8) and (11.9) into (11.7),

$$|V| \leq (l+1)(\delta_0 \lceil \log_2 \Delta \rceil)n = O(n\lceil \log_{\rho_0} \Delta \rceil). \tag{11.10}$$

11.4   Handling Small Input Angles

To build a practical general-use mesh generator using Delaunay refinement, an essential task is determining a method for extending Ruppert's algorithm to domains with small input angles. Small input angles cause an unavoidable conflict between the two objectives of Ruppert's algorithm: (1) creating a mesh conforming to the input and (2) eliminating tetrahedra with poor radius-edge ratio. The necessary compromise between these two goals is that poorly-shaped tetrahedra are allowed to remain if they are near small input angles. This section will describe the algorithm for producing a conforming tetrahedralization which terminates despite small input angles and precisely define which poor quality tetrahedra near small angles may remain in the final mesh.

Algorithms for generating a conforming Delaunay tetrahedralization of a general PLC have relied on creating a "collar" region in acutely adjacent input faces to produce a consistent triangulation strategy in each face near this acute angle [26, 27]. These ideas were extended into theoretical quality mesh generation algorithms by Cheng and Poon [28] and Pav and Walkington [15]. The collar construction which is now described simplifies these initial approaches into a manageable theory which has been implemented.

Definitions. The *local feature size* [10] is the function lfs : $\mathbb{R}^d \to \mathbb{R}^+$ which at each $\mathbf{x} \in |\mathcal{X}|$ is equal to the radius of the smallest ball centered at $\mathbf{x}$ that intersects at least two disjoint polytopes of $\mathcal{X}$. When the PLC under consideration (which is usually the input $\mathcal{X}$) is unclear, the PLC is given a second argument to remove ambiguity, lfs$(\mathbf{x}; \mathcal{X})$. Local feature size is 1-Lipschitz, i.e., for any $\mathbf{x}, \mathbf{y} \in |\mathcal{X}|$, lfs$(\mathbf{x}) \leq$ lfs$(\mathbf{y}) + \|\mathbf{x} - \mathbf{y}\|$.

The 0-*local feature size* at $\mathbf{x}$, lfs$_0(\mathbf{x})$ is the distance from $\mathbf{x}$ to the second nearest vertex in $\mathcal{X}$. This is the local feature size with respect to the 0-dimensional features (vertices) of $\mathcal{X}$. We consider an additional function for measuring the size of the mesh over a subsegment of an input segment. The 1-*feature size* of segment s, fs$_1(s)$, is the radius of the smallest closed ball centered at a point $\mathbf{x} \in s$ intersecting a segment or input vertex of $\mathcal{X}$ which does not intersect s. A *piecewise smooth complex* (abbreviated as PSC) generalizes the PLC definition to allow non-polyhedral manifolds. Namely, a PSC $\mathcal{Y}$ is a collection of manifolds such that: *(i)* $P \in \mathcal{Y}$ implies that the boundary of P is a union of objects in $\mathcal{Y}$, and *(ii)* the intersection of any two manifolds in $\mathcal{Y}$ is a patch of the boundary in both of them. Carefully defining and refining general PSCs involves a number of issues beyond the analysis of PLCs. However, we will only consider a very special class of PSCs: complexes containing straight-line segments and circular arcs. By using such a restricted class of PSCs we avoid the need to delve further into the subtleties of meshing smooth complexes in full generality. The definition of local feature size with respect to a PSC is identical to that of a PLC.

Hypothesis. Before protecting acute input angles by forming an auxiliary PSC for refinement, input segments should be subdivided to appropriately estimate the 1-feature size of each subsegment. Specifically, we must assume that all input segments $\mathcal{S}$ can be subdivided into a new set of segments $\mathcal{S}_1$ such that for some constants $b > 0$, $c_0 \in (0, .5)$, and $c_1 \in (0, 1)$, the following hold.

$H1$: For any input vertex $\mathbf{v}_0$, all $s_1 \in \mathcal{S}_1$ containing $\mathbf{v}_0$ have equal length satisfying $|s_1| \leq c_0 \cdot$ lfs$_0(\mathbf{v}_0; \mathcal{X})$.

$H2$: For all $s_1 \in \mathcal{S}_1$,
$$b \cdot \min(\text{fs}_1(s_1; \mathcal{X}), \text{lfs}(s_1; \mathcal{X})) < |s_1| < c_1 \cdot \min(\text{fs}_1(s_1; \mathcal{X}), \text{lfs}(s_1; \mathcal{X})).$$

Several approaches have been developed to compute a refinement satisfying $H1$ and $H2$ [28, 29, 30]. Determining the best approach to computing this subdivision is a topic of ongoing research.

11.4.1 Collar Construction

With an estimate of the 1-feature size on each segment, we are able to isolate acute input angles from the remainder of the input complex. Vertices are inserted according to the following three rules to separate each face into two regions: a protected collar region near the boundaries and a non-protected interior region suitable for Ruppert's algorithm.

$P1$: If s and s' are adjacent non-end segments which meet at vertex $\mathbf{v}$, then a vertex $\mathbf{u}$ is inserted at distance $\frac{\max(|s|,|s'|)}{2}$ from $\mathbf{v}$, in any direction in the facet which is perpendicular to s.

$P2$: If s is an end segment and s' is an adjacent non-end segment, both containing vertex $\mathbf{v}$, then insert vertex $\mathbf{u}$ at the intersection of any line parallel to s in the face at distance $\frac{|s'|}{2}$ away from s and the circle of radius $|s|$ around the input vertex on $\mathbf{v}_0$.

$P3$: For any input vertex $\mathbf{v}_0$ on a segment s, insert collar vertices such that the sphere of radius $|s|$ around $\mathbf{v}_0$ restricted to the face has no arcs of angle larger than $90°$.

With these vertices, the collar region and new features describing its boundary can be defined. Any vertex inserted during collar construction is called a *collar vertex*. A *collar segment* is a segment between collar vertices corresponding to adjacent vertices on an input segment, and the circular arc between adjacent collar vertices corresponding to the same input vertex is called a *collar arc*. The *collar region* of a face is the subset of that face between input segments and collar segments and arcs. In the Delaunay triangulation of a face, any simplex which lies inside the collar region is called a *collar simplex*. See Figure 11.9 for a depiction of these objects.

Let $\mathcal{Y}$ be the PSC containing all input and collar features. This includes all the input vertices of $\mathcal{X}$, all segments and endpoints of $\mathcal{S}_1$, all collar vertices, collar segments, collar arcs, the collar region of each face, and the non-collar region of each face. The following lemma demonstrates that $H1$ and $H2$ are sufficient conditions to ensure that the local feature size with respect to $\mathcal{Y}$ is comparable to the local feature size with respect to the original complex $\mathcal{X}$. As in Theorem 11.3.1, let θ_m denote the smallest input angle in $\mathcal{X}$.

Lemma 11.4.1 *There exists a constant $K > 0$ depending only upon b, c_0, and c_1 such that*

$$\text{lfs}(\mathbf{x}; \mathcal{Y}) \leq \text{lfs}(\mathbf{x}; \mathcal{X}) \leq \frac{K}{\sin^2 \theta_m} \text{lfs}(\mathbf{x}; \mathcal{Y}).$$

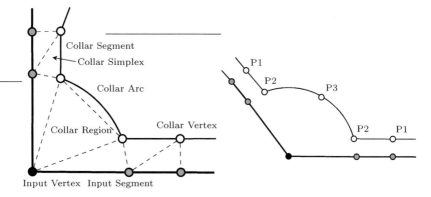

FIGURE 11.9: Left: examples of vertices, simplices, segments, and arcs associated with the collar region. Right: Collar vertices labeled based on the insertion rule applied to add them.

Lemma 11.4.1 allows the upcoming algorithm to be analyzed using $\mathrm{lfs}(\cdot\,;\mathcal{Y})$, i.e., to treat the complex containing the collar as the input.

11.4.2   Protected Delaunay Refinement

The algorithm DELAUNAYREFINEMENT given in Figure 11.5 can be modified to ensure termination given a collared PLC. The resulting algorithm PROTECTEDDELAUNAYREFINEMENT is given in Figure 11.10.

PROTECTEDDELAUNAYREFINEMENT $(\mathcal{X}, \rho_0)$
// $\mathcal{X}$ is a 3D PLC; ρ_0 is a radius-edge ratio bound.

0.   insert collar features to PLC $\mathcal{X}$ to produce PSC $\mathcal{Y}$;
1.   initialize a DT $\mathcal{D}$ of the vertex set of $\mathcal{Y}$;
2.   **while** ($\exists$ encroached subsegment, subarc, or subface)
3.   **or** ($\exists\ \tau \in \mathcal{D} \setminus \mathcal{D}_p$ such that $\rho(\tau) > \rho_0$), **do**
4.   create a new point $\mathbf{v}$ by rule i, $i \in \{1, 1a, 2, 3\}$;
5.   update $\mathcal{D}$ to be the DT of $\mathrm{vert}(\mathcal{D}) \cup \{\mathbf{v}\}$;
6.   **endwhile**
7. $\mathcal{T} := \mathcal{D} \setminus \{\tau \,|\, \tau \in \mathcal{D} \text{ and } \tau \not\subseteq |\mathcal{X}|\}$
8.   **return** $\mathcal{T}$;

FIGURE 11.10: The modified Delaunay refinement algorithm which terminates in the presence of small input angles. It generates a conforming Delaunay mesh $\mathcal{T}$ of a 3D PLC $\mathcal{X}$ such that all tetrahedra of $\mathcal{T}$ with radius-edge ratio larger than ρ_0 occur near acute input angles.

Performing Delaunay refinement on a complex containing circular arcs requires a careful definition of encroachment for these arcs as well as a decision on where a vertex is to be inserted when an encroached subarc is processed. An arc is encroached if the diametral ball defined by its endpoints contains no vertices of the triangulation. (Note: others have used alternative encroachment rules for curved objects [28, 31]. Using the diametral ball of the endpoints is often the simplest solution in practice since this yields essentially the same encroachment check as for segments. Moreover, this rule does not lead to any additional restrictions on the input or algorithm parameters.) A new point generating rule is given for subarcs:

$R1a$: If a subarc is encroached, $\mathbf{v}$ is its midpoint. For the purpose of the analysis $\mathbf{v}$ is treated as an $R1$-vertex.

Finally, let $\mathcal{D}_p$ be the set of tetrahedra with circumcenters that encroach upon a collar simplex. By prohibiting the insertion of circumcenters of tetrahedra in $\mathcal{D}_p$, the infinite encroachment sequences caused by acute angles are avoided; recall Figure 11.3. Then the key estimate, relating local feature size to the insertion radius, can be shown as stated in the lemma below. Recall that $r(\mathbf{v})$ denotes the insertion radius of vertex $\mathbf{v}$.

Lemma 11.4.2 ([32]) *For any $\rho_0 > 2$, there exists $K > 0$ depending only upon ρ, b, c_0, and c_1 such that for each vertex $\mathbf{v}$ inserted by* PROTECTEDDELAUNAYREFINEMENT$(\mathcal{X}, \rho_0)$

$$\mathrm{lfs}(\mathbf{v}; \mathcal{X}) \leq \frac{K}{(\sin \theta_m)^2} r(\mathbf{v}).$$

The proof of the lemma follows Ruppert's original arguments [10] considering local feature size with respect to the augmented complex $\mathcal{Y}$. Then Lemma 11.4.1 completes the proof. The fact that $\mathcal{Y}$ is a smooth complex creates only minor technical difficulties. With this key lemma in place, the algorithm is guaranteed to terminate (since local feature size is bounded away from zero for a given PLC) and produce a conforming quality tetrahedralization as summarized in the following theorem.

Theorem 11.4.3 ([32]) PROTECTEDDELAUNAYREFINEMENT *terminates producing a conforming Delaunay tetrahedralization of $\mathcal{X}$. The circumcenter of any remaining tetrahedra with radius-edge ratio larger than ρ_0 lies in the circumball of a collar simplex.*

11.5 Implementation and Examples

Two public codes, `TetGen` [33] and `DIR3` [34], have respectively implemented the algorithms discussed in this chapter.

TetGen is a Delaunay tetrahedral mesh generator. It can handle arbitrary 3D polyhedral domains and can generate good quality adaptive tetrahedral meshes for numerical methods.

Given a 3D PLC $\mathcal{X}$ described by its boundary facets, TetGen first generates an initial tetrahedral mesh $\mathcal{T}$ of $\mathcal{X}$ by inserting few additional points in the boundary of $\mathcal{X}$. $\mathcal{T}$ contains no interior points of $|\mathcal{X}|$. It usually consists of badly-shaped tetrahedra. The algorithm DELAUNAYREFINEMENT (Figure 11.5) is used to refine $\mathcal{T}$ into a good quality mesh. The main advantage of this approach is that it can detect acute input angles during the mesh refinement. TetGen avoids inserting vertices in places near small input angles. Hence it terminates on arbitrary input geometry. However, the resulting mesh may not be Delaunay if there are small acute input angles.

The DELAUNAYREFINEMENT algorithm does not remove sliver tetrahedra, but it is natural to include sliver tetrahedra in the queue of tetrahedra to be removed via circumcenter insertion. While lacking a theoretical proof, the modified algorithm has been observed to terminate in practice [14, 5]. Both TetGen and DIR3 have included this heuristic for producing meshes with a prescribed minimum dihedral angle.

DIR3 is also a Delaunay tetrahedral mesh generator which admits an arbitrary 3D PLC as input. Using the PROTECTEDDELAUNAYREFINEMENT algorithm (Figure 11.10), it produces a conforming Delaunay tetrahedralization even in the presence of small input angles. Again, slivers can typically be removed (without any theoretical guarantee) by inserting their circumcenters during refinement. However, sliver circumcenters (just like those of poor radius-edge tetrahedra) cannot be inserted if they encroach protected collar simplices as this will prevent termination of the algorithm.

The number of vertices in the mesh produced by DIR3 is typically most dependent on the size of the smallest angle in the mesh and the size of the constants in the hypothesis $H2$. If the collar region is smaller, the local feature size with respect to the PSC $\mathcal{Y}$ is smaller leading to larger meshes due to estimate (11.6). Finding the best size for the collar region remains a topic of ongoing research.

Figure 11.11 contains an example meshes produced by TetGen which demonstrates the DELAUNAYREFINEMENT algorithm on a PLC which contains no input facet-facet angle less than $71°$. Several example meshes produced by DIR3 are given in Figure 11.12 and Figure 11.13. The results of the PROTECTEDDELAUNAYREFINEMENT algorithm on several PLCs with small input angles are shown.

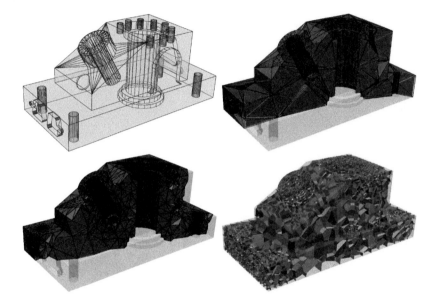

FIGURE 11.11: (See color insert.) Example meshes produced with TetGen. Top left: The input PLC. The minimum facet-facet angle is about 76°. Top right: The initial coarse mesh; i.e., the input to DELAUNAYREFINEMENT. Bottom left: The refined Delaunay mesh by DELAUNAYREFINEMENT. Bottom right: The dual Voronoi partition (Voronoi faces are randomly colored for visualization).

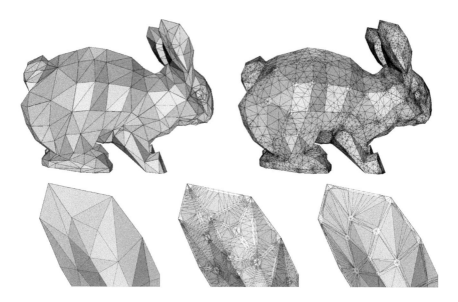

FIGURE 11.12: An example mesh produced with DIR3. Top left: The input PLC with minimum input angle of about 2.5°. Top right: Resulting Delaunay mesh using the DELAUNAYREFINEMENT algorithm. Slivers with angles smaller than 5° have also been removed during refinement. Only 0.1% of the resulting tetrahedra contain angles of less than 5°. Bottom left: An enlarged view of one ear of the rabbit. Bottom center: Collar region protecting all input segments. Bottom right: Collar region protecting only input angles that could prevent the termination of the DELAUNAYREFINEMENT algorithm.

FIGURE 11.13: Several meshes produced with DIR3. In each case, small input angles requiring a collar region lead to the most mesh refinement.

Bibliography

[1] Edelsbrunner, H., *Algorithms in Combinatorial Geometry*, Springer-Verlag, Heidelberg, 1987.

[2] Aurenhammer, F., "Voronoi diagrams – a study of fundamental geometric data structures," *ACM Comput. Surveys*, Vol. 23, 1991, pp. 345–405.

[3] Edelsbrunner, H., *Geometry and Topology for Mesh Generation*, Cambridge University Press, England, 2001.

[4] Eymard, R., Gallouët, T., and Herbin, R., "The finite volume method," *Handb. Numer. Anal.*, edited by P. Ciarlet and J. Lions, North Holland, 2000, pp. 715–1022.

[5] Si, H., *Three dimensional boundary conforming Delaunay mesh generation*, Ph.D. thesis, Institute für Mathematik, Technische Universität Berlin, 2008.

[6] Gärtner, K., "Existence of bounded discrete steady state solutions of the van Roosbroeck system on boundary conforming Delaunay grids," *SIAM J. Sci. Comput.*, Vol. 31, 2009, pp. 1347–1362.

[7] Glitzky, A. and Gärtner, K., "Existence of bounded steady state solutions to spin-polarized drift-diffusion systems," *SIAM J. Math. Anal.*, Vol. 41, 2010, pp. 2489–2513.

[8] Glitzky, A. and Griepentrog, J., "Discrete Sobolev–Poincaré inequalities for Voronoi finite volume approximations," *SIAM J. Num. Anal.*, Vol. 48, 2010, pp. 372–391.

[9] Chew, P. L., "Guaranteed-Quality Triangular Meshes," Tech. Rep. TR 89-983, Dept. of Comp. Sci., Cornell University, Ithaca, NY, 1989.

[10] Ruppert, J., "A Delaunay Refinement Algorithm for Quality 2-Dimensional Mesh Generation," *J. Algorithms*, Vol. 18, No. 3, 1995, pp. 548–585.

[11] Babuška, I. and Aziz, A. K., "On the Angle Condition in the Finite Element Method," *SIAM J. Num. Anal.*, Vol. 13, No. 2, 1976, pp. 214–226.

[12] Miller, G. L., Talmor, D., Teng, S., and Walkington, N. J., "A Delaunay Based Numerical Method For Three Dimensions: generation, formulation and partition," *Proc. 27th Symp. Theory Comput.*, ACM Press, New York, 1995.

[13] Miller, G. L., Pav, S. E., and Walkington, N. J., "An Incremental Delaunay Meshing Algorithm," Tech. Rep. 02-CNA-023, Center for Nonlinear Analysis, Carnegie Mellon University, Pittsburgh, PA, 2002.

[14] Shewchuk, J. R., *Delaunay refinement mesh generation*, Ph.D. thesis, Department of Computer Science, Carnegie Mellon University, Pittsburgh, PA, 1997.

[15] Pav, S. E. and Walkington, N. J., "Robust Three Dimensional Delaunay Refinement," *Proc. 13th Int. Meshing Roundtable*, 2004, pp. 145–156.

[16] Jamet, P., "Estimations d'erreur pour des elements finis droits presque degeneres," *RAIRO Anal. Numér.*, Vol. 10, No. 3, 1976, pp. 43–61.

[17] Shenk, N. A., "Uniform error estimates for certain narrow Lagrange finite elements," *Math. Comput.*, Vol. 63, No. 207, 1994, pp. 105–119.

[18] Pav, S. E., *Delaunay refinement algorithms*, Ph.D. thesis, Department of Mathematical Sciences, Carnegie Mellon University, Pittsburgh, PA, May 2003.

[19] Miller, G. L., Pav, S. E., and Walkington, N. J., "When and Why Ruppert's Algorithm Works," *Proc. 12th Int. Meshing Roundtable*, 2003, pp. 91–102.

[20] Si, H., "An analysis of Shewchuk's Delaunay refinement algorithm," *Proc. 18th Int. Meshing Roundtable*, 2009, pp. 499–518.

[21] Miller, G. L., Phillips, T., and Sheehy, D., "Linear-size meshes," *Proc. 20th Can. Conf. Comput. Geom.*, 2008.

[22] Hudson, B., Miller, G. L., Phillips, T., and Sheehy, D., "Size Complexity of Volume Meshes vs. Surface Meshes," *Proc. 20th Symp. Discrete Algorithms*, 2009, pp. 1041–1047.

[23] Cheng, S.-W., Dey, T., Edelsbrunner, H., Facello, M. A., and Teng, S.-H., "Sliver Exudation," *J. Assoc. Comput. Mach.*, Vol. 47, 2000, pp. 883–904.

[24] Erickson, J., "Nice point sets can have nasty Delaunay triangulations," *Discrete Comput. Geom.*, Vol. 30, No. 1, 2003, pp. 109–132.

[25] Talmor, D., *Well-spaced points for numerical methods*, Ph.D. thesis, Department of Computer Science, Carnegie Mellon University, Pittsburgh, PA,1997.

[26] Murphy, M., Mount, D. M., and Gable, C. W., "A Point-Placement Strategy for Conforming Delaunay Tetrahedralization," *Int. J. Comput. Geom. Appl.*, Vol. 11, No. 6, 2001, pp. 669–682.

[27] Cohen-Steiner, D., Colin de Verdière, E., and Yvinec, M., "Conforming Delaunay triangulations in 3D," *Comput. Geom.*, Vol. 28, No. 2-3, 2004, pp. 217–233.

[28] Cheng, S.-W. and Poon, S.-H., "Three-Dimensional Delaunay Mesh Generation," *Discrete Comput. Geom.*, Vol. 36, No. 3, 2006, pp. 419–456.

[29] Rand, A. and Walkington, N., "3D Delaunay Refinement of Sharp Domains Without a local feature size oracle," *Proc. 17th Int. Meshing Roundtable*, 2008, pp. 37–54.

[30] Miller, G. L., Phillips, T., and Sheehy, D., "Fast sizing calculations for meshing," *18th Fall Workshop Comput. Geom.*, 2008.

[31] Hudson, B., "Safe Steiner Points for Delaunay Refinement," *17th Int. Meshing Roundtable Res. Notes*, Pittsburgh, PA, 2008.

[32] Rand, A. and Walkington, N., "Collars and Intestines: Practical Conforming Delaunay Refinement," *Proc. 18th Int. Meshing Roundtable*, 2009, pp. 481–497.

[33] TetGen: A Quality Tetrahedral Mesh Generator and a 3D Delaunay Triangulator, 2009, `http://tetgen.berlios.de/`.

[34] DIR3: Delauany Incremental Refinement in 3D, 2009, `http://math.cmu.edu/~arand/DIR3/`.

Chapter 12

Two-Dimensional Approaches to Sparse Matrix Partitioning

Rob H. Bisseling, Bas O. Fagginger Auer, A. N. Yzelman

Utrecht University

Tristan van Leeuwen

University of British Columbia

Ümit V. Çatalyürek

Ohio State University

12.1 Introduction

Parallel computers are appearing on our desktops, as the multicore revolution is driving technology towards parallelism. In our applications, this causes the need for partitioning of the computational work and the corresponding data. In scientific computing, these data are often represented by a sparse matrix, which has many zero elements but stores the interesting information in its nonzero elements.

To partition well, we need to balance the amount of work among the processors and minimize the amount of communication caused by having nonzeros in different processors. In this chapter, we focus on an important sparse matrix computation: the multiplication of a sparse matrix A by a (dense) input vector $\mathbf{v}$, which yields as output a (dense) vector $\mathbf{u} = A\mathbf{v}$. Here, A is of size $m \times n$, $\mathbf{v}$ is of length n, and $\mathbf{u}$ is of length m. This multiplication is the most time-consuming operation of most iterative solvers for linear systems [1] and eigensystems [2].

The load balance of the computation is commonly expressed by the following constraint:

$$\text{nz}(A_s) \leq (1 + \epsilon)\frac{\text{nz}(A)}{p}, \quad \text{for } s = 0, 1, \ldots, p - 1. \tag{12.1}$$

Here, $\text{nz}(A)$ denotes the number of nonzeros of the matrix A and $\text{nz}(A_s)$ the number of nonzeros of matrix part A_s. The parts are numbered from 0 to $p - 1$, where p is the number of processors (or processor cores) among which A needs to be distributed; to avoid confusion, we will always call the basic processing element a *processor*. The allowed load imbalance $\epsilon > 0$ is a user-specified parameter. The proper choice of ϵ depends on the characteristics of the parallel computer: for a machine with faster communication, the load imbalance becomes more important, and hence ϵ should be reduced. In our exposition, we assume that the parallel computer has a distributed memory, so that the partitioning implies an actual physical distribution of the data. Still, in case of a shared memory, a good partitioning is also needed, primarily to divide the work, but also to reduce memory conflicts between processors trying to access the same location in the shared memory.

Communication arises because the nonzeros in a row or column may not all reside at the same processor, having been assigned to different parts by the partitioning. Since the aim of a matrix–vector multiplication is to compute $\mathbf{u} = A\mathbf{v}$, i.e.,

$$u_i = \sum_{j=0}^{n-1} a_{ij} v_j, \quad \text{for } i = 0, 1, \ldots, m - 1, \tag{12.2}$$

nonzeros of a row assigned to different parts cause communication, as products $a_{ij} v_j$ or sums of such products have to be sent to the owner of vector component u_i. In the worst case, all p processors have nonzeros in row i, and $p - 1$ processors need to send a contribution to u_i. In general, the number of processors having part of row i equals p_i with $p_i \leq p$, and the number of data words to be sent for computing u_i will be at most $p_i - 1$. (The owner of u_i is equal to one of the owners of a nonzero in row i, thus preventing an unnecessary communication.) For columns, similar considerations hold: the vector component v_j must be made known to all q_j processors that have a nonzero in column j, giving rise to $q_j - 1$ communications of a data word.

The aim of matrix partitioning is to find an assignment of the nonzeros to parts such that Equation (12.1) is satisfied and the *communication volume*, the total number of data words communicated, is minimized. The volume is defined by

$$V = \sum_{i=0}^{m-1}(p_i - 1) + \sum_{j=0}^{n-1}(q_j - 1), \tag{12.3}$$

where we assume that matrix rows and columns are not empty.

One-dimensional (1D) partitioning has been common practice for decades, usually assigning complete rows to a processor; of course assigning complete columns is also possible. Sometimes this is natural because a row has a physical meaning, like the connections of a grid point in a regular computational grid. Often, all the information concerning a grid point is kept together, and assigning the corresponding matrix row to one processor is natural. In other situations, there is not much meaning beyond the matrix entries themselves and the nonzeros of a row could be assigned to different processors. An example is web matrices, where rows/columns represent webpages, and nonzeros represent links between them. An advantage of 1D partitioning is that the parallel matrix–vector multiplication becomes slightly simpler; e.g., in the case of row partitioning (assigning complete rows to parts) we do not need to send contributions to the output vector, saving one phase of the computation (and, depending on the type of parallel computing model used, saving one global synchronization). The disadvantage is that the remaining phase may become more expensive. As a solution to an optimization problem, minimizing the communication in a 1D fashion is less general, and hence could be suboptimal.

Two-dimensional (2D) partitioning of sparse matrices is more recent, and it has been shown to be highly effective in distributing the work and minimizing communication for parallel computations in a wide range of applications [4, 5, 6, 7, 8].

Two types of 2D methods are particularly promising: (i) the *fine-grain approach* by Çatalyürek and Aykanat [4, 5] which partitions the nonzero matrix elements looking at individual elements, thereby allowing for the most general solution; (ii) the *Mondriaan approach* by Vastenhouw and Bisseling [8] which recursively bipartitions the sparse matrix into two (not necessarily contiguous) submatrices, trying splits in both the row and column directions and each time choosing the best. This orthogonal recursive bisection method is of the coarse-grain type; each split of a submatrix keeps either the nonzeros of the rows together as an atomic unit, or those of the columns. Both approaches use a multilevel hypergraph partitioner as their splitting engine, which minimizes the communication volume in the exact metric given by Equation (12.3).

A natural question to ask is whether combining the two approaches in a hybrid method would lead to savings in communication volume. This chapter explains the different approaches and also presents a new hybrid fine-grain/Mondriaan approach which tries to split the current subset of nonzeros in three possible ways: by rows, by columns, and by individual nonzeros, and then chooses the best split. Thus, the partitioning algorithm automatically detects the best method for the type of matrix involved, and adjusts this decision to the current submatrix as the partitioning progresses, at the cost of an additional splitting attempt for each bipartitioning.

As a final note, partitioning methods for parallelizing the sparse matrix–vector multiplication can easily be combined with methods for accelerating sequential sparse matrix–vector multiplication, e.g., autotuning by using OSKI [9], or matrix reordering to improve cache use [3], as these can be applied directly to the local part of the matrix–vector multiplication.

12.2 Sparse Matrices and Hypergraphs

We will now reformulate the matrix partitioning problem of the previous section as a hypergraph partitioning problem. A *hypergraph* $\mathcal{H} = (\mathcal{V}, \mathcal{N})$ consists of a set of vertices $\mathcal{V}$ and a set of *hyperedges* or *nets* $\mathcal{N}$, which are subsets of $\mathcal{V}$. A hypergraph is a generalization of an *undirected graph*, which represents the special case where the subsets are pairs of vertices. A *bipartitioning* of a hypergraph is a splitting of the vertices into two disjoint nonempty sets $\mathcal{V}_0$ and $\mathcal{V}_1$ with $\mathcal{V}_0 \cup \mathcal{V}_1 = \mathcal{V}$.

A sparse matrix can be converted to a hypergraph in various ways. Since the numerical value of the matrix nonzeros is irrelevant for minimizing the communication volume given by Equation (12.3), we only consider the sparsity pattern of the nonzeros. The first way of converting is by the *column-net model* [10], where each row i corresponds to a vertex v_i and each column j to a net $n_j = \{v_i : 0 \leq i < m \wedge a_{ij} \neq 0\}$. This hypergraph is denoted by $\mathcal{H}_c$; it contains all the information on the sparsity pattern of the matrix. An example is given by Figure 12.1. A different way is by the *row-net model*, where the roles of rows and columns are reversed: rows correspond to nets, and columns to vertices. In this case, the hypergraph is denoted by $\mathcal{H}_r$. Both ways are 1D in nature. A different, two-dimensional way of conversion will be discussed in Section 12.5.

12.3 Parallel Sparse Matrix–Vector Multiplication

A sparse matrix can be multiplied in parallel by a vector using either a 1D or a 2D partitioning. As the algorithm for a 1D partitioning is simpler, we will explain this algorithm first.

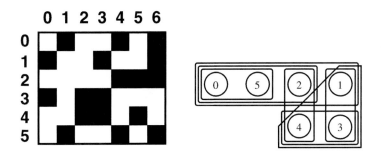

FIGURE 12.1: Example sparse matrix of size 6×7 with 18 nonzeros. Left: the matrix, with nonzeros depicted in black. Right: the column-net hypergraph created from the matrix, with vertices (circles) representing the rows of the matrix, and nets (sets containing the vertices) representing the columns. For example, column 6 of the matrix has nonzeros $a_{06}, a_{16}, a_{26}, a_{56}$, thus creating the net $\{0, 1, 2, 5\}$ of the hypergraph.

12.3.1   Using a One-Dimensional Partitioning

The natural parallel algorithm for sparse matrix–vector multiplication using a 1D row partitioning can be formulated as follows. Let $P(x)$ denote the processor that owns variable x. We write a_{i*} for row i, and we assume that $P(a_{i*}) = P(u_i)$. Furthermore, we assume that the vector component v_j is assigned to one of the processors with nonzeros in column j. Let I_s be the set of matrix rows of processor s and let J_s be the set of nonempty matrix columns of processor s, i.e.,

$$J_s = \{j : 0 \leq j < n \wedge (\exists i : 0 \leq i < m \wedge a_{ij} \neq 0 \wedge P(a_{ij}) = s)\}. \quad (12.4)$$

Algorithm 4 presents the natural row-based parallel 1D sparse matrix–vector multiplication algorithm. This algorithm consists of two phases, also called *supersteps* in bulk-synchronous parallel (BSP) language [11, 12]. Superstep (0) communicates vector components v_j and superstep (1) computes the vector components u_i. In superstep (1), the nonzeros a_{ij} are multiplied with the corresponding v_j and added into the output component u_i. The total communication volume of the algorithm equals the number of vector components v_j communicated in superstep (0).

The communication volume of parallel sparse matrix–vector multiplication based on a 1D matrix partitioning can be modeled by using a hypergraph where the vertices are partitioned. Assume we have a row partitioning and a corresponding partitioning of the vertices. If λ_j is the number of parts of the partitioning that have vertices in net j, the *cost* of the hypergraph partitioning is defined as

$$C = \sum_{j=0}^{n-1} c_j(\lambda_j - 1), \quad (12.5)$$

Algorithm 4 1D sparse matrix–vector multiplication for processor s.

(0)	{ Communication of $\mathbf{v}$ }	

for all $j \in J_s$ **do**
  get v_j from $P(v_j)$;

(1)   { Local sparse matrix–vector multiplication }
  for all $i \in I_s$ **do**
    $u_i := 0$;
    for all $j : 0 \leq j < n \wedge a_{ij} \neq 0$ **do**
      $u_i := u_i + a_{ij}v_j$;

---

where c_j is a scaling constant reflecting the importance of net j. This cost metric is commonly called the $(\lambda - 1)$-*metric*.

Let $\mathcal{H} = (\mathcal{V}, \mathcal{N})$ be a hypergraph with vertex weights $W : \mathcal{V} \to \mathbf{R}_{\geq 0}$ and net costs $c : \mathcal{N} \to \mathbf{R}_{\geq 0}$. Let p be the desired number of parts and $\epsilon > 0$ the allowed imbalance. Then the *hypergraph partitioning problem* is defined as finding a partitioning $\mathcal{V} = \mathcal{V}_0 \cup \cdots \cup \mathcal{V}_{p-1}$ of the vertices into pairwise disjoint nonempty subsets, such that the *load balance constraint*

$$W(\mathcal{V}_s) \leq (1 + \epsilon)\frac{W(\mathcal{V})}{p}, \quad \text{for } s = 0, 1, \ldots, p - 1, \qquad (12.6)$$

is satisfied, and the cost, given by Equation (12.5), is minimal.

By the translation of a matrix into a hypergraph for the column-net model, we have that $\lambda_j = q_j$, the number of different processors present in matrix column j. If we now define $c_j = 1$, the cost C becomes exactly equal to the communication volume V of the parallel sparse matrix–vector multiplication, defined by Equation (12.3). If we define vertex weight W_i as the number of nonzeros in matrix row i, the balance constraint becomes equal to the constraint for matrix partitioning, Equation (12.1). This is illustrated by Figure 12.2.

The total communication volume may not be the only important metric. On parallel architectures with high latencies, where there is a large cost for starting up a message, and a much smaller cost for each word of the message, it may be important to keep the number of messages small as well. In particular this holds for smaller problem sizes, and for a large number of processors p, since then the number of send–receive pairs can become $p(p - 1)$. In this chapter, however, we will be exclusively concerned with the metric of communication volume.

The translation from a matrix partitioning problem to a hypergraph partitioning problem has made it possible to use existing software such as ML-Part [13] and hMETIS [14], developed for the area of VLSI design where hypergraph partitioning is needed for obtaining good circuit layouts. But even more, the importance of hypergraph partitioning for parallel matrix computations has driven the development of new hypergraph partitioners, such as PaToH [15], Zoltan [16], Parkway [17], and also the native hypergraph partitioner of Mondriaan [8]. Of these, the Zoltan hypergraph partitioner (see Chapter 13) and Parkway are themselves parallel, which is important for larger problems that cannot be stored on a single processor in the case of distributed-memory parallel architectures. All these partitioners are based on the multilevel idea [18]: first the hypergraph is coarsened by merging vertices, doing this repeatedly at several levels, then an initial partitioning is obtained of the coarse hypergraph, and finally this partitioning is further refined during an uncoarsening phase. A detailed explanation of the multilevel approach, within the context of graph partitioning, can be found in Chapter 14.

The hypergraph bipartitioning problem is NP-hard [19], so all partitioning approaches for large problems will have to be based on heuristics. Even finding good approximate solutions with a guaranteed quality bound is NP-hard, as was shown for graphs by Bui and Jones [20].

Graph partitioning is used with much success in the area of parallel mesh computations, which are often based on finite elements. Software such as Chaco [21], Metis [22], ParMetis [23], Jostle [24], Scotch [25], PT-Scotch [26], can be employed to partition meshes and, more in general, graphs, typically aiming at reducing the total number of cut edges. This is not the same as reducing the total communication volume in a parallel mesh computation, as two cut edges (i,j) and (i,j') with $j \neq j'$ may cause only one communication leaving the source processor $P(i)$ owning vertex i, since it might happen that $P(j) = P(j')$, i.e., the destination processors are the same. Therefore, minimizing the edge cut amounts to minimizing an upper bound on the communication, not the actual communication, and hence is an approximation. Since graph partitioning is in general faster than hypergraph partitioning, it may still be the method of choice for problems where the approximation is good. The packages ParMetis [23] and PT-Scotch [26] are parallel, and Jostle [24] also has a parallel version. Since a symmetric sparse matrix can be viewed as the adjacency matrix of a graph, with $a_{ij} \neq 0$ if and only if (i,j) is an edge of the graph, graph partitioners can also be used to partition such sparse matrices. Historically, graph partitioners were even the first to be used, before the hypergraph partitioners came to the field [27]. For more on graph partitioning, see Chapter 14. An overview of existing software packages ordered by problem type and by the possibility to run the package in parallel is given in Table 12.1.

12.3.2 Using a Two-Dimensional Partitioning

The natural parallel algorithm for sparse matrix–vector multiplication with the most general (2D) matrix distribution can be formulated as follows. We assume that all data elements, i.e., the matrix nonzeros a_{ij} and the vector components u_i and v_j, have been assigned to arbitrary processors. Let I_s now

TABLE 12.1: Overview of available software for partitioning graphs and hypergraphs.

Name	Graph/ hypergraph	Sequential/ parallel	Reference
Chaco	graph	sequential	[21]
Metis	graph	sequential	[22]
Scotch	graph	sequential	[25]
Jostle	graph	parallel	[24]
ParMetis	graph	parallel	[23]
PT-Scotch	graph	parallel	[26]
hMETIS	hypergraph	sequential	[14]
ML-Part	hypergraph	sequential	[13]
Mondriaan	hypergraph	sequential	[8]
PaToH	hypergraph	sequential	[15]
Parkway	hypergraph	parallel	[17]
Zoltan	hypergraph	parallel	[16]

be the set of nonempty matrix rows of processor s, i.e.,

$$I_s = \{i : 0 \le i < m \ \wedge \ (\exists j : 0 \le j < n \ \wedge \ a_{ij} \ne 0 \ \wedge \ P(a_{ij}) = s)\} \quad (12.7)$$

and, as before, let J_s be the set of nonempty matrix columns of processor s, see Equation (12.4). Algorithm 5, adapted from [11], presents the natural parallel 2D sparse matrix–vector multiplication algorithm. This algorithm consists of four supersteps. These supersteps alternate between communication and computation: first we communicate, then compute, then communicate, and finally compute again. Here, the central superstep is (1), where nonzeros a_{ij} are multiplied with the corresponding v_j and added to the local contribution u_{is} of processor s to the output component u_i. The total communication volume of the algorithm is the sum of the number of vector components v_j communicated in superstep (0) and the number of contributions u_{is} communicated in superstep (2). For convenience, we take as the total computation cost the cost of superstep (1) measured in multiplications $a_{ij} \cdot v_j$. This cost is defined as the maximum over the processors of the number of local nonzeros. We ignore the cost of summing the contributions in superstep (3); this cost is much less than that of sending them in the preceding superstep.

The communication requirements of parallel sparse matrix–vector multiplication based on a 2D matrix partitioning is illustrated by Figure 12.3. Using both dimensions gives more flexibility, for instance to better use an allowed imbalance to reduce communication, as happens in the example.

In this chapter, we only discuss matrix partitioning, and not the vector partitioning, aiming at balancing the computational work load and minimizing the total communication volume. We assume that the vector is partitioned after the matrix partitioning in such a way that this does not cause additional

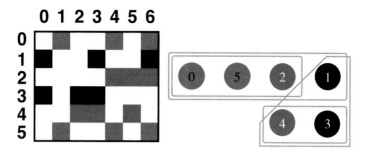

FIGURE 12.2: (See color insert.) Sparse matrix from Figure 12.1 partitioned into 3 parts by a 1D row partitioning. Rows 0 and 5 are assigned to processor 0 (red), rows 1 and 3 to processor 1 (black), and rows 2 and 4 to processor 2 (blue). Left: the matrix, with nonzeros painted by the color of their processor. Right: the column-net hypergraph created from the matrix, with only the cut nets shown (in green). The vertices are painted in the same color as their corresponding row. Column 6 has $\lambda_6 = 3$ processors, causing $\lambda_6 - 1 = 2$ communications during sparse matrix–vector multiplication, and the corresponding net $\{0, 1, 2, 5\}$ has three parts. This means that vector component v_6 has to become known to all three processors, and hence it has to be sent by the source processor to the other two processors. The other cut columns are columns 2, 3, 4; they each have two parts and cause one communication. Columns 0, 1, and 5 do not cause communication. The load balance happens to be perfect. The communication volume as defined by Equation (12.3) equals $V = 2 + 1 + 1 + 1 = 5$, which is the sum of the number of communications for the separate columns.

communication, meaning that $P(u_i)$ is one of the processors with nonzeros in row i, and $P(v_j)$ is one of the processors with nonzeros in column j. This leaves some freedom that can be used to achieve a secondary objective, e.g., balancing the communication load [28], reducing the number of messages [29], or balancing the number of vector components. The latter objective may be important in an iterative solver with many vector operations such as GMRES [30].

12.4 Coarse-Grain Partitioning

Coarse-grain approaches to partitioning keep naturally occurring sets of nonzeros such as rows or columns (or parts of rows or columns) together during the splits of the partitioning process; this is in contrast with the fine-grain approach, where nonzeros are handled individually. In the present section, we

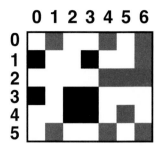

FIGURE 12.3: (See color insert.) Sparse matrix from Figure 12.1 partitioned into 3 parts by a 2D partitioning, using the Mondriaan package with the hybrid approach of this chapter. Rows 1 and 4, and columns 4 and 6 each cause one communication. The load imbalance is one nonzero, i.e., $\epsilon = 1/6$. The total communication volume is 4.

review various types of coarse-grain partitioning. In the next section, we will discuss the fine-grain approach.

12.4.1 Cartesian Partitioning and Its Variants

Cartesian partitioning, also known as *checkerboard partitioning* and *rectilinear partitioning*, is one of the frequently used matrix partitioning methods for dense matrices or sparse matrices with structured patterns that are difficult to exploit [31, 32, 33, 34, 35, 36]. A Cartesian partitioning is obtained by partitioning the rows of the matrix into P subsets, the columns into Q subsets, with $p = PQ$, and then assigning the Cartesian product of a row subset and a column subset to a processor, for each of the p possible combinations of row subset and column subset. This is a desirable partitioning due to its advantage of naturally limiting the number of processors communicating in each communication superstep of parallel 2D matrix–vector multiplication.

For sparse matrices, one can also achieve Cartesian partitioning by intelligently combining row-net and column-net hypergraph models with multi-constraint partitioning [4, 6, 7]. In this two-step method, for a $P \times Q$ processor mesh with $p = PQ$, the matrix is first partitioned into P horizontal strips (blocks of rows, which may have been permuted) using the column-net hypergraph model. Then, in the second step, the same matrix is partitioned into Q vertical strips using the row-net hypergraph model with multi-constraint partitioning, such that each vertex (representing a matrix column) will have P weights corresponding to the number of nonzeros in the P different blocks of the first partitioning. Here, for each of the P blocks of rows, a constraint of the form (12.1) must be satisfied, with Q parts instead of p. It is, of course, also possible to start the process in the vertical direction.

Despite the obvious communication advantages, Cartesian partitioning is

---

Algorithm 5 2D sparse matrix–vector multiplication for processor s.

---

(0)   { Communication of **v** }
        for all $j \in J_s$ **do**
            get v_j from $P(v_j)$;

(1)   { Local sparse matrix–vector multiplication }
        for all $i \in I_s$ **do**
            $u_{is} := 0$;
            for all $j : 0 \leq j < n \wedge a_{ij} \neq 0 \wedge P(a_{ij}) = s$ **do**
                $u_{is} := u_{is} + a_{ij}v_j$;

(2)   { Communication of contributions to **u** }
        for all $i \in I_s$ **do**
            put u_{is} in $P(u_i)$;

(3)   { Summation of contributions }
        for all $i : 0 \leq i < n \wedge P(u_i) = s$ **do**
            $u_i := 0$;
            for all $t : 0 \leq t < p \wedge u_{it} \neq 0$ **do**
                $u_i := u_i + u_{it}$;

---

very restrictive and for sparse matrices with very irregular patterns, it may yield a higher communication volume. Another partitioning method, *jagged partitioning* [35, 37], is a variant of Cartesian partitioning. In the jagged partitioning for a $P \times Q$ processor mesh, the matrix is first partitioned into P horizontal strips, just as for Cartesian partitioning, but with contiguous rows in each strip. In the second step, however, every horizontal strip is independently partitioned into Q submatrices, each containing a set of contiguous columns. This yields a partitioning that is not Cartesian, since splits span the entire matrix in one dimension but they are jagged in the other dimension. The resulting submatrices are contiguous blocks in the original matrix. Although efficient algorithms exist for producing jagged partitions with optimal load balance [38, 39, 40], they do not consider the minimization of communication volume explicitly. The *jagged-like partitioning* method [4, 7] utilizes 1D row-net and column-net hypergraph models to achieve communication-aware jagged matrix partitionings. The reason this method is called jagged-like instead of jagged is that it no longer enforces the rows and columns of the submatrices to be contiguous.

12.4.2 Mondriaan Partitioning by Orthogonal Recursive Bisection

The Mondriaan approach [8] to partitioning consists of a sequence of splits, each time trying both the row and column direction, and then taking the best of the two (i.e., the one with the lowest communication volume). Each split is thus 1D, and the overall result is a 2D matrix partitioning. The motivation for this approach is that by keeping the nonzeros of a row or column together, communication in one direction is prevented by construction, while we still benefit from the flexibility of using both directions in the overall process. The 1D splits are a hypergraph-based bipartitioning of the current largest part in a prescribed ratio of numbers of nonzeros, such as 1:2 for the first split in the case of $p = 3$. An advantage of 1D splittings is that they are relatively fast, since they consider hypergraphs with a limited number of vertices, namely m or n for the first split. For $p = 2$, the Mondriaan method takes the best of two 1D partitionings, and hence the result is a 1D partitioning. Only for $p > 2$, it potentially exploits both dimensions.

A nice property of the sparse matrix–vector multiplication algorithm, Algorithm 5, is that the total communication volume for a p-way matrix partitioning obtained by successive bipartitionings is just the sum of the communication volumes of the separate splits, where each split can be seen as acting on a sparse matrix of the same size as A but containing only those nonzeros of the subset to be split. Thus, split k acts on an $m \times n$ matrix $A^{(k)}$, for $k = 0, \ldots, p - 2$, where $A^{(0)} = A$. This enables looking at every split in isolation. This property was shown for a 1D partitioning in [10] and for an arbitrary partitioning in [8, Theorem 2.2].

The result of the Mondriaan splitting is not necessarily Cartesian, as the parts created by a bipartitioning are split independently later on. Still, the partitioning has internal structure: the nonzeros a_{ij} of processor s have indices in the set $I_s \times J_s$, cf. Equations (12.4) and (12.7), which represents a (scattered) submatrix of A; all these submatrices are disjoint, and by construction they fit in the original matrix, as the colored rectangles in the paintings by the Dutch painter Piet Mondriaan (1872–1944). This is illustrated by Figure 12.4. Such structure will generally not be present in an arbitrary partitioning of the nonzeros, where the submatrices enclosing the nonzeros of a processor may overlap.

12.5 Fine-Grain Partitioning

The *fine-grain approach* [4, 5] to partitioning translates the matrix to a hypergraph $\mathcal{H}_f$, as follows. Every vertex represents a matrix nonzero. There are two types of nets: row nets and column nets. A *row net* i contains all

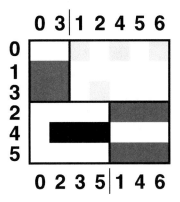

FIGURE 12.4: (See color insert.) Sparse matrix from Figure 12.1 partitioned into 4 parts by a Mondriaan partitioning. The first split is in the row direction, whereas the second and third splits are in the column direction. The matrix is shown in the local view, displaying the local submatrices of the processors. The column indices are different for the top and the bottom part. The submatrix $I_0 \times J_0$ in the upper left corner contains the nonzeros assigned to processor 0 (red), where $I_0 = \{0, 1, 3\}$ and $J_0 = \{0, 3\}$.

the vertices corresponding to nonzeros in matrix row i, and a *column net j* contains all the vertices corresponding to nonzeros in matrix column j. The total number of vertices of $\mathcal{H}_f$ is $\mathrm{nz}(A)$ and the number of nets is $m + n$, assuming that no row or column is empty.

The hypergraph can then be partitioned into p parts, and the corresponding cost given by Equation (12.5) with $c_j = 1$ is exactly equal to the communication volume of the sparse matrix–vector multiplication, as every column net j that is cut into λ_j parts corresponds to a column that is cut into the same number of parts, so that $q_j = \lambda_j$, and similarly for cut row nets. Choosing vertex weights of 1 makes the load balance constraint of the hypergraph identical to that of the matrix, given by Equation 12.1. As a result, any hypergraph partitioner can be used, either splitting the hypergraph directly into p parts, or using recursive bipartitioning.

A different view of the fine-grain representation of a sparse matrix can be obtained by first translating the matrix A into its corresponding fine-grain hypergraph $\mathcal{H}_f = \mathcal{H}_f(A)$, and then translating it back into a matrix, but not by the same model, using either the row-net model, yielding the matrix $F = F(A)$ shown in Figure 12.5, or the column-net model, yielding the matrix F^T. The 2D partitioning of the matrix A then corresponds to a 1D column partitioning of the matrix F. Note that the matrix F, which is of size $(m + n) \times \mathrm{nz}(A)$, has a special form: for instance, in the figure every column has exactly two

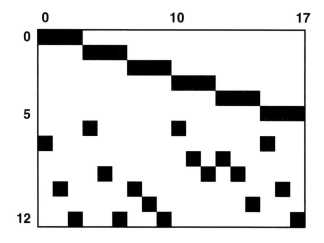

FIGURE 12.5: Fine-grain incidence matrix $F = F(A)$ of size $(6 + 7) \times 18$ created from the 6×7 matrix A with 18 nonzeros of Figure 12.1 by translating the fine-grain hypergraph $\mathcal{H}_f = \mathcal{H}_f(A)$ back into a matrix by the row-net model. The kth nonzero a_{ij} in A corresponds to column k in F, which has 2 nonzeros, f_{ik} and $f_{(m+j),k}$, where $m = 6$ is the number of row nets.

nonzeros, one corresponding to a row net from $\mathcal{H}_f$ and the other to a column net.

12.6 The Hybrid Partitioning Algorithm

The Mondriaan method tries to bipartition the hypergraphs $\mathcal{H}_r$ and $\mathcal{H}_c$. It is a straightforward extension of this method to bipartition the fine-grain hypergraph $\mathcal{H}_f$ as well. This can be done within the same framework of the Mondriaan p-way matrix partitioner, and using the same hypergraph splitter. We call the resulting method the *hybrid fine-grain/Mondriaan method* for sparse matrix partitioning, or in short the *hybrid method*.

This hybrid method will cost an additional hypergraph bipartitioning run for every split, which may be expensive because the fine-grain bipartitioning, which is 2D, needs more computation time and memory than the row and column bipartitionings, which are 1D. Therefore, an important implementation issue in making the hybrid method work in practice is exploitation of the special structure that every hypergraph $\mathcal{H}_f$ possesses.

Algorithm 6 presents a recursive formulation of the hybrid method. The algorithm splits the current subset A of nonzeros into two parts, with a fraction

p_0/p of the nonzeros assigned to part 0 and a fraction $1 - p_0/p = p_1/p$ to part 1. The values of p_0 and p_1 are chosen as close as possible to $p/2$, to minimize the number of split levels. Following the adaptive strategy proposed in [8], the imbalance allowed in the first split is based on the assumption that each subsequent split in a path down the binary splitting tree adds the same imbalance. For part r, with $r = 0, 1$, the first split thus allows an imbalance of ϵ/q_r, with $q_r = \lceil \log_2 p_r \rceil + 1$. The imbalances are recomputed when calling the function recursively, based on the actual current division of nonzeros and on the corresponding number of parts.

The function split$(A, \text{method}, \delta_0, \delta_1, f)$ is not given in detail here; an outline of how it works follows. The aim is to split A into two parts, with part 0 having at most a fraction $(1 + \delta_0)f$ of the nonzeros and part 1 at most a fraction $(1+\delta_1)(1-f)$. The matrix is first translated into a hypergraph for the chosen splitting method (row, column, or fine-grain), and then this hypergraph is split into two parts using multilevel partitioning, see Chapter 14. This is done by coarsening the hypergraph repeatedly, merging similar vertices until the hypergraph is small enough, and then running a heuristic Kernighan–Lin/ Fiduccia–Mattheyses (KLFM) algorithm [41, 42] which assigns merged vertices to the two parts trying to minimize communication while balancing the nonzeros within the specification. Then the hypergraph is uncoarsened while a simpler version of KLFM is run to refine the partitioning at each level.

In Algorithm 6, all three hypergraph-based splitting methods are tried, i.e., by rows, by columns, and by the fine-grain method, and the best is chosen. This is motivated by the idea that often taking rows or columns as atomic units is beneficial, as this completely prevents communication in one direction and requires the solution of a smaller hypergraph partitioning problem (in terms of number of vertices), but that sometimes a true 2D bipartitioning is required. Since it cannot be easily predicted when this situation occurs, this should be automatically detected by the algorithm. Figure 12.6 illustrates the hybrid method.

12.7   Time Complexity

The time complexity of the complete hybrid p-way partitioning procedure of an $m \times n$ matrix is

$$T = \mathcal{O}((m + n) \cdot \text{nz}_{\max}^{\text{r}} \cdot \text{nz}_{\max}^{\text{c}} \cdot \log_2 p), \tag{12.8}$$

where $\text{nz}_{\max}^{\text{r}}$ is the maximum number of nonzeros in a matrix row and $\text{nz}_{\max}^{\text{c}}$ the maximum in a column, which is briefly explained as follows. Consider a 1D column partitioning. The first merging step of the coarsening computes inner products between pairs of matrix columns, which can be done in at most $\mathcal{O}(n \cdot \text{nz}_{\max}^{\text{r}} \cdot \text{nz}_{\max}^{\text{c}})$ operations, since for each of the n columns at most

---

Algorithm 6 Recursive adaptive algorithm for the computation of a p-way partitioning of a sparse matrix A with an allowed load imbalance of ϵ. The number of parts p can be any number ≥ 1, not necessarily a power of two. The output is a sequence $(A_0, A_1, \ldots, A_{p-1})$ of mutually disjoint subsets of nonzeros, each containing the nonzeros of one processor, with $\cup_{s=0}^{p-1} A_s = A$.

---

    function MatrixPartition(A, p, ϵ)

        if $p > 1$ **then**

$$p_0 := \lfloor p/2 \rfloor;$$
$$p_1 := \lceil p/2 \rceil;$$
$$q_0 := \lceil \log_2 p_0 \rceil + 1;$$
$$q_1 := \lceil \log_2 p_1 \rceil + 1;$$
$$(A_0^{\mathrm{r}}, A_1^{\mathrm{r}}) := \mathrm{split}(A, \mathrm{row}, \epsilon/q_0, \epsilon/q_1, p_0/p);$$
$$(A_0^{\mathrm{c}}, A_1^{\mathrm{c}}) := \mathrm{split}(A, \mathrm{col}, \epsilon/q_0, \epsilon/q_1, p_0/p);$$
$$(A_0^{\mathrm{f}}, A_1^{\mathrm{f}}) := \mathrm{split}(A, \mathrm{fine}, \epsilon/q_0, \epsilon/q_1, p_0/p);$$
$$(A_0, A_1) := \mathrm{best\ of}\ (A_0^{\mathrm{r}}, A_1^{\mathrm{r}}),\ (A_0^{\mathrm{c}}, A_1^{\mathrm{c}}),\ (A_0^{\mathrm{f}}, A_1^{\mathrm{f}});$$

$$\mathrm{nz}_{\max} := \frac{\mathrm{nz}(A)}{p}(1 + \epsilon);$$
$$\epsilon_0 := \frac{\mathrm{nz}_{\max}}{\mathrm{nz}(A_0)} \cdot p_0 - 1;$$
$$\epsilon_1 := \frac{\mathrm{nz}_{\max}}{\mathrm{nz}(A_1)} \cdot p_1 - 1;$$

            MatrixPartition(A_0, p_0, ϵ_0);
            MatrixPartition(A_1, p_1, ϵ_1);

        else output A;

$\mathrm{nz}_{\max}^{\mathrm{c}}$ nonzeros are handled, each requiring access to a row of at most $\mathrm{nz}_{\max}^{\mathrm{r}}$ nonzeros. The next step has only half the number of columns. If the first merge was successful, meaning that columns merge mainly with columns that have a very similar sparsity pattern, then the number of nonzeros per column hardly increases, while the number of nonzeros per row halves. Thus the next coarsening step requires less than half the work of the first step, and so on. The complete coarsening thus takes at most twice the time of its first step. The KLFM algorithm used in the uncoarsening is somewhat cheaper, but still of the same order. Overall, $\log_2 p$ stages of splitting are needed. The first stage is the most expensive, as the complete matrix is handled. The second stage works on two submatrices with smaller $m, n, \mathrm{nz}_{\max}^{\mathrm{r}}, \mathrm{nz}_{\max}^{\mathrm{c}}$ and because of the form of the complexity formula each submatrix costs less than half the work of the first stage. The overall cost of the second stage is thus at most that of the first stage. For other methods than 1D column partitioning, the complexity formula stays the same.

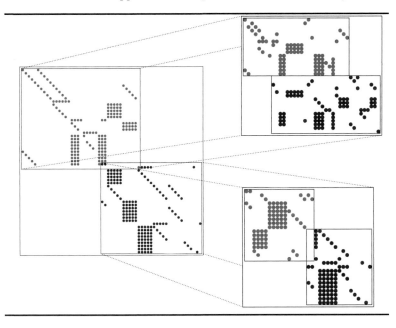

FIGURE 12.6: (See color insert.) Split of the 59×59 matrix `impcol_b` with 312 nonzeros into four parts. The first split (left) is by the fine-grain method; the second split (top right) is by rows; and the third split (bottom right) is again by the fine-grain method. The matrix is permuted correspondingly by moving, after each split, the newly cut rows and columns to the middle and the uncut ones to the left or right, or top or bottom. The matrix structure seen on the left is the (doubly) separated block-diagonal (SBD) structure, see [3].

12.8 Experimental Results

To compare the different 1D and 2D approaches to partitioning, we performed numerical experiments on a set of 10 test matrices, originating in various fields, see Table 12.2. All experiments were performed on an Intel Core i7 860 2.8 GHz processor, under the Ubuntu Linux 10.10 operating system (kernel 2.6.35, x86_64), with 8 Gbyte of RAM.

Table 12.3 presents a comparison of 1D and 2D methods, showing in the 1D case both a row-wise and a column-wise partitioning, and in the 2D case three methods, namely the Mondriaan approach based on orthogonal recursive bisection, the fine-grain approach, and the hybrid approach. For this table, the native hypergraph bipartitioner of Mondriaan was used. Table 12.4 presents a similar comparison, but now with PaToH [15] as a bipartitioner. The bottom line of the tables shows an average performance, obtained by dividing the

TABLE 12.2: Test set of sparse matrices.

Matrix	m	n	nz	Field
df1001	6071	12230	35632	linear programming
cre_b	9648	77137	260785	linear programming
tbdmatlab	19859	5979	430171	information retrieval
nug30	52260	379350	1567800	linear programming
c98a	56243	56274	2075889	cryptography
tbdlinux	112757	20167	2157675	information retrieval
stanford	281903	281903	2312497	web search
polyDFT	46176	46176	3690048	polymer self-assembly
cage13	445315	445315	7479343	DNA electrophoresis
stanford_berkeley	683446	683446	7583376	web search

communication volume by the volume for the 2D Mondriaan method using its own hypergraph bipartitioner, as given in the '2D Mon' column of Table 12.3. The allowed load imbalance defined by Equation (12.1) was set to $\epsilon = 0.03$. Communication volumes in the table represent averages over 25 runs with different random number seeds, where the volume of a run is influenced to some extent by random choices made by the algorithm. An example of randomness appearing in the algorithm is in the initial partitioning, where eight random starting partitionings of the coarsened hypergraph are generated, and the best resulting partitioning is chosen. Furthermore, ties in the algorithm are usually broken by 'flipping a coin.' For debugging purposes, the random number seed can be fixed, but for comparing results of two different methods we retain the randomness and perform a large number of runs.

The 2D methods perform much better than the 1D methods. In Table 12.3, the best 1D method, which is 1D column partitioning, yields on average 21% more communication than the 2D Mondriaan method, and in 24 out of 30 *problem instances*, i.e., pairs (Matrix, p), it beats the 1D row method, and in three instances it performs equally. This preference for column partitioning is a bias of the test set, since partitioning A^T instead of A would have given the opposite conclusion. Taking for each problem instance the best of the 1D row and 1D column results still gives on average 23% more communication than with the 2D Mondriaan method. Similar differences between 1D and 2D results can be seen in Table 12.4.

The 2D Mondriaan method with its own bipartitioner is better than the fine-grain method for all matrices except cage13, see Table 12.3. The hybrid method beats both methods in 20 out of 30 problem instances, and the difference with the best of both is small in the other cases. We may conclude that in most cases the hybrid method succeeds in picking the best of both, exactly what it was designed for.

With the PaToH implementation of the hypergraph bipartitioner, we find

TABLE 12.3: Communication volume of p-way partitioning using the Mondriaan package (version 3.11) with its own hypergraph bipartitioner. Average over 25 runs.

Matrix	p	1D row	1D column	2D Mon	2D fine	2D hybrid
dfl001	4	2974	1411	1402	1646	1410
	16	6086	3650	3660	3939	3618
	64	9124	6175	6142	6439	6044
cre_b	4	33293	1402	1379	3498	1393
	16	59910	4142	4153	8889	4122
	64	86092	9353	9312	17876	9276
tbdmatlab	4	13784	15061	10373	13987	10340
	16	56117	41769	26569	44122	26553
	64	152760	93667	52603	96504	58058
nug30	4	284683	67065	65926	109410	65491
	16	355935	150653	148256	258382	145881
	64	376809	250778	246508	439015	244726
c98a	4	162188	117532	100232	140069	100250
	16	652151	351879	239427	379300	239257
	64	1349998	647273	452397	665934	449727
tbdlinux	4	41275	54438	30159	35999	33945
	16	167039	152469	69971	120256	87672
	64	486500	328067	143767	298505	172510
stanford	4	22074	1309	1313	2462	1246
	16	96563	4662	4774	7106	4389
	64	339045	13324	13112	19631	12391
polyDFT	4	8813	8744	8501	12670	8496
	16	39066	39135	37855	51094	37707
	64	83527	83237	75034	113611	74966
cage13	4	118701	118701	112532	97957	99504
	16	267701	267701	245730	204808	208823
	64	495286	495286	428818	357141	359062
stanford_berkeley	4	35276	1626	1673	2367	1169
	16	230519	7219	10108	15184	6307
	64	917105	20473	23789	43413	19373
Avg normalized value		7.80	1.21	1.00	1.53	0.97

TABLE 12.4: Communication volume of p-way partitioning using the Mondriaan package (version 3.11) with PaToH as hypergraph bipartitioner. Average over 25 runs. Note that the 'Mon' column refers to the Mondriaan partitioning method used, based on orthogonal recursive bisection, but the splits of this method are carried out using PaToH.

Matrix	p	1D row	1D column	2D Mon	2D fine	2D hybrid
dfl001	4	2953	1374	1391	1386	1358
	16	5888	3608	3599	3552	3522
	64	8406	6049	6039	6060	5872
cre_b	4	33097	1206	1195	1225	1151
	16	57110	3481	3483	3585	3412
	64	73974	7811	7831	8097	7576
tbdmatlab	4	14078	15052	10666	10841	10391
	16	54397	40554	26511	31329	26219
	64	134093	81681	49238	64680	50810
nug30	4	270049	56809	55810	62323	52675
	16	384826	117883	116083	133336	106562
	64	542912	197808	198609	231081	185720
c98a	4	161633	117691	100096	124396	96455
	16	643899	351613	228064	325374	226643
	64	1341322	646249	417233	578384	408051
tbdlinux	4	42773	52710	35180	24921	24967
	16	166555	147091	77038	81754	73274
	64	443921	308329	145350	186478	149748
stanford	4	8755	934	896	921	826
	16	50281	3330	3169	3345	3004
	64	217976	9801	9643	9066	8359
polyDFT	4	9448	9228	8785	8794	8615
	16	35427	35177	34009	37024	34381
	64	78560	78615	73531	82879	73290
cage13	4	120883	121737	116349	89953	90534
	16	273048	273487	250559	188854	190640
	64	496963	498482	437986	333481	334123
stanford_berkeley	4	3321	1432	1139	1406	1031
	16	134868	5287	4650	5046	4080
	64	579112	15489	14958	14083	12810
Avg normalized value		5.32	1.11	0.90	0.95	0.83

again that 2D Mondriaan performed better than the fine-grain method, although by a much smaller margin. Apparently, the advantage of the more general solution is less important than the positive effect, as a solution heuristic, of forcing the nonzeros of a row (or column) to stay together, which is built into the Mondriaan approach. Here too, the hybrid method usually (i.e., in 24 out of 30 instances) beats the best of both, with at most 3% difference between hybrid and best of both in the other instances.

Comparing Tables 12.3 and 12.4, we see that PaToH is significantly better optimized for the fine-grain method, achieving an average normalized volume of 0.95 vs. 1.53 for the native Mondriaan bipartitioner. For the Mondriaan method, it is also better, 0.90 vs. 1.00. Overall, the best results are obtained with the hybrid method using PaToH as the bipartitioner, with a normalized volume of 0.83. Note that all these results are obtained within the same Mondriaan software framework, so that they can be compared with each other. Using PaToH for the whole partitioning process, i.e., not only for the bipartitioning process, would give slightly different results; see [16] for results using only PaToH, and also for results using Zoltan.

Tables 12.5 and 12.6 present a comparison of the run time of the five methods, using both the native hypergraph bipartitioner of Mondriaan and PaToH. Our main observation here is that the time of the hybrid method is almost the time of the Mondriaan and fine-grain method combined. This can be expected since both methods are tried within the hybrid partitioner. Another observation from both tables is that the time of the Mondriaan method is about the time of the 1D row and 1D column method combined. This means that 1D methods are only useful in terms of computing time if there is a quick way of determining which of the two directions, row or column, has to be chosen, see e.g., the recipe given in [7]. Remarkable is the speed of the 2D Mondriaan method using the PaToH bipartitioner; it is almost three times as fast compared to using the native Mondriaan bipartitioner. It is also 2.4 times faster than the fine-grain method using PaToH. Examining the timings for $p = 4, 16, 64$ in Tables 12.5 and 12.6, we see that the scaling with p in practice is often better than the theoretical upper bound of $\mathcal{O}(\log_2 p)$ given by the time complexity, Equation (12.8).

Table 12.7 presents the actual run time of a parallel sparse matrix–vector multiplication for the largest two test matrices after partitioning using Mondriaan with PaToH as bipartitioner. The purpose of the table is to check whether lower communication volumes indeed lead to lower run times of the parallel program. For cage13 and $p = 64$, the savings of one third in communication volume shown in Table 12.4 between 1D row and 2D hybrid leads to a decrease in run time of about 26% as shown in Table 12.7. The large savings for stanford_berkeley between 1D row and 1D column show up as a decrease of a factor of 2.5 in run time. The overall gain in speed is not directly proportional to the gain in communication volume because there are other factors involved as well, such as computation time and synchronization time; for smaller problems, the latter may actually be dominant. For larger p,

TABLE 12.5: Time (in s) needed to compute a p-way partitioning using the Mondriaan package (version 3.11) with its own hypergraph bipartitioner. Average over 25 runs.

Matrix	p	1D row	1D column	2D Mon	2D fine	2D hybrid
df1001	4	0.29	0.19	0.47	0.35	0.84
	16	0.38	0.36	0.76	0.64	1.38
	64	0.53	0.53	1.10	0.90	1.94
cre_b	4	1.63	1.58	3.36	2.76	6.17
	16	2.17	2.46	5.23	4.32	9.56
	64	4.04	3.21	6.78	5.69	12.38
tbdmatlab	4	2.84	2.45	5.41	10.09	16.59
	16	4.72	4.38	9.25	14.63	25.86
	64	5.58	5.75	11.24	17.59	30.15
nug30	4	17.45	14.71	40.68	23.71	63.07
	16	15.86	23.27	58.57	42.64	100.10
	64	21.59	29.36	68.28	54.21	116.30
c98a	4	25.30	82.15	104.40	391.60	493.00
	16	38.70	112.10	162.20	482.90	667.70
	64	53.10	123.60	193.90	517.10	745.50
tbdlinux	4	35.73	21.41	61.23	142.50	199.50
	16	52.24	32.30	94.13	190.70	290.30
	64	60.14	39.12	111.90	214.60	336.90
stanford	4	7.93	70.58	79.16	249.60	310.70
	16	13.19	105.20	118.60	390.50	479.90
	64	17.52	123.20	141.70	455.80	565.50
polyDFT	4	7.23	7.11	14.55	49.13	64.13
	16	15.42	15.45	29.15	87.43	117.10
	64	21.25	20.90	39.90	117.50	157.50
cage13	4	54.14	54.24	109.80	121.80	233.20
	16	95.52	95.72	169.00	195.70	372.20
	64	116.60	115.80	227.10	255.70	491.10
stanford_berkeley	4	16.96	288.20	300.50	1454.00	1746.00
	16	31.84	484.10	523.60	2788.00	3140.00
	64	51.03	574.60	632.60	3393.00	3797.00
Avg normalized value		0.39	0.57	1.00	2.19	3.17

TABLE 12.6: Time (in s) needed to compute a p-way partitioning using the Mondriaan package (version 3.11) with PaToH as hypergraph bipartitioner. Average over 25 runs.

Matrix	p	1D row	1D column	2D Mon	2D fine	2D hybrid
df1001	4	0.06	0.07	0.14	0.11	0.24
	16	0.11	0.20	0.30	0.24	0.50
	64	0.17	0.26	0.44	0.35	0.72
cre_b	4	0.31	1.07	1.36	1.85	3.19
	16	0.50	1.60	2.11	2.75	4.87
	64	0.83	1.99	2.77	3.37	6.16
tbdmatlab	4	1.61	1.30	2.93	6.47	9.28
	16	2.50	2.26	5.07	9.67	15.12
	64	2.88	2.79	6.46	11.56	18.62
nug30	4	4.00	3.89	8.52	8.47	16.65
	16	5.99	6.76	15.05	14.78	29.48
	64	19.65	8.65	22.87	17.84	38.20
c98a	4	9.74	12.83	19.71	50.39	67.54
	16	13.97	25.35	38.83	84.48	112.60
	64	19.87	34.10	52.48	106.40	145.30
tbdlinux	4	16.83	8.96	24.43	75.48	99.05
	16	23.87	14.90	42.14	109.50	149.80
	64	27.66	17.67	53.64	125.30	179.20
stanford	4	3.27	5.77	8.68	28.26	36.64
	16	5.79	11.17	16.07	48.64	62.69
	64	8.27	15.86	22.73	66.05	86.15
polyDFT	4	6.04	6.05	12.16	57.23	68.41
	16	11.85	12.32	23.33	93.86	116.60
	64	15.87	16.41	33.32	120.60	152.10
cage13	4	12.54	12.43	25.19	43.27	67.04
	16	26.57	28.31	44.53	76.49	117.80
	64	36.58	41.24	66.67	102.10	161.70
stanford_berkeley	4	5.84	10.08	15.53	49.34	63.99
	16	12.57	22.37	31.34	102.20	123.80
	64	21.10	36.90	49.00	160.20	186.30
Avg normalized value		0.17	0.19	0.35	0.83	1.16

TABLE 12.7: Time (in ms) of parallel sparse matrix–vector multiplication using the educational program bsp_mv from BSPedupack [11] on the Huygens IBM Power6+ parallel computer in Amsterdam, which consists of 1664 dual-core processors running at 4.7 GHz. Here, p denotes the number of cores used. This program has not been optimized; for comparison, the highly optimized sequential program from [3] takes only 18.08 ms for cage13 and 19.61 ms for stanford_berkeley. The matrices were partitioned by using Mondriaan with PaToH as bipartitioner for $\epsilon = 0.03$.

Matrix	p	1D row	1D column	2D Mon	2D fine	2D hybrid
cage13	1	372.2	372.2	372.2	372.2	372.2
	4	125.0	124.2	120.7	123.0	124.2
	16	39.5	38.8	37.1	36.6	35.9
	64	18.3	17.0	16.1	13.5	13.5
stanford_berkeley	1	552.6	552.6	552.6	552.6	552.6
	4	222.7	171.5	169.3	182.5	171.9
	16	74.4	49.3	71.2	52.9	59.1
	64	49.5	21.2	21.4	22.6	21.6

communication increases and computation decreases, making communication gains more pronounced. Note that partitioning with the 2D hybrid method takes the same time as about 8625 sparse matrix–vector multiplications would take for stanford_berkeley with $p = 64$, where we ignore the fact that we use different hardware in our timings. For the simpler 1D row method, the break-even point is already at 1030 iterations. Still, in most cases the extra time spent on partitioning by using a 2D method would pay off.

12.9 Conclusions and Outlook

The main conclusions of this chapter are

- 2D partitioning is superior to 1D partitioning, as it leads to significantly smaller communication volumes in sparse matrix–vector multiplication.

- Both the Mondriaan approach, based on orthogonal recursive bisection, and the fine-grain approach, based on assigning individual nonzeros to processors, have their strong points and will perform better on some classes of matrices. The hybrid approach combines the best of the two and almost always achieves the lowest communication volume.

- The hybrid approach takes longer to compute, about the same time as the Mondriaan and the fine-grain approach together.

- The fastest 2D partitioner is the 2D Mondriaan method with PaToH as a bipartitioner.

Open questions that deserve further investigation are

- In addition to reducing the communication volume, how can we achieve other objectives as well, such as reducing the total number of messages? Note that all data words to be sent from the same source processor to the same destination processor can be packaged into one message. The number of messages is important in certain types of parallel computation, e.g., where message-passing is employed as a model, in particular if the latency for sending a message is high. The aim would be to keep the number of messages far below the theoretical maximum of $2p(p-1)$ for the 2D case.

- Can we partition the matrix and the vectors simultaneously while modeling the computation and communication requirements exactly? In our approach, the vectors are partitioned after the matrix partitioning, but then it may already be too late to do this well. For instance, if balancing the communication is desired, current methods [28] in practice achieve the best solution that can be obtained after the matrix partitioning, but still there can be severe communication imbalance. Uçar and Aykanat [43] perform the two partitionings at the same time, but do not model the communication cost exactly.

- Can we find a recipe for choosing automatically when to use 2D Mondriaan or fine-grain partitioning, based on a few problem statistics? Such a recipe is given for the choice between 1D row, 1D column, 2D fine-grain, and jagged-like methods in [7], based on $m, n, \mathrm{nz}(A)$ and the average and median nonzero counts of the rows and columns. For instance, for rectangular matrices that are far from square, splitting the largest dimension is prescribed, i.e., 1D column partitioning if $m \ll n$. This choice can be made again at every split. Having such a recipe will save partitioning time, while still keeping the same quality as the hybrid partitioning.

Acknowledgments

This work was supported in part by the DOE grant De-FC02-06ER2775 and by the NSF grants CNS-0643969, OCI-0904809, and OCI-0904802.

We thank the Dutch National Computer Facilities foundation and SARA in Amsterdam for providing access to the supercomputer Huygens.

Bibliography

[1] Barrett, R., Berry, M., Chan, T. F., Demmel, J., Donato, J., Dongarra, J., Eijkhout, V., Pozo, R., Romine, C., and van der Vorst, H., *Templates for the Solution of Linear Systems: Building Blocks for Iterative Methods*, SIAM, Philadelphia, PA, 1994.

[2] Bai, Z., Demmel, J., Dongarra, J., Ruhe, A., and van der Vorst, H., editors, *Templates for the Solution of Algebraic Eigenvalue Problems: A Practical Guide*, SIAM, Philadelphia, PA, 2000.

[3] Yzelman, A. N. and Bisseling, R. H., "Cache-oblivious sparse matrix–vector multiplication by using sparse matrix partitioning methods," *SIAM Journal on Scientific Computing*, Vol. 31, No. 4, 2009, pp. 3128–3154.

[4] Çatalyürek, Ü. V., *Hypergraph Models for Sparse Matrix Partitioning and Reordering*, Ph.D. thesis, Bilkent University, Computer Engineering and Information Science, Nov 1999, Available at `http://www.cs.bilkent.edu.tr/tech-reports/1999/ABSTRACTS.1999.html`.

[5] Çatalyürek, Ü. V. and Aykanat, C., "A Fine-Grain Hypergraph Model for 2D Decomposition of Sparse Matrices," *Proceedings Eighth International Workshop on Solving Irregularly Structured Problems in Parallel (Irregular 2001)*, IEEE Press, Los Alamitos, CA, 2001, p. 118.

[6] Çatalyürek, Ü. V. and Aykanat, C., "A Hypergraph-Partitioning Approach for Coarse-Grain Decomposition," *Proceedings Supercomputing 2001*, ACM Press, New York, 2001, p. 42.

[7] Çatalyürek, Ü. V., Aykanat, C., and Uçar, B., "On Two-Dimensional Sparse Matrix Partitioning: Models, Methods, and a Recipe," *SIAM Journal on Scientific Computing*, Vol. 32, No. 2, 2010, pp. 656–683.

[8] Vastenhouw, B. and Bisseling, R. H., "A two-dimensional data distribution method for parallel sparse matrix-vector multiplication," *SIAM Review*, Vol. 47, No. 1, 2005, pp. 67–95, Available at `http://www.staff.science.uu.nl/~bisse101/Mondriaan`.

[9] Vuduc, R., Demmel, J. W., and Yelick, K. A., "OSKI: A library of automatically tuned sparse matrix kernels," *J. Phys. Conf. Series*, Vol. 16, 2005, pp. 521–530.

[10] Çatalyürek, Ü. V. and Aykanat, C., "Hypergraph-Partitioning-Based Decomposition for Parallel Sparse-Matrix Vector Multiplication," *IEEE Transactions on Parallel and Distributed Systems*, Vol. 10, No. 7, 1999, pp. 673–693.

[11] Bisseling, R. H., *Parallel Scientific Computation: A Structured Approach using BSP and MPI*, Oxford University Press, Oxford, UK, 2004.

[12] Valiant, L. G., "A Bridging Model for Parallel Computation," *Communications of the ACM*, Vol. 33, No. 8, 1990, pp. 103–111.

[13] Caldwell, A. E., Kahng, A. B., and Markov, I. L., "Improved Algorithms for Hypergraph Bipartitioning," *Proceedings Asia and South Pacific Design Automation Conference*, ACM Press, New York, 2000, pp. 661–666.

[14] Karypis, G. and Kumar, V., "Multilevel k-way hypergraph partitioning," *Proceedings 36th ACM/IEEE Conference on Design Automation*, ACM Press, New York, 1999, pp. 343–348.

[15] Çatalyürek, Ü. V. and Aykanat, C., *PaToH: A Multilevel Hypergraph Partitioning Tool, Version 3.0*, Bilkent University, Department of Computer Engineering, Ankara, 06533 Turkey. PaToH is available at `http://bmi.osu.edu/~umit/software.htm`, 1999.

[16] Devine, K. D., Boman, E. G., Heaphy, R., Bisseling, R. H., and Catalyurek, U. V., "Parallel Hypergraph Partitioning for Scientific Computing," *Proceedings IEEE International Parallel and Distributed Processing Symposium 2006*, IEEE Press, 2006.

[17] Trifunović, A. and Knottenbelt, W. J., "Parallel multilevel algorithms for hypergraph partitioning," *Journal of Parallel and Distributed Computing*, Vol. 68, No. 5, 2008, pp. 563–581.

[18] Bui, T. and Jones, C., "A heuristic for reducing fill-in in sparse matrix factorization," *Proceedings Sixth SIAM Conference on Parallel Processing for Scientific Computing*, SIAM, Philadelphia, PA, 1993, pp. 445–452.

[19] Lengauer, T., *Combinatorial algorithms for integrated circuit layout*, John Wiley and Sons, Chichester, UK, 1990.

[20] Bui, T. N. and Jones, C., "Finding good approximate vertex and edge partitions is NP-hard," *Information Processing Letters*, Vol. 42, 1992, pp. 153–159.

[21] Hendrickson, B. and Leland, R., "An Improved Spectral Graph Partitioning Algorithm for Mapping Parallel Computations," *SIAM Journal on Scientific Computing*, Vol. 16, No. 2, 1995, pp. 452–469.

[22] Karypis, G. and Kumar, V., "A Fast and High Quality Multilevel Scheme for Partitioning Irregular Graphs," *SIAM Journal on Scientific Computing*, Vol. 20, No. 1, 1998, pp. 27–58, 359–392.

[23] Karypis, G. and Kumar, V., "Parallel Multilevel k-Way Partitioning Scheme for Irregular Graphs," *SIAM Review*, Vol. 41, No. 2, 1999, pp. 278–300.

[24] Walshaw, C. and Cross, M., "JOSTLE: Parallel Multilevel Graph-Partitioning Software – An Overview," *Mesh Partitioning Techniques and Domain Decomposition Techniques*, edited by F. Magoules, Civil-Comp Ltd., 2007, pp. 27–58.

[25] Pellegrini, F. and Roman, J., "SCOTCH: A Software Package for Static Mapping by Dual Recursive Bipartitioning of Process and Architecture Graphs," *Proceedings High Performance Computing and Networking Europe*, Vol. 1067 of *Lecture Notes in Computer Science*, Springer, Berlin, 1996, pp. 493–498.

[26] Chevalier, C. and Pellegrini, F., "PT-Scotch: a tool for efficient parallel graph ordering," *Parallel Computing*, Vol. 34, No. 6-8, 2008, pp. 318–331.

[27] Hendrickson, B. and Kolda, T. G., "Graph Partitioning Models for Parallel Computing," *Parallel Computing*, Vol. 26, No. 12, 2000, pp. 1519–1534.

[28] Bisseling, R. H. and Meesen, W., "Communication balancing in parallel sparse matrix-vector multiplication," *Electronic Transactions on Numerical Analysis*, Vol. 21, 2005, pp. 47–65, Special Issue on Combinatorial Scientific Computing.

[29] Uçar, B. and Aykanat, C., "Encapsulating Multiple Communication-Cost Metrics in Partitioning Sparse Rectangular Matrices for Parallel Matrix-Vector Multiplies," *SIAM Journal on Scientific Computing*, Vol. 25, No. 6, 2004, pp. 1837–1859.

[30] Saad, Y. and Schultz, M. H., "GMRES: A Generalized Minimal Residual Algorithm for Solving Nonsymmetric Linear Systems," *SIAM Journal on Scientific and Statistical Computing*, Vol. 7, No. 3, 1986, pp. 856–869.

[31] Bisseling, R. H., "Parallel Iterative Solution of Sparse Linear Systems on a Transputer Network," *Parallel Computation*, edited by A. E. Fincham and B. Ford, Vol. 46 of *The Institute of Mathematics and its Applications Conference Series. New Series*, Oxford University Press, Oxford, UK, 1993, pp. 253–271.

[32] Hendrickson, B., Leland, R., and Plimpton, S., "An efficient parallel algorithm for matrix-vector multiplication," *International Journal of High Speed Computing*, Vol. 7, No. 1, 1995, pp. 73–88.

[33] Lewis, J. G., Payne, D. G., and van de Geijn, R. A., "Matrix-Vector Multiplication and Conjugate Gradient Algorithms on Distributed Memory Computers," *Proceedings IEEE Scalable High Performance Computing Conference*, IEEE Press, 1994, pp. 542–550.

[34] Lewis, J. G. and van de Geijn, R. A., "Distributed Memory Matrix-Vector Multiplication and Conjugate Gradient Algorithms," *Proceedings Supercomputing'93*, ACM Press, New York, 1993, pp. 484–492.

[35] Nicol, D. M., "Rectilinear partitioning of irregular data parallel computations," *Journal of Parallel and Distributed Computing*, Vol. 23, No. 2, 1994, pp. 119–134.

[36] Ogielski, A. T. and Aiello, W., "Sparse matrix computations on parallel processor arrays," *SIAM Journal on Scientific Computing*, Vol. 14, No. 3, 1993, pp. 519–530.

[37] Saltz, J. H., Petition, S. G., Berryman, H., and Rifkin, A., "Performance Effects of Irregular Communication Patterns on Massively Parallel Multiprocessors," *Journal of Parallel and Distributed Computing*, Vol. 13, No. 2, 1991, pp. 202–212.

[38] Kutluca, H., Kurc, T. M., and Aykanat, C., "Image-Space Decomposition Algorithms for Sort-First Parallel Volume Rendering of Unstructured Grids," *Journal of Supercomputing*, Vol. 15, 2000, pp. 51–93.

[39] Manne, F. and Sørevik, T., "Partitioning an Array onto a Mesh of Processors," *PARA '96: Proceedings Third International Workshop on Applied Parallel Computing, Industrial Computation and Optimization*, Springer-Verlag, London, UK, 1996, pp. 467–477.

[40] Pınar, A. and Aykanat, C., "Sparse Matrix Decomposition with Optimal Load Balancing," *Proceedings International Conference on High Performance Computing*, Dec 1997, pp. 224–229.

[41] Fiduccia, C. M. and Mattheyses, R. M., "A Linear-Time Heuristic for Improving Network Partitions," *Proceedings of the 19th IEEE Design Automation Conference*, IEEE Press, Los Alamitos, CA, 1982, pp. 175–181.

[42] Kernighan, B. W. and Lin, S., "An efficient heuristic procedure for partitioning graphs," *Bell System Technical Journal*, Vol. 29, 1970, pp. 291–307.

[43] Uçar, B. and Aykanat, C., "Revisiting Hypergraph Models for Sparse Matrix Partitioning," *SIAM Review*, Vol. 49, No. 4, 2007, pp. 595–603.

Chapter 13

Parallel Partitioning, Coloring, and Ordering in Scientific Computing

E. G. Boman

Sandia National Laboratories[1]

Ü. V. Çatalyürek

Ohio State University

C. Chevalier

CEA, DAM, DIF, Arpajon, France

K. D. Devine

Sandia National Laboratories

[1]Sandia is a multi-program laboratory managed and operated by Sandia Corporation, a wholly owned subsidiary of Lockheed Martin Corporation, for the U.S. Department of Energy's National Nuclear Security Administration under contract DE-AC04-94AL85000.

13.1 Introduction

Combinatorial problems often arise within the context of scientific computing. As a result, combinatorial scientific computing (CSC) has become a fruitful research area, enabling the development of efficient and effective numerical simulations for a variety of scientific applications. In this chapter, we focus on three important CSC problems: partitioning (load balancing), graph coloring, and sparse matrix ordering. Through examples, we show how these problems occur in applications. Our target audience is computational scientists who may have little prior knowledge of CSC.

Much software has been written to address these CSC problems. In addition to implementations of CSC algorithms directly in applications, many libraries have been written to provide CSC services. For example, many libraries provide serial partitioning (e.g., Chaco [42, 43], METIS [48], Scotch [56], Pa-ToH [21], Mondriaan [71]) and parallel partitioning and load balancing (e.g., ParMETIS [49], PT-Scotch [57], Jostle [72]). The ColPack [33] library provides a suite of graph coloring algorithms; parallel graph coloring is available in the Parallel Boost Graph Library [37]. Matrix ordering is available in Scotch, PT-Scotch, METIS, and ParMETIS, as well as in the Boost Graph Library [64] and AMD and its variants [1, 28] in SuiteSparse [27].

In this chapter, we focus on the Zoltan toolkit [29, 30, 6], which provides algorithms for these three CSC problems through a common interface shared by all of the algorithms. Its "data structure neutral" design allows it to work with a wide variety of applications. A short tutorial can be found in [31].

In the rest of this chapter, we first discuss partitioning and load-balancing in the context of mesh-based and matrix-based applications. Then, we turn to coloring for a Jacobian matrix. We discuss ordering with focus on fill-reducing ordering for sparse direct solvers. Finally, we describe Zoltan and how it addresses these CSC problems.

13.2 Partitioning and Load Balancing

For efficient parallel computing, it is essential to balance the work (load) among processors (or processes). On many architectures, especially distributed-memory computers, it is also important to minimize the communication among processors. In this section we assume the "owner computes" model, where computations are performed by the processor (core) that owns the data. This gives rise to the *partitioning* problem: How to partition data among processors such that all processors have roughly equal load (work) and communication is minimized? Many algorithms have been proposed for this problem. Mainly, they fall into two categories: geometric, or combinatorial (based on graph or hypergraph models). Zoltan implements several methods of each type.

From the application point of view, it is important to first identify the data objects to be distributed among processors. For mesh-based computations, this can be elements (regions) or nodes (vertices). For matrix-based computations, this is typically matrix rows or columns, but other partitioning of nonzeros could also be used [18, 71]. For molecular dynamics, this is typically particles.

Geometric methods, such as Recursive Coordinate Bisection [3], Recursive Inertial Bisection [65, 66], and Space-Filling Curve Partitioning [73, 58, 55], are simple to use since they just require the coordinates of each data point and not connectivity among them. Implicitly these methods rely on the fact that objects that are closer in space are relevant and may have dependencies to each other; hence communication may incur if they are assigned to different parts. Therefore, they try to partition the space such that objects that are closer in space are assigned to same part. So informally, these methods' main objective is to partition space such that each part have almost equal number of objects to balance the work.

On the other side, connectivity-based methods (i.e., methods that use graph and hypergraph) explicitly model the objects and their dependencies. Vertices represent objects and edges (hyperedges) represent dependencies in a graph (hypergraph). In these methods, the main objective is minimizing the cost function defined on the edges (hyperedges) [41] that corresponds to communication volume, while partitioning the vertices into balanced disjoint subsets. Therefore, since these methods directly aim at minimizing the communication cost, they usually partitions with lower communication cost.

For applications that have only binary dependencies, a graph model with edges only connecting two vertices is usually sufficient. However, in many applications, interactions between objects are more complex and multi-way. In such cases, a hypergraph model with hyperedges connecting more than two vertices provides a more natural, elegant computational model for load balancing [13, 14, 41].

Below, we will examine how to partition meshes and sparse matrices using connectivity based methods.

13.2.1 Partitioning for Mesh Computations

A mesh is a discretization of the computational domain, mainly used to solve partial differential equations (PDEs). We focus on unstructured meshes commonly used in finite element and finite volume methods. The mesh is a set of entities called elements or regions that are geometrically limited by nodes and by the edges between nodes, as shown in Figure 13.1(a).

The goal of mesh partitioning is to divide the elements into p subsets (domains) such that the work per processor is approximately constant and that the communication cost is minimized. Typically, communication cost is approximated by the total volume, though other factors (e.g., number of messages) also play a role.

13.2.1.1 Mesh Models

In the following models, we partition the elements. This is usually preferred because many computations are internal to the elements, so parallel codes require each element to be owned by a single processor. However, for a node-centric code, much of the discussion below still applies by simply switching elements with nodes.

Dual graph models: In these models, the mesh is modeled as a graph in order to use a graph partitioning tool. These approaches consist in looking only at the elements of the mesh and their adjacencies. The more common relation of adjacency for two elements is to share a face and, in this case, we can talk about face-based dual graph (Figure 13.1(c)). It is also possible to have a relation of adjacency between two elements based on the fact that they share a node and, in this case, we speak about node-based dual graph (Figure 13.1(d)).

Performing graph partitioning on these dual graphs consists in dividing the elements in p disjoint subsets (parts, domains) of the same size, while minimizing the cut edges. In the case of a faced-based dual graph, the partitioning will minimize the number of faces shared between elements of different parts.

With a node-based dual graph, we tend to minimize the number of nodes shared between the parts. However, this model is not exact in terms of minimizing the number of nodes, due to the fact that each node correspond to a clique in the graph to partition and, therefore, the edge cut is not directly a measurement of the number of nodes, even if the two are closely correlated.

Hypergraph Model: The number of nodes shared is particularly important for the widely used continuous Galerkin finite elements methods because the communications are on the nodes, not on the faces. As the graph-based approach does not model this accurately, we introduce an hypergraph based approach which can resolve this issue.

This model is perhaps easier to describe first with a bipartite graph, which is a graph with two kinds of vertices. One kind will be the elements of the mesh, the other kind being the nodes of the mesh. Edges are only between two vertices of different types and their meaning in this case is that the element needs this node.

With the hypergraph terminology, this model is defined by the set of vertices which correspond to the elements in the mesh, and a set of hyperedges which correspond to the nodes in the mesh (Figure 13.1(b)). Hypergraph offers mainly two metrics that are relevant in our case: the number of cut hyperedges, which is the number of shared nodes in the mesh, and the connectivity metric (also called λ-1 cut metric), which is sum of the number of proces-

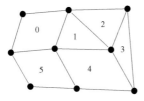

(a) A 2D mesh with 6 elements (3 quadrangles (0,5,4) and 3 triangles(1,2,3)) and 9 nodes.

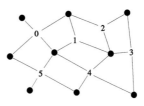

(b) A hypergraph model corresponding to the previous mesh, where mesh elements are vertices in the graph. Note elements and nodes of the mesh are both present.

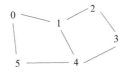

(c) Dual graph defined by the faces. Note that the nodes in the mesh are not explicitly represented.

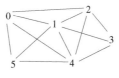

(d) Dual graph defined by the nodes.

FIGURE 13.1: A mesh and four associated models: (a) graph, (b) hypergraph, and (c, d) dual graph.

sors (minus one) for each shared node (cut hyperedge). The later one exactly represents the communication volume for a node-based mesh method.

13.2.1.2 Observations and Conclusions

There is, in general, no direct correlation between the minimization of the number of shared faces and the number of shared nodes implied by the mesh partitioning. Nevertheless, in our experience, the partitioning model does not impact the results much in most cases. Indeed, when all the faces of all the elements are the same type, as the separators are mostly contiguous, it is almost the same to compute a face separation or a node separation of the mesh.

However, for meshes with mixed types of elements, the hypergraph model will be more accurate to model the number of nodes shared by different parts. For example, for meshes with tetrahedrons, hexahedrons, and pyramids, faces can be of different types, like triangle, quadrangle, or any type of polygons, and this information does not appear with the dual graph models.

Moreover, the hypergraph model is intrinsically richer than the dual graph models because it keeps information for both elements and nodes. This will allow more complex objectives for partitioning, like minimizing the number of ghost vertices or elements, or having a better load balancing of the computation and of the memory. Indeed, when a code uses ghosting of elements, the ghost elements are not modeled during the partitioning and, thus, are not balanced. This is an important phenomenon because the number of parts become larger but the size of locally owned elements stay constant, like on massive parallel computers with limited local memory (e.g., BlueGene).

Another point is that the structure of the hypergraph is exactly the same as the one of the mesh, thus it will be very easy to update the hypergraph in order to take into account mesh changes to do dynamic load balancing.

Hypergraphs, thus, offer a more accurate model for mesh partitioning, and this model can be more easily extended to handle the new challenges of petascale computing. Parallel software for hypergraph partitioning is currently available [68, 30] (see Section 13.5.1 for available hypergraph partitioning methods in Zoltan).

13.2.2 Partitioning for Sparse Matrices

In scientific community, sparse matrix partitioning problems have been usually modeled using graphs and hypergraphs [13, 18, 41, 71]. One of the most commonly used and, hence, optimized, kernel is sparse-matrix vector multiplication, and the computation structure in this kernel involves multi-way interactions, as well as non-symmetric dependencies; hence, it can be naturally modeled using hypergraphs [13, 14, 41]. Therefore, during the last decade, hypergraph-based models and methods have gained wide acceptance in the literature for applications like parallelization of sparse matrix-vector multiplications [18, 71, 70]. There are three basic hypergraph models, namely, the column-net model [14], the row-net model [14], and the column-row-net (fine-grain) model [15, 18]. The column-net and row-net models are respectively used for one-dimensional (1D) rowwise and 1D columnwise partitioning. The column-row-net model is used for two-dimensional(2D) nonzero-based fine-grain partitioning. In all these models, partitioning constraint of maintaining a balance on part weights corresponds to maintaining a computational load balance, and the partitioning objective of minimizing the cutsize exactly corresponds to minimizing the total communication volume.

There are other methods such as the orthogonal recursive bisection [71], the jagged-like [18], the checkerboard [16], and the hybrid [4] (see Chapter 12 for details) partitioning methods, which utilize various combination of the basic hypergraph models to achieve different 2D partitionings.

13.3 Coloring

A distance-k coloring of a graph is an assignment of colors and integer number, to vertices, such that any two vertices connected by a path consisting of at most k edges receive different colors, and the goal of the associated problem is to use as few colors as possible.

Finding a distance-k coloring using the fewest colors is an NP-hard problem [53], but greedy heuristics are effective in practice, as they run fast and provide solutions of acceptable quality [24, 34]. They are, however, inherently sequential and, thus, challenging to parallelize.

Distance-1 coloring is used in numerous applications, including in discovering concurrency in parallel scientific computing [46, 47, 62]. Distance-2 coloring finds applications in other areas, such as in channel assignment problems in radio networks [50, 51] as well as in the efficient computation of sparse Jacobian and Hessian matrices, whether derivatives are calculated using automatic differentiation or estimated using finite differencing [34]. Below we outline how coloring is relevant to the finite difference approach and, then, briefly present parallel coloring efforts for distance-1 and distance-2 coloring problems and their variants.

13.3.1 Jacobians by Finite Differences

Typically, nonlinear problems are solved by a sequence of linear systems, e.g., with Newton's method. Examples are time-dependent PDEs used in weather and climate simulation. A significant amount of work (and memory) can be involved in computing the Jacobian matrix $J(x)$ for a smooth, nonlinear function $f(x)$. When f is highly complex, for example, it may involve solving a PDE, then forming exact (analytic) derivatives is often impossible.

Two common options in this case is to estimate J by finite differences or to compute it exactly using automatic differentiation (AD). The AD approach typically involves using AD software and is described in other chapters.

Suppose we need to explicitly form the Jacobian J. This is typically the case if one plans to solve the linear system $Jx = b$ using a direct solver. We note that using iterative methods like Jacobi-Free Newton-Krylov (JFNK), one only needs J implicitly, e.g., a black-box to compute matrix-vector products. By using finite differences, we can form the kth column of J by the formula $J_k(x) = J(x)e_k \approx (f(x + \epsilon e_k) - f(x))/\epsilon$, where e_k is the kth unit vector and ϵ is a small real number. Since $f(x)$ is typically already known, this costs one function evaluation. Thus, n function evalutions are needed to form the whole $n \times n$ Jacobian this way.

Fortunately, for sparse Jacobians there is a better way. The key idea is to form several columns of the Jacobian at once [25]. Columns that are *structurally orthogonal*, that is, their sparsity patterns do not overlap, can be computed in a single step. Let d be a binary vector with ones in a set of indices corresponding to a group of structurally orthogonal columns. Then, we can estimate Jd using a single function evaluation along the direction d. The process to form all of J works as follows:

1. Derive the sparsity pattern of J.

2. Compute independent groups of columns of J (e.g., by coloring).

3. Compute the compressed Jacobian: $\hat{J} = JD$, where D is a binary matrix, corresponding to the grouping of the columns.

4. Uncompress $\hat{J}$ to recover J.

We wish to minimize the number of columns in the compressed Jacobian, which corresponds to the number of function evaluations. This combinatorial problem is a coloring problem (see [35]). One approach [23] is to form the column intersection graph, G_C, and color the vertices such that no two adjacent vertices have the same color (distance-1 coloring). Here, in a column intersection graph, each column is represented by a vertex, and vertices are connected by an edge if the respective columns have a nonzero in the same row. A second approach, which uses less memory, is to compute a partial distance-2 coloring directly on the structure of J (see [35] for details). Both approaches can reduce the number of function evaluations by an order of magnitude (or more) compared to the naive method (one evaluation per column).

13.3.2 Preconditioning for Iterative Solvers

In the previous section, we assumed we required an explicit Jacobian to be used in a direct solver. However, for large problems iterative solvers may be more efficient. In this case, J is only needed implicitly as an operator.

Nevertheless, finite differencing and coloring can be very useful to form a pre-conditioner for J. One example is the parallel ocean code POP (part of the Community Climate System Model), which uses a JFNK method. Still, to form a preconditioner, an explicit Jacobian is needed as input. An interesting point is that an approximation to the Jacobian is often sufficient in this case, since the preconditioner only need solve the problem inexactly. Thus, we can form $\tilde{J} \approx J$, where $\tilde{J}$ contains only a subset of the entries in the true Jacobian J. Generally, $\tilde{J}$ requires less colors than J and, therefore, fewer function evaluations.

13.3.3   Parallel Coloring

In parallel applications where a distance-k coloring is needed, the graph is either already partitioned and mapped or needs to be partitioned and mapped onto the processors of a distributed-memory parallel machine. Under such circumstances, gathering the graph on one processor to perform the coloring sequentially is prohibitively time consuming or infeasible due to memory constraints. Hence the graph needs to be colored in parallel.

In a recent work, Bozdağ et al. [9] presented a computational framework that provides efficient parallelization of the distance-1 greedy coloring heuristics on distributed memory machines. The framework is an iterative, data-parallel, algorithmic scheme that proceeds in two-phased rounds. In the first phase of each round, processors concurrently color the vertices assigned to them in a speculative manner, communicating at a course granularity using bulk-syncronous communication. In the second phase, processors concurrently check the validity of the colors assigned to their respective vertices and identify a set of vertices that needs to be recolored in the next round to resolve any detected inconsistencies. The scheme terminates when every vertex has been colored correctly.

In a more recent work [8], this framework is extended to provide distance-2 coloring. In addition to speculative coloring, processors proactively help coloring of the vertices colored in the neighbour processors by constructing partial *forbidden-colors* lists that are later combined and used by the processor that owns the vertex to be colored.

13.4   Ordering

The *ordering* of data can affect performance in several ways. First, a re-ordering of data in an application can give better locality and, thus, improved performance due to better use of cache or fast memory. This has become increasingly important as memory hierarchies have grown deeper and become more complex. However, we will rather focus on a more specific problem: How to reorder a sparse matrix such as to minimize fill in a sparse direct solver?

13.4.1   Fill-Reducing Ordering of Sparse Matrices

In many scientific computing applications, it is necessary to solve the system $Ax = b$. Either direct or iterative methods can be used. When A is ill-conditioned and no good preconditioner is known, direct solvers are preferred since iterative solvers will converge slowly or not at all. The work and memory required to solve $Ax = b$ depends on the fill in the factorization. A standard technique is, therefore, to permute (reorder) the matrix A such that one solves the equivalent system $(PAQ)(Q^T x) = Pb$, where P and Q are permutation matrices. This reordering is usually performed as a preprocessing step, and depends only on the nonzero structure not on the numerical values. In the symmetric positive definite case, a symmetric permutation is used to preserve symmetry and definiteness, i.e., $Q = P^T$. This gives the symmetric ordering problem:

Problem 13.4.1 (Symmetric Ordering) *Given a symmetric positive definite, sparse matrix A, find a permutation matrix P that minimizes the fill in the Cholesky factor L where $PAP^T = LL^T$.*

In the nonsymmetric case, numerical pivoting is required for stability during the factorization phase. Generally, only a column permutation is, therefore, performed in the setup phase to reduce fill.

Problem 13.4.2 (Nonsymmetric Ordering) *Given a sparse matrix A, find a permutation matrix Q that minimizes the fill in the LU factors of PAQ, where P is determined by partial pivoting.*

There are two broad categories of fill-reducing ordering methods: *minimum degree* [67] and *nested dissection* [36] methods. The minimum-degree class takes a local perspective and the algorithms are greedy. The algorithms are inherently sequential but fast to compute. The nested dissection methods, on the other hand, take a global perspective. Typically, recursive bisection is used, which is suited for parallel processing. Nested dissection methods usually give less fill for large problems, and are better suited for parallel factorization. In practice, hybrid methods that combine nested dissection with a local ordering are most effective.

13.4.2   Symmetric Positive Definite Case

Most of the state-of-the-art symmetric ordering tools [22, 40, 48] are hybrid methods that combine multilevel graph partitioning for computing nested dissections with a variant of minimum degree for local ordering. The main idea of nested dissection is as follows. Consider a partitioning of vertices (V) into three sets: V_1, V_2 and V_S, such that the removal of V_S, called *separator*, decouples two parts V_1 and V_2. If we order the vertices of V_S after the vertices of V_1 and V_2, i.e., if the elimination process starts with vertices of either V_1 or V_2, no fill will occur between the vertices of V_1 and V_2. Furthermore, the elimination process in V_1 and V_2 are independent from each other and their elimination only incurs fill to themselves and V_S. In the nested dissection, the ordering of the vertices of V_1 and V_2 is simply computed by applying the same process recursively. In the hybrid methods, recursion is stopped when a part becomes smaller than predetermined size, and minimum degree-based local ordering is used to order the vertices in that part.

Çatalyürek et al. [20, 17] proposed use of hypergraphs for ordering sparse matrices. Use of hypergraphs in this context has two advantages. First, although multilevel paradigm is a perfect fit in partitioning, in the ordering context it has a small flaw: A vertex separator found in the coarser graphs may not be minimal when separator projected to a finer graph in the multilevel hierarchy. Use of hypergraphs eliviates this problem. Second, in many applications, like the solution of linear programming problems using an interior point method, symmetric matrices are in the form of $M = AA^T$ (or $M = AD^2A^T$) where A is a rectangular matrix. In such cases, one can achieve better and faster orderings for M by simply running nested dissection on A using hypergraph models [14, 2].

13.4.3 Unsymmetric Case

In unsymmetric direct solvers, partial pivoting is required for numerical stability and to avoid breakdown. Partial pivoting is usually performed by looking down a column for a large entry and then swapping rows. This defines a row permutation but it is determined during the numerical factorization. Therefore, fill-reducing ordering has focused on column ordering. Consider $\hat{A} = PAQ$. The column permutation Q can be computed in a preprocessing step based on the structure of A. Note that P is not known at this point, so Q should reduce fill for *any* P.

There are two popular approaches to column ordering. First, one can simply symmetrize A and compute a nested dissection ordering for either $A + A^T$ or $A^T A$. This is attractive because one can use the same algorithms and software as in the symmetric case. However, ordering based on $A + A^T$ does not necessarily give good quality ordering for A. Ordering based on $A^T A$ generally works better and has a theoretical foundation because the fill for LU factors of A is contained within the Cholesky factor of $A^T A$. Unfortunately, $A^T A$ is often much denser than A and takes too much memory to form explicitly. The second approach is, therefore, to order the columns of A to reduce fill in the Cholesky factor of $A^T A$ but without forming $A^T A$. Several versions of minimum degree have been adapted in this way; the most popular is COLAMD [26]. A limitation of COLAMD is that it is inherently sequential and not suited for parallel computations.

A third approach, the HUND method, was recently proposed by Grigori et al. [38]. The idea is to use hypergraphs to perform a unsymmetric version of nested dissection, then switch to CCOLAMD on small subproblems. Hypergraph partitioning is used to obtain a singly bordered block diagonal (SBBD) form. It is shown in [38] that the fill in LU factorization is contained within the blocks in the SBBD form, even with partial pivoting. Experiments with serial LU factorization show that HUND compares favorably to other methods both in terms of fill and flops. Perhaps, the greatest advantage of HUND, though, is that it can both be computed in parallel and produces orderings well suited for parallel factorization.

We remark that several parallel direct solvers (e.g., MUMPS) currently cannot perform column reordering but only apply symmetric permutations. In this case, HUND ordering can be applied symmetrically. The row reordering is not necessary but also does not destroy sparsity. In fact, with classical partial pivoting, the row permutation does not matter but with threshold pivoting the row permutation will make a difference (though we expect the effect to be small).

13.5   The Zoltan Toolkit for CSC

The Zoltan toolkit is a collection of algorithms for parallel partitioning, load balancing, matrix ordering, and coloring. Most algorithms are implemented natively within Zoltan, but Zoltan also provides the option to use several third-party libraries, e.g., ParMETIS or PT-Scotch. All algorithms are implemented in parallel using the MPI Message Passing Interface [54]. Algorithms use a common application interface, so users can access partitioning, matrix ordering, and coloring algorithms through a single interface. Zoltan's design is "data-structure neutral"; that is, an application does not have to use or construct any particular data structure in order to use Zoltan. Instead, Zoltan accesses application data through callback functions that provide information such as unique identifiers for objects to be partitioned, ordered, or colored; geometric coordinates of objects; and dependencies between objects (i.e., graph or hypergraph edges). Typically, these callback functions are easy for an application developer to write, and only 4-6 callbacks are required for any particular algorithm. More details on the Zoltan interface are provided in the Zoltan tutorial [31] and User's Guide [6].

13.5.1   Zoltan Partitioning

Zoltan began as a dynamic load-balancing library for parallel, unstructured, and/or adaptive applications. It includes a suite of native partitioning algorithms as well as interfaces to the popular partitioning libraries PT-

Scotch [57], ParMETIS [49], and PaToH [21]. Users can select different partitioning algorithms through a simple parameter setting, making it easy for users to compare different methods in their applications.

Zoltan's suite of partitioning algorithms includes geometric, hypergraph-based, and graph-based methods. In addition, Zoltan provides utilities for moving data from an old distribution to the new one; use of these utilities is optional and separate from the partitioning algorithm.

Geometric methods partition data based on their geometric locality. Objects that are physically close to each other are assigned to the same process. Geometric methods in Zoltan include Recursive Coordinate Bisection [3], Recursive Inertial Bisection [65, 66], and Space-Filling Curve partitioning [73, 58, 55]. Zoltan callback functions that return the dimensionality of the domain and coordinates of objects are required for geometric partitioning. Geometric methods are fast and easy to use, making them appropriate for dynamic applications requiring frequent repartitioning. Because they preserve geometric locality of objects, they are ideal for contact detection and particle simulations. However, because they do not explicitly model communication, they can produce partitions with relatively high communication costs.

A unique feature of Zoltan is its parallel hypergraph partitioner, PHG [30]. PHG is a fully parallel multilevel partitioner that uses the hypergraph model. Combined with a repartitioning model [19] that reduces total communication costs (application communication costs plus data movement costs that arise in redistributing data), it is effective for redistributing data during dynamic load balancing.

Zoltan provides graph-based partitioning natively through its parallel hypergraph partitioner, as well as through interfaces to PT-Scotch and ParMETIS. In general, graph partitioning produces lower quality partitions (with respect to communication costs) than hypergraph partitioning, but graph partitioning is faster to compute. Zoltan's graph and hypergraph partitioners can use either hypergraph-based callbacks (that return, in effect, a compressed-sparse-row representation of the graph or hypergraph) or graph-based callbacks (that return edge lists for each object).

Applications that use Zoltan for partitioning and load-balancing include the finite element simulation framework SIERRA [32], the ACME contact detection library [11], the circuit simulator Xyce [45], the Maxwell particle-in-cell code Pic3P [12], and the CFD code PHASTA [63]. Zoltan has been used to partition meshes with more than one billion elements on more than 32 thousand processors.

13.5.2 Zoltan Coloring

Over time, support for matrix ordering and coloring was added to Zoltan. Zoltan's parallel coloring algorithms are based on the framework described in [9]. The coloring algorithms use the same graph-based callback functions used in Zoltan's graph and hypergraph partitioners. The output of the coloring algorithms is a vector of integers, corresponding to colors assigned to the objects. Zoltan currently supports several common coloring problems: distance-1 [5, 9], distance-2 [10, 8], and partial distance-2 [8] coloring. Coloring problems arising from Hessian computations (star and acyclic) are planned for the future.

A unique feature of Zoltan is the ability to compute coloring in parallel on distributed memory platforms. The parallel coloring algorithms in Zoltan have been shown scalable up to thousands of processors [7]. Zoltan's coloring algorithms are used by the NOX non-linear solver package in the Trilinos solver toolkit [44].

13.5.3   Zoltan Matrix Ordering

Zoltan provides ordering algorithms to improve locality in explicit methods or reduce fill in direct linear solvers. The locality-enhancing ordering is based on space-filling curves; the suggested numbering and storage of objects within a processor or core is determined by the objects' position along a space-filling curve through the objects. This ordering uses the same callback functions returning coordinates as the geometric partitioning algorithms. This ordering has been shown to reduce execution time in the parallel FV-MAS climate modeling code [61] by 20%-25% on a mesh with 2.6 million cells. Future work includes implementation of the graph-based Reverse Cuthill McKee ordering algorithm.

Zoltan's fill-reducing orderings are intended for large matrices in direct solvers and Cholesky factorizations. The focus in Zoltan is on nested-dissection methods [36], since nested-dissection methods generally work best on large matrices. Classic nested dissection for symmetric problems is provided through interfaces to the third-party libraries PT-Scotch [57, 22] and ParMETIS [49]. For nonsymmetric problems, Zoltan contains a native parallel implementation of HUND. The HUND algorithm can be applied to symmetric matrices, but it works best for nonsymmetric ones. Zoltan's fill-reducing orderings require the graph or hypergraph callbacks as in graph and hypergraph partitioning.

Zoltan returns a permutation vector that can be provided as input to the application's direct solver. Most direct solvers allow symmetric permutations, but only a few allow nonsymmetric permutations ($P \neq Q$). Zoltan ordering has been tested with several versions of the direct solver library SuperLU [52, 39].

13.6   Conclusions and Future Work

We have described three important problems in CSC (partitioning, coloring, and ordering). The Zoltan toolkit provides a common interface to solve all of these problems. It is the only software that can do so in parallel for large-scale problems. There is also a matrix-based front end to Zoltan in the Trilinos software, called Isorropia.

There are of course other CSC problems that could be added to Zoltan in the future. A natural candidate is *matching*, where one wants to match pairs of vertices to maximize some objective. This problem has applications to sparse matrix computations (see Chapter 2).

Although Zoltan provides a rich set of algorithms, the problems solved are often simplified abstractions of real problems. For example, most partitioning algorithms focus on minimizing the communication *volume*, but this is only a rough approximation of the communication cost on a parallel system. In reality, factors like the number of messages and the maximum communication load over all processors, are also important. Although some progress has been made in these directions [69], much work remains to be done (also see Chapter 12 for a similar discussion on sparse matrix partitioning).

Our current models (graph and hypergraph partitioning) assume that the objective to minimize is a simple linear sum of edge cuts. However, there are situations where such models are insufficient. One example is to balance both computation and communication. Another example is to perform domain decomposition such that each subdomain has approximately the same memory requirement after the matrix on each subdomain is factored (complete or incomplete factorization). Typically, one cannot evaluate the memory usage until after the symbolical factorization phase (certainly not during the partitioning). An iterative approach to such *complex objectives* was proposed in [60, 59].

Bibliography

[1] P. R. Amestoy, T. A. Davis, and I. S. Duff. Algorithm 837: Amd, an approximate minimum degree ordering algorithm. *ACM Trans. Math. Softw.*, 30(3):381–388, 2004.

[2] C. Aykanat, A. Pınar, and Ü. V. Çatalyürek. Permuting sparse rectangular matrices into block-diagonal form. *SIAM Journal on Scientific Computing*, 26(6):1860–1879, 2004.

[3] M. J. Berger and S. H. Bokhari. A partitioning strategy for nonuniform problems on multiprocessors. *IEEE Trans. Computers*, C-36(5):570–580, 1987.

[4] R. Bisseling, T. van Leeuwen, and U. V. Çatalyürek. Combinatorial problems in high-performance computing: Partitioning. In U. Naumann, O. Schenk, H. D. Simon, and S. Toledo, editors, *Abstract, Combinatorial Scientific Computing*, number 09061 in Dagstuhl Seminar Proceedings,

Dagstuhl, Germany, 2009. Schloss Dagstuhl - Leibniz-Zentrum fuer Informatik.

[5] E. G. Boman, D. Bozdag, U. Çatalyürek, A. Gebremedhin, and F. Manne. A scalable parallel graph coloring algorithm for distributed memory computers. In *Euro-Par 2005*, volume 3648 of *LNCS*, pages 241–251. Springer, 2005.

[6] E. Boman, K. Devine, R. Heaphy, B. Hendrickson, V. Leung, L. A. Riesen, C. Vaughan, U. Çatalyürek, D. Bozdag, W. Mitchell, and J. Teresco. *Zoltan 3.0: Parallel Partitioning, Load Balancing, and Data-Management Services; User's Guide*. Sandia National Laboratories, Albuquerque, NM, 2007. Tech. Report SAND2007-4748W; http://www.cs.sandia.gov/Zoltan/ug_html/ug.html.

[7] D. Bozdag, U. Çatalyürek, F. Dobrian, A. Gebremedhin, M. Halappanavar, and A. Pothen. Poster abstract - combinatorial algorithms enabling scientific computing: Petascale algorithms for graph coloring and matching. In *Proc. of 21st Supercomputing Conference*, 2009.

[8] D. Bozdag, U. V. Çatalyürek, A. H. Gebremedhin, F. Manne, E. G. Boman, and F. Ozguner. Distributed-memory parallel algorithms for distance-2 coloring and their application to derivative computation. *SIAM Journal of Scientific Computing*, 32(4):2418–2446, 2010.

[9] D. Bozdag, A. H. Gebremedhin, F. Manne, E. G. Boman, and U. V. Çatalyürek. A framework for scalable greedy coloring on distributed-memory parallel computers. *J. Parallel and Distributed Computing*, 68(4):515–535, 2008.

[10] D. Bozdağ, U. Çatalyürek, A. H. Gebremedhin, F. Manne, E. G. Boman, and F. Özgüner. A parallel distance-2 graph coloring algorithm for distributed memory computers. In *The 2005 International Conference on High Performance Computing and Communications (HPCC-05)*, 2005.

[11] K. H. Brown, M. W. Glass, A. S. Gullerud, M. W. Heinstein, R. E. Jones, and T. E. Voth. *ACME Algorithms for Contact in a Multiphysics Environment, API Version 1.3*. Sandia National Laboratories, Albuquerque, NM, 2003. Tech. Report SAND2003-1470.

[12] A. Candel, A. Kabel, L. Lee, Z. Li, C. Ng, G. Schussman, K. Ko, I. Ben-Zvii, and J. Kewisch. Parallel 3D finite element particle-in-cell simulations with Pic3P. Technical Report SLAC-PUB-13671, Stanford Linear Accelerator Center, Menlo Park, CA, 2009.

[13] Ü. Çatalyürek and C. Aykanat. Decomposing irregularly sparse matrices for parallel matrix-vector multiplications. *Lecture Notes in Computer Science*, 1117:75–86, 1996.

[14] Ü. Çatalyürek and C. Aykanat. Hypergraph-partitioning-based decomposition for parallel sparse-matrix vector multiplication. *IEEE Trans. Parallel Dist. Systems*, 10(7):673–693, 1999.

[15] Ü. Çatalyürek and C. Aykanat. A fine-grain hypergraph model for 2d decomposition of sparse matrices. In *Proc. IPDPS 8th Int'l Workshop on Solving Irregularly Structured Problems in Parallel (Irregular 2001)*, April 2001.

[16] Ü. Çatalyürek and C. Aykanat. A hypergraph-partitioning approach for coarse-grain decomposition. In *Proc. Supercomputing 2001*. ACM, 2001.

[17] U. V. Çatalyürek, C. Aykanat, and E. Kayaaslan. Hypergraph partitioning-based fill-reducing ordering. Technical Report OSUBMI-TR-2009-n02 and BU-CE-0904, The Ohio State University, Department of Biomedical Informatics and Bilkent University, Computer Engineering Department, 2009. submitted for publication.

[18] Ü. V. Çatalyürek, C. Aykanat, and B. Uçar. On two-dimensional sparse matrix partitioning: Models, methods, and a recipe. *SIAM Journal on Scientific Computing*, 32(2):656–683, 2010.

[19] U. V. Çatalyürek, E. G. Boman, K. D. Devine, D. Bozdag, R. T. Heaphy, and L. A. Riesen. Hypergraph-based dynamic load balancing for adaptive scientific computations. In *Proc. of 21th International Parallel and Distributed Processing Symposium (IPDPS'07)*. IEEE, 2007.

[20] Ü. V. Çatalyürek. *Hypergraph Models for Sparse Matrix Partitioning and Reordering*. PhD thesis, Bilkent University, Computer Engineering and Information Science, Nov 1999.

[21] Ü. V. Çatalyürek and C. Aykanat. *PaToH: A Multilevel Hypergraph Partitioning Tool, Version 3.0*. Bilkent University, Department of Computer Engineering, Ankara, 06533 Turkey. PaToH is available at `http://bmi.osu.edu/~umit/software.htm`, 1999.

[22] C. Chevalier and F. Pellegrini. PT-SCOTCH: A tool for efficient parallel graph ordering. *Parallel Computing*, 34(6–8):318–331, 2008.

[23] T. F. Coleman and J. J. Moré. Estimation of sparse Jacobian matrices and graph coloring. *SIAM J. Numer. Anal.*, 20(1):187–209, 1983.

[24] T. F. Coleman and J. J. Moré. Estimation of sparse Jacobian matrices and graph coloring problems. *SIAM J. Numer. Anal.*, 1(20):187–209, 1983.

[25] A. R. Curtis, M. J. D. Powell, and J. K. Reid. On the estimation of sparse Jacobian matrices. *J. Inst. Math. Appl.*, 13:117–119, 1974.

[26] T. A. Davis, J. R. Gilbert, S. I. Larimore, and E. G. Ng. A column approximate minimum degree ordering algorithm. *ACM Trans. on Mathematical Software*, 30(3):353–376, 2004.

[27] T. A. Davis. SuiteSparse home page. http://www.cise.ufl.edu/research/ sparse/SuiteSparse/.

[28] T. A. Davis, J. R. Gilbert, S. I. Larimore, and E. G. Ng. Algorithm 836: Colamd, a column approximate minimum degree ordering algorithm. *ACM Trans. Math. Softw.*, 30(3):377–380, 2004.

[29] K. Devine, E. Boman, R. Heaphy, B. Hendrickson, and C. Vaughan. Zoltan data management services for parallel dynamic applications. *Computing in Science and Engineering*, 4(2):90–97, 2002.

[30] K. D. Devine, E. G. Boman, R. T. Heaphy, R. H. Bisseling, and U. V. Çatalyürek. Parallel hypergraph partitioning for scientific computing. In *Proc. of 20th International Parallel and Distributed Processing Symposium (IPDPS'06)*. IEEE, 2006.

[31] K. D. Devine, E. G. Boman, L. A. Riesen, U. V. Çatalyürek, and C. Chevalier. Getting started with zoltan: A short tutorial. In U. Naumann, O. Schenk, H. D. Simon, and S. Toledo, editors, *Combinatorial Scientific Computing*, number 09061 in Dagstuhl Seminar Proceedings, page 10 pages, Dagstuhl, Germany, 2009. Schloss Dagstuhl - Leibniz-Zentrum fuer Informatik.

[32] H. Carter Edwards. SIERRA framework version 3: Core services theory and design. Technical Report SAND2002-3616, Sandia National Laboratories, Albuquerque, NM, 2002.

[33] A. Gebremedhin, A.Tarafdar, A. Walther, and A. Pothen. ColPack home page. http://www.cscapes.org/coloringpage/software.htm.

[34] A. H. Gebremedhin, F. Manne, and A. Pothen. What color is your Jacobian? Graph coloring for computing derivatives. *SIAM Rev.*, 47(4):629–705, 2005.

[35] A. H. Gebremedhin, F. Manne, and A. Pothen. What color is your Jacobian? Graph coloring for computing derivatives. *SIAM Review*, 47(4):629–705, 2005.

[36] J. A. George. Nested dissection of a regular finite element mesh. *SIAM Journal on Numerical Analysis*, 10:345–363, 1973.

[37] D. Gregor and A. Lumsdaine. The parallel BGL: A generic library for distributed graph computations. *Parallel Object-Oriented Scientific Computing*, July 2005.

[38] L. Grigori, E. G. Boman, S. Donfack, and T. A. Davis. Hypergraph-based unsymmetric nested dissection ordering for sparse LU factorization. *SIAM J. Sci. Comp.*, 32(6):3426–3446, 2010.

[39] L. Grigori, J. W. Demmel, and X. S. Li. Parallel symbolic factorization for sparse LU with static pivoting. *SIAM J. Scientific Computing*, 29(3):1289–1314, 2007.

[40] B. Hendrickson and E. Rothberg. Effective sparse matrix ordering: just around the bend. In *Proc. Eighth SIAM Conf. Parallel Processing for Scientific Computing*, 1997.

[41] B. Hendrickson and T. G. Kolda. Graph partitioning models for parallel computing. *Parallel Computing*, 26:1519–1534, 2000.

[42] B. Hendrickson and R. Leland. The chaco user's guide, version 1.0. Technical Report SAND93-2339, Sandia National Laboratories, 1993.

[43] B. Hendrickson and R. Leland. The chaco user's guide, version 2.0. Technical Report SAND94-2692, Sandia National Laboratories, 1994.

[44] M. Heroux, R. Bartlett, V. Howle, R. Hoekstra, J. Hu, T. Kolda, R. Lehoucq, K. Long, R. Pawlowski, E. Phipps, A. Salinger, H. Thornquist, R. Tuminaro, J. Willenbring, A. Williams, and K. S. Stanley. An overview of the Trilinos project. *ACM TOMS*, 31(3):397–423, 2005.

[45] S. A. Hutchinson, E. R. Keiter, R. J. Hoekstra, L. J. Waters, T. V. Russo, E. L. Rankin, and S. D. Wix. *Xyce Parallel Electronic Simulator User's Guide, Version 1.0*. Sandia National Laboratories, Albuquerque, NM, 2002. Tech. Report SAND2002-3790.

[46] D. Hysom and A. Pothen. A scalable parallel algorithm for incomplete factor preconditioning. *SIAM J. Sci. Comput.*, 22:2194–2215, 2001.

[47] M. T. Jones and P. E. Plassmann. Scalable iterative solution of sparse linear systems. *Parallel Computing*, 20(5):753–773, 1994.

[48] G. Karypis and V. Kumar. METIS: Unstructured graph partitioning and sparse matrix ordering system. Technical report, Dept. Computer Science, University of Minnesota, 1995. *http://glaros.dtc.umn.edu/ gkhome/metis/metis/download*.

[49] G. Karypis, K. Schloegel, and V. Kumar. ParMETIS: Parallel graph partitioning and sparse matrix ordering library, version 3.1. Technical report, Dept. Computer Science, University of Minnesota, 2003. `http:// www-users.cs.umn.edu/~karypis/metis/parmetis/download.html`.

[50] S. O. Krumke, M. V. Marathe, and S. S. Ravi. Models and approximation algorithms for channel assignment in radio networks. *Wireless Networks*, 7(6):575–584, 2001.

[51] V. S. A. Kumar, M. V. Marathe, S. Parthasarathy, and A. Srinivasan. End-to-end packet-scheduling in wireless ad-hoc networks. In *SODA 2004: Proceedings of the Fifteenth Annual ACM-SIAM Symposium on Discrete Algorithms*, pages 1021–1030, 2004.

[52] X. S. Li and J. W. Demmel. SuperLU_DIST: A scalable distributed-memory sparse direct solver for unsymmetric linear systems. *ACM Trans. Mathematical Software*, 29(2):110–140, June 2003.

[53] S. T. McCormick. Optimal approximation of sparse Hessians and its equivalence to a graph coloring problem. *Math. Programming*, 26:153 – 171, 1983.

[54] Message Passing Interface Forum, University of Tennessee, Knoxville, Tennessee. *MPI: A Message-Passing Interface Standard*, 1995. http://www.mpi-forum.org/docs/mpi-11-html/mpi-report.html.

[55] A. Patra and J. Tinsley Oden. Problem decomposition for adaptive *hp* finite element methods. *J. Computing Systems in Engg.*, 6(2), 1995.

[56] F. Pelligrini. SCOTCH 3.4 user's guide. Research Rep. RR-1264-01, LaBRI, Nov. 2001.

[57] F. Pelligrini. PT-SCOTCH 5.1 user's guide. Research rep., LaBRI, 2008.

[58] J. R. Pilkington and S. B. Baden. Partitioning with spacefilling curves. CSE Technical Report CS94–349, Dept. Computer Science and Engineering, University of California, San Diego, CA, 1994.

[59] A. Pınar and B. Hendrickson. Communication support for adaptive computation. In *Proc. 10th SIAM Conf. Parallel Processing for Scientific Computing*, Portsmouth, VA, March 2001.

[60] A. Pınar and B. Hendrickson. Partitioning for complex objectives. In *Proc. Irregular 2001*, 2001.

[61] T. D. Ringler, J. Thuburn, J. B. Klemp, and W. C. Skamarock. A unified approach to energy conservation and potential vorticity dynam ics for arbitrarily-structured c-grids. *Journal of Computational Physics*, 229(9):3065–3090, 2010.

[62] Y. Saad. ILUM: A multi-elimination ILU preconditioner for general sparse matrices. *SIAM J. Sci. Comput.*, 17:830–847, 1996.

[63] O. Sahni, M. Zhou, M. S. Shephard, and K. E. Jansen. Scalable implicit finite element solver for massively parallel processing with demonstration to 160k cores. In *Proceedings of the 2009 ACM/IEEE Conference on Supercomputing*, November 2009. 2009 ACM Gordon Bell Finalist.

[64] J. G. Siek, L.-Q. Lee, and A. Lumsdaine. *The Boost Graph Library: User Guide and Reference Manual.* Addison-Wesley Professional, 2001.

[65] H. D. Simon. Partitioning of unstructured problems for parallel processing. *Comp. Sys. Engng.*, 2:135–148, 1991.

[66] V. E. Taylor and B. Nour-Omid. A study of the factorization fill-in for a parallel implementation of the finite element method. *Int. J. Numer. Meth. Engng.*, 37:3809–3823, 1994.

[67] W. F. Tinney and J. W. Walker. Direct solution of sparse network equations by optimally ordered triangular factorization. In *Proc. IEEE*, volume 55, pages 1801–1809, 1967.

[68] A. Trifunovic and W. J. Knottenbelt. Parkway 2.0: A parallel multilevel hypergraph partitioning tool. In *Proc. 19th International Symposium on Computer and Information Sciences (ISCIS 2004)*, volume 3280 of *LNCS*, pages 789–800. Springer, 2004.

[69] B. Uçar and C. Aykanat. Encapsulating multiple communication-cost metrics in partitioning sparse rectangular matrices for parallel matrix-vector multiplies. *SIAM Journal on Scientific Computing*, 25(6):1837–1859, 2004.

[70] B. Uçar, Ü. V. Çatalyürek, and C. Aykanat. A matrix partitioning interface to patoh in matlab. *Parallel Computing*, 36(5-6):254 – 272, 2010.

[71] B. Vastenhouw and R. H. Bisseling. A two-dimensional data distribution method for parallel sparse matrix-vector multiplication. *SIAM Review*, 47(1):67–95, 2005.

[72] C. Walshaw. *The Parallel JOSTLE Library User's Guide, Version 3.0.* University of Greenwich, London, UK, 2002.

[73] M. S. Warren and J. K. Salmon. A parallel hashed oct-tree n-body algorithm. In *Proc. Supercomputing '93*, Portland, OR, Nov. 1993.

Chapter 14

Scotch and PT-Scotch Graph Partitioning Software: An Overview

François Pellegrini

Institut Polytechnique de Bordeaux, LaBRI & INRIA Bordeaux—Sud-Ouest

14.1   Introduction

Graph partitioning is a ubiquitous technique which has applications in many fields of computer science and engineering, especially among the CSC community. It is mostly used to help solve domain-dependent optimization problems modeled in terms of weighted or unweighted graphs, where finding good solutions amounts to computing, possibly recursively according to a divide-and-conquer framework, small vertex or edge cuts that balance the weights of the graph parts.

Save for very constrained subproblems, the computation of a balanced partition with minimal cut is NP-hard, even in the bipartitioning case [1]. It is also the case for some of its variants, such as the static mapping problem which will be described below. Consequently, for problems of sizes above several thousands of vertices, heuristics are the only way to provide acceptable solutions in reasonable time.

The SCOTCH [2] project is yet another attempt to devise a versatile software toolbox for graph partitioning. This project, which is currently developed within the Bacchus team of INRIA Bordeaux Sud-Ouest, started in late 1992 at the *Laboratoire Bordelais de Recherche en Informatique* (LaBRI) of Université Bordeaux 1, roughly at the same time as competing software CHACO [3], JOSTLE [4] and METIS [5]. At that time, members of our *"Parallel Programming and Environments"* team were working on a software environment for a multicomputer which was part of the team's plans, and a tool was required to adequately map communicating processes onto the processors of the machine, so as to minimize inter-processor communication and network congestion. This problem, referred to as static mapping in the literature, is a generalization of the graph partitioning problem, since the latter is equivalent to computing a static mapping of the graph to partition onto an unweighted complete graph. It is also NP-hard in its general form [1].

Shortly after the sequential static mapping features of SCOTCH were developed [6], the team started working on a parallel supernodal direct sparse linear solver called PASTIX [7], which required the computation of fill-reducing reorderings of the unknowns of the matrix. To fulfill this request, the existing recursive edge bipartitioning scheme was transposed to vertex bipartitioning, so as to compute high-quality vertex separators for the ordering of sparse matrices by nested dissection [8]. In a few years, a complete software chain for solving structural mechanics problems was designed and implemented [9], under contract from the CESTA center of the French CEA atomic energy agency. Its scientists, who had access to very large machines, began generating problems which were so large that the sequential ordering phase soon became the main bottleneck because these large graphs did not fit in memory and resulted in heavy disk swapping. In order to overcome this difficulty, a first solution was to extend the ordering capabilities of SCOTCH to handle native mesh structures, in the form of bipartite graphs. These node-element bipartite graphs are a representation of hypergraphs, for which vertex separation algorithms were designed [10]. For the type of elements which were considered, memory usage was divided by a factor of three, at the expense of an equivalent increase in run time. Yet, in the meantime, graph sizes went on increasing, so that the memory problem remained.

For these reasons, the PT-SCOTCH project (for "*Parallel Threaded SCOTCH*") was started. Several parallel graph partitioning tools had already been developed by other teams in the meantime [4, 5], but serious concerns existed about their scalability, both in terms of running time and of the quality of the results produced. In order to anticipate the expected growth rate of problem and machine sizes, our initial design goal was to be able to partition graphs of above one billion vertices, distributed over a thousand processors. Scalability issues were to receive considerable attention, all the more now because the growth in machine sizes leads us to expect parallel architectures comprising millions of processing elements in the years to come.

This chapter gives an insight on the way SCOTCH and PT-SCOTCH were designed so as to fulfill their user's needs. We will specifically focus on parallel processing issues, because their proper handling is essential to the solving of today's large size problems. In addition to describing the algorithms which we devised and implemented within our software, we will use this chapter to present relevant state of the art, in order to give a bigger picture of how CSC is intimately linked with graph partitioning. Readers interested in a thorough description of the algorithms used both in SCOTCH and in the state of the art may refer to [11]. We will not provide precise performance measurements, as they obsolete quickly (a new revision of PT-SCOTCH, most often comprising new algorithms, is released about every six months); instead, we will describe the observed behavior of our software in terms of scalability, which derives from the nature of the algorithms that are currently implemented, and refer to papers where these measurements are shown whenever necessary.

In the next section, we will briefly present the two main problems for which SCOTCH has been designed, namely process to processor allocation and sparse matrix ordering. Then we will introduce the multilevel framework which is commonly used for partitioning graphs, and show how this framework can be parallelized. The next two sections will be devoted to the study of two critical issues for the efficient execution of the multilevel framework, which are its coarsening and refinement phases, respectively; we will describe the main features of the novel algorithms which have been implemented in SCOTCH to tackle these issues, which comprise a probabilistic matching algorithm for graph coarsening and a diffusion-based optimization algorithms that takes advantage of problem space reduction by working on band graphs. We will conclude by discussing some paths for current and future research in parallel graph partitioning.

14.2 The Problems to Solve

Scotch and PT-Scotch were designed to solve two main problems: the distribution of processes and/or data to the processing elements of parallel machines, and the computation of fill-reducing orderings.

14.2.1 Static Mapping for Parallel Processing

The efficient execution of a parallel program on a parallel machine requires that the communicating processes of the program be assigned to the processors of the machine so as to minimize its overall running time. When processes are assumed to coexist simultaneously for the entire duration of the program, this optimization problem is referred to as *mapping*. It amounts to balancing the computational weight of the processes among the processors of the machine, while reducing the cost of communication by keeping intensively inter-communicating processes on nearby processors. A mapping is called *static* if it is computed prior to the execution of the program and is never reconsidered at run-time. Else, dynamic scheduling techniques have to be employed, which are outside the scope of this chapter.

In most cases, the underlying computational structure of the parallel programs to map can be conveniently modeled as a graph in which vertices correspond to processes that handle distributed pieces of data, and edges reflect data dependencies. The mapping problem can then be addressed by assigning processor labels to the vertices of the graph, so that all processes assigned to some processor are loaded and run on it. In a SPMD[1] context, this is equivalent to the *distribution* across processors of the data structures of parallel programs; in this case, all pieces of data assigned to some processor are handled by a single process located on this processor. This is why, in this context, we will talk about *processes* rather than about *processors* or *processing elements* to represent the entities across which data will be distributed.

When the architecture of the target machine is assumed to be fully homogeneous in term of communication, this problem is called the *partitioning* problem, which has been intensely studied [12, 13, 14, 15]. Graph partitioning can be seen as a subproblem of mapping where the target architecture is modeled as a complete graph with identical edge weights, while vertex weights can differ according to the compute power of the processing elements running the processes of the SPMD program. On the other hand, mapping can be seen as the inclusion of additional constraints to the standard graph partitioning problem, turning it into a more general optimization problem termed *skewed graph partitioning* by some authors [16].

---

[1] *Single Process, Multiple Data.* This is a programming paradigm in which all processes of a parallel program have the very same code, the individual behavior of the processes being conditioned by the data they own.

The computation of efficient static mappings requires an *a priori* knowledge of the dynamic behavior of the target machine with respect to the programs which are run on it. This knowledge is synthesized in a *cost function*, the nature of which determines the characteristics of the desired optimal mappings. The main criteria which can intervene in the computation of cost functions comprise estimated overall running time, load imbalance across processing elements, inter-process communication minimization, input and output bandwidth imbalance across processing elements, etc. Two classes of formulae may be used to evaluate these criteria [17, 18]: *global sum* methods add up values across processing elements, while *minimax* methods aim at minimizing the maximum, over all processing elements, of a given quantity (load imbalance, communication cost, etc.). While, in some methods, all criteria are placed at the same priority level, being aggregated in a unique cost function by means of weighting coefficients, load imbalance takes precedence in most cases: the goal of the algorithms is then to minimize some communication cost function, while keeping load balance within a specified tolerance.

One of the most widely used cost functions for communication minimization is the sum, for all edges of the source graph to be mapped onto the target graph modeling the target architecture, of their dilation[2] multiplied by their weight [13, 19]; we will denote it as f_C in the following. In the case of plain graph partitioning, where all edge dilations are equal to 1, this function simplifies into the sum of the weights of all cut edges, that is, edges the ends of which belong to different processing elements. This global sum-type function has several interesting properties. First, it is easy to compute, and it allows for incremental updates performed by iterative algorithms, unlike minimax cost functions. Second, regardless of the type of routing implemented on the target machine (store-and-forward or cut-through), it can model the traffic on the interconnection network, and its minimization favors the mapping of intensively intercommunicating processes onto nearby processors [19, 20]. However, it may be inaccurate to model data exchange for parallel computations. For instance, in the case of hybrid[3] or iterative solving of sparse linear systems, the main operation to perform is a parallel matrix-vector product. Data borne by frontier vertices have to be sent to neighboring processes, but only once per destination process, while the edge cut may account for several edges incident to the same vertex, as illustrated in Figure 14.1. This is why many authors have advocated for a hypergraph-based communication cost model, and use hypergraph partitioning tools such as PARKWAY, PATOH or ZOLTAN, instead of graph partitioning tools, to assign vertices to subdomains [22, 23, 24, 25]

---

[2]The dilation of a source graph edge is the length of the shortest path, within the target graph, linking the two target graph vertices onto which are placed the two source graph vertices which are the ends of the considered edge.

[3]Hybrid solving is a paradigm which tries to combine the stability of direct methods and the speed and small memory footprint of iterative methods for solving large sparse systems of equations. For instance, some hybrid methods use a Schur complement decomposition to perform direct linear system solving on the separated subdomains distributed across the processes, and to perform iterative solving on the separators, as in [21].

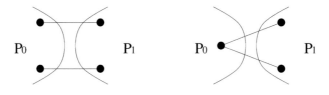

FIGURE 14.1: Inaccuracy of the graph partitioning model for representing data exchange. For both of the above bipartitions, $f_C = 2$, but 4 vertex data have to be exchanged in the leftmost case, while it is 3 in the rightmost case.

(see Section 13.2, p. 352, for a thorough description of this model, including further references).

Although hypergraph partitioning represents data exchange in an exact way, yielding an average decrease of communication cost of 5 to 10 percent on mesh graphs, and up to 60 percent for some non-mesh graphs, compared to plain graph partitioning [22], it is several times slower than the latter, which is therefore still of economical interest for scientific computing.

14.2.2 Sparse Matrix Reordering

Many scientific and engineering problems can be modeled by sparse linear systems, which are solved either by iterative or direct methods. Direct methods are popular because of their generality and robustness, especially in some application fields such as linear programming and structural mechanics, where iterative methods may fail to achieve convergence for specific classes of problems. However, to achieve efficiency with direct methods, one must minimize the amount of fill-in induced by factorization. As described above, fill-in is a direct consequence of the order in which the unknowns of the linear system are numbered, and its effects are critical both in terms of memory and computation costs (see Section 2.3, p. 25, for a detailed explanation of the fill-in process).

An efficient way to compute fill reducing orderings of symmetric sparse matrices is to use recursive nested dissection [26]. It amounts to computing a vertex set S which separates the adjacency graph associated with some matrix into two parts A and B, ordering S last, and proceeding recursively on parts A and B until their sizes become smaller than some threshold value. This ordering guarantees that no non-zero term can appear in the factorization process between unknowns of A and unknowns of B.

The main issue of the nested dissection ordering algorithm is thus to find small vertex separators which balance the remaining subgraphs as evenly as possible. Vertex separators are commonly computed by means of vertex-based graph algorithms [27, 28], or from edge separators [29, and included references] using minimum cover techniques [30, 31], but other techniques such as spectral vertex partitioning have also been used [15]. The divide-and-conquer nature of nested dissection makes this algorithm easy to parallelize at a coarse-grain level: once a separator has been computed on a graph distributed on p processes, each of the separated parts can be redistributed onto $\frac{p}{2}$ processes, such that the nested dissection algorithm can recursively operate independently on each of the process subsets. Moreover, the elimination trees induced by nested dissection are broad, short, and balanced, and therefore exhibit high concurrency when factorizing and solving these linear systems on parallel architectures [32].

14.3 General Architecture of the Scotch Library

Great care has been taken in the design of the library since the inception of the sequential version. In particular, it has been designed in a highly modular way, based on structured data types, in order to allow for extensibility without sacrificing algorithmic efficiency. This approach eased the design of the parallel extension of the library, on top of the existing sequential modules. The key aspects of the library design are exposed below.

14.3.1 Design Choices

While the sequential version of the library makes very few assumptions on the nature of graphs to be handled, the only limitation regarding the fact that vertex and edge weights should be strictly positive integers, the parallel version required some early design decisions to be taken, which condition the ability of the software to handle some types of graphs.

In particular, we assumed that distributed graphs are of reasonably small degree, that is, that graph adjacency matrices have sparse rows and columns. Unlike more robust approaches [33] (see Section 12.1, p. 321), we presume that vertex adjacencies can be stored locally on every process, without incurring too much memory imbalance or even memory shortage.

Also, we wanted our software to run on any number of processes p, and produce any number of parts k, to avoid limitations which existed in other packages. For instance, PARMETiS can compute orderings only on a number of processes which is a power of two ($k = 2^i$), and JOSTLE produces only as many domains as the number of processes it is running on ($k = p$). These limitations were introduced, in each of the above cases, to ease software development, and tackling these issues in PT-SCOTCH required additional data structures, for instance so as to decouple the target and execution architectures: a process can handle several target domains, and conversely a given target domain can be managed concurrently by several processes. This decoupling was quite straightforward, because the management of abstract target architectures is already part of the static mapping computation process.

In terms of execution architecture, we considered a distributed memory model, since large machines with hundred thousands of processing elements are not likely to ever implement efficiently a shared-memory paradigm, even of NUMA[4] type. The distributed memory paradigm has a strong impact on algorithms, because mutual exclusion and locking issues, which are implicitly resolved by doing memory accesses in the sequential case, have to be reformulated in terms of non-deterministic, latency-inducing, message exchanges.

Since we wanted to depend as less as possible on underlying hardware and system constraints, the basic integer type used in the SCOTCH and PT-SCOTCH libraries to represent vertex and edge indices, called SCOTCH_Num, is completely independent from machine word size, and can be set at compile time to be either a 32-bit or a 64-bit quantity. Also, while the PT-SCOTCH library, Application Programmer Interface (API) explicitly refers to the MPI API for handling communication in the distributed memory environment (for instance, initializing a PT-SCOTCH distributed graph structure requires the passing of a MPI communicator which describes the group of processes across which graph data is distributed), we did not base on MPI limitations regarding message lengths and displacements, which are restricted to int values by definition of the MPI API, to perform internal index computations. PT-SCOTCH is therefore fully operational in 64-bit mode, and can handle graphs well above 2^{32} vertices and edges, distributed on any number of processing elements, provided there are less than 2^{32} elements to be exchanged between any pair of processes. This is most likely to be the case since, even on very large machines, the available memory per core is still less than a few gigabytes, so that the number of vertices and edges per process cannot be higher than a few tens of millions, well below the two billion limit.

In order for parallel partitioning algorithms to be able to scale on an arbitrary number of processing elements, graph data must never be duplicated, across all of the processing elements, more than a (small) factor of the vertex and edge set sizes $|V|$ and $|E|$. The distributed graph structure which we implemented, and which we describe below, fulfills this goal. Also, all of our algorithms should never use data structures that are super-linear in k, p and/or $|V|$, such as in $k|V|$, k^2, kp, $p|V|$ or p^2, as doing so precludes scalability and can quickly lead to memory shortage (it is a known problem that the current version of ParMeTiS uses an array of size p^2).

---

[4]Non Uniform Memory Access. The machines implementing this model are categorized by the fact that, while their memory can be accessed from all of their processing elements, the cost of access may differ depending on the location of the memory with respect to the processing element which makes the request.

14.3.2 Distributed Graph Structure

According to the above assumptions, in our PT-SCOTCH library, like in other packages, distributed graphs are represented by means of adjacency lists (the PT-SCOTCH distributed graph data structure is extensively described in [34]). Vertices are distributed across processes along with their adjacency lists and with some duplicated global data. In order to allow users to create and destroy vertices without needing any global renumbering, every process is assigned a user-defined range of global vertex indices, in which vertex indices may not all be assigned. Range arrays are duplicated across all processes in order to allow each of them to determine the owner process of any non-local vertex by dichotomy search, whenever necessary.

Since many algorithms require that local data be attached to every vertex, and since using global indices extensively would be too expensive in our opinion, all vertices owned by some process are also assigned local indices, suitable for the indexing of compact local data arrays. This local indexing is extended so as to encompass all non-local vertices which are neighbors of local vertices, which are referred to as *ghost* or *halo* vertices. A low-level halo exchange routine is provided by PT-SCOTCH, to diffuse data borne by local vertices to the ghost copies possessed by all of their neighboring processes. This low-level routine is internally used by many algorithms of PT-SCOTCH, for instance to share matching data in the coarse graph building routine.

Because global and local indexings coexist, two adjacency arrays are in fact maintained on every process. The first one, provided by the user, holds the global indices of the neighbors of any given vertex. The second one, which is internally maintained by PT-SCOTCH, holds the local and ghost indices of the neighbors, and can be made available to users willing to perform local computations on vertex data. Since only local vertices are processed by the distributed algorithms, the adjacency of ghost vertices is never stored on the processes, which guarantees the scalability of the data structure as no process will store information of a size larger than its number of local outgoing arcs.

14.3.3 Library Architecture

The main data structures of SCOTCH concern the handling of unoriented graphs, represented as adjacency lists. It is also in this form that they should be passed by users, through both the centralized [35] and distributed [36] library API. Internally, the basic centralized and distributed graph data structures, called `Graph` and `Dgraph`, respectively, are encapsulated into several extended, module dependent, structures. These extended data types serve to store internal working data such as the current state of a bipartition, of a vertex separator or of a static mapping. For instance, the centralized bipartition graph `Bgraph` contains fields which represent the ideal load in part 0 (which may be different from half of the overall graph load if unequal parts are desired in the course of the recursive bipartitioning process), the current load in part 0, an array holding the current part of each vertex, an array holding the indices of all frontier vertices which accelerates the initialization of boundary optimization algorithms, etc.

The modularity of the API of SCOTCH matches the modularity of its coding. The source code of SCOTCH is made of more than 450 files written in C, 410 of which belong to the `libscotch` library itself. Although it is coded in an imperative language, its design follows the principles of object-oriented programming. A module is a set of files, the functions of which all operate on instances of a common data type. In that respect, module functions represent the private and public methods of the class the abstract type of which is defined by the structure. This enhanced modularity enables external contributors to use SCOTCH as a testbed for new partitioning and ordering methods, which can easily be integrated into the existing code base. A thorough description of the internals of SCOTCH, including the instructions for extending its features, can be found in [37].

14.4   Multilevel Framework

Experience has shown that, for many graphs used in scientific computations[5], best partition quality is achieved when using a multilevel framework. This method, which derives from the multigrid algorithms originally used in numerical physics, repeatedly reduces the size of the graph to partition by clustering vertices and edges, computes an initial partition for the coarsest graph obtained, and prolongs[6] the result back to the original graph [12, 13], as illustrated in Figure 14.2.

14.4.1   Refinement

Multilevel methods are most often combined with partition refinement methods, in order to smooth the prolonged partitions at every level so that the granularity of the solution is that of the original graph and not that of the coarsest graph.

In the sequential case, the most popular refinement heuristics in the literature are the local optimization algorithms of Kernighan-Lin [38] (KL) and Fiduccia-Mattheyses [39] (FM), either on edge-based [13] or vertex-based [27]

---

[5]These graphs have in common good locality and separation properties: their maximum degree is small, and their diameter evolves as a fractional power of their size, e.g., in $O\left(n^{\frac{1}{d}}\right)$ for d-dimensional mesh graphs of size n.

[6]While a *projection* is an application to a space of lower dimension, a *prolongation* refers to an application to a space of higher dimension. Yet, the term projection is also commonly used to refer to the propagation of the values borne by coarse graph vertices to the finer vertices the coarsening of which they result from.

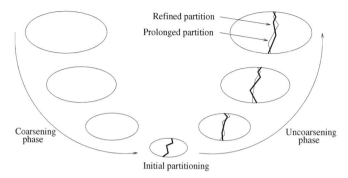

FIGURE 14.2: The multilevel partitioning process, here in the bipartitioning case. In the uncoarsening phase, the light and bold lines represent for each level the prolonged partition obtained from the coarser graph, and the partition obtained after refinement, respectively.

forms depending on the nature of the desired separators. All of these algorithms are based on successive moves of frontier vertices to decrease the current value of the prescribed cost function. Moves which adversely affect the cost function may be accepted, if they are further compensated by moves which result in an overall gain, therefore allowing these algorithms to perform *hill-climbing* from local minima of the cost function.

Vertices to be moved are chosen according to their *gain value*, denoted by ∇f_C in the following, which is the amount by which the current value of the cost function would be modified if the given vertex were moved to some other part; vertices of negative gain values are consequently the most interesting to move. For that purpose, vertex gain values are computed in advance, and vertices of interest are inserted into data structures which implement some flavor of sorting, such that vertices with lowest gain values can be retrieved at small cost.

In the case of bipartitioning, there is only one part to move the vertex to, and candidate vertices can be stored in only one such data structure when individual vertex movements are considered (for FM-like algorithms), or two twin structures when vertex swaps are considered (for KL-like algorithms). In the case of k-way partitioning, while most frontier vertices also have only one neighbor part (e.g., the vertices belonging to subdomain interface faces for 3D meshes), others may have more (e.g., vertices representing subdomain interface edges and points in 3D). Handling frontier vertices with multiple potential destination parts requires either the use of enhanced data structures which can record several gains per vertex, that is, one per target subdomain, or some pre-selection of the most appropriate destination part if simpler data structures are sought [40].

While, in theory, all of the $(k-1)$ parts other than the current part of any vertex might be potential destinations for it, the set of destinations considered by the algorithms is always much more reduced. Since the refinement process is expected to be local by the very nature of the multilevel framework, the only vertices for which gains are computed are frontier vertices, an optimization often referred to as *boundary refinement* in the literature [13, 41]. Also, the only parts to be considered for destination of a frontier vertex are, in the context of graph partitioning, the ones which are currently adjacent to it [40], which coerces the local optimization algorithm to behave in the same way as a diffusion-like method. For static mapping, the set of potential destination parts may be extended to the union of all of the parts adjacent to the part of the vertex [42]. This extension with respect to the graph partitioning case comes from the potential need to insert some "buffer" vertices of intermediate subdomains between two subdomains mapped onto remote processing elements, so as to improve communication locality on the target machine (see [11] for a thorough description of this issue).

14.4.2 Coarsening

Coarsening and refinement are intimately linked. By reducing the size of large-scale, graph-wide topological structures into sets of a few tens of vertices, coarsening allows these structures to be wholly handled by local optimization algorithms, conversely broadening the scope of the latter and turning them into global-like optimization algorithms at the coarsest levels. Then, as uncoarsening takes place, partition refinement only focuses on smaller and smaller details of the finer graphs, as large-scale structures regain their original sizes and become out of the scope of local optimization.

However, the quality of partitions produced by a multilevel approach may not be as good as the one of long-running, global optimization algorithms. Indeed, the critical issue in multilevel methods is the preservation, by the coarsening algorithm, of the topological properties of the original graphs. Otherwise, solutions provided by the initial partitioning method would not be relevant approximations of the solutions to the finest problems to solve. While some of the topological bias induced by the coarsening can be removed by refinement methods, the local nature of the latter most often hinders their ability to perform the massive changes that would be required to alleviate any significant bias. Coarsening artifacts, as well as the topology of the original graphs, can trap local optimization algorithms into local optima of their cost function, such that refined frontiers are often made of non-optimal sets of locally optimal segments, as illustrated in Figure 14.3.

Coarsening algorithms are based on the computation of vertex clusters in the fine graph, which define a partition from which the coarse graph will be built by quotienting. Clusters will become the vertices of the coarser graph, while edges of this latter will be created from the edges linking fine vertices belonging to different clusters, so that the weight of a coarse edge is equal to the sum of the weights of the fine edges it replaces.

FIGURE 14.3: Partition of graph BUMP into 8 parts with SCOTCH 4.0, using the multilevel framework with a Fiduccia-Mattheyses refinement algorithm. The cut is equal to 714 edges. Segmented frontiers are clearly evidenced, compared to Figure 14.8, p. 397.

Save for some promising work on multiple and weighted aggregation [43] which aims at removing further the impact of coarsening artefacts, all of the existing coarsening schemes are based on edge matchings to mate the vertices of the fine graph, so that any coarse vertex contains at most two fine vertices. Matchings are an essential tool for the solving of many combinatorial scientific computing problems [44]. In the context of graph coarsening, matchings do not need to be maximal, which allows for the use of fast heuristics.

The most popular matching scheme in the literature is heavy-edge random matching [14], which is very easy to implement and provides matchings of good quality on average [45]. It operates by considering all graph vertices in random order, and by mating each of them with one of its non already matched neighbors linked by an edge of heaviest possible weight. This scheme favors the preservation of the large-scale topological structure of graphs because edge weights increase with the desirability to match in a dimension which has not yet been explored.

14.4.3   Initial Partitioning

Initial partitioning algorithms are in charge of computing, on the coarsest graph, the partition which will be prolonged back to finer graphs. Since initial partition quality is mandatory, expensive algorithms may be used, all the more that they operate on small graphs.

Many global methods have been investigated to compute graph partitions from scratch, either as is or within a multilevel framework: evolutionary algorithms (comprising simulated annealing [46], genetic algorithms [47], ant colonies [48], greedy iterative algorithms [49]), graph algorithms [14, 50], geometry-based inertial methods [51], spectral methods [12], region growing [52], diffusion-based methods [53], etc. While taking their inspiration from radically different fields such as genetics or statistical physics, analogies between these methods are numerous, as well as cross-fertilization in their implementations, so that it is sometimes difficult to categorize them unambiguously.

Region growing algorithms have received much attention, as they result in connected parts, which is a most desirable feature in numerous application domains [52, 54]. They consist in selecting as many seed vertices in the graph as the number of desired parts, and growing regions from these seeds until region boundaries meet. The most basic versions of this approach sequentially grow regions one by one [14, 50], while more elaborate algorithms, such as *bubble-growing* algorithms, grow all regions simultaneously in order to yield interfaces of higher quality [52, 55].

We will not discuss, in this chapter, the implementation in PT-SCOTCH of the computation of initial partitions, as we implemented it in a quite straightforward way. At the time being, distributed graphs are coarsened down to a few thousands of vertices, and coarsest distributed graphs are centralized and duplicated on all of the processes, where traditional, sequential initial partitioning algorithms are used. Yet, this method may no longer work in the future, when the desired number of parts becomes larger than a few millions. Since initial partitioning algorithms require at least as many vertices as there are parts to be computed, so that all part indices can be propagated back to the finer graphs, coarsest graphs may no longer fit in the memory of single processing elements. Parallel initial partitioning methods will have to be designed. We plan to overcome this difficulty by setting up a multilevel initial parallel static mapping method, where partial static mappings will be computed on the available coarsest vertices, these coarse mappings being locally completed when enough fine vertices are available in the course of the uncoarsening process.

14.5   Parallel Graph Coarsening Algorithms

Multilevel schemes are extremely powerful tools for graph partitioning, provided that unbiased coarsening and local optimization algorithms are defined for the targeted classes of graphs.

In this section, we will present the ways in which the parallel graph partitioning problem has been addressed in the literature, with respect to each of its key issues (i.e., matching, folding, and centralization of the distributed data for the computation of the initial partition on the coarsest graph), as well as the solutions which have been implemented in PT-SCOTCH.

14.5.1   Matching

The process of creating a coarser distributed graph from some finer distributed graph is quite straightforward, once every fine vertex has been assigned to the cluster representing its future coarse vertex. By performing a halo exchange of the cluster indices borne by fine vertices, each of the latter can know the indices of the clusters to which its fine neighbors belong, and create its coarse neighbors list. In PT-SCOTCH, like in the sequential SCOTCH software, vertices are clustered by pairs, by means of edge matching. The key issue for parallel graph coarsening is therefore to devise parallel matching algorithms which are both efficient and unbiased, that is, which provide coarse graphs the topological properties of which are independent of the distribution of fine vertices across the processes of the parallel program.

Parallel matching algorithms are a very active research topic [56, and included references]. In our context, this problem is relaxed, because matchings do not need to be maximal. Instead, they should exhibit randomness properties suitable to the preservation, by the coarsened graphs, of the topological properties of the finer graphs to which they are applied. Otherwise, the prolongation to the finer graphs of partitions computed on the coarser graphs would not be likely to represent good solutions of the original problem, because of the biases introduced by coarsening artifacts.

The critical issue for the computation of matchings on distributed graphs is the breaking of mating decision ties for edges spanning across processes. Any process willing to mate one of its local vertices to a remote adjacent vertex has to perform some form of query-reply communication with the process holding the remote vertex. Because of communication latency and overhead, mating requests are usually aggregated per neighbor process, resulting in a two-phase algorithm. In the first phase, processes send mating requests to their neighbors. In the second phase, neighbors answer positively or negatively depending whether the requested vertices are free, have already been matched in a previous phase, or are temporarily unavailable because they have themselves requested another remote vertex for mating. The asynchronicity between requests and replies creates ties and dependency chains which hinder the convergence of the algorithm. For instance, when several vertices located on different processes request the same remote vertex, at most only one of the requests can be satisfied. Moreover, when the requested vertex has itself asked for another vertex, none of the incoming requests will be satisfied, because the solicited vertex cannot know whether its own request will be satisfied or not before replying to its applicants. This phenomenon increases along with the probability of neighbor vertices to be remote: when graphs are arbitrarily distributed (which is usually the case before they are partitioned for the first time), when they have high degrees, and/or when the number of processes is high.

Several solutions to the tie-breaking problem have been proposed in the literature, which differ in the granularity of the remote dependency exclusion mechanism.

The first approach is to handle tie-breaking at the vertex level [57, 58, 59]. These algorithms base on a parallel formulation of Luby's graph coloring algorithm [60] to compute a maximal independent set of vertices. The rationale behind the use of a graph coloring is to completely avoid dependency chains, by preventing potential sought-after vertices from being applicants themselves. In the matching algorithm, each of the colors is considered in turn, and only vertices of the current color can perform mating requests. While this algorithm does not guarantee that collisions between requests will not happen, since two vertices of the same color can still send a request to a same common neighbor vertex of another color, it significantly reduces the number of such collisions.

The second approach is to handle tie breaking at the subdomain level. The goal of these methods is to avoid as much communication as possible during the matching process, so as to reduce the impact of dependency chains on the convergence of the matching algorithm. This result is sought for by privileging local matings over remote ones, that is, by restricting matings of vertices to their local subdomains [61, 62]. In this case, like in the fully sequential case, tie breaking is naturally resolved by the sequentiality of computations within each subdomain. Depending on the distribution of graph vertices, privileging local matching can induce heavy topological bias in the coarsened graphs.

The solution to the parallel matching problem which we have implemented in PT-SCOTCH follows a Monte-Carlo approach. This idea, first exposed by Chevalier in [63], consists in allowing a vertex to send a request only when the random number it draws is above some threshold. In [63] and [34], this threshold depended on the numbers of local and remote neighbors of the vertex. In [64], we opted for a more straightforward algorithm, which does not take into consideration any parameter linked to the graph distribution, and is consequently less susceptible to distribution biases: every unmatched vertex can send a mating request with probability 0.5, or remain silent. The insensitivity to data distribution makes our algorithm more robust than the greedy matching algorithm currently implemented in PARMETIS [58, Section 4], which bases on vertex and process indices to determine whether a matching request will be sent or not be sent.

According to our experiments [64], five collective rounds of our probabilistic algorithm are enough to match more than 80% of the graph vertices, with a low resulting coarsening ratio of 53.7%. This ratio indicates that remote mating is efficient, and that no topological biases, due to initial graph data distribution across processes, are likely to occur.

14.5.2 Folding

In the course of the coarsening phase of a multilevel framework, folding refers to the process of redistributing the vertices of a coarse graph onto a smaller number of processes than those holding the finer graph. Performing

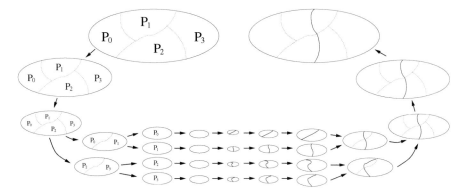

FIGURE 14.4: Diagram of the parallel computation of the separator of a graph distributed across four processes, by parallel coarsening with folding-with-duplication, multi-sequential computation of initial partitions that are locally prolonged back and refined on every process, and then parallel un-coarsening of the best partition encountered.

such a folding has several benefits. First, it can reduce communication over-head. As coarsening goes on, the number of vertices per process is bound to decrease, and communication time tends to be dominated by start-up time and latency. Using too many processes to handle so little data makes collective communication prone to operating system hazards, and increases the proba-bility of increased latency. Second, concentrating vertices on fewer processes is most likely to decrease the number of remote neighbors per process, resulting in shorter messages. Third, for algorithms biased towards local computations (see the above section), increasing the number of local vertices can improve the quality of the output they produce [65].

One of the original features of our multilevel framework is its ability to duplicate coarse graphs during the folding process, so that no process remains idle. The coarsening phase starts once the matching phase has completed. It can be parametrized so as to allow users to choose between two options. In the first case, all coarsened vertices are kept on their local processes (that is, processes that hold at least one of the ends of the coarsened edges), as shown in the first steps of Figure 14.4, which decreases the number of vertices owned by every process and speeds-up future computations. In the second case, coarsened graphs are folded and duplicated, as shown in the next steps of Figure 14.4, which increases the number of working copies of the graph and can thus reduce communication and improve the final quality of the separators.

In fact, separator computation algorithms, which are local heuristics, heav-ily depend on the quality of the coarsened graphs, and we have observed with the sequential version of SCOTCH that taking every time the best parti-tion among two, obtained from two fully independent multilevel runs, usually improves overall ordering quality. By enabling the folding-with-duplication routine (which will be referred to as "fold-dup" in the following) in the first coarsening levels, one can implement this approach in parallel, every subgroup of processes that hold a working copy of the graph being able to perform an almost-complete independent multilevel computation, save for the very first level which is shared by all subgroups, for the second which is shared by half of the subgroups, and so on. The problem with the "fold-dup" approach is that it consumes a lot of memory, because the amount of memory per process does not decrease. Consequently, as in [58], our strategy is to resort to folding only when the number of vertices of the graph to be considered reaches some minimum threshold.

14.5.3 Multi-Centralization for Initial Partitioning

Once the distributed graph is coarsened down to a prescribed, small size, an initial partition has to be computed. While fully parallel methods for an initial partitioning could be used, such as parallel genetic algorithms (see Section 14.6.1), all implementations we are aware of resort to state-of-the-art sequential methods, because of the very high quality of the partitions they produce, at reasonable cost for such small graphs. Since the initial partition will only be subject to refinement and will never be reconsidered globally, quality concerns dominate at this stage.

When folding is used, the coarsest graphs are already concentrated on a small number of processes, and centralizing them on a single process is not much work. Yet, all the remaining processes are idle, and may be used to speed-up the computation of the initial partition. In [59], the centralized graph is duplicated on all processes, and each of them computes sequentially the branch of the recursive bipartitioning tree leading to the subdomain it is in charge of. Doing so results only in a limited gain in run time, as it only concerns small graphs. Moreover, this approach can only be used when successive bipartitioning tasks are fully independent, which prevents this method from being used in the context of static mapping [19].

In PT-SCOTCH, after the final stage of our fold-dup method, no more than two processes may have the same coarse graph, since the two copies of a graph folded at some stage undergo afterward an independent coarsening process, using a different pseudo-random seed. Unlike in [59], where all centralized graphs are identical and the same bipartitions must be computed on all processes, we favor the computation of different initial partitions by each of the processes.

This multi-sequential phase is illustrated at the bottom of Figure 14.4: the routines of the sequential SCOTCH library are used on every process to complete the coarsening process, compute an initial partition, and prolong it back up to the largest centralized coarsened graph stored on each of the processes. Then, the partitions are prolonged back in parallel to the finer distributed graphs, selecting the best partition between the two available when prolonging to a level where fold-dup had been performed. This distributed prolongation process is repeated until we obtain a partition of the original graph. The benefits of producing different initial partitions have been shown in [63].

14.6   Parallel Partition Refinement Algorithms

While satisfactory parallel graph coarsening methods have been devised at early stages of our project, the design of efficient and scalable partition refinement algorithms is still an open problem, so that most of the research work in PT-SCOTCH is devoted to this topic.

14.6.1   Partition Refinement

Because the state-of-the-art local optimization methods used for partition refinement in sequential multilevel frameworks were so robust and cost-effective, much effort has been put on their parallelization. Yet, it is common knowledge in the parallel computing community that the best parallel algorithm to solve a given problem is most often not the parallel transposition of the best sequential algorithm for this purpose. Partition refinement by no means escapes this rule, as state-of-the-art sequential methods bear strong sequentiality constraints which prevent their direct transposition into scalable parallel formulations, so that global methods also have to be considered.

As seen in Section 14.4.1, local optimization algorithms for partition refinement are based on successive moves of frontier vertices to decrease the current value of the prescribed cost function. In the sequential case, vertices are moved one by one, and the gains of the selected vertex and of its neighbors are recomputed on the fly, such that one always knows exactly the outcome of a sequence of moves. However, this no longer holds if moves are to be performed independently in parallel, as illustrated by the infamous "neighbors swap" case of Figure 14.5.

If the two vertices had been far away one from one another, the problem described above would not have shown up. What causes it is that the two vertices are neighbors, and that the mutual impact of their moves cannot be accounted for to avoid performing the second move as soon as the first one takes place. Indeed, if the two vertices are distributed across separate processors, there is no immediate means for each of the processes running the algorithm to inform its peers; this is not desirable either, as taking into account the constraint of having to update the gain values of neighbor vertices would create time dependencies which would dramatically hinder the scalability of the algorithm.

It follows from the previous example that, for any parallel version of some

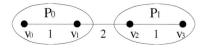

(a) Current bipartition of a weighted graph, with $f_C = 2$, $\nabla f_C(v_0) = +1$, $\nabla f_C(v_1) = -1$, $\nabla f_C(v_2) = -1$, $\nabla f_C(v_3) = +1$.

(b) Resulting partition after moving simultaneously v_1 and v_2. While each of them had negative gain, $f_C = 4$.

FIGURE 14.5: Adverse impact on the cost function of the simultaneous move of two vertices which, moved individually, would have improved it.

FM-like local optimization algorithm to work, no two neighbor vertices must ever be considered at the same time by distinct processes. This is indeed a very strong constraint, which imposes a heavy burden on the programmer. Since dynamic remote locking mechanisms would generate too much communication, this constraint must be enforced a priori, with certainty, within the core of the algorithm itself. Several solutions have been proposed in the literature, which differ by the granularity of the exclusion mechanism, in a way very similar to the matching problem (see Section 14.5.1). All of them are based on rounds within which the exclusion constraint will be enforced, with some synchronization taking place at the end of the rounds so as to recompute the gains of remote neighbors.

The first class of solutions to this problem handles collision exclusion at the vertex level [59]. Much like for matching, it is based on a coloring of the distributed graph, computed by means of a distributed variant of Luby's algorithm (see Section 14.5.1), to split vertices into groups such that no two neighbor vertices belong to the same group. The groups associated with the colors are then processed in a round-robin fashion, and within each round all vertices of negative gain are moved to their preferred destination part. Then, vertex movement information is exchanged between neighboring processes, so that the gains of their neighbors can be recomputed before the next round begins. While this method is very scalable, as computations are most likely to be evenly spread across processes, it loses the hill-climbing capabilities of the original sequential algorithm, resulting in poorer partition quality. Indeed, moves that increase the cost function can no longer be accepted, else many processes could decide simultaneously, within the same round, to select vertices of positive gain, which might not be of overall minimum positive gain, without any chance for the algorithm to compensate later for these moves.

The second class of solutions enforces neighbor exclusion at the subdomain interface level. Here again, the algorithms take the form of two nested loops. In JOSTLE [66], at the outer level, interface regions between different subdomains are separated into disjoint areas, which are centralized on one of their two owner processes. These subgraphs are treated as independent problems, onto which sequential optimization algorithms are applied. This algorithm preserves, in each interface subgraph, the hill-climbing capabilities of the sequential local optimization algorithm. However, what is preserved in quality is lost in runtime efficiency and scalability, because interface areas may be of very different sizes [66, page 1653]. In the method actually implemented in PARMETIS [58, Section 4], interfaces are not considered explicitly. Instead, for each pass, all vertex moves which would improve the cut are performed provided that some condition, depending on the pass number and on a randomized function of the numbers of the origin and destination subdomains, is satisfied. Collisions are therefore excluded at the subdomain level, but because individual vertex moves are performed in parallel, only gradient-like improvements are allowed.

Since the road to parallel local optimization methods was so full of hurdles, several authors have investigated the path of global methods. These methods, which are too expensive to be applied to large graphs in the sequential case (see Section 14.4.3 for an overview of the most popular ones), most often have a strong potential for parallelization. Yet, due to their global nature, their cost may be high even in parallel, and their use within a multilevel framework somehow goes against the nature of this framework, which is designed to use fast algorithms that operate on reduced problem spaces.

The class of global methods which has been used the most extensively in the literature is evolutionary algorithms, and more specifically genetic algorithms [47, 67, 68]. Genetic algorithms (GA) are meta-heuristics used to solve multi-criteria optimization problems using an evolutionary method. They consist in simulating iteratively, generation after generation, the evolution of a population of individuals which represent solutions to the given problem, selecting best-fitting individuals as candidates for breeding the next generation. GA are known to converge very slowly and cannot therefore be applied to large graphs [47]. Talbi and Bessière [68] were among the first to use parallel GA to solve the graph partitioning problem, using a multi-deme[7] parallel implementation.

Also, many authors noticed that partitions yielded by local optimization algorithms were not optimal. One of the most vocal communities is users of iterative linear system solving methods [52, 54], who found that such partitions were not fitted for their purpose, because subdomains with longer frontiers or irregular shapes resulted in a larger number of iterations to achieve convergence. To measure the quality of each of the parts, several authors refer to a metric called *aspect ratio*, which can be thought in d dimensions as the ratio between the size of the interface of a part with respect to the $\left(\frac{d-1}{d}\right)^{th}$ power of the size of its contents[8] [52, 67, 69]. The more compact a part is, the smaller its aspect ratio value is.

---

[7]In this approach, processes handle almost independent populations, between which "champions" are exchanged, from time to time, across neighboring processing elements. It is also called the "island" model.

[8]For instance, for a 2D mesh, the aspect ratio of a part is measured as the ratio between

In [52], Diekmann *et al.* demonstrated such a behavior, and proposed both a measure of the aspect ratio of the parts, as well as a set of heuristics to create and refine the partitions, with the objective of decreasing their aspect ratio. Among these algorithms is a bubble-growing algorithm, the principle of which was presented in Section 14.4.3. An important drawback of this method is that it does not guarantee that all parts will hold the same number of vertices, which means that afterward other heuristics have to be called on to restore the load balance. Also, all of the graph vertices must be visited many times, which makes this algorithm quite expensive, the more so because it is combined with costly algorithms such as simulated annealing, and the computation of the aspect ratio requires some knowledge on the geometry of the graphs, which is not always available. In [55], Meyerhenke and Schamberger further explore the bubble model, and devise a way to grow the bubbles by solving, possibly in parallel, systems of linear equations, instead of iteratively computing bubble centers. This method yields partitions of high quality too, but is very slow, even in parallel [70], and the load balancing problem is also not addressed. Speed concerns were partially resolved by their integration of our band graph technique [71], which we are going to describe below.

14.6.2 Band Graphs

Since the strong sequentiality constraint of local refinement algorithms seems too hard to overcome, resulting either in a loss of partition quality or offering poor scalability, the most straightforward path towards efficient parallel refinement is to try to reduce the cost of global algorithms without losing in partition quality. The only solution is therefore to reduce problem space.

When looking at the essence of the multilevel framework, it can be assumed that, because of the local nature of both the FM and the uncoarsening algorithms, the refined partition computed at any level should not differ much from the partition that is prolonged back to this level, as this latter is itself the prolongation of a partition that is a local optimum in the coarser levels. Therefore, to refine a partition, FM-like algorithms should not need to know more of the graph topology than a small "band" around the boundary of the prolonged partition. We have therefore implemented a band graph extraction algorithm, which only keeps vertices that are at small distance from the prolonged separators, such that our local optimization algorithms are applied to these band graphs rather than to the whole graphs, as illustrated in Figure 14.6 in the sequential case. Vertices which do not belong to the band graph are merged into two *anchor vertices* of weights equivalent to those of the merged vertices.

The experiments which we carried out on band graphs, either in the edge

---

the length of the perimeter of the part, with respect to the square root of its surface. The lower bound for this metric is achieved for parts of circular shape in the 2D Euclidean space.

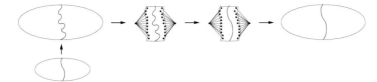

FIGURE 14.6: Multilevel banded refinement scheme. A band graph of small width is created around the prolonged finer separator, with anchor vertices representing all of the removed vertices in each part. After some optimization algorithm (whether local to the frontier or global to the band graph) is applied, the refined band separator is prolonged back to the full graph, and the uncoarsening process continues.

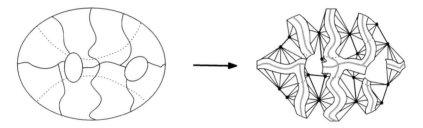

FIGURE 14.7: Computation of a k-way distributed band graph from a 8-partitioned graph distributed across three processes. The solid line is the current partition frontier, while dotted lines represent the separation between process domains. Merged vertices in each part are now represented by a clique of local anchor vertices, one per process and per part. Anchor vertices which would not be connected to vertices in their part are not created.

or vertex separation cases [34, 72, 73], consistently showed that the overwhelming majority of refined separator vertices is not located at a distance greater than three from the vertices of the prolonged separators. This value is significant: it is the maximum distance at which two vertices can be in some graph when the coarse vertices to which they belong are neighbors in the coarser graph at the next level. Therefore, keeping more layers of vertices in the band graph is not useful, because allowing the fine separator to move a distance greater than three in the fine graph to reach some local optimum means that it could already have moved to this local optimum in the coarser graph, save for coarsening artefacts, which are exactly what we do not want to be influenced by.

The computation and use of distributed band graphs does not differ much from their sequential counterparts [64]. However, while sequential band graphs have only one anchor vertex per part, in the parallel k-way case, there are as many interconnected anchor vertices as there are parts in the partition, as illustrated in Figure 14.7.

14.6.3   Multi-Centralization for Sequential Refinement

The implementation of distributed band graphs in PT-SCOTCH was accompanied by the coding of a multi-centralized refinement algorithm. At every distributed uncoarsening step of the multilevel framework, a distributed band graph is created. Centralized copies of this band graph are then gathered on every participating process, which can run fully independent instances of sequential local optimization algorithms such as FM, using different pseudo-random seeds so as to explore slightly different solution spaces. The best refined band separator is prolonged back to the distributed graph, and the uncoarsening process continues.

This method is clearly not scalable at all. There will always be a point where centralized graphs will no longer fit in the memory of the processors of the parallel machine, and where the execution of so many local optimization algorithms on the multi-centralized band graphs will be a waste of run time because, in spite of the perturbations of their initial state, most of the obtained local optima will be identical.

Yet, since the band graphs which we introduced in the previous section result in a dramatic reduction of problem space, global methods could well take advantage of this reduction. This is why we also implemented global refinement methods, such as genetic algorithms (we will not describe our own implementation, as it does not differ much from the literature [72]), as well as a new flavor of diffusion algorithm.

14.6.4   Diffusion Algorithms

As demonstrated above, GAs have interesting characteristics in terms of scalability and quality. However, their convergence rate is slow, because they are not specialized enough with respect to the problem at stake. In order to save time and compute power, it appeared worthwhile to us to turn to methods which specifically address our requirements for small and smooth frontiers, while still being of global and scalable nature. This is why we investigated diffusion-based methods instead of developing a fully parallel version of our GA method.

Our diffusion scheme, referred to as the *"jug of the Danaides"* [73], can apply to an arbitrary number of parts, but for the sake of clarity we will describe it in the context of graph bipartitioning, that is, with two parts only. We modeled the graph to bipartition in the following way. Nodes are represented as barrels of infinite capacity, which leak so that one unit of liquid at most drips per unit of time. When graph vertices are weighted, always with integer weights, the maximum quantity of liquid to be lost per unit of time is equal to the weight of the vertex. Graph edges are modeled by pipes of cross-section equal to their weight. In both parts, a source vertex is chosen, to which a source pipe is connected, which flows in a number of units of liquid per unit of time, equal to half the overall vertex weight. Two sorts of liquids are

FIGURE 14.8: Partition of graph BUMP into 8 parts with SCOTCH 5.0.6, using our diffusion-based method in a multilevel context. The cut is equal to 713 edges. This picture is to be compared with Figure 14.3, p. 385.

injected in the system: scotch in the first pipe, and anti-scotch in the second pipe, such that when some quantity of scotch mixes with the same quantity of anti-scotch, both vanish.

Unlike all of the methods presented in Section 14.6.1, our diffusion algorithm privileges load balancing over cut minimization. Yet, as such, it presents two weaknesses: nothing is said about the selection of the seed vertices, and performing such iterations over all of the graph vertices is very expensive compared to local optimization algorithms which only consider vertices in the immediate vicinity of the frontiers. However, both issues can be solved by the use of the band graphs presented in Section 14.6.2.

Anchor vertices represent a natural choice to be taken as seed vertices. Indeed, the most important problem for bubble-growing algorithms is the determination of the seed vertices from which bubbles are grown, which requires expensive processes involving all of the graph vertices [52, 55]. Since anchor vertices are connected to all of the vertices of the last layers, the diffused liquids flow as a front as if they originated from the farthest vertices from the frontier, which is indeed what would happen if they flowed from the center of a bubble having the frontier as its perimeter.

Our experiments show that, while our diffusion-based refinement method is much slower than its FM counterpart (from 3 to 20 times, according to the diffusion algorithm used), it helps provide partitions of good quality [73], as illustrated in Figure 14.8. Yet, because of the recursive bipartitioning nature of the algorithm, running times increase along with the number of parts, while partition quality tends to decrease, as analyzed in [74]. A full k-way diffusion algorithm is therefore necessary to preserve quality when the number of parts increases.

Our diffusion algorithm has the additional benefit of being highly scalable, and its parallel version does not substantially differ from its sequential counterpart [64] in the bipartitioning case. In the k-way case, our preliminary results show that the behavior of this version does not differ much from the one of the bipartitioning version when parts are sufficiently large, as the majority of interface vertices has only two neighbors. It bases on the structure of the k-way distributed band graph presented in Figure 14.7, to flow k sorts of liquids from the existing anchor vertices. Yet, load imbalance tends to increase along with the number of parts, apparently because the approximation of fronts by anchor vertices is more subject to artifacts [75]. These preliminary findings have to be confirmed by further experiments.

Basing on a k-way formulation of our band graphs, Meyerhenke *et al.* [71] have devised a parallel version of their bubble-growing algorithm, which exhibits a much faster convergence speed because of this reduced problem space. Their implementation provides very good results in term of quality, but the time they take is still long, so we expect that our solution will bring an improvement in this respect.

14.7   Performance Issues

We will briefly present in this section the results that we have obtained with revision 5.1.6 of PT-SCOTCH. This revision includes all of the contributions that were discussed in the previous sections. In these tests, PT-SCOTCH has been used to compute graph partitions in parallel by means of recursive bisection, using probabilistic matching, fold-dup on the final coarsening stages, and diffusion on distributed band graphs.

Regarding graph partitioning, in [64], we demonstrated that, on most of our test cases, the partitions computed by PT-SCOTCH compare favorably to those produced by PARMETIS. The decrease in cut size can be as high as 20% on some graphs, irrespective of the number of processes. PT-SCOTCH always computes better results for small numbers of parts, reflecting the intrinsic quality of its parallel graph bipartitioning framework. When the number of parts increases, the quality of the partitions produced by PT-SCOTCH degrades, and matches those produced by PARMETIS for about a thousand parts; it may eventually become slightly poorer for some graphs. This is due to the fact that, to date, PT-SCOTCH performs k-way partitioning by means of recursive bipartitioning, while PARMETIS uses a direct k-way algorithm. For large numbers of parts, the greedy nature of the recursive bipartitioning scheme, which prevents reconsidering earlier choices, negatively impacts partition quality. It is not surprising that the graphs for which PARMETIS gains over PT-SCOTCH are those of higher degree, for which bad decisions in the earlier bipartitioning stages produce higher penalties in terms of cut.

However, this phenomenon is most often compensated by the improvement in quality yielded by folding-with-duplication and multi-sequential phases (see Section 14.5.2). Indeed, for most of the test graphs, both PT-SCOTCH and

PARMETIS exhibit stable partition quality for all numbers of parts and up to 384 processes, which was the largest number of processes in these experiments.

The negative impact of recursive bipartitioning is of course even more perceptible for run time. While PT-SCOTCH can be more than three times faster than PARMETIS in the bipartitioning case, its execution time increases along with the number of parts, and can be several times higher than the one of PARMETIS for large number of parts.

Regarding sparse matrix ordering, the results which were obtained for a representative set of test matrices, using from 2 to 64 processes, are very satisfactory [34]. The quality of the orderings computed by PT-SCOTCH, measured as the overall number of floating-point operations required to factor the reordered sparse matrix (also called "operation count"), is equivalent to the one of the sequential SCOTCH software. Also, it is not significantly impacted by the number of processes involved in the computation of the orderings, while the operation counts of PARMETIS orderings increase dramatically along with their number. While PT-SCOTCH is about two to three times slower on average than PARMETIS, it can yield factorization operation counts that are as much twice as small as those from PARMETIS. This is of great interest as factorization times are more than one order of magnitude higher than ordering times.

14.8   Conclusion and Future Works

The SCOTCH software package is distributed at no charge, under the CeCILL-C free software license, to all interested parties, whether they are simple users, researchers working on the same subjects, or potential contributors. We plan to improve SCOTCH regularly, according to the new developments of our research.

The goals of our initial roadmap have been fulfilled, as SCOTCH has been able to partition 3D graphs of above 2 billion vertices, distributed on 2048 processes, on the *Platine* machine of CCRT. Weak scalability studies were also carried out up to 8192 processing elements on the *Hera* machine at LLNL.

To date, PT-SCOTCH can only compute graph partitions in parallel by means of recursive bipartitioning, but thanks to the upcoming parallel k-way partitioning scheme based on distributed k-way band graphs, we are about to release a new version providing parallel static mapping features. The latter are likely to gain popularity, as large parallel systems become more and more heterogeneous.

Yet, large heterogeneous machines may pose a synchronicity problem. Most algorithms that we presented here make use of halo exchanges, that is, some form of all-to-all communication, as well as of parallel reduction. With the advent of machines having more than a million processing elements, and in spite of the continuous improvement of communication subsystems, the demand for more asynchronicity in parallel algorithms is likely to increase. Yet, devising an asynchronous version of the jug of the Danaides algorithm is not as easy as for the original diffusion algorithms. The latter is based on solving a linear system, which may be performed by means of iterative algorithms for which there exist asynchronous formulations [76], while our requirement to remove one unit of liquid per time step creates a threshold effect which cannot be rolled back so easily. In this respect, GAs may offer a better alternative, as computations can take place asynchronously within independent demes. When enough of them exist, convergence is likely to be achieved even if some of them may lag behind. The study of parallel asynchronous graph partitioning algorithms is therefore, in our opinion, to become an active way of research in the years to come.

Acknowledgments

We thank Cédric Chevalier, Jun-Ho Her and all the past SCOTCH team interns for their contributions to the materials presented in this chapter. Thanks also to Jean Roman, former head of the *"Parallel Programming and Environments"* team at LaBRI, who welcomed our very first work on graph partitioning algorithms, making possible all of the above.

Bibliography

[1] Garey, M. R. and Johnson, D. S., *Computers and Intractablility: A Guide to the Theory of NP-completeness*, W. H. Freeman, San Francisco, CA, 1979.

[2] "SCOTCH: Static mapping, graph partitioning, and sparse matrix block ordering package," http://www.labri.fr/~pelegrin/scotch/.

[3] "Chaco: Software for Partitioning Graphs," http://www.sandia.gov/~bahendr/chaco.html.

[4] "JOSTLE: Graph partitioning software," http://staffweb.cms.gre.ac.uk/~c.walshaw/jostle/.

[5] "METIS: Family of Multilevel Partitioning Algorithms," http://glaros.dtc.umn.edu/gkhome/views/metis.

[6] Pellegrini, F. and Roman, J., "SCOTCH: A Software Package for Static Mapping by Dual Recursive Bipartitioning of Process and Architecture Graphs," *Proc. HPCN'96, Brussels*, Vol. 1067 of *LNCS*, April 1996, pp. 493–498.

[7] Hénon, P., Ramet, P., and Roman, J., "PASTIX: A High-Performance Parallel Direct Solver for Sparse Symmetric Definite Systems," *Parallel Computing*, Vol. 28, No. 2, Jan. 2002, pp. 301–321.

[8] Pellegrini, F., Roman, J., and Amestoy, P., "Hybridizing Nested Dissection and Halo Approximate Minimum Degree for Efficient Sparse Matrix Ordering," *Concurrency: Practice and Experience*, Vol. 12, 2000, pp. 69–84.

[9] Goudin, D., Hénon, P., Pellegrini, F., Ramet, P., Roman, J., and Pesqué, J.-J., "Parallel Sparse Linear Algebra and Application to Structural Mechanics," *Numerical Algorithms*, Vol. 24, 2000, pp. 371–391.

[10] Pellegrini, F. and Goudin, D., "Using the native mesh partitioning capabilities of Scotch 4.0 in a parallel industrial electromagnetics code," *Eleventh SIAM Conference on Parallel Processing for Scientific Computing*, San Francisco, CA, USA, Feb. 2004.

[11] Pellegrini, F., *Contributions to parallel multilevel graph partitioning*, Habilitation à Diriger des Recherches, Université Bordeaux I, Dec. 2009.

[12] Barnard, S. T. and Simon, H. D., "A fast multilevel implementation of recursive spectral bisection for partitioning unstructured problems," *Concurrency: Practice and Experience*, Vol. 6, No. 2, 1994, pp. 101–117.

[13] Hendrickson, B. and Leland, R., "A multilevel algorithm for partitioning graphs," *Proc. ACM/IEEE conference on Supercomputing (CDROM)*, Dec. 1995, 28 pages.

[14] Karypis, G. and Kumar, V., "Multilevel Graph Partitioning Schemes," *Proc. 24th Intern. Conf. Par. Proc., III*, CRC Press, 1995, pp. 113–122.

[15] Pothen, A., Simon, H. D., and Liou, K.-P., "Partitioning sparse matrices with eigenvectors of graphs," *SIAM J. Matrix Analysis*, Vol. 11, No. 3, July 1990, pp. 430–452.

[16] Hendrickson, B., Leland, R., and Van Driessche, R., "Skewed Graph Partitioning," *Proceedings of the 8th SIAM Conference on Parallel Processing for Scientific Computing*, IEEE, March 1997.

[17] Lo, V. M., "Heuristic algorithms for task assignment in distributed systems," *International Conference on Distributed Computer Systems*, IEEE, 1984, pp. 30–39.

[18] Sadayappan, P. and Ercal, F., "Cluster partitioning approaches to mapping parallel programs onto a hypercube," *Proc. International Conference on Supercomputing*, Vol. 297 of *LNCS*, June 1987, pp. 475–497.

[19] Pellegrini, F., "Static Mapping by Dual Recursive Bipartitioning of Process and Architecture Graphs," *Proc. SHPCC'94*, IEEE, May 1994, pp. 486–493.

[20] Walshaw, C., Cross, M., Everett, M. G., Johnson, S., and McManus, K., "Partitioning & Mapping of Unstructured Meshes to Parallel Machine Topologies," *Proc. Irregular'95*, No. 980 in LNCS, 1995, pp. 121–126.

[21] Gaidamour, J. and Hénon, P., "A parallel direct/iterative solver based on a Schur complement approach," *Proc. 11th Int. Conf. on Comp. Sci. and Eng.*, Sao Paulo, 2008, pp. 98–105.

[22] Çatalyurek, U. and Aykanat, C., "Hypergraph-Partitioning-Based Decomposition for Parallel Sparse-Matrix Vector Multiplication," *IEEE Trans. on Par. and Dist. Syst.*, Vol. 10, No. 7, 1999, pp. 673–693.

[23] Devine, K., Boman, E., Heaphy, R., Bisseling, R., and Catalyurek, U., "Parallel hypergraph partitioning for scientific computing," *Proc. 20th International Parallel and Distributed Processing Symposium*, IEEE, 2006.

[24] Hendrickson, B., "Graph partitioning and parallel solvers: Has the emperor no clothes?" *IRREGULAR'98: solving irregularly structured problems in parallel*, No. 1457 in LNCS, Aug. 1998, pp. 218–225.

[25] Trifunovic, A. and Knottenbelt, W., "Parallel multilevel algorithms for hypergraph partitioning," *Journal of Parallel and Distributed Computing*, Vol. 68, No. 5, 2008, pp. 563–581.

[26] George, A. and Liu, J. W.-H., *Computer solution of large sparse positive definite systems*, Prentice Hall, 1981.

[27] Hendrickson, B. and Rothberg, E., "Improving the Runtime and Quality of Nested Dissection Ordering," *SIAM J. Sci. Comput.*, Vol. 20, No. 2, 1998, pp. 468–489.

[28] Leiserson, C. and Lewis, J., "Orderings for parallel sparse symmetric factorization," *Third SIAM Conference on Parallel Processing for Scientific Computing*, 1987.

[29] Pothen, A. and Fan, C.-J., "Computing the Block Triangular Form of a Sparse Matrix," *ACM Trans. Math. Software*, Vol. 16, No. 4, Dec. 1990, pp. 303–324.

[30] Duff, I., "On Algorithms for Obtaining a Maximum Transversal," *ACM Trans. Math. Soft.*, Vol. 7, No. 3, Sept. 1981, pp. 315–330.

[31] Hopcroft, J. and Karp, R., "An $n^{5/2}$ Algorithm for Maximum Matchings in Bipartite Graphs," *SIAM J. Comput.*, Vol. 2, No. 4, Dec. 1973, pp. 225–231.

[32] Gupta, A., Karypis, G., and Kumar, V., "Highly scalable parallel algorithms for sparse matrix factorization," *IEEE Trans. Parallel Distrib. Syst.*, Vol. 8, No. 5, 1997, pp. 502–520.

[33] Vastenhouw, B. and Bisseling, R. H., "A Two-Dimensional Data Distribution Method for Parallel Sparse Matrix-Vector Multiplication," *SIAM Rev.*, Vol. 47, No. 1, 2005, pp. 67–95.

[34] Chevalier, C. and Pellegrini, F., "PT-SCOTCH: A tool for efficient parallel graph ordering," *Parallel Computing*, Vol. 34, 2008, pp. 318–331.

[35] Pellegrini, F., SCOTCH *and* LIBSCOTCH 5.1 *User's Guide*, LaBRI, Université Bordeaux I, Aug. 2008, Available from `http://www.labri.fr/~pelegrin/scotch/`.

[36] Pellegrini, F., PT-SCOTCH *and* LIBSCOTCH 5.1 *User's Guide*, LaBRI, Université Bordeaux I, Aug. 2008.

[37] Pellegrini, F., "Distillating knowledge about Scotch," *Combinatorial Scientific Computing*, No. 09061 in Dagstuhl Seminar Proceedings, Feb. 2009, 12 pages.

[38] Kernighan, B. W. and Lin, S., "An efficient heuristic procedure for partitionning graphs," *BELL System Technical Journal*, Feb. 1970, pp. 291–307.

[39] Fiduccia, C. M. and Mattheyses, R. M., "A linear-time heuristic for improving network partitions," *Proc. 19th Design Automation Conference*, IEEE, 1982, pp. 175–181.

[40] Karypis, G. and Kumar, V., "Multilevel k-way Partitioning Scheme for Irregular Graphs," *Journal of Parallel and Distributed Computing*, Vol. 48, 1998, pp. 96–129.

[41] Karypis, G. and Kumar, V., METIS- *A Software Package for Partitioning Unstructured Graphs, Partitioning Meshes, and Computing Fill-Reducing Orderings of Sparse Matrices*, University of Minnesota, Sept. 1998.

[42] Walshaw, C. and Cross, M., "Multilevel Mesh Partitioning for Heterogeneous Communication Networks," *Future Generation Comput. Syst.*, Vol. 17, No. 5, 2001, pp. 601–623.

[43] Chevalier, C. and Safro, I., "Weighted aggregation for multi-level graph partitioning," *Combinatorial Scientific Computing*, No. 09061 in Dagstuhl Seminar Proceedings, Schloss Dagstuhl - Leibniz-Zentrum für Informatik, Germany, 2009.

[44] Hendrickson, B. and Pothen, A., "Combinatorial Scientific Computing: The Enabling Power of Discrete Algorithms in Computational Science," *High Performance Computing for Computational Science—VECPAR 2006*, No. 4395 in LNCS, 2007, pp. 260–280.

[45] Karypis, G. and Kumar, V., "Analysis of Multilevel Graph Partitioning," *Proc. ACM/IEEE conference on Supercomputing (CDROM)*, Dec. 1995, 19 pages.

[46] Bollinger, S. W. and Midkiff, S. F., "Processor and link assignment in multicomputers using simulated annealing," *Proc. 11th Int. Conf. on Parallel Processing*, The Penn. State Univ. Press, Aug. 1988, pp. 1–7.

[47] Bui, T. N. and Moon, B. R., "Genetic Algorithm and Graph Partitioning," *IEEE Trans. Comput.*, Vol. 45, No. 7, 1996, pp. 841–855.

[48] Langham, A. E. and Grant, P. W., "Using Competing Ant Colonies to Solve k-way Partitioning Problems with Foraging and Raiding Strategies," *ECAL'99: Proc. 5th European Conference on Advances in Artificial Life*, No. 1674 in LNCS, 1999, pp. 621–625.

[49] Battiti, R. and Bertossi, A. A., "Greedy, Prohibition, and Reactive Heuristics for Graph Partitioning," *IEEE Trans. Comput.*, Vol. 48, No. 4, 1999, pp. 361–385.

[50] Fahrat, C., "A simple and efficient automatic FEM domain decomposer," *Computers and Structures*, Vol. 28, No. 5, 1988, pp. 579–602.

[51] Nour-Omid, B., Raefsky, A., and Lyzenga, G., "Solving finite element equations on concurrent computers," *Parallel Computations and Their Impact on Mechanics*, edited by A. K. Noor, ASME Press, 1986, pp. 209–227.

[52] Diekmann, R., Preis, R., Schlimbach, F., and Walshaw, C., "Aspect Ratio for Mesh Partitioning," *Proc. Euro-Par'98*, Vol. 1470 of *LNCS*, 1998, pp. 347–351.

[53] Wan, Y., Roy, S., Saberi, A., and Lesieutre, B., "A Stochastic Automaton-based Algorithm for Flexible and Distributed Network Partitioning," *Proc. Swarm Intelligence Symposium*, IEEE, 2005, pp. 273–280.

[54] Vanderstraeten, R., Keunings, R., and Farhat, C., "Beyond conventional mesh partitioning algorithms," *SIAM Conf. on Par. Proc.*, 1995, pp. 611–614.

[55] Meyerhenke, H. and Schamberger, S., "Balancing Parallel Adaptive FEM Computations by Solving Systems of Linear Equations," *Proc. Euro-Par'05*, Vol. 3648 of *LNCS*, 2005, pp. 209–219.

[56] Chan, A., Dehne, F. Bose, P., and Latzel, M., "Coarse grained parallel algorithms for graph matching," *Parallel Computing*, Vol. 34, No. 1, 2008, pp. 47–62.

[57] Barnard, S. T., "PMRSB: Parallel multilevel recursive spectral bisection," *Proc. ACM/IEEE Conference on Supercomputing (CDROM)*, Dec. 1995.

[58] Karypis, G. and Kumar, V., "A Coarse-Grain Parallel Formulation of Multilevel k-way Graph-partitioning Algorithm," *Proc. 8th SIAM Conference on Parallel Processing for Scientific Computing*, 1997.

[59] Karypis, G. and Kumar, V., "Parallel multilevel k-way partitioning scheme for irregular graphs," *SIAM Review*, Vol. 41, No. 2, 1999, pp. 278–300.

[60] Luby, M., "A simple parallel algorithm for the maximal independent set problem," *SIAM J. Comput.*, Vol. 15, No. 4, 1986, pp. 1036–1055.

[61] Walshaw, C., Cross, M., and Everett, M. G., "Parallel Dynamic Graph Partitioning for Unstructured Meshes," Tech. Rep. 97/IM/20, University of Greenwich, London SE18 6PF, UK, March 1997.

[62] Walshaw, C., Cross, M., and Everett, M. G., "Parallel Dynamic Graph Partitioning for Adaptive Unstructured Meshes," *J. Parallel Distrib. Comput.*, Vol. 47, No. 2, 1997, pp. 102–108.

[63] Chevalier, C., *Conception et mise en œuvre d'outils efficaces pour le partitionnement et la distribution parallèles de problèmes numériques de très grande taille*, Thèse de Doctorat, LaBRI, Université Bordeaux I, Sept. 2007.

[64] Her, J.-H. and Pellegrini, F., "Efficient and scalable parallel graph partitioning by recursive bipartitioning," *Parallel Computing*, 2010, Submitted. `http://www.labri.fr/~pelegrin/papers/scotch_parallelbipart_parcomp.pdf`.

[65] Karypis, G. and Kumar, V., "A Parallel Algorithm for Multilevel Graph Partitioning and Sparse Matrix Ordering," *Journal of Parallel and Distributed Computing*, Vol. 48, 1998, pp. 71–95.

[66] Walshaw, C. and Cross, M., "Parallel Optimisation Algorithms for Multilevel Mesh Partitioning," *Parallel Computing*, Vol. 26, No. 12, 2000, pp. 1635–1660.

[67] Rama Mohan Rao, A., "Distributed evolutionary multi-objective mesh-partitioning algorithm for parallel finite element computations," *Computers & Structures*, Vol. 87, No. 23–24, 2009, pp. 1461–1473.

[68] Talbi, E.-G. and Bessière, P., "A parallel genetic algorithm for the graph partitioning problem," *ICS'91: Proceedings of the 5th international conference on Supercomputing*, ACM, 1991, pp. 312–320.

[69] Diekmann, R., Meyer, D., and Monien, B., "Parallel Decomposition of Unstructured FEM-Meshes," *Concurrency: Practice & Experience*, Vol. 10, No. 1, 1998, pp. 53–72.

[70] Meyerhenke, H. and Schamberger, S., "A Parallel Shape Optimizing Load Balancer," *Proc. Euro-Par'06*, Vol. 4128 of *LNCS*, 2006, pp. 232–242.

[71] Meyerhenke, H., Monien, B., and Sauerwald, T., "A new diffusion-based multilevel algorithm for computing graph partitions of very high quality," *Proc. 22nd IPDPS*, 2008, pp. 1–13.

[72] Chevalier, C. and Pellegrini, F., "Improvement of the Efficiency of Genetic Algorithms for Scalable Parallel Graph Partitioning in a Multi-Level Framework," *Proc. Euro-Par'06, Dresden*, Vol. 4128 of *LNCS*, Sept. 2006, pp. 243–252.

[73] Pellegrini, F., "A parallelisable multi-level banded diffusion scheme for computing balanced partitions with smooth boundaries," *Proc. Euro-Par'07*, Vol. 4641 of *LNCS*, Springer, Rennes, Aug. 2007, pp. 191–200.

[74] Simon, H. D. and Teng, S.-H., "How good is recursive bisection," *SIAM J. Sci. Comput.*, Vol. 18, No. 5, Sept. 1997, pp. 1436–1445.

[75] Her, J.-H. and Pellegrini, F., "Towards efficient and scalable parallel static mapping," *CSC'2009*, SIAM, Oct. 2009.

[76] Tongxiang, G., "Asynchronous relaxed iterative methods for solving linear systems of equations," *Applied Mathematics and Mechanics*, Vol. 18, No. 8, Aug. 1997, pp. 801–806.

Chapter 15

Massively Parallel Graph Partitioning: A Case in Human Bone Simulations

C. Bekas, A. Curioni

IBM Research

P. Arbenz, C. Flaig, G. H. van Lenthe, R. Müller, A. J. Wirth

ETH Zurich

15.1 Introduction

High resolution *in vivo* peripheral quantitative computed tomography (HR-pQCT) provides detailed information on bone structure (see Figure 15.1). In particular, bone is readily modeled by a direct mapping of HR-pQCT voxels to bone elements. Obviously, bone density estimates are easily computed from this voxel model. In fact, the analysis of bone density is today's standard approach of predicting bone strength and fracture risk in diseases like osteoporosis, that is, according to the World Health Organization (WHO), second only to cardiovascular disease as a leading health care problem.

Observe that such a quantitative analysis of bone density does not take into account the microarchitectural structure of the bone. Indeed, it has been shown that the correlation of osteoporotic fracture risk with bone density alone can give poor results with respect to accurately assessing this risk. This

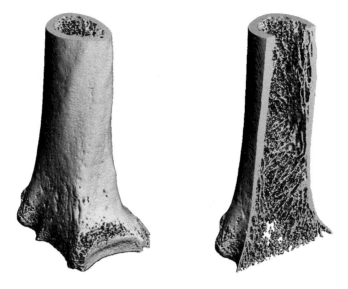

FIGURE 15.1: Bone specimens from human subjects. Left: mostly cortical bone is shown. Right: Trabecular (osteoporotic) trabecular bone is shown.

is not surprising: bone density, bone geometry, and bone microarchitecture are all components that determine bone strength, hence, all these aspects will affect bone fracture risk [1]

Recent advances in imaging capabilities have given us unprecedented capabilities in understanding bone architecture. In addition, by coupling these new imaging capabilities with finite element modeling allows us to perform highly accurate simulations (at the micro scale level) of bone behavior to external stress that allow us to efficiently determine bone stiffness and strength. The target is to accurately model bone microarchitecture, because of its important contribution to bone stiffness and strength. The novel approach shows high potential to improve individual prediction, a tool much needed in the diagnosis and treatment of osteoporosis, because it can be done in vivo. This means that the finite element models are derived from image data taken from living subjects. This is a property that clearly sets this methodology apart from in vitro studies, where bone specimens from animal subjects or diseased individuals are subjected to mechanical testing in a laboratory.

Figure 15.2 graphically illustrates the proposed methodology. A patient undergoes a high-resolution pQCT scan from which an accurate 3D representation of the bone structure is obtained. For example, Melton *et al* [1] have taken scans of patient wrists resulting in stacks of 116 slices with a resolution of 1024-by-1024 pixels per slice. In the sequel, the underlying voxel model is transformed into a finite element model, using a direct voxel-to-element conversion. The resulting micro-Finite Element model is used to simulate how

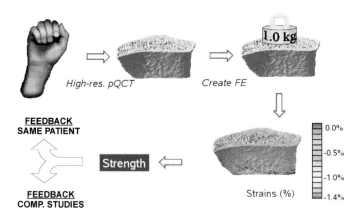

FIGURE 15.2: Simulation model.

the bone withstands external forces. In this work we have used the pure displacement model of linear elasticity, which has been shown to give results in good agreement with normal physiological bone loading. With the computational model at hand, a large array of external loads can be simulated, giving researchers and clinicians unprecedented, in-depth understanding of bone behavior. Based on the computed bone displacements a "heat" map of strains is created which provides a very useful graphical aid.

It is not difficult to see that the new approach allows for a systematic study of bone behavior, that targets to allow only minimal (and controlled) empirical bias. Of course, this much desired property comes at a cost. Due to the intricate microarchitecture of human bone, hundreds of millions of voxels and thus elements will be required (e.g., see Figure 15.2 for the mentioned wrist study). Since the bone's structural details are of the micrometer scale, we call these studies micro-Finite Element analyses. As mentioned earlier, it is absolutely essential to include the micro scale in the model. Failing to do so, would provide model predictions that are quickly deviating from the true physical behavior. For these reasons, the resulting computer models involve hundreds of millions or even billions of degrees of freedom. In particular, as we will see in the following sections, a positive definite linear system will need to be solved, where its dimension will reach 1.5×10^9 in the largest ever simulation of this kind.

As we will see in the following sections, accurately studying bone specimens will require a huge number of degrees of freedom in the underlying linear equations that are to be solved. This number will vary from tens of millions (such as in the aforementioned wrist study) to billions when biopsy vertebrae specimens are under investigation. In the former case, we envision a wide clinical adoption of our methodologies where on on-line diagnosis can be rendered possible by the use of multicore platforms. In parallel work to the present one, we have been able to verify the validity of the on-line approach on for the wrist specimens [2]. In fact, we were able to show that less than 32 Gigabytes of main memory are required while run times did will not exceed a few minutes. A further implication, has to do with comparison studies of a current simulation result with previous ones, both from the same as well as form other patients. (see feedback loop arrow in Figure 15.2.

These problems stretch our numerical algorithms and computational platforms to their limits. Indeed, for the success of the simulations, state-of-the-art numerical and non-numerical algorithms need to be implemented using state-of-the-art software engineering environments and deployed on the latest massively parallel computing platforms. In this work we target to exploit the massive computational power of the Blue Gene/L Supercomputer. We have used up to 8192 computing cores which allowed us to simulate the largest ever human bone specimen in under 20 minutes of wall-clock time, including input/output operations. Our endeavors in these exciting simulations have highlighted kernels that we believe will be particularly crucial in the era of petascale computing.

In particular, applications that involve irregular domains, rely on graph (or hypergraph) partitioning in order to be distributed on parallel processors. In our case the mesh that models the bone is distributed to processors in two phases. First, a geometric partitioning distributes horizontal voxel slices to the processors. Then, a multilevel k-way repartitioning is utilized with a goal to evenly distribute voxels to processors while at the same time keeping inter-processor communication as small as possible. The problem sizes we encounter in bone simulations are so large (billions of voxels) that parallel graph partitioners need to be used. We will showcase the importance of parallel graph partitioning, in identifying it as the single most significant bottleneck for massive scale-out of such applications.

15.2   Computational Model

The mathematical modeling of the problem is given by the Lamé equations of elasticity [3, 4]. Find $\mathbf{u} \in \mathbf{V}$ such that,

$$\int_\Omega [2\mu\epsilon(\mathbf{u}) : \epsilon(\mathbf{v}) + \lambda \div \mathbf{u} \div \mathbf{v}]d\Omega = \int_\Omega \mathbf{f}^\top \mathbf{v}d\Omega + \int_{\Gamma_N} \mathbf{g}_S^\top \mathbf{v} \ d\Gamma, \quad \forall \mathbf{v} \in \mathbf{V} (15.1)$$

where $V \subset (H^1_{\Gamma_D}(\Omega))^3$ is a Sobolev space, Γ_D and Γ_N are the Dirichlet and Neumann part of $\partial\Omega$, $\overline{\Gamma_D \cup \Gamma_N} = \overline{\partial\Omega}$, $\Gamma_D \cap \Gamma_N = \emptyset$, and $\mathbf{u}$ describes displacements, λ and μ are the Lamé constants, f the volume forces, and g the boundary tractions. The symmetric strains in 15.1 are given by:

$$\epsilon(\mathbf{u}) = \frac{1}{2}(\nabla\mathbf{u} + (\nabla\mathbf{u})^\top). \tag{15.2}$$

The intricate geometric structure of the bone is approximated by voxels, which drives the discretization of the linear elasticity equation. Each displacement component is a continuous piecewise trilinear function that is determined by its values at the vertices of the voxels. The discretization of (15.1) leads to the solution of a very large and sparse linear system

$$Au = f, \tag{15.3}$$

where matrix A is symmetric and positive definite as long as $\Gamma_D \neq \emptyset$.

The large number of voxels entails large symmetric positive-definite systems of linear equations. The method of choice to solve (15.3) is the preconditioned conjugate gradient (PCG) algorithm. PCG has been applied earlier to large problems from bone structure analysis with success [5].

The conjugate gradient method for linear systems is an iterative, Krylov subspace technique which restricts the linear system (15.3) on the linear subspace

$$\mathcal{K}_m(A, r_0) = \text{span}\{r_0, Ar_0, \ldots, A^{m-1}r_0\}, \tag{15.4}$$

where, $r_0 = f - Au^{(0)}$ is the residual for some initial guess $u^{(0)}$ (see for example [6]). In particular, the method constructs successive approximations $u^{(k)}$ from the subspace $\mathcal{K}_m(A, r_0)$ such that the $A-$norm of the forward error, $\|u^* - u^{(k)}\|_A$ is minimized at each step k.

Unlike direct (Gaussian elimination like) methods, such as the ones implemented in the popular LAPACK[7] library, PCG does not alter the elements of the matrix A. Instead, the key computational kernel is a matrix-vector product with matrix A at every step of the method. Thus, since matrix A is extremely large and very sparse, parallel sparse matrix-vector products play a central role in our method. We will illustrate this in the sections that follow.

The Conjugate Gradient (CG) algorithm can converge very fast, i.e., requiring a small dimension m for the Krylov subspace, in the case that the matrix involved is not too ill-conditioned, or more precisely, if the matrix does not have many (and possibly clustered) eigenvalues close to 0. Thus, CG can be extremely competitive in comparison to direct methods which impose a cost that in general grows cubically with the size of the matrix. We note here that in the case of matrices with special (e.g., banded) structure, sparse direct methods, such as the ones implemented in popular software packages as PARDISO and MUMPS [8, 9, 10, 11], can offer a quite competitive alternative to iterative methods. However, irregular sparsity structure and large

TABLE 15.1: The Preconditioned Conjugate Gradient method for SPD linear systems. $<,>$ depicts inner products between vectors.

Conjugate Gradient: $u := cg(A, u^{(0)}, f, cg\_iter)$

Input	SPD matrix: A, right hand side: f, preconditioner B
	Initial solution $u^{(0)}$, CG iterations: cg_iter
Output	Approximate solution u of linear system $Au = f$

1.   Compute initial residual $r^{(1)} = f - Au^{(0)}$, $z^{(1)} = B^{-1}r^{(1)}$, $p^{(1)} = z^{(1)}$

2.   **do** $k = 1, \ldots, cg\_iter$

3. $t := Ap^{(k)}$

4. $\alpha_k = \dfrac{<r^{(k)}, z^{(k)}>}{<t, p^{(k)}>}$

5. $u^{(k)} = u^{(k-1)} + \alpha_k t$

6. $t = \alpha_k t$

7. $r^{(k+1)} = r^{(k)} - t$, $z^{(k+1)} = B^{-1}r^{(k+1)}$

8. $\beta_k = \dfrac{<r^{(k+1)}, z^{(k+1)}>}{<r^{(k)}, p^{(k)}>}$

9. $p^{(k+1)} = r^{(k+1)} + \beta_k p^{(k)}$

10.   **enddo**

---

bandwidth in 3D problems, causes huge fill-in (i.e., new non-zero entries in the matrix factorization). These are problems that can significantly negatively affect the applicability of such techniques.

It is precisely this case that is posed in our simulations. The matrix dimension can easily reach the order of a billion and bone architecture will impose a large bandwidth on the matrix. In this case, it is straightforward to see that the CG method is a natural choice for our simulations. However, as we mentioned above, the convergence properties of CG are linked with the condition number $\kappa(A)$ of the matrix A (or more generally with its spectral properties [6]). It is well known that $\kappa(A) \approx O(h^{-2})$, where h is the mesh size [3]. In order to improve the spectral properties of matrix A, the original linear system (15.3) is replaced by the linear system

$$AB^{-1}x = f, \qquad u = B^{-1}x, \qquad (15.5)$$

where the matrix B is known as a preconditioner. It is designed in such a way such that AB^{-1} has favorable spectral properties, while at the same time the linear system $Bu = x$ can be solved cheaply. Thus, the PCG method is the CG method with the application of a suitable preconditioner. Table 15.1 provides an algorithmic description.

We employ an aggregation-based Algebraic Multigrid (AMG) preconditioner, that is known to perform very well in the kind of linear systems that we study [5]. Indeed, in all of the bones that we have studied so far, we were able to reduce the residual norm by 6-7 orders magnitude in about one hundred PCG iteration steps. It is important to point out that in order to render our method memory efficient, we have developed a matrix-free variant of PCG that does not require building the system matrix but is still able to construct an aggregation-based AMG preconditioner [12]. This preconditioner is incorporated in the Trilinos package ML [13]. It consumes only about a forth of the memory of ML's standard AMG preconditioner. From now on we will be referring to this preconditioner.

As the following sections will show, efficient partitioning of the finite element/volume mesh is crucial in achieving high scalability of the simulations. In particular, the matrix-vector products in the CG method, as well as the preconditioner construction are crucially affected by the partitioning quality. In parallel sparse matrix-vector multiplications, which is the core of the current application, the de facto standard is to distribute the work of each matrix-vector product to the available processors by means of graph partitioning. That is, every sparse matrix defines an associated graph by means of the adjacency information of every matrix row. Let each unknown (or finite element or voxel) correspond to a vertex in the graph. Since each row of the sparse matrix corresponds to an unknown (or a finite element, or voxel), the non zeros of this row determine the neighboring nodes in the graph. Thus, when we seek to perform an efficient and scalable parallel matrix-vector product, we can model it as follows: We seek to partition the graph nodes in groups, such that the number of edges that connect different groups (known as the partition cut) is as small as possible. At the same time, we seek the population of each group to be as even as possible. The cut models communication, as surely splitting a matrix row between two or more processors will require some form of data exchange among these processors in order to compute the dot product for that matrix row. The graph partitioning problem is known to be NP-complete. However, efficient and powerful heuristic methods have been suggested and used with great success. In what follows we will demonstrate that in view of massively parallel simulations of very large problems, such as the ones on bone simulations, this model has reached its limits.

15.2.1   Software Implementation Environment

Large scale scientific applications, that are targeted to be deployed on massively parallel architectures, require the use of sophisticated software engineering techniques and libraries. Our simulation framework is implemented in the software package ParFE [14] that is based on the public-domain object-oriented software framework Trilinos [13]. The latter provides the backbone sparse linear algebra kernels, such as distributed data-structures and sparse matrix-vector multiplications. In addition, Trilinos implements the ML software suite for multigrid preconditioners. Trilinos is parallelized using MPI [15] and data is distributed by means of ParMETIS [16] which has been shown to exhibit good scaling in several parallel applications. Data is stored in the

HDF5 data format [17]. In [12] the authors presented results obtained on the Cray XT/3 at the Swiss National Supercomputing Centre (CSCS) in Manno, Switzerland[1], that validated the correctness and usefulness of this approach.

In the present work we carry this important research over to the next level. In particular, the accurate study of realistic simulations of large human bone specimens requires extreme scale-out that stretches the limits of both algorithms and computational platforms. Here, we report our progress as well as the challenges that we face towards this goal. We focused on the IBM BG/L Supercomputer with the goal of exploiting its excellent scale-out potential that allows us to attempt realistic simulations involving a computational loads that are one order of magnitude larger than in previous computations. Indeed, the BG/L Supercomputer, coupled with algorithmic advances introduced into our solver, made possible the largest simulation of its kind of a human vertebra bone specimen.

15.3 The Study

We conducted a study of artificial as well as human bone specimens resulting in very large sparse systems of up to about 1.5 billions of unknowns. These runs always required less than half an hour, using up to 8 racks (8192 nodes) of the BG/L system at the IBM T.J. Watson Research Center. We note that this run-time includes input as well as output. In fact, the largest input meshes occupied more than 90 gigabytes on disk, while output was also worth of a few tens of gigabytes. Input and output I/O was always a few percentage (%1–%2) of the total run time.

We have conducted two series of experiments. In the first, we measured the so called weak scalability of the code on a sequence of carefully constructed artificial bone samples that increase in size proportional to the increase of the cores employed. The second series involves strong scalability (the number of BG/L cores increases while the problem size stays fixed) on a real human bone specimen which (to the best of our knowledge) is the largest simulation of its kind so far, the results of which provide valuable information about the strength of osteoporotic bone. A second important target of this experiment was to identify the parts of the code that exhibit the best as well as the worst scalability. In particular, we focused on measuring the performance of the graph repartitioning, the construction of the preconditioner and the solution phase (application of the preconditioned CG algorithm).

---

[1]http://www.cscs.ch

TABLE 15.2: Run times in seconds for the weak scalability test. The last column indicates the number of PCG iterations.

number of nodes	time to repartition	preconditioner setup time	solution time	total time	iteration count
1	2.50	27.5	113	149	94
8	6.60	45.2	116	179	86
27	7.10	51.5	113	185	80
64	7.10	53.6	124	199	86
125	7.60	55.7	122	202	81
216	8.00	65.6	119	207	79
343	8.60	55.0	119	211	77
512	9.10	67.5	118	214	75
729	10.4	70.5	118	216	74
1000	12.0	87.0	126	248	77
1728	18.5	185.0	145	376	81

15.3.1   Weak Scalability Test

The problems were obtained by mirroring a small cube of human trabecular bone scanned with a high-resolution microCT system. This cubical piece is a small portion of the interior of the bone displayed in Figure 15.1. We have tested weak scalability using a series of 12 cubes (see Table 15.2), starting from 1 to 1728 BG/L nodes, where the k-th cube required k^3 BG/L nodes. The smallest cube ($k = 1$) involved about 300.000 degrees of freedom, the largest cube ($k = 12$) about $300.000k^3 \approx 520$ million. In weak scalability tests the problem size is increased in proportion to the number of processors. This means that we solve increasingly larger problems and we suitably increase the number of processors employed in such a way that the total work per processor remains the same. Of course, parallel AMG preconditioners behave differently in terms of convergence properties as processor counts increase. However, in our case we observed a very steady iteration count (last column of Table 15.2). A code passes the test if the execution time does not depend on the processor number.

All runs were done in co-processor mode in which the second PPC440 core of the BG/L nodes handles MPI communication. We first note again that we observed that the number of iterations that PCG required to be fairly constant. We observed satisfactory scalability up to 1000 CPUs. However, load imbalance caused by poor repartitioning started to manifest in the largest cases entailing the increase of the time for constructing the preconditioner. In particular, what happened is that the work per processor was not even for large processor counts. This is because the graph partitioner was not able to distribute the finite volume mesh evenly. In addition, we observed a communication imbalance as well. That is, a poor partition will also lead to some processors requiring significantly more communication than others. Notice that

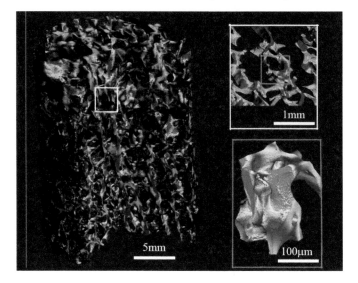

FIGURE 15.3: Effective strain on a human bone (vertebra) specimen.

preconditioner construction was the most severely affected component. This is because load and communication imbalance cause some processors to spent much more time than others in the multigrid cycle. We were able to address this problem in part by limiting the number of AMG levels. In particular, we did not allow them to exceed ten levels. The drawback of such an approach is that the complexity of the solver at the coarsest level can increase significantly. To remedy this, we opted not to use sparse direct solvers but rather a few steps (not larger than four) of simple polynomial smoothers, such as the Chebyshev iteration. We followed this practice for all simulations in this code, with minimal increase in the number of PCG steps (i.e., the quality of the AMG preconditioner was not noticeably affected).

15.3.2 Strong Scalability Test

We conducted the largest simulation of its kind so far (1.5 billions degrees of freedom) and calculated the local effective strains of a vertebral bone specimen. This enabled a highly detailed analysis of bone deformation under load, and calculation of bone stiffness and strength. Figure 15.3 illustrates the effective strain of the bone specimen. For this specimen we conducted a strong scalability test (see the run times in Table 15.3). No less than 4800 BG/L nodes were required to meet the memory requirements of the problem (ca. 2.4 Terabytes). We scaled our runs up to eight BG/L racks correspondig to 8192 BG/L dual core nodes.

The solution phase scales quite well and uniformly (column 'solution') as

TABLE 15.3: Run times in seconds for the strong scalability test. The last column indicates the number of PCG iterations.

number of nodes	time to repartition	preconditioner setup time	solution time	total time	iteration count
4800	157	614	482	1341	71
5000	187	420	447	1143	68
5400	230	577	431	1323	67
5800	304	362	417	1165	69
6200	397	430	425	1332	69
6800	497	368	416	1359	70
7900	749	427	380	1632	71
8100	775	418	365	1635	70

the number of employed BG/L cores increases. The construction of the pre-conditioner exhibits similar overall scalability (column 'precond'). However, this is not as uniform as before. The reason is the load imbalance that is caused by the graph repartitioner (ParMETIS). Ideally, each node accommodates an (almost) equal number of nodes. Figure 15.4 displays the maximum, median and minimum numbers of finite element per MPI process distribution, for the employed processing units (BG/L cores in this case). It is clear that we are experiencing significant load imbalance. In the ideal case, the load of each processor should be clustered around the median value. We note that in our case only a few cores are assigned many fewer mesh nodes than the average, while several (close to 5%) cores are assigned a significantly larger node load than the average. This causes cores with a light load to wait at synchronization points and thus to register much larger communication times than the average. Moreover, it is important to notice that the cases for which the mesh-node distribution disparity is the smallest are those that achieve the best scalability in the preconditioner construction phase (i.e., the case of 6800 BG/L cores).

A second important observation with respect to load imbalance is that on massively parallel platforms its effects can be more challenging to address. The reason is that computing cores have access to a limited amount of memory. Of course, one needs to understand that this is not a limitation of massively parallel machines, but rather a feature. Indeed, it is becoming common knowledge that only well balanced machines, with respect to processor clock, interconnect, and main memory amount and bandwidth per chip, can scale to hundreds of thousands of computing cores. Thus, severe load imbalance can not only significantly hamper performance, but even halt the program altogether. As we pointed out in the previous section, in order to address this important problem, we have made a series of changes in the AMG preconditioner as compared to its original design. The most important change was to reduce the number of levels (grids) of the multigrid preconditioner. Of course,

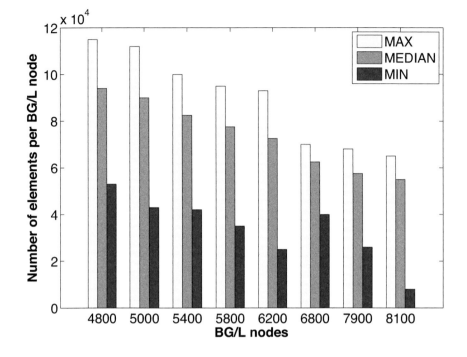

FIGURE 15.4: Maximum, median and minimum number of mesh elements per MPI process achieved by graph repartitioning (ParMETIS) for the strong scalability test.

this leads to larger coarsest linear systems. Therefore, we opt for executing a few steps of the Chebyshev or Gauss-Seidel iteration instead of a sparse direct solver [18].

Returning to the scalability of the solution phase we note that the preconditioned conjugate gradient method is dominated by the sparse matrix-vector products (including the application of the preconditioner) and the global reductions (`MPI_ALLREDUCE`) for the calculation of vector inner products. For this purpose we utilize the global tree network available on the BG/L platform that achieves excellent latency as well as bandwidth which is important (remember that global reduction involves more that 10 gigabytes of data since we have more than 1.5 billion degrees of freedom). However, fast communication times are achieved thanks to the large bandwidth of the TREE network.

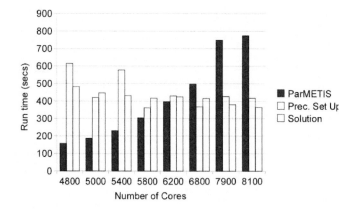

FIGURE 15.5: Strong scalability test on the large bone specimen (c.a. 1.5×10^9 degrees of freedom). Timings for the repartition (ParMETIS), construction of the preconditioner, and solution phase.

15.3.3   Repartitioning Scalability

In parallel sparse matrix computations, graph partitioning is the de facto standard of obtaining parallelization in terms of data storage as well as computations. Although more advanced models based on hypergraph partitioning (see [19, 20]) have shown excellent potential to give better quality partitions than standard graph partitioning, these come with a significantly increased cost that can become prohibitive for very large partition counts such as the ones that we are interested in this work.

We observed that the major bottleneck in extreme scale-out of ParFE is caused by graph repartitioning. Figure 15.5 displays the run times for the construction of the preconditioner, the solution phase, and the time required for graph repartitioning as listed in Table 15.3. It is evident that the latter dominates the overall runtimes.

In order for ParMETIS to be suitable to run on 8 racks of a BG/L system (with 512 Megabytes of memory per node) we introduced a number of algorithmic modifications. The main change involved the geometric partitioning algorithm which relies on a serial version of the quicksort algorithm. Geometric partitioning is used to achieve an initial partitioning of CT scan data to the available cores on which k-way multilevel repartitioning is applied at a second step. Figure 15.7 illustrates the geometric and the subsequent multilevel partitioning achieved by ParMETIS for a small bone specimen on 16 processors.

We implemented a fully parallel mergesort algorithm to replace the serial quicksort, since using the latter on 8192 BG/L nodes (8 racks) requires 512 MBbytes of memory per node leaving no memory for any other calculations

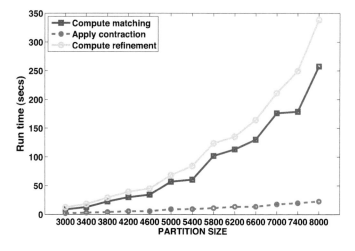

FIGURE 15.6: Timing analysis for repartitioning on a smaller bone sample. Notice the log-log scales.

(or even for the light weight Linux OS kernel on the compute cores), thus causing the program to halt. A second important modification was to replace asynchronous all-to-all communications with MPI collective communications. The former floods the network with messages, immediately overflowing the MPI buffers, while the latter is very efficiently implemented on the BG/L communication layer.

In achieving scale-out on thousands of compute cores the challenges we face are aligned towards two main directions.

15.3.3.1   Load Imbalance.

Complicated geometry of the application domain, such as the intricate structure of trabecular bone (cf. Figures 15.1, 15.3, and 15.7) can entail significantly imbalanced partitions that have a strong negative impact when thousands of processors are used (see Figure 15.4). It is well known that graph partitioning is an extremely challenging combinatorial problem for which the best algorithms utilize complicated heuristics. Without doubt during the last years multilevel coarsening-decoarsening algorithms (such as the ones implemented in popular software such as ParMETIS [16], Zoltan [21], or SCOTCH [22] have dominated the scene. However, our experiments indicate that in the era of massively parallel machines with tens (or even hundreds) of thousands of computing cores this model of graph partitioning has reached its limits.

15.3.3.2 Scalability.

The scalability of parallel graph partitioning tools on tens of thousands of processors appears to be a formidable task. Figure 15.6 illustrates a run time breakdown for ParMETIS on a smaller real human bone sample. There are two main observations:

(i) There are two phases that clearly do not scale at all. These are the coarsening and decoarsening (partition refinement) phases. They involve the solution of a series of graph maximal-matching and matching refinement problems for which many point to point communications are needed. In ParMETIS, these are implemented in an asynchronous manner (in order to avoid dead-locks and race conditions among other targets) that pose great difficulties when tens of thousands of computing cores are involved. Finding the optimal scheduling for these communications, so that they can be safely performed concurrently, is a formidable task on massively parallel platforms.

(ii) Observe that the run times of the phases that appear to cause much less scalability problems, in comparison with the above, increase with a slope that suggests they will also not scale when several tens of thousands of cores will be used. Thus, we claim that in order to achieve extreme scale-out on real world applications with intricate geometric characteristic, we will have to use either an algorithmic approach, that naturally scales on massively parallel machines, or we will need to consider a mapping of applications to available resources that does not rely on graph partitioning.

15.4 Conclusion

We have presented our progress in very large scale simulations of human bones that can provide crucial information and insight for appreciating the risk of fractures due to osteoporosis. Since osteoporosis is a leading health care problem, the development of a software tool that can render this kind of risk analysis to be a routine procedure is very important. In previous work we showed the validity of the approach and its ability to solve large problems [23]. In the scope of this book, we adopt this work in the context of combinatorial scientific computing and illustrate the central role of graph partitioning in large scale bone simulations. We demonstrate that it is possible to achieve simulations of unprecedented size in a few minutes. We were able to achieve this by careful algorithm adaptation and code re-engineering. As a result we were able to harness the excellent scale-out potential of the Blue Gene Supercomputer.

FIGURE 15.7: (See color insert.) Left: Geometric partitioning on 16 processors. Right: K-way multilevel partitioning on 16 processors. Both partitionings have been computed using ParMETIS.

Our work has led us to pinpoint an issue of utmost importance, that we believe will become the challenge for the solution of unstructured problems in the next years. At the dawn of the Petaflop/s era, it is inevitable that the problem on how to efficiently map real world applications on hundreds of thousands or even millions of processing elements will need to be solved efficiently. Today's most successful approaches are graph based. However, we shaw that the NP-complete nature of this model requires for heuristics that, at least currently, present immense obstacles for their massively parallel deployment. We believe that the solution will require next generation algorithms and efficient mapping models. We furthermore believe that the mapping models such as graph partitioning or the more advanced hypergraph partitioning ([24]) will need to be extended or modified so that particularities of the underlying machine architecture are seriously taken into consideration. Preliminary results ([25]) indicate that this is indeed possible. These mapping tools will need to be naturally suited for extreme scale-out as well. Otherwise they will remain the bottleneck of the whole process.

Acknowledgment

We would like to thank the Swiss Supercomputing Center (CSCS) for help and facilities that made pre- and post-processing of data (creation of meshes and visualizations) possible.

Bibliography

[1] Melton, L. J., Riggs, B. L., Harry van Lenthe, G., Achenbach, S. J., Müller, R., L. Bouxsein, M., Amin, S., Atkinson, E. J., and Khosla, S., "Contribution of In Vivo Structural Measurements and Load/Strength Ratios to the Determination of Forearm Fracture Risk in Postmenopausal Women," *Journal of Bone and Mineral Research*, Vol. 22, No. 9, 2007, pp. 1442–1448.

[2] Bekas, C., Curioni, A., and Müller, R., "Multicore Computers Can Protect Your Bones: Rendering the Computerized Diagnosis of Osteoporosis a Routine Clinical Practice," Poster. Supercomputing 09, Nov. 14-20, 2009, Portland, OR, USA.

[3] Ciarlet, P. G., *Three-dimensional elasticity*, Vol. 20 of *Studies in mathematics and its applications*, North Holland, Amsterdam, 1988.

[4] Arbenz, P., van Lenthe, G., Mennel, U., Mller, R., and Sala, M., "Multilevel μ-Finite Element Analysis for Human Bone Structures," *Applied Parallel Computing. State of the Art in Scientific Computing*, edited by B. Kgstrm, E. Elmroth, J. Dongarra, and J. Wasniewski, Vol. 4699 of *Lecture Notes in Computer Science*, Springer, Berlin/Heidelberg, 2007, pp. 240–250.

[5] Adams, M. F., Bayraktar, H. H., Keaveny, T. M., and Papadopoulos, P., "Ultrascalable Implicit Finite Element Analyses in Solid Mechanics with over a Half a Billion Degrees of Freedom," *SC '04: Proceedings of the 2004 ACM/IEEE conference on Supercomputing*, IEEE Computer Society, Washington, DC, USA, 2004, p. 34.

[6] Saad, Y., *Iterative Methods for Sparse Linear Systems*, SIAM, Philadelphia, 2nd ed., 2003.

[7] Anderson, E., Bai, Z., Bischof, C., Blackford, S., Demmel, J., Dongarra, J., Croz, J. D., Greenbaum, A., Hammerling, S., McKenney, A., and Sorensen, D., *LAPACK Users' Guide*, SIAM, Philadelphia, 3rd ed., 1999.

[8] Schenk, O., Wächter, A., and Hagemann, M., "Matching-based Preprocessing Algorithms to the Solution of Saddle-Point Problems in Large-Scale Nonconvex Interior-Point Optimization," *Comput. Optim. Appl.*, Vol. 36, No. 2-3, April 2007, pp. 321–341.

[9] Schenk, O. and Gärtner, K., "Solving Unsymmetric Sparse Systems of Linear Equations with PARDISO," *Journal of Future Generation Computer Systems*, Vol. 20, No. 3, 2004, pp. 475–487.

[10] Amestoy, P. R., Duff, I. S., Koster, J., and L'Excellent, J.-Y., "A Fully Asynchronous Multifrontal Solver Using Distributed Dynamic Scheduling," *SIAM Journal on Matrix Analysis and Applications*, Vol. 23, No. 1, 2001, pp. 15–41.

[11] Amestoy, P. R., Guermouche, A., L'Excellent, J.-Y., and Pralet, S., "Hybrid scheduling for the parallel solution of linear systems," *Parallel Computing*, Vol. 32, No. 2, 2006, pp. 136–156.

[12] Arbenz, P., van Lenthe, G. H., Mennel, U., Müller, R., and Sala, M., "A Scalable Multi-level Preconditioner for Matrix-Free μ-Finite Element Analysis of Human Bone Structures," *Internat. J. Numer. Methods Eng.*, Vol. 73, No. 7, 2007, pp. 927–947.

[13] Heroux, M., Bartlett, R., Hoekstra, V. H. R., Hu, J., Kolda, T., Lehoucq, R., Long, K., Pawlowski, R., Phipps, E., Salinger, A., Thornquist, H., Tuminaro, R., Willenbring, J., and Williams, A., "An Overview of Trilinos," Tech. Rep. SAND2003-2927, Sandia National Laboratories, 2003.

[14] Sala, M., Mennel, U., Arbenz, P., C., F., van Lenthe G. H., and A., W., "The PARFE Project Home Page," http://parfe.sourceforge.net, 2010.

[15] Snir, M., Otto, S., Huss-Lederman, S., Walker, D., and Dongarra, J., *MPI-The Complete Reference, Volume 1: The MPI Core*, MIT Press, Cambridge, MA, USA, 2nd ed., 1998.

[16] Karypis, G. and Kumar, V., "MeTis: Unstructured Graph Partitioning and Sparse Matrix Ordering System, Version 4.0," `http://www.cs.umn.edu/~metis`, 2009.

[17] Group, T. H., "HDF5: Hierarchical Data Format," Reference manual and user's guide available online at: http://www.hdfgroup.org/HDF5/doc/, 2009.

[18] Gee, M., Siefert, C., Hu, J., Tuminaro, R., and Sala, M., "ML 5.0 Smoothed Aggregation User's Guide," Tech. Rep. SAND2006-2649, Sandia National Laboratories, 2006.

[19] Çatalyürek, U. V., Aykanat, C., and Uçar, B., "On Two-Dimensional Sparse Matrix Partitioning: Models, Methods, and a Recipe," *SIAM J. Sci. Comput.*, Vol. 32, February 2010, pp. 656–683.

[20] Çatalyürek, U. V., Aykanat, C., and Uçar, B., "On Two-Dimensional Sparse Matrix Partitioning: Models, Methods, and a Recipe," *SIAM Journal on Scientific Computing*, Vol. 32, No. 2, 2010, pp. 656–683.

[21] Devine, K., Boman, E., Heaphy, R., Hendrickson, B., and Vaughan, C., "Zoltan Data Management Services for Parallel Dynamic Applications," *Computing in Science and Engineering*, Vol. 4, No. 2, 2002, pp. 90–97.

[22] Pellegrini, F., "Graph partitioning based methods and tools for scientific computing," *Parallel Computing*, Vol. 23, No. 1-2, 1997, pp. 153 – 164, Environment and tools for parallel scientific computing.

[23] Bekas, C., Curioni, A., Arbenz, P., Flaig, C., Van Lenthe, G. H., Müller, R., and Wirth, A. J., "Extreme scalability challenges in micro-finite element simulations of human bone," *Concurrency and Computation: Practice and Experience*, Vol. 22, No. 16, 2010, pp. 2282–2296.

[24] Catalyurek, U. and Aykanat, C., "A hypergraph-partitioning approach for coarse-grain decomposition," *Supercomputing '01: Proceedings of the 2001 ACM/IEEE conference on Supercomputing (CDROM)*, ACM, New York, USA, 2001, pp. 28–28.

[25] Stathopoulos, A., Bekas, C., and Curioni, A., "Architecture Aware Graph Partitioning," SIAM Workshop on Combinatorial Scientific Computing, May 10-11, 2011, Darmstadt, Germany.

Chapter 16

Algorithmic and Statistical Perspectives on Large-Scale Data Analysis

Michael W. Mahoney

Stanford University

16.1   Introduction

In recent years, motivated in large part by technological advances in both scientific and Internet domains, there has been a great deal of interest in what may be broadly termed *large-scale data analysis*. In this chapter, I will focus on what seems to me to be a remarkable and inevitable trend in this increasingly-important area. This trend has to do with the convergence of two very different perspectives or worldviews on what are appropriate or interesting or fruitful ways to view the data. At the risk of oversimplifying a large body of diverse work, I would like to draw a distinction between what I will loosely term the *algorithmic perspective* and the *statistical perspective* on large-scale data analysis problems. By the former, I mean roughly the approach that one trained in computer science might adopt. From this perspective, primary concerns include database issues, algorithmic questions such as models of data access, and the worst-case running time of algorithms for a given objective function. By the latter, I mean roughly the approach that one trained in statistics (or some application area where strong domain-specific assumptions about the data may be made) might adopt. From this perspective, primary concerns include questions such as how well the objective functions being considered conform to the phenomenon under study and whether one can make reliable predictions about the world from the data at hand.

Although very different, these two approaches are certainly not incompatible. Moreover, in spite of peoples' best efforts *not* to forge the union between these two worldviews, and instead to view the data from one perspective or another, depending on how one happened to be trained, the large-scale data analysis problems that we are increasingly facing—in scientific, Internet, financial, etc. applications—are so important and so compelling—either from a business perspective, or an academic perspective, or an intellectual perspective, or a national security perspective, or from whatever perspective you find to be most compelling—that we are being forced to forge this union.

Thus, *e.g.*, if one looks at the SIG-KDD meeting, ACM's flagship meeting on data analysis, now (in 2010) versus 10 or 15 years ago, one sees a substantial shift away from more database-type issues toward topics that may be broadly termed statistical, *e.g.*, that either involve explicit statistical modeling or involve making generative assumptions about data elements or network participants. And vice-versa—there has been a substantial shift in statistics, and especially in the natural and social sciences more generally, toward thinking in great detail about computational issues. Moreover, to the extent that statistics as an academic discipline is deferring on this matter, the relatively new area of machine learning is filling in the gap.

I should note that I personally began thinking about these issues some time ago. My Ph.D. was in computational statistical mechanics, and it involved a lot of Monte Carlo and molecular dynamics computations on liquid water and

DNA-protein-water interactions. After my dissertation, I switched fields to theoretical computer science, where I did a lot of work on the theory (and then later on the application) of randomized algorithms for large-scale matrix problems. One of the things that struck me during this transition was the deep conceptual disconnect between these two areas. In computer science, we have a remarkable infrastructure of machinery—from complexity classes and data structuring and algorithmic models, to database management, computer architecture, and software engineering paradigms—for solving problems. On the other hand, it seems to me that there tends to be a remarkable lack of appreciation, and thus associated cavalierness, when it comes to understanding how the data can be messy and noisy and poorly-structured in ways that adversely affect how one can be confident in the conclusions that one draws about the world as a result of the output of one's fast algorithms.

In this chapter, I would like to describe some of the fruits of this thought. To do so, I will focus on two very particular very applied problems in large-scale data analysis on which I have worked. The solution to these two problems benefited from taking advantage of the complementary algorithmic versus statistical perspectives in a novel way. In particular, we will see that, by understanding the statistical properties *implicit* in worst-case algorithms, we can make very strong claims about very applied problems. These claims would have been *much* more difficult to make and justify if one chose to view the problem from just one perspective or the other. The first problem has to do with selecting good features from a data matrix representing DNA microarray or DNA Single Nucleotide Polymorphism data. This is of interest, *e.g.*, to geneticists who want to understand historical trends in population genetics, as well as to biomedical researchers interested in so-called personalized medicine. The second problem has to do with identifying, or certifying that there do not exist, good clusters or communities in a data graph representing a large social or information network. This is of interest in a wide range of applications, such as finding good clusters or micro-markets for query expansion or bucket testing in Internet advertising applications.

The applied results for these two problems have been reported previously in appropriate domain-specific (in this case, genetics and Internet) publications [1, 2, 3, 4, 5, 6], and I am indebted to my collaborators with whom I have discussed some of these ideas in preliminary form. Thus, rather than focusing on the genetic issues *per se* or the Internet advertising issues *per se* or the theoretical analysis *per se*, in this chapter I would like to focus on what was going on "under the hood" in terms of the interplay between the algorithmic and statistical perspectives. The hope is that these two examples can serve as a model for exploiting the complementary aspects of the algorithmic and statistical perspectives in order to solve very applied large-scale data analysis problems more generally.[1] As we will see, in neither of these two applications

---

[1] This chapter grew out of an invited talk at a seminar at Schloss Dagstuhl on Combinatorial Scientific Computing (a research area that sits at the interface between algorithmic computer science and traditional scientific computing). While the connection between "com-

did we *first* perform statistical modeling, independent of algorithmic considerations, and *then* apply a computational procedure as a black box. This
approach of more closely coupling the computational procedures used with
statistical modeling or statistical understanding of the data seems particularly appropriate more generally for very large-scale data analysis problems,
where design decisions are often made based on computational constraints but
where it is of interest to understand the implicit statistical consequences of
those decisions.

16.2   Diverse Approaches to Modern Data Analysis Problems

Before proceeding, I would like in this section to set the stage by describing
in somewhat more detail some of diverse approaches that have been brought
to bear on modern data analysis problems [7, 8, 9, 10, 11, 12, 13, 14]. In particular, although the significant differences between the algorithmic perspective
and the statistical perspective have been highlighted previously [11], they are
worth reemphasizing.

A common view of the data in a database, in particular historically among
computer scientists interested in data mining and knowledge discovery, has
been that the data are an accounting or a record of everything that happened
in a particular setting. For example, the database might consist of all the
customer transactions over the course of a month, or it might consist of all
the friendship links among members of a social networking site. From this
perspective, the goal is to tabulate and process the data at hand to find
interesting patterns, rules, and associations. An example of an association rule
is the proverbial "People who buy beer between 5 p.m. and 7 p.m. also buy
diapers at the same time." The performance or quality of such a rule is judged
by the fraction of the database that satisfies the rule exactly, which then boils
down to the problem of finding frequent itemsets. This is a computationally

binatorial" and "algorithmic" in my title is hopefully obvious, the reader should also note
a connection between the statistical perspective I have described and approaches common
in traditional scientific computing. For example, even if a formal statistical model is not
specified, the fact that hydrated protein-DNA simulations or fluid dynamics simulations
"live" in two or three dimensions places very strong regularity constraints on the ways in
which information can propagate from point to point and thus on the data sets derived.
Moreover, in these applications, there is typically a great deal of domain-specific knowledge
that can be brought to bear that informs the questions one asks, and thus one clearly observes a tension between worst-case analysis versus field-specific assumptions. This implicit
knowledge should be contrasted with the knowledge available for matrices and graphs that
arise in much more unstructured data analysis applications such as arise in certain genetics
and in many Internet domains.

hard problem, and much algorithmic work has been devoted to its exact or approximate solution under different models of data access.

A very different view of the data, more common among statisticians, is one of a particular random instantiation of an underlying process describing unobserved patterns in the world. In this case, the goal is to extract information about the world from the noisy or uncertain data that are observed. To achieve this, one might posit a model: $data \sim F_\theta$ and $mean(data) = g(\theta)$, where F_θ is a distribution that describes the random variability of the data around the deterministic model $g(\theta)$ of the data. Then, using this model, one would proceed to analyze the data to make inferences about the underlying processes and predictions about future observations. From this perspective, modeling the noise component or variability well is as important as modeling the mean structure well, in large part since understanding the former is necessary for understanding the quality of predictions made. With this approach, one can even make predictions about events that have yet to be observed. For example, one can assign a probability to the event that a given user at a given web site will click on a given advertisement presented at a given time of the day, even if this particular event does not exist in the database.

Although these two perspectives are certainly not incompatible, they are very different, and they lead one to ask very different questions of the data and of the structures that are used to model data. Recall that in many applications, graphs and matrices are common ways to model the data [13, 14]. For example, a common way to model a large social or information network is with an interaction graph model, $G = (V, E)$, in which nodes in the vertex set V represent "entities" and the edges (whether directed, undirected, weighted or unweighted) in the edge set E represent "interactions" between pairs of entities. Alternatively, these and other data sets can be modeled as matrices, since an $m \times n$ real-valued matrix A provides a natural structure for encoding information about m objects, each of which is described by n features. Thus, in the next two sections, I will describe two recent examples—one having to do with modeling data as matrices, and the other having to do with modeling data as a graph—of particular data analysis problems that benefited from taking advantage in novel ways of the respective strengths of the algorithmic and statistical perspectives.

16.3 Genetics Applications and Novel Matrix Algorithms

In this section, I will describe an algorithm for selecting a "good" set of exactly k columns (or, equivalently, exactly k rows) from an arbitrary $m \times n$ matrix A. This problem has been studied extensively in scientific computing and Numerical Linear Algebra (NLA), often motivated by the goal of finding a good basis with which to perform large-scale numerical computations. In addition, variants of this problem have recently received a great deal of attention in Theoretical Computer Science (TCS). More generally, problems of this sort arise in many data analysis applications, often in the context of finding a good set of features with which to describe the data or to perform tasks such as classification or regression.

16.3.1 Motivating Genetics Application

Recall that "the human genome" consists of a sequence of roughly 3 billion base pairs on 23 pairs of chromosomes, roughly 1.5% of which codes for approximately 20,000–25,000 proteins. A DNA microarray is a device that can be used to measure simultaneously the genome-wide response of the protein product of each of these genes for an individual or group of individuals in numerous different environmental conditions or disease states. This very coarse measure can, of course, hide the individual differences or polymorphic variations. There are numerous types of polymorphic variation, but the most amenable to large-scale applications is the analysis of Single Nucleotide Polymorphisms (SNPs), which are known locations in the human genome where two alternate nucleotide bases (or alleles, out of A, C, G, and T) are observed in a non-negligible fraction of the population. These SNPs occur quite frequently, ca. 1 b.p. per thousand, and thus they are effective genomic markers for the tracking of disease genes (i.e., they can be used to perform classification into sick and not sick) as well as population histories (i.e., they can be used to infer properties about population genetics and human evolutionary history).

In both cases, $m \times n$ matrices A naturally arise, either as a people-by-gene matrix, in which A_{ij} encodes information about the response of the jth gene in the ith individual/condition, or as people-by-SNP matrices, in which A_{ij} encodes information about the value of the jth SNP in the ith individual. Thus, matrix computations have received attention in these applications [15, 16, 17, 18, 19]. A common *modus operandi* in applying NLA and matrix techniques such as PCA and the SVD to to DNA microarray, DNA SNPs, and other data problems is

- Model the people-by-gene or people-by-SNP data as an $m \times n$ matrix A.

- Perform the SVD (or related eigen-methods such as PCA or recently-popular manifold-based methods [20, 21, 22] that boil down to the SVD) to compute a small number of eigengenes or eigenSNPs or eigenpeople that capture most of the information in the data matrix.

- Interpret the top eigenvectors as meaningful in terms of underlying biological processes; or apply a heuristic to obtain actual genes or actual

SNPs from the corresponding eigenvectors in order to obtain such an interpretation.

In certain cases, such reification may lead to insight and such heuristics may be justified. (For instance, if the data happen to be drawn from a Guassian distribution, then the eigendirections tend to correspond to the axes of the corresponding ellipsoid, and there are many vectors that, up to noise, point along those directions.) In such cases, however, the justification comes from domain knowledge and not the mathematics [23, 16, 3]. The reason is that the eigenvectors themselves, being mathematically defined abstractions, can be calculated for any data matrix and thus are not easily understandable in terms of processes generating the data: eigenSNPs (being linear combinations of SNPs) cannot be assayed; nor can eigengenes be isolated and purified; nor is one typically interested in how eigenpatients respond to treatment when one visits a physician.

For this and other reasons, a common task in genetics and other areas of data analysis is the following: given an input data matrix A and a parameter k, find the best subset of exactly k *actual* DNA SNPs or *actual* genes, i.e., *actual* columns or rows from A, to use to cluster individuals, reconstruct biochemical pathways, reconstruct signal, perform classification or inference, etc.

16.3.2   A Formalization of and Prior Approaches to This Problem

Unfortunately, common formalizations of this algorithmic problem—including looking for the k actual columns that capture the largest amount of information or variance in the data or that are maximally uncorrelated—lead to intractable optimization problems [24, 25]. In this chapter, I will consider the so-called Column Subset Selection Problem (CSSP): given as input an arbitrary $m \times n$ matrix A and a rank parameter k, choose the set of exactly k columns of A s.t. the $m \times k$ matrix C minimizes (over all $\binom{n}{k}$ sets of such columns) the error

$$\min ||A - P_C A||_\xi = \min ||A - CC^+ A||_\xi \quad (\xi = 2, F) \qquad (16.1)$$

where $\xi = 2, F$ represents the spectral or Frobenius norm[2] of A and where $P_C = CC^+$ is the projection onto the subspace spanned by the columns of C.

Within NLA, a great deal of work has focused on this CSSP problem [26, 27, 28, 29, 30, 31, 32, 33, 34, 35, 36]. Several general observations about the NLA approach include

---

[2] Recall that the spectral norm is the largest singular value of the matrix, and thus it is a "worst case" norm in that it measures the worst-case stretch of the matrix, while the Frobenius norm is more of an "averaging" norm, since it involves a sum over every singular direction. The former is of greater interest in scientific computing and NLA, where one is interested in actual columns for the subspaces they define and for their good numerical properties, while the latter is of greater interest in data analysis and machine learning, where one is more interested in actual columns for the features they define.

- The focus in NLA is on *deterministic algorithms*. Moreover, these algorithms are greedy, in that at each iterative step, the algorithm makes a decision about which columns to keep according to a pivot-rule that depends on the columns it currently has, the spectrum of those columns, etc. Differences between different algorithms often boil down to how deal with such pivot rules decisions, and the hope is that more sophisticated pivot-rule decisions lead to better algorithms in theory or in practice.

- There are deep *connections with QR factorizations* and in particular with the so-called Rank Revealing QR factorizations. Moreover, there is an emphasis on optimal conditioning questions, backward error analysis issues, and whether the running time is a large or small constant multiplied by n^2 or n^3.

- Good *spectral norm bounds* are obtained. A typical spectral norm bound is

$$||A - P_C A||_2 \leq O\left(\sqrt{k(n-k)}\right)||A - P_{U_k}A||_2,$$

and these results are algorithmic, in that the running time is a low-degree polynomial in m and n [33]. On the other hand, the strongest results for the Frobenius norm in this literature is

$$||A - P_C A||_F \leq \sqrt{(k+1)(n-k)}||A - P_{U_k}A||_2,$$

but it is only an existential result, i.e., the only known algorithm essentially involves exhaustive enumeration [31]. (In these two expressions, U_k is the $m \times k$ matrix consisting of the top k left singular vectors of A, and P_{U_k} is a projection matrix onto the span of U_k.)

Within TCS, a great deal of work has focused on the related problem of choosing good columns from a matrix [37, 38, 39, 40, 41, 42]. Several general observations about the TCS approach include

- The focus in TCS is on *randomized algorithms*. In particular, with these algorithms, there exists some nonzero probability, which can typically be made extremely small, say 10^{-20}, that the algorithm will return columns that fail to satisfy the desired quality-of-approximation bound.

- The algorithms select *more than k columns*, and the best rank-k projection onto those columns is considered. The number of columns is typically a low-degree polynomial in k, most often $O(k \log k)$, where the constants hidden in the big-O notation are quite reasonable.

- Very good *Frobenius norm bounds* are obtained. For example, the algorithm (described below) that provides the strongest Frobenius norm bound achieves

$$||A - P_{C_k}A||_F \leq (1+\epsilon)||A - P_{U_k}A||_F, \qquad (16.2)$$

while running in time of the order of computing an exact or approximate basis for the top-k right singular subspace [42]. The TCS literature also demonstrates that there exists a set of k columns that achieves a constant-factor approximation

$$||A - P_{C_k}A||_F \leq \sqrt{k}||A - P_{U_k}A||_F,$$

but note that this is an existential result [43]. (Here, C_k is the best rank-k approximation to the matrix C, and P_{C_k} is the projection matrix onto this k-dimensional space.)

Much of the early work in TCS focused on randomly sampling columns according to an importance sampling distribution that depended on the Euclidean norm of those columns [37, 38, 39, 40, 41]. This had the advantage of being "fast," in the sense that it could be performed in a small number of "passes" over that data from external storage, and also that additive-error quality-of-approximation bounds could be proved. This had the disadvantage of being less immediately-applicable to scientific computing and large-scale data analysis applications. For example, columns are often normalized during data preprocessing; and even when not normalized, column norms can still be uninformative, as in heavy-tailed graph data[3] where they often correlate strongly with simpler statistics such as node degree.

The algorithm from the TCS literature that achieves the strongest Frobenius norm bounds of the form (16.2) is the following.[4] Given an $m \times n$ matrix A and a rank parameter k,

- Compute the importance sampling probabilities $\{p_i\}_{i=1}^n$, where $p_i = \frac{1}{k}||V_k^{T(i)}||_2^2$, where V_k^T is *any* $k \times n$ orthogonal matrix spanning the top-k right singular subspace of A. (Note that these quantities are proportional to the diagonal elements of the projection matrix onto the span of V_k^T.)

- Randomly select and rescale $c = O(k \log k/\epsilon^2)$ columns of A according to these probabilities.

A more detailed description of this algorithm may be found in [42, 3], where it is proven that (16.2) holds with extremely high probability. The computational bottleneck for this algorithm is computing $\{p_i\}_{i=1}^n$, for which it suffices to

---

[3]By *heavy-tailed graph*, I mean a graph (or equivalently the adjacency matrix of such a graph) such as the social and information networks described below, in which quantities such as the degree distribution or eigenvalue distribution decay in a heavy-tailed or power law manner.

[4]Given a matrix A and a rank parameter k, one can express the SVD as $A = U\Sigma V^T$, in which case the best rank-k approximation to A can be expressed as $A_k = U_k\Sigma_k V_k^T$. In the text, I will sometimes overload notation and use V_k^T to refer to any $k \times n$ orthonormal matrix spanning the space spanned by the top-k right singular vectors. The reason is that this basis is used only to compute the importance sampling probabilities; since those probabilities are proportional to the diagonal elements of the projection matrix onto the span of this basis, the particular basis does not matter.

compute *any* $k \times n$ matrix V_k^T that spans the top-k right singular subspace of A. (That is, it suffices to compute any orthonormal basis spanning V_k^T, and it is not necessary to compute the full SVD.) It is an open problem whether these importance sampling probabilities can be approximated more rapidly.

To motivate the importance sampling probabilities used by this algorithm, recall that if one is looking for a worst-case relative-error approximation of the form (16.2) to a matrix with $k-1$ large singular values and one much smaller singular value, then the directional information of the kth singular direction will be hidden from the Euclidean norms of the matrix. Intuitively, the reason is that, since $A_k = U_k \Sigma_k V_k^T$, the Euclidean norms of the columns of A are convolutions of "subspace information" (encoded in U_k and V_k^T) and "size-of-A information" (encoded in Σ_k). This suggests deconvoluting subspace information and size-of-A information by choosing importance sampling probabilities that depend on the Euclidean norms of the columns of V_k^T. Thus, this importance sampling distribution defines a nonuniformity structure over $\mathbb{R}^n$ that indicates *where* in the n-dimensional space the information in A is being sent, independent of *what* that (singular value) information is. As we will see in the next two subsections, by using these importance sampling probabilities, we can obtain novel algorithms for two very traditional problems in NLA and scientific computing.

16.3.3 An Aside on Least Squares and Statistical Leverage

The analysis of the relative-error algorithm described in the previous subsection and of the algorithm for the CSSP described in the next subsection boils down to a least-squares approximation result. Intuitively, these algorithms find columns that provide a space that is good in a least-squares sense, when compared to the best rank-k space, at reconstructing every row of the input matrix [42, 44].

Thus, consider the problem of finding a vector x such that $Ax \approx b$, where the rows of A and elements of b correspond to constraints and the columns of A and elements of x correspond to variables. In the very overconstrained case where the $m \times n$ matrix A has $m \gg n$,[5] there is in general no vector x such that $Ax = b$, and it is common to quantify "best" by looking for a vector x_{opt} such that the Euclidean norm of the residual error is small, i.e., to solve the least-squares (LS) approximation problem

$$x_{opt} = \mathrm{argmin}_x ||Ax - b||_2.$$

This problem is ubiquitous in applications, where it often arises from fitting the parameters of a model to experimental data, and it is central to theory. Moreover, it has a natural statistical interpretation as providing the best estimator within a natural class of estimators, and it has a natural geometric interpretation as fitting the part of the vector b that resides in the column

---

[5]In this section only, we will assume that $m \gg n$.

space of A. From the viewpoint of low-rank matrix approximation and the CSSP, this LS problem arises since measuring the error with a Frobenius or spectral norm, as in (16.1), amounts to choosing columns that are "good" in a least squares sense.

From an algorithmic perspective, the relevant question is: how long does it take to compute x_{opt}? The answer here is that is takes $O(mn^2)$ time [45]—*e.g.*, depending on numerical issues, condition numbers, etc., this can be accomplished with the Cholesky decomposition, a variant of the QR decomposition, or by computing the full SVD.

From a statistical perspective, the relevant question is: when is computing this x_{opt} the right thing to do? The answer to this is that this LS optimization is the right problem to solve when the relationship between the "outcomes" and "predictors" is roughly linear and when the error processes generating the data are "nice" (in the sense that they have mean zero, constant variance, are uncorrelated, and are normally distributed; or when we have adequate sample size to rely on large sample theory) [46].

Of course, in practice these assumptions do not hold perfectly, and a prudent data analyst will check to make sure that these assumptions have not been too violated. To do this, it is common to assume that $b = Ax + \varepsilon$, where b is the response, the columns $A^{(i)}$ are the carriers, and ε is the nice error process. Then $x_{opt} = (A^T A)^{-1} A^T b$, and thus $\hat{b} = Hb$, where the projection matrix onto the column space of A, $H = A(A^T A)^{-1} A^T$, is the so-called *hat matrix*. It is known that H_{ij} measures the influence or statistical leverage exerted on the prediction $\hat{b}_i$ by the observation b_j [47, 46, 48]. Relatedly, if the ith diagonal element of H is particularly large then the ith data point is particularly sensitive or influential in determining the best LS fit, thus justifying the interpretation of the elements H_{ii} as *statistical leverage scores* [3].[6]

To gain insight into these statistical leverage scores, consider the so-called "wood beam data" example [51, 47], which is visually presented in Figure 16.1(a), along with the best-fit line to that data. In Figure 16.1(b), the leverage scores for these ten data points are shown. Intuitively, data points that "stick out" have particularly high leverage—*e.g.*, the data point that has the most influence or leverage on the best-fit line to the wood beam data is the point marked "4", and this is reflected in the relative magnitude of the corresponding statistical leverage score. Indeed, since Trace$(H) = n$, a rule of thumb that has been suggested in diagnostic regression analysis to identify errors and outliers in a data set is to investigate the ith data point if $H_{ii} > 2n/m$ [52, 53], i.e., if H_{ii} is larger that 2 or 3 times the "average" size. On the other hand, of course, if it happens to turn out that such a point is a legitimate data point, then one might expect that such an outlying data point will be a particularly important or informative data point.

Returning to the algorithmic perspective, consider the following random

---

[6]This concept has also proven useful under the name of graph resistance [49] and also of coherence [50].

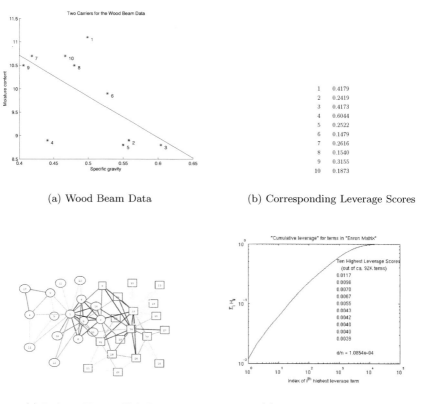

(a) Wood Beam Data

(b) Corresponding Leverage Scores

(c) Zachary Karate Club Data

(d) Cumulative Leverage

FIGURE 16.1: (a) The Wood Beam Data described in [47] is an example illustrating the use of statistical leverage scores in the context of least-squares approximation. Shown are the original data and the best least-squares fit. (b) The leverage scores for each of the ten data points in the Wood Beam Data. Note that the data point marked "4" has the largest leverage score, as might be expected from visual inspection. (c) The so-called Zachary karate club network [54], with edges color-coded such that leverage scores for a given edge increase from yellow to red. (d) Cumulative leverage (with $k = 10$) for a $65,031 \times 92,133$ term-document matrix constructed Enron electronic mail collection, illustrating that there are a large number of data points with very high leverage score. (e) The normalized statistical leverage scores and information gain score—information gain is a mutual information-based metric popular in the application area [2, 3]—for each of the $n = 5520$ genes, a situation in which the data cluster well in the low-dimensional space defined by the maximum variance axes of the data [3]. Red stars indicate the 12 genes with the highest leverage scores, and the red dashed line indicates the average or uniform leverage scores. Note the strong correlation between the unsupervised leverage score metric and the supervised information gain metric.

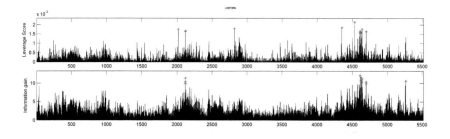

(e) Leverage Score and Information Gain for DNA Microarray Data

FIGURE 16.1: (*Continued.*)

sampling algorithm for the LS approximation problem [55, 42]. Given a very overconstrained least-squares problem, where the input matrix A and vector b are *arbitrary*, but $m \gg n$:

- Compute normalized statistical leverage scores $\{p_i\}_{i=1}^m$, where $p_i = ||U_A^{(i)}||_2^2/k$, where U is the $m \times n$ matrix consisting of the left singular vectors of A.[7]

- Randomly sample and rescale $r = O(n \log n/\epsilon^2)$ constraints, i.e., rows of A and the corresponding elements of b, using these scores as an importance sampling distribution.

- Solve (using any appropriate LS solver as a black box) the induced subproblem to obtain a vector $\tilde{x}_{opt}$.

Since this algorithm samples constraints and not variables, the dimensionality of the vector $\tilde{x}_{opt}$ that solves the subproblem is the same as that of the vector x_{opt} that solves the original problem. This algorithm is described in more detail in [55, 42], where it is shown that relative error bounds of the form

$$||b - A\tilde{x}_{opt}||_2 \leq (1 + \epsilon)||b - Ax_{opt}||_2 \quad \text{and}$$
$$||x_{opt} - \tilde{x}_{opt}||_2 \leq O(\epsilon)||x_{opt}||_2$$

hold. Even more importantly, this algorithm highlights that the essential nonuniformity structure for the worst-case analysis of the LS (and, as we will see, the related CSSP) problem is defined by the statistical leverage scores!

---

[7]More generally, recall that if U is *any* orthogonal matrix spanning the column space of A, then $H = P_U = UU^T$ and thus $H_{ii} = ||U^{(i)}||_2^2$, i.e., the statistical leverage scores equal the Euclidean norm of the rows of any such matrix U [3]. Clearly, the columns of U are orthonormal, but the rows of U in general are not—they can be uniform if, *e.g.*, U consists of columns from a truncated Hadamard matrix; or extremely nonuniform if, *e.g.*, the columns of U come from a truncated Identity matrix; or anything in between.

That is, the same "outlying" data points that the diagnostic regression analyst tends to investigate are those points that are biased toward by the worst-case importance sampling probability distribution.

Clearly, for this LS algorithm, which holds for arbitrary input A and b, $O(mn^2)$ time suffices to compute the sampling probabilities; in addition, it has recently been shown that one can obtain a nontrivial approximation to them in $o(mn^2)$ time [56]. For many applications, such as those described in subsequent subsections, spending time on the order of computing an exact or approximate basis for the top-k singular subspace is acceptable, in which case immediate generalizations of this algorithm are of interest. In other cases, one can preprocess the LS system with a "randomized Hadamard" transform (as introduced in the "fast" Johnson-Lindenstrauss lemma [57, 58]). Application of such a Hadamard transform tends to "uniformize" the leverage scores, intuitively for the same reason that a Fourier matrix delocalizes a localized δ-function, in much the same way as application of a random orthogonal matrix or a random projection does. This has led to the development of relative-error approximation algorithms for the LS problem that run in $o(mn^2)$ time in theory [59, 60]—essentially $O(mn \log(n/\epsilon) + \frac{n^3 \log^2 m}{\epsilon})$ time, which is much less than $O(mn^2)$ when $m \gg n$—and whose numerical implementation performs faster than traditional deterministic algorithms for systems with as few as thousands of constraints and hundreds of variables [61, 62].

16.3.4   A Two-Stage Hybrid Algorithm for the CSSP

In this subsection, I will describe an algorithm for the CSSP that uses the concept of statistical leverage to combine the NLA and TCS approaches, that comes with worst-case performance guarantees, and that performs well in practical data analysis applications. I should note that, prior to this algorithm, it was not immediately clear how to combine these two approaches. For example, if one looks at the details of the pivot rules in the deterministic NLA methods, it isn't clear that keeping more columns will help at all in terms of reconstruction error. Similarly, since there is a version of the "coupon collecting" problem at the heart of the usual TCS analysis, keeping fewer than $\Omega(k \log k)$ will fail with respect to this worst-case analysis. Moreover, the obvious hybrid algorithm of first randomly sampling $O(k \log k)$ columns and then using a deterministic QR procedure to select exactly k of those columns does not seem to perform so well (either in theory or in practice).

Consider the following more sophisticated version of a two-stage hybrid algorithm. Given an arbitrary $m \times n$ matrix A and rank parameter k:

- (Randomized phase) Let V_k^T be *any* $k \times n$ orthogonal matrix spanning the top-k right singular subspace of A. Compute the importance sampling probabilities $\{p_i\}_{i=1}^n$, where

$$p_i = \frac{1}{k}||V_k^{T(i)}||_2^2. \qquad (16.3)$$

Randomly select and rescale $c = O(k \log k)$ columns of V_k^T according to these probabilities.

- (Deterministic phase) Let $\tilde{V}^T$ be the $k \times O(k \log k)$ non-orthogonal matrix consisting of the down-sampled and rescaled columns of V_k^T. Run a deterministic QR algorithm on $\tilde{V}^T$ to select exactly k columns of $\tilde{V}^T$. Return the corresponding columns of A.

In particular, note that both the original choice of columns in the first phase, as well as the application of the QR algorithm in the second phase, involve the matrix V_k^T, i.e., the matrix defining the relevant non-uniformity structure over the columns of A in the $(1 + \epsilon)$-relative-error algorithm [42, 3], rather than the matrix A itself, as is more traditional.[8] A more detailed description of this algorithm may be found in [44], were it is shown that with extremely high probability the following spectral[9] and Frobenius norm bounds hold:

$$
\begin{array}{rcl}
||A - P_C A||_2 & \leq & O(k^{3/4} \log^{1/2}(k) n^{1/2}) ||A - P_{U_k} A||_2 \\
||A - P_C A||_F & \leq & O(k \log^{1/2} k) ||A - P_{U_k} A||_F
\end{array}
$$

Interestingly, the analysis of this algorithm makes critical use of the importance sampling probabilities (16.3), which are a generalization of the concept of *statistical* leverage described in the previous subsection, for its worst-case *algorithmic* performance guarantees. Moreover, it is critical to the success of this algorithm that the QR procedure in the second phase be applied to the randomly-sampled version of V_k^T, i.e., the matrix defining the worst-case nonuniformity structure in A, rather than of A itself.

With respect to running time, the computational bottleneck for this algorithm is computing $\{p_i\}_{i=1}^n$, for which it suffices to compute *any* $k \times n$ matrix V_k^T that spans the top-k right singular subspace of A. (In particular, a full SVD computation is *not* necessary.) Thus, this running time is of the same order as the running time of the QR algorithm used in the second phase when applied to the original matrix A. Moreover, this algorithm easily scales up to matrices with thousands of rows and millions of columns, whereas existing off-the-shelf implementations of the traditional algorithm may fail to run at all. With respect to the worst-case quality of approximation bounds, this algorithm selects columns that are comparable to the state-of-the-art algorithms for constant k (i.e., $O(k^{1/4} \log^{1/2} k)$ worse than previous work) for the spectral norm and only a factor of at most $O((k \log k)^{1/2})$ worse than the best previously-known existential result for the Frobenius norm.

---

[8]Note that QR (as opposed to the SVD) is *not* performed in the second phase to speed up the computation of a relatively cheap part of the algorithm, but instead it is performed since the goal of the algorithm is to return actual columns of the input matrix.

[9]Note that to establish the spectral norm bound, [44] used slightly more complicated (but still depending only on information in V_k^T) importance sampling probabilities, but this may be an artifact of the analysis.

16.3.5   Data Applications of the CSSP Algorithm

In the applications we have considered [1, 2, 3, 63, 64], the goals of DNA microarray and DNA SNP analysis include the reconstruction of untyped genotypes, the evaluation of tagging SNP transferability between geographically-diverse populations, the classification and clustering into diseased and non-diseased states, and the analysis of admixed populations with non-trivial ancestry; and the goals of selecting good columns more generally include diagnostic data analysis and unsupervised feature selection for classification and clustering problems. Here, I will give a flavor of when and why and how the CSSP algorithm of the previous subsection might be expected to perform well in these and other types of data applications.

To gain intuition for the behavior of leverage scores in a typical application, consider Figure 16.1(c), which illustrates the so-called Zachary karate club network [54], a small but popular network in the community detection literature. Given such a network $G = (V, E)$, with n nodes, m edges, and corresponding edge weights $w_e \geq 0$, define the $n \times n$ Laplacian matrix as $L = B^T W B$, where B is the $m \times n$ edge-incidence matrix and W is the $m \times m$ diagonal weight matrix. The effective resistance between two vertices is given by the diagonal entries of the matrix $R = B L^\dagger B^T$ (where $L^\dagger$ denotes the Moore-Penrose generalized inverse) and is related to notions of "network betweenness" [65]. For many large graphs, this and related betweenness measures tend to be strongly correlated with node degree and tend to be large for edges that form articulation points between clusters and communities, i.e., for edges that "stick out" a lot. It can be shown that the effective resistances of the edges of G are proportional to the statistical leverage scores of the m rows of the $m \times n$ matrix $W^{1/2} B$—consider the $m \times m$ matrix

$$P = W^{1/2} R W^{1/2} = \Phi (\Phi^T \Phi)^+ \Phi^T,$$

where $\Phi = W^{1/2} B$, and note that if U_Φ denotes any orthogonal matrix spanning the column space of Φ, then

$$P_{ii} = (U_\Phi U_\Phi^T)_{ii} = ||(U_\Phi)_{(i)}||_2^2.$$

Figure 16.1(c) presents a color-coded illustration of these scores for Zachary karate club network.

Next, to gain intuition for the (non-)uniformity properties of statistical leverage scores in a typical application, consider a term-document matrix derived from the publicly-released Enron electronic mail collection [66], which is an example of social or information network we will encounter again in the next section and which is also typical of the type of data set to which SVD-based latent semantic analysis (LSA) methods [67] have been applied. I constructed a $65,031 \times 92,133$ matrix, as described in [66], and I chose the rank parameter as $k = 10$. Figure 16.1(d) plots the cumulative leverage, i.e., the running sum of top t statistical leverage scores, as a function of increasing t. Since $\frac{k}{n} = \frac{10}{92,133} \approx 1.0854 \times 10^{-4}$, we see that the highest leverage term has a leverage score nearly two orders of magnitude larger than this "average" size scale, that the second highest-leverage score is only marginally less than the first, that the third highest score is marginally less than the second, etc. Thus, by the traditional metrics of diagnostic data analysis [52, 53], which suggests flagging a data point if

$$(P_{U_k})_{ii} = (H_k)_{ii} > 2k/n,$$

there are a *huge* number of data points that are *extremely* outlying. In retrospect, of course, this might not be surprising since the Enron email corpus is extremely sparse, with nowhere on the order of $\Omega(n)$ nonzeros per row. Thus, even though LSA methods have been successfully applied, plausible generative models associated with these data are clearly not Gaussian, and the sparsity structure is such that there is no reason to expect that nice phenomena such as measure concentration occur.

Finally, note that DNA microarray and DNA SNP data often exhibit a similar degree of nonuniformity, although for somewhat different reasons. To illustrate, Figure 16.1(e) presents two plots. First, it plots the normalized statistical leverage scores for a data matrix, as was described in [3], consisting of $m = 31$ patients with 3 different cancer types with respect to $n = 5520$ genes. A similar plot illustrating the remarkable nonuniformity in statistical leverage scores for DNA SNP data was presented in [2]. Empirical evidence suggests that two phenomena may be responsible. First, as with the term-document data, there is no domain-specific reason to believe that nice properties like measure concentration occur—on the contrary, there are reasons to expect that they do not. Recall that each DNA SNP corresponds to a single mutational event in human history. Thus, it will "stick out," as its description along its one axis in the vector space will likely not be well-expressed in terms of the other axes, i.e., in terms of the other SNPs, and by the time it "works its way back" due to population admixing, etc., other SNPs will have occurred elsewhere. Second, the correlation between statistical leverage and supervised mutual information-based metrics is particularly prominent in examples where the data cluster well in the low-dimensional space defined by the maximum variance axes. Considering such data sets is, of course, a strong selection bias, but it is common in applications. It would be of interest to develop a model that quantifies the observation that, conditioned on clustering well in the low-dimensional space, an unsupervised measure like leverage scores should be expected to correlate well with a supervised measure like informativeness [2] or information gain [3].

With respect to some of the more technical and implementational issues, several observations [63, 44] shed light on the inner workings of the CSSP algorithm and its usefulness in applications. Recall that an important aspect of QR algorithms is how they make so-called pivot rule decisions about which columns to keep [45] and that such decisions can be tricky when the columns are not orthogonal or spread out in similarly nice ways.

- We looked at several versions of the QR algorithm, and we compared each version of QR to the CSSP using that version of QR in the second phase. One observation we made was that different QR algorithms behave differently—*e.g.*, some versions such as the Low-RRQR algorithm of [68] tend to perform much better than other versions such as the qrxp algorithm of [34, 69]. Although not surprising to NLA practitioners, this observation indicates that some care should be paid to using "off the shelf" implementations in large-scale applications. A second less-obvious observation is that preprocessing with the randomized first phase tends to improve more poorly-performing variants of QR more than better variants. Part of this is simply that the more poorly-performing variants have more room to improve, but part of this is also that more sophisticated versions of QR tend to make more sophisticated pivot rule decisions, which are relatively less important after the randomized bias toward directions that are "spread out."

- We also looked at selecting columns by applying QR on V_k^T and then keeping the corresponding columns of A, i.e., just running the classical deterministic QR algorithm with no randomized first phase on the matrix V_k^T. Interestingly, with this "preprocessing" we tended to get better columns than if we ran QR on the original matrix A. Again, the interpretation seems to be that, since the norms of the columns of V_k^T define the relevant nonuniformity structure with which to sample with respect to, working directly with those columns tends make things "spread out," thereby avoiding (even in traditional deterministic settings) situations where pivot rules have problems.

- Of course, we also observed that randomization further improves the results, assuming that care is taken in choosing the rank parameter k and the sampling parameter c. In practice, the choice of k should be viewed as a "model selection" question. Then, by choosing $c = k, 1.5k, 2k, \ldots$, we often observed a "sweet spot," in bias-variance sense, as a function of increasing c. That is, for a fixed k, the behavior of the deterministic QR algorithms improves by choosing somewhat more than k columns, but that improvement is degraded by choosing too many columns in the randomized phase.

16.3.6   Some General Thoughts on Leverage Scores and Matrix Algorithms

I will conclude this section with two general observations raised by these theoretical and empirical results having to do with using the concept of statistical leverage to obtain columns from an input data matrix that are good both in worst-case analysis and also in large-scale data applications.

One high-level question raised by these results is: why should statistical leverage, a traditional concept from regression diagnostics, be useful to obtain improved worst-case approximation algorithms for traditional NLA matrix problems? The answer to this seems to be that, intuitively, if a data point has a high leverage score and is not an error then it might be a particularly important or informative data point. Since worst-case analysis takes the input matrix as given, each row is assumed to be reliable, and so worst-case guarantees are obtained by focusing effort on the most informative data points. It would be interesting to see if this perspective is applicable more generally in the design of matrix and graph algorithms.

A second high-level question is: why are the statistical leverage scores so nonuniform in many large-scale data analysis applications. Here, the answer seems to be that, intuitively, in many very large-scale applications, statistical models are *implicitly* assumed based on computational and not statistical considerations. In these cases, it is not surprising that some interesting data points "stick out" relative to obviously inappropriate models. This suggests the use of these importance sampling scores as cheap signatures of the "inappropriateness" of a statistical model (chosen for algorithmic and not statistical reasons) in large-scale exploratory or diagnostic applications. It would also be interesting to see if this perspective is applicable more generally.

16.4   Internet Applications and Novel Graph Algorithms

In this section, I will describe a novel perspective on identifying good clusters or communities in a large graph. The general problem of finding good clusters in (or good partitions of) a data graph has been studied extensively in a wide range of applications. For example, it has been studied for years in scientific computation (where one is interested in load balancing in parallel computing applications), machine learning and computer vision (where one is interested in segmenting images and clustering data), and theoretical computer science (where one is interested in it as a primitive in divide-and-conquer algorithms). More recently, problems of this sort have arisen in the analysis of large social and information networks, where one is interested in finding communities that are meaningful in a domain-specific context.

16.4.1   Motivating Internet Application

Sponsored search is a type of contextual advertising where Web site owners pay a fee, usually based on click-throughs or ad views, to have their Web site search results shown in top placement position on search engine result pages. For example, when a user enters a term into a search box, the search engine typically presents not only so-called "algorithmic results," but it also presents text-based advertisements that are important for revenue generation. In this context, one can construct a so-called *advertiser-bidded-phrase graph*, the simplest variant of which is a bipartite graph $G = (U, V, E)$, in which U consists of some discretization of the set of advertisers, V consists of some discretization of the set of keywords that have been bid upon, and an edge $e = (u, v) \in E$ is present if advertiser $u \in U$ had bid on phrase $v \in V$. It is then of interest to perform data mining on this graph in order to optimize quantities such as the user click-through-rate or advertiser return-on-investment.

Numerous community-related problems arise in this context. For example, in *micro-market identification*, one is interested in identifying a set of nodes that is large enough that it is worth spending an analyst's time upon, as well as coherent enough that it can be usefully thought about as a "market" in an economic sense. Such a cluster can be useful for A/B bucket testing, as well as for recommending to advertisers new queries or sub-markets. Similarly, in *advanced match*, an advertiser places bids not only when an exact match occurs to a set of advertiser-specified phrases, but also when a match occurs to phrases "similar to" the specified bid phrases. Ignoring numerous natural language and game theoretic issues, one can imagine that if the original phrase is in the middle of a fairly homogeneous concept class, then there may be a large number of similar phrases that are nearby in the graph topology, in which case it might be advantageous to both the advertiser and the search engine to include those phrases in the bidding. On the other hand, if the original phrase is located in a locally very unstructured part of the graph, then there may be a large number of phrases that are nearby in the graph topology but that have a wide range of meanings very different than the original bid phrase, in which case performing such an expansion might not make sense.

As in many other application areas, in these clustering and community identification applications, a common *modus operandi* in applying data analysis tools is

- Define an objective function that formalizes the intuition that one has as a data analyst as to what constitutes a good cluster or community.

- Since that objective function is almost invariably intractable to optimize exactly, apply some approximation algorithm or heuristic to the problem.

- If the set of nodes that are thereby obtained look plausibly good in an application-dependent sense, then declare success. Otherwise, modify the objective function or heuristic or algorithm, and iterate the process.

Such an approach can lead to insight—*e.g.*, if the data are "morally" low-dimensional, as might be the case, as illustrated in Figure 16.2(a), when the Singular Value Decomposition, manifold-based machine learning methods, or k-means-like statistical modeling assumptions are appropriate; or if the data have other "nice" hierarchical properties that conform to one's intuition, as suggested by the schematic illustration in Figure 16.2(b); or if the data come from a nice generative model such as a mixture model. In these cases, a few steps of such a procedure will likely lead to a reasonable solution.

On the other hand, if the size of the data is larger, or if the data arise from an application where the sparsity structure and noise properties are more adversarial and less intuitive, then such an approach can be problematic. For example, if a reasonable solution is not readily obtained, then it is typically not clear whether one's original intuition was wrong; whether this is due to an improper formalization of the correct intuitive concept; whether insufficient computational resources were devoted to the problem; whether the sparsity and noise structure of the data have been modeled correctly, etc. That common visualization algorithms applied to large networks lead to largely non-interpretable figures, as illustrated in Figure 16.2(c), which reveal more about the inner workings of the visualization algorithm than the network being visualized, *suggests* that this more problematic situation is more typical of large social and information network data. In such cases, it would be of interest to have principled tools, the algorithmic and statistical properties of which are well-understood, to "explore" the data.

16.4.2 A Formalization of and Prior Approaches to This Problem

One of the most basic question that one can ask about a data set (and one which is intimately related to questions of community identification) is: what does the data set "look like" if it is cut into two pieces? This is of interest since, *e.g.*, one would expect very different properties from a data set that "looked like" a hot dog or a pancake, in that one could split it into two roughly equally-sized pieces, each of which was meaningfully coherent in and of itself; as opposed to a data set that "looked like" a moderately-dense expander-like random graph, in that there didn't exist any good partitions of any size of the data into two pieces; as opposed to a data set in which there existed good partitions, but those involved nibbling off just 0.1% of the nodes of the data and leaving the rest intact.

A common way to formalize this question of qualitative connectivity is via the *graph partitioning* problem [70, 71, 72, 73, 74, 75, 76]. Graph partitioning refers to a family of objective functions and associated approximation algorithms that involve cutting or partitioning the nodes of a graph into two sets with the goal that the cut has good quality (i.e., not much edge weight crosses the cut) as well as good balance (i.e., each of the two sets has a lot of the node weight). There are several standard formalizations of this bi-criterion. In this chapter, I will be interested in the quotient cut formulations,[10] which require the small side to be large enough to "pay for" the edges in the cut.

Given an undirected, possibly weighted, graph $G = (V, E)$, the *expansion*

---

[10]Other formulations include the graph bisection problem, which requires perfect $\frac{1}{2} : \frac{1}{2}$ balance and the β-balanced cut problem, with β set to a fraction such as $\frac{1}{10}$, which requires at least a $\beta : (1 - \beta)$ balance.

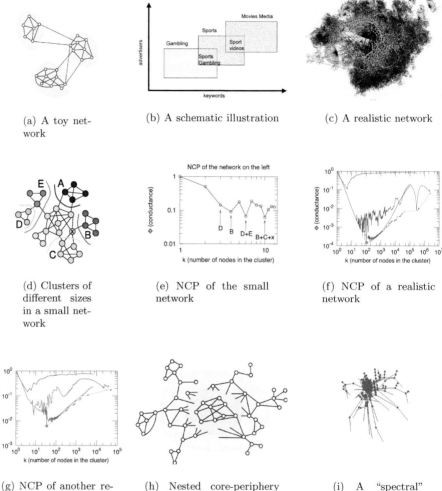

(a) A toy network

(b) A schematic illustration

(c) A realistic network

(d) Clusters of different sizes in a small network

(e) NCP of the small network

(f) NCP of a realistic network

(g) NCP of another realistic network

(h) Nested core-periphery structure

(i) A "spectral" community

FIGURE 16.2: (a) A toy network that is "nice" and easily-visualizable. (b) A schematic of an advertiser-bidded-phrase graph that is similarly "nice." (c) A more realistic depiction of a large network. (d) Illustration of conductance clusters in a small network. (e) The downward-sloping NCP common to small networks, low-dimensional graphs, and common hierarchical models. (f) The NCP of a "LiveJournal" network [5]. (g) The NCP of an "Epinions" network [5]. (h) The nested core-periphery structure responsible for the upward-sloping NCP. (i) A typical 500-node "community" found with a spectral method. (j) Another 500-node "community" found with a spectral method. (k) A typical 500-node "community" found with a flow-based method. (l) Another 500-node "community" found with a flow-based method.

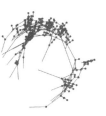

(j) Another "spectral" community

(k) A "flow" community

(l) Another "flow" community

FIGURE 16.2: (*Continued.*)

$\alpha(S)$ *of a set of nodes* $S \subseteq V$ is

$$\alpha(S) = \frac{|E(S, \overline{S})|}{\min\{|S|, |\overline{S}|)\}}, \qquad (16.4)$$

where $E(S, \overline{S})$ denotes the set of edges having one end in S and one end in the complement $\overline{S}$, and where $| \cdot |$ denotes cardinality (or weight); and the *expansion of the graph G* is

$$\alpha(G) = \min_{S \subseteq V} \alpha(S). \qquad (16.5)$$

Alternatively, if there is substantial variability in node degree, then normalizing by a related quantity is of greater interest. The *conductance* $\phi(S)$ *of a set of nodes* $S \subset V$ is

$$\phi(S) = \frac{|E(S, \overline{S})|}{\min\{A(S), A(\overline{S})\}}, \qquad (16.6)$$

where $A(S) = \sum_{i \in S} \sum_{j \in V} A_{ij}$, where A is the adjacency matrix of a graph. In this case, the *conductance of the graph G* is

$$\phi(G) = \min_{S \subseteq V} \phi(S). \qquad (16.7)$$

In either case, one could replace the "min" in the denominator with a "product," *e.g.*, replace $\min\{A(S), A(\overline{S})\}$ in the denominator of Equation (16.6) with $A(S) \cdot A(\overline{S})$.[11] The product formulation provides a slightly greater reward to a cut for having a big big-side, and it so has a slightly weaker preference for balance than the minimum formulation. Both formulations are equivalent, though, in that the objective function value of the set of nodes achieving the minimum with the one is within a factor of 2 of the objective function value

---

[11]Note that, up to scaling, this is simply the popular Normalized Cuts metric [72].

of the (in general different) set of nodes achieving the minimum with the other. Generically, the Minimum Conductance Cut Problem refers to solving any of these formulations; and importantly, all of these variants of the graph partitioning problem lead to intractable combinatorial optimization problems.

Within scientific computing [70, 77, 78, 79], and more recently within statistics and machine learning [72, 80, 81, 82], a great deal of work has focused on the conductance (or normalized cut) versions of this graph partitioning problem. Several general observations about the approach adopted in these literatures include

- The focus is on *spectral approximation algorithms*. Spectral algorithms use an exact or approximate eigenvector of the graph's Laplacian matrix to find a cut that achieves a "quadratic approximation," in the sense that the cut returned by the algorithm has conductance value no bigger than ϕ if the graph actually contains a cut with conductance $O(\phi^2)$ [83, 84, 85, 86, 87].

- The focus is on *low-dimensional graphs, e.g.*, bounded-degree planar graphs and finite element meshes (for which quality-of-approximation bounds are known that depend on just the number of nodes and not on structural parameters such as the conductance value). Moreover, there is an acknowledgement that *spectral methods are inappropriate for expander graphs* that have constant expansion or conductance (basically, since for these graphs any clusters returned are not meaningful in applications).

- Since the algorithm is typically applied to find partitions of a graph that are useful in some downstream application, there is a *strong interest in the actual pieces returned by the algorithm*. Relatedly, there is interest in the robustness or consistency properties of spectral approximation algorithms under assumptions on the data.

Recursive bisection heuristics—recursively divide the graph into two groups, and then further subdivide the new groups until the desired number of clusters groups is achieved—are also common here. They may be combined with local improvement methods [88, 89], which when combined with multi-resolution ideas leads to programs such as Metis [78, 79], Cluto [90], and Graclus [91].

Within TCS, a great deal of work has also focused on this graph partitioning problem [92, 93, 94, 95, 96, 97]. Several general observations about the traditional TCS approach include

- The focus is on *flow-based approximation algorithms*. These algorithms use multi-commodity flow and metric embedding ideas to find a cut whose conductance is within an $O(\log n)$ factor of optimal [92, 93, 95], in the sense that the cut returned by the algorithm has conductance no bigger than $O(\log n)$, where n is the number of nodes in the graph, times the conductance value of the optimal conductance set in the graph.

- The focus is on worst-case analysis of *arbitrary graphs*. Although flow-based methods achieve their worst-case $O(\log n)$ bounds on expanders, it is observed that *spectral methods are appropriate for expander graphs* (basically, since the quadratic of a constant is a constant, which implies a worst-case constant-factor approximation).

- Since the algorithmic task is simply to approximate the value of the objective function of Equation (16.5) or Equation (16.7), *there is no particular interest in the actual pieces returned by the algorithm*, except insofar as those pieces have objective function value that is close to the optimum.[12]

Most of the TCS work is on the expansion versions of the graph partitioning problem, but it is noted that most of the results obtained "go through" to the conductance versions by considering appropriately-weighted functions on the nodes.

There has also been recent work within TCS that has been motivated by achieving improved worst-case bounds and/or being appropriate in data analysis applications where very large graphs, say with millions or billions or nodes, arise. These methods include

- *Local Spectral Methods.* These methods [98, 99, 100, 101, 102] take as input a seed node and a locality parameter and return as output a "good" cluster "nearby" the seed node. Moreover, they often have computational cost that is proportional to the size of the piece returned, and they have roughly the same kind of quadratic approximation guarantees as the global spectral method.

- *Cut-Improvement Algorithms.* These methods [103, 104, 105, 106], *e.g.*, MQI [104, 105], take as input a graph and an initial cut and use spectral or flow-based methods to return as output a "good" cut that is "within" or "nearby" the original cut. As such, they can be combined with a good method (say, from spectral or Metis) for initially splitting the graph into two pieces to obtain a heuristic method for finding low conductance cuts in the whole graph [104, 105].

- *Combining Spectral and Flow.* These methods can be viewed as combinations of spectral and flow-based techniques which exploit the complementary strengths of these two classes of techniques. They include an algorithm that used semidefinite programming to find a solution that is within a multiplicative factor of $O(\sqrt{\log n})$ of optimal [96, 97], as well as several related algorithms [107, 108, 109, 110, 111] that are more amenable to medium- and large-scale implementation.

---

[12]That is not to say that those pieces are not sometimes then used in a downstream data application. Instead, it is to say simply that the algorithmic problem takes into account only the objective function value of those pieces and not how close those pieces might be to pieces achieving the optimum.

16.4.3 A Novel Approach to Characterizing Network Structure

Most work on community detection in large networks begins with an observation such as: "It is a matter of common experience that communities exist in networks. Although not precisely defined, communities are usually thought of as sets of nodes with better connections amongst its members than with the rest of the world." (I have not provided a reference for this quotation since variants of it could have been drawn from any one of scores or hundreds of papers on the topic. Most of this work then typically goes on to apply the *modus operandi* described previously.) Although far from perfect, conductance is probably the combinatorial quantity that most closely captures this intuitive bi-criterial notion of what it means for a set of nodes to be a good community.

Even more important from the perspective of this chapter, the use of conductance as an objective function has the following benefit. Although exactly solving the combinatorial problem of Equation (16.5) or Equation (16.7) is intractable, there exists a wide range of heuristics and approximation algorithms, the respective strengths and weaknesses of which are well-understood in theory and/or in practice, for approximately optimizing conductance. In particular, recall that spectral methods and multi-commodity flow-based methods are complementary in that

- The worst-case $O(\log n)$ approximation factor is obtained for flow-based methods on expander graphs [92, 95], a class of graphs which does not cause problems for spectral methods (to the extent than any methods are appropriate).

- The worst-case quadratic approximation factor is obtained for spectral methods on graphs with "long stringy" pieces [112, 77], basically since spectral methods can confuse "long path" with "deep cuts," a difference that does not cause problems for flow-based methods.

- Both methods perform well on "low-dimensional graphs"—*e.g.*, road networks, discretizations of low-dimensional spaces as might be encountered in scientific modeling, graphs that are easily-visualizable in two dimensions, and other graphs that have good well-balanced cuts—although the biases described in the previous two bullets still manifest themselves.

I should note that empirical evidence [4, 5, 6] clearly demonstrates that large social and information networks have all of these properties—they are expander-like when viewed at large size scales; their sparsity and noise properties are such that they have structures analogous to stringy pieces that are cut off or regularized away by spectral methods; and they often have structural regions that at least locally are meaningfully low-dimensional.

All of this suggests a rather novel approach—that of using approximation algorithms for the intractable graph partitioning problem (or other intractable problems) as "experimental probes" of the structure of large social and information networks. By this, I mean using these algorithms to "cut up" or "tear apart" a large network in order to provide insights into its structure. From this perspective, although we are defining an edge-counting metric and then using it to perform an optimization, we are not particularly interested *per se* in the clusters that are output by the algorithms. Instead, we are interested in what these clusters, coupled with knowledge of the inner workings of the approximation algorithms, tell us about network structure. Relatedly, given the noise properties and sparsity structure of large networks, we are not particularly interested *per se* in the solution to the combinatorial problem. For example, if we were provided with an oracle that returned the solution to the intractable combinatorial problem, it would be next to useless to us. Instead, much more information is revealed by looking at ensembles of output clusters in light of the artifactual properties of the heuristics and approximation algorithms employed. That is, from this perspective, we are interested in the *statistical properties implicit in worst-case approximation algorithms* and what these properties tell us about the data.

The analogy between approximation algorithms for the intractable graph partitioning problem and experimental probes is meant to be taken more seriously rather than less. Recall, *e.g.*, that it is a non-trivial exercise to determine what a protein or DNA molecule "looks like"—after all, such a molecule is very small, it can't easily be visualized with traditional methods, it has complicated dynamical properties, etc. To determine the structure of a protein, an experimentalist puts the protein in solution or in a crystal and then probes it with X-ray crystallography or a nuclear magnetic resonance signal. In more detail, one sends in a signal that scatters off the protein; one measures a large quantity of scattering output that comes out the other side; and then, using what is measured as well as knowledge of the physics of the input signal, one reconstructs the structure of the hard-to-visualize protein. In an analogous manner, we can analyze the structure of a large social or information by tearing it apart in numerous ways with one of several local or global graph partitioning algorithms; measuring a large number of pieces output by these procedures; and then, using what was measured as well as knowledge of the artifactual properties of the heuristics and approximation algorithms employed, reconstructing structural properties of the hard-to-visualize network.

By adopting this approach, one can hope to use approximation algorithms to test commonly-made data analysis assumptions, *e.g.*, that the data meaningfully lie on a manifold or some other low-dimensional space, or that the data have structures that are meaningfully-interpretable as communities.

16.4.4 Community-Identification Applications of This Approach

As a proof-of-principle application of this approach, let me return to the question of characterizing the small-scale and large-scale community structure in large social and information networks. Given the conductance (or some other) community quality score, define the *network community profile* (NCP) as the conductance value, as a function of size k, of the minimum conductance set of cardinality k in the network:

$$\Phi(k) = \min_{S \subset V, |S|=k} \phi(S).$$

Intuitively, just as the "surface-area-to-volume" aspect of conductance captures the "gestalt" notion of a good cluster or community, the NCP measures the score of "best" community as a function of community size in a network. See Figure 16.2(d) for an illustration of small network and Figure 16.2(e) for the corresponding NCP. Operationally, since the NCP is NP-hard to compute exactly, one can use approximation algorithms for solving the Minimum Conductance Cut Problem in order to compute different approximations to it. Although we have experimented with a wide range of procedures on smaller graphs, in order to scale to very large social and information networks,[13] we used

- *Metis+MQI*—this flow-based method consists of using the popular graph partitioning package Metis [78] followed by a flow-based MQI cut-improvement post-processing step [104], and it provides a surprisingly strong heuristic method for finding good conductance cuts;

- *Local Spectral*—although this spectral method [99] is worse than Metis+MQI at finding very good conductance pieces, the pieces it finds tend to be "tighter" and more "community-like"; and

- *Bag-of-Whiskers*—this is a simple heuristic to paste together small barely-connected sets of nodes that exert a surprisingly large influence on the NCP of large informatics graphs;

and we compared and contrasted the outputs of these different procedures.

The NCP behaves in a characteristic downward-sloping manner [4, 5, 6] for graphs that are well-embeddable into an underlying low-dimensional geometric structure, *e.g.*, low-dimensional lattices, road networks, random geometric graphs, and data sets that are well-approximatable by low-dimensional spaces

---

[13]We have examined a large number of networks with these methods. The networks we studied range in size from tens of nodes and scores of edges up to millions of nodes and tens of millions of edges. The networks were drawn from a wide range of domains and included large social networks, citation networks, collaboration networks, web graphs, communication networks, citation networks, internet networks, affiliation networks, and product co-purchasing networks. See [4, 5, 6] for details.

and non-linear manifolds. (See, *e.g.*, Figure 16.2(e).) Relatedly, a downward sloping NCP is also observed for small commonly-studied networks that have been widely-used as testbeds for community identification algorithms, while the NCP is roughly flat at all size scales for well-connected expander-like graphs such as moderately-dense random graphs. Perhaps surprisingly, common generative models, including preferential attachment models, as well as "copying" models and hierarchical models specifically designed to reproduce community structure, also have flat or downward-sloping NCPs. Thus, when viewed from the perspective of 10,000 meters, all of these graphs "look like" either a hot dog (in the sense that one can split the graph into two large and meaningful pieces) or a moderately-dense expander-like random graph (in the sense that there are no good partitions of any size in the graph). This fact is used to great advantage by common machine learning and data analysis tools.

From the same perspective of 10,000 meters, what large social and information networks "look like" is very different. In Figure 16.2(f) and Figure 16.2(g), I present the NCP computed several different ways for two representative large social and information networks. Several things are worth observing: first, up to a size scale of roughly 100 nodes, the NCP goes down; second, it then achieves its global minimum; third, above that size scale, the NCP gradually increases, indicating that the community-quality of possible communities gets gradually worse and worse as larger and larger purported communities are considered; and finally, even at the largest size scales there is substantial residual structure above and beyond what is present in a corresponding random graph. Thus, good network communities, at least as traditionally conceptualized, tend to exist only up to a size scale of roughly 100 nodes, while at larger size scales network communities become less and less community-like.

The explanation for this phenomenon is that large social and information networks have a *nested core-periphery structure*, in which the network consists of a large moderately-well connected core and a large number of very well-connected communities barely connected to the core, as illustrated in Figure 16.2(h). If we define a notion of a *whisker* to be maximal sub-graph detached from network by removing a single edge and the associated notion of the *core* to be the rest of the graph, i.e., the 2-edge-connected core, then whiskers contain (on average) 40% of nodes and 20% of edges. Moreover, in nearly every network, the global minimum of the NCP is a whisker. Importantly, though, even if *all* the whiskers are removed, then the core itself has a downward-then-upward-sloping NCP, indicating that the core itself has a nested core-periphery structure. In addition to being of interest in community identification applications, this fact explains the inapplicability of many common machine learning and data analysis tools for large informatics graphs. For example, it implies that recursive-bisection heuristics are largely inappropriate for this class of graphs (basically since the recursion depth will be very large).

There are numerous ways one can be confident in these domain-specific conclusions.[14] Most relevant for the interplay between the algorithmic and statistical perspectives on data that I am discussing in this chapter are

- *Modeling Considerations.* Several lines of evidence point to the role that sparsity and noise have in determining the large-scale clustering structure of large networks.

  - Extremely sparse Erdős-Rényi random graphs with connection probability parameter roughly $p \in (1/n, \log(n)/n)$—i.e., in the parameter regime where measure fails to concentrate sufficiently to have a fully-connected graph or to have observed node degrees close to expected node degrees—provide the simplest mechanism to reproduce the qualitative property of having very-imbalanced deep cuts and no well-balanced deep cuts.

  - So-called Power Law random graphs [113, 114] with power law parameter roughly $\beta \in (2,3)$—which, due to exogenously-specified node degree variability, have an analogous failure of measure concentration—also exhibit this same qualitative property.

  - A model in which new edges are added randomly but with an iterative "forest fire" burning mechanism provides a mechanism to reproduce the qualitative property of a relatively *gradually-increasing* NCP.[15]

- *Algorithmic-Statistical Considerations.* Ensembles of clusters returned by different approximation algorithms, *e.g.*, Local Spectral versus the flow-based methods such as Metis+MQI versus the Bag-of-Whiskers heuristic, have very different properties, in a manner consistent with known artifactual properties of how those algorithms operate.

  - For example, Metis+MQI finds sets of nodes that have very good conductance scores—at very small size scales, these are similar to the pieces from Local Spectral and these sets of nodes could plausibly be interpreted as good communities; but at size scales larger

---

[14]For example, *structural considerations* (*e.g.*, small barely connected whiskers, discoverable with pretty much any heuristic or algorithm, are responsible for minimum); *lower-bound considerations* (*e.g.*, SDP-based lower bounds for large partitions provide a strong nonexistence results for large communities); *domain-specific considerations* (*e.g.*, considering the properties of "ground truth" communities and the comparing results from networks in different application areas with very different degrees of homogeneity to their edge semantics); and *alternate-formalization considerations* (*e.g.*, other formalizations of the community concept that take into account the inter-connectivity versus intra-connectivity bi-criterion tend to have similar NCPs, while formalizations that take into account one or the other criteria behave very differently).

[15]As an aside, these considerations suggest that the data look like *local structure on top of a global sparse quasi-random scaffolding*, much more than the more commonly-assumed model of *local noise on top of global geometric or hierarchical structure*, an observation with significant modeling implications for this class of data.

than roughly 100 nodes, these are often tenuously-connected (and in some cases unions of disconnected) pieces, for which such an interpretation is tenuous at best.

- Similarly, the NCP of a variant of flow-based partitioning that permits disconnected pieces to be output mirrors the NCP from the Bag-of-Whiskers heuristic that combines disconnected whiskers, while the NCP of a variant of flow-based partitioning that requires connected pieces more closely follows that of Local Spectral, at least until much larger size scales where flow has problems since the graph becomes more expander-like.

- Finally, Local Spectral finds sets of nodes with good (but not as good as Metis+MQI) conductance value that that are more "compact" and thus more plausibly community-like than those pieces returned by Metis+MQI. For example, since spectral methods confuse long paths with deep cuts, empirically one obtains sets of nodes that have worse conductance scores but are internally more coherent in that, *e.g.*, they have substantially smaller diameter and average diameter. This is illustrated in Figures 16.2(i) through 16.2(l): Figures 16.2(i) and 16.2(j) show two ca. 500-node communities obtained from Local Spectral, which are thus more internally coherent than the two ca. 500-node communities illustrated in Figures 16.2(k) and 16.2(l) which were obtained from Metis+MQI and which thus consist of two tenuously-connected smaller pieces.

16.4.5 Some General Thoughts on Statistical Issues and Graph Algorithms

I will conclude this section with a few general thoughts raised by these empirical results. Note that the modeling considerations suggest, and the algorithmic-statistical considerations verify, that statistical issues, and in particular *regularization issues*, are particularly important for extracting domain-specific understanding from approximation algorithms applied this class of data.

By regularization, of course, I mean the traditional notion that the computed quantities—clusters and communities in this case—are empirically "nicer," or more "smooth" or "regular," in some useful domain-specific sense [115, 116, 117]. Recall that regularization grew out of the desire to find meaningful solutions to ill-posed problems in integral equation theory, and it may be viewed as adding a small amount of bias to reduce substantially the variance or variability of a computed quantity. It is typically implemented by adding a "loss term" to the objective function being optimized. Although this manner of implementation leads to a natural interpretation of regularization as a trade-off between optimizing the objective and avoiding over-fitting the data, it typically leads to optimization problems that are harder (think of

ℓ_1-regularized ℓ_2-regression) or at least no easier (think of ℓ_2-regularized ℓ_2-regression) than the original problem, a situation that is clearly unacceptable in the very large-scale applications I have been discussing.

Interestingly, though, at least for very large and extremely sparse social and information networks, intuitive notions of cluster quality tend to fail as one aggressively optimizes the community-quality score [6]. For instance, by aggressively optimizing conductance with Metis+MQI, one obtains disconnected or barely-connected clusters that do not correspond to intuitive communities. This suggests the rather interesting notion of implementing *implicit regularization via approximate computation*—approximate optimization of the community score in-and-of-itself introduces a systematic bias in the properties of the extracted clusters, when compared with much more aggressive optimization or computation of the intractable combinatorial optimum. Of course, such a bias may in fact be preferred since, as in case of Local Spectral, the resulting "regularized communities" are more compact and correspond to more closely to intuitive communities. Explicitly incorporating the tradeoff between the conductance of the bounding cut of the cluster and the internal cluster compactness into a new modified objective function would not have led to a more tractable problem; but implicitly incorporating it in the approximation algorithm had both algorithmic and statistical benefits. It is clearly of interest to formalize this approach more generally.

16.5   Conclusions and Future Directions

As noted above, the algorithmic and statistical perspectives on data and what might be interesting questions to ask of the data are quite different, but they are not incompatible. Some of the more prominent recent examples of this include

- Statistical, randomized, and probabilistic ideas are central to much of the recent work on developing improved worst-case approximation algorithms for matrix problems.

- Otherwise intractable optimization problems on graphs and networks yield to approximation algorithms when assumptions are made about the network participants.

- Much recent work in machine learning draws on ideas from both areas; similarly, in combinatorial scientific computing, implicit knowledge about the system being studied often leads to good heuristics for combinatorial problems.

- In boosting, a statistical technique that fits an additive model by minimizing an objective function with a method such as gradient descent, the computation parameter, i.e., the number of iterations, also serves as a regularization parameter.

In this chapter, I have focused on two examples that illustrate this point in a somewhat different way. For the first example, we have seen that using a concept fundamental to statistical and diagnostic regression analysis was crucial in the development of improved algorithms that come with worst-case performance guarantees and that, for related reasons, also perform well in practice. For the second example, we have seen that an improved understanding of worst-case algorithms (*e.g.*, when they perform well, when they perform poorly, and what regularization is implicitly implemented by them) suggested methods for using those approximation algorithms as "experimental probes" of large informatics graphs and for better inference and prediction on those graphs.

To conclude, I should re-emphasize that in neither of these applications did we *first* perform statistical modeling, independent of algorithmic considerations, and *then* apply a computational procedure as a black box. This approach of more closely coupling the computational procedures used with statistical modeling or statistical understanding of the data seems particularly appropriate more generally for very large-scale data analysis problems, where design decisions are often made based on computational constraints but where it is of interest to understand the implicit statistical consequences of those decisions.

Acknowledgments

I would like to thank Schloss Dagstuhl and the organizers of Dagstuhl Seminar 09061 for an enjoyable and fruitful meeting; Peristera Paschou, Petros Drineas, and Christos Boutsidis for fruitful discussions on DNA SNPs, population genetics, and matrix algorithms; Jure Leskovec and Kevin Lang for fruitful discussions on social and information networks and graph algorithms; and my co-organizers of as well as numerous participants of the MMDS meetings, at which many of the ideas described here were formed.

Bibliography

[1] Paschou, P., Mahoney, M. W., Javed, A., Kidd, J. R., Pakstis, A. J., Gu, S., Kidd, K. K., and Drineas, P., "Intra- and interpopulation genotype

reconstruction from tagging SNPs," *Genome Research*, Vol. 17, No. 1, 2007, pp. 96–107.

[2] Paschou, P., Ziv, E., Burchard, E., Choudhry, S., Rodriguez-Cintron, W., Mahoney, M., and Drineas, P., "PCA-Correlated SNPs for Structure Identification in Worldwide Human Populations," *PLoS Genetics*, Vol. 3, 2007, pp. 1672–1686.

[3] Mahoney, M. and Drineas, P., "CUR Matrix Decompositions for Improved Data Analysis," *Proc. Natl. Acad. Sci. USA*, Vol. 106, 2009, pp. 697–702.

[4] Leskovec, J., Lang, K., Dasgupta, A., and Mahoney, M., "Statistical Properties of Community Structure in Large Social and Information Networks," *WWW '08: Proceedings of the 17th International Conference on World Wide Web*, 2008, pp. 695–704.

[5] Leskovec, J., Lang, K., Dasgupta, A., and Mahoney, M., "Community Structure in Large Networks: Natural Cluster Sizes and the Absence of Large Well-Defined Clusters," Vol. arXiv:0810.1355, October 2008.

[6] Leskovec, J., Lang, K., and Mahoney, M., "Empirical Comparison of Algorithms for Network Community Detection," *WWW '10: Proceedings of the 19th International Conference on World Wide Web*, 2010, pp. 631–640.

[7] Fayyad, U., Piatetsky-Shapiro, G., and Smyth, P., "From data mining to knowledge discovery in databases," *AI Magazine*, Vol. 17, 1996, pp. 37–54.

[8] Smyth, P., "Data mining: data analysis on a grand scale?" *Statistical Methods in Medical Research*, Vol. 9, No. 4, 2000, pp. 309–327.

[9] Donoho, D., "Aide-Memoire. High-Dimensional Data Analysis: The Curses and Blessings of Dimensionality," 2000, `http://www-stat.stanford.edu/ donoho/Lectures/AMS2000/Curses.pdf` Accessed December, 17 2008.

[10] Breiman, L., "Statistical Modeling: The Two Cultures (with comments and a rejoinder by the author)," *Statistical Science*, Vol. 16, No. 3, 2001, pp. 199–231.

[11] Lambert, D., "What use is statistics for massive data?" *Crossing boundaries: statistical essays in honor of Jack Hall*, edited by J. E. Kolassa and D. Oakes, Institute of Mathematical Statistics, 2003, pp. 217–228.

[12] Poggio, T. and Smale, S., "The Mathematics of Learning: Dealing with Data," *Notices of the AMS*, Vol. 50, No. 5, May 2003, pp. 537–544.

[13] Golub, G. H., Mahoney, M. W., Drineas, P., and Lim, L.-H., "Bridging the Gap Between Numerical Linear Algebra, Theoretical Computer Science, and Data Applications," *SIAM News*, Vol. 39, No. 8, October 2006.

[14] Mahoney, M. W., Lim, L.-H., and Carlsson, G. E., "Algorithmic and Statistical Challenges in Modern Large-Scale Data Analysis are the Focus of MMDS 2008," *SIAM News*, Vol. 42, No. 1 & 2, January/February and March 2009.

[15] Alter, O., Brown, P., and Botstein, D., "Singular value decomposition for genome-wide expression data processing and modeling," *Proc. Natl. Acad. Sci. USA*, Vol. 97, No. 18, 2000, pp. 10101–10106.

[16] Kuruvilla, F., Park, P., and Schreiber, S., "Vector algebra in the analysis of genome-wide expression data," *Genome Biology*, Vol. 3, 2002, pp. research0011.1–0011.11.

[17] Meng, Z., Zaykin, D., Xu, C., Wagner, M., and Ehm, M., "Selection of genetic markers for association analyses, using linkage disequilibrium and haplotypes," *American Journal of Human Genetics*, Vol. 73, No. 1, 2003, pp. 115–130.

[18] Horne, B. and Camp, N., "Principal component analysis for selection of optimal SNP-sets that capture intragenic genetic variation," *Genetic Epidemiology*, Vol. 26, No. 1, 2004, pp. 11–21.

[19] Lin, Z. and Altman, R., "Finding Haplotype Tagging SNPs by Use of Principal Components Analysis," *American Journal of Human Genetics*, Vol. 75, 2004, pp. 850–861.

[20] Tenenbaum, J., de Silva, V., and Langford, J., "A global geometric framework for nonlinear dimensionality reduction," *Science*, Vol. 290, 2000, pp. 2319–2323.

[21] Roweis, S. and Saul, L., "Nonlinear dimensionality reduction by local linear embedding," *Science*, Vol. 290, 2000, pp. 2323–2326.

[22] Saul, L. K., Weinberger, K. Q., Ham, J. H., Sha, F., and Lee, D. D., "Spectral methods for dimensionality reduction," *Semisupervised Learning*, edited by O. Chapelle, B. Schoelkopf, and A. Zien, MIT Press, Cambridge, MA, 2006, pp. 293–308.

[23] Gould, S., *The Mismeasure of Man*, W. W. Norton and Company, New York, 1996.

[24] Civril, A. and Magdon-Ismail, M., "On Selecting A Maximum Volume Sub-matrix of a Matrix and Related Problems," *Theoretical Computer Science*, Vol. 410, 2009, pp. 4801–4811.

[25] Civril, A. and Magdon-Ismail, M., "Column Based Matrix Reconstruction via Greedy Approximation of SVD," Unpublished manuscript, 2009.

[26] Businger, P. and Golub, G., "Linear least squares solutions by Householder transformations," *Numerische Mathematik*, Vol. 7, 1965, pp. 269–276.

[27] Foster, L. V., "Rank and null space calculations using matrix decomposition without column interchanges," *Linear Algebra and Its Applications*, Vol. 74, 1986, pp. 47–71.

[28] Chan, T., "Rank revealing QR factorizations," *Linear Algebra and Its Applications*, Vol. 88/89, 1987, pp. 67–82.

[29] Chan, T. and Hansen, P., "Computing Truncated Singular Value Decomposition Least Squares Solutions by Rank Revealing QR-Factorizations," *SIAM Journal on Scientific and Statistical Computing*, Vol. 11, 1990, pp. 519–530.

[30] Bischof, C. H. and Hansen, P. C., "Structure-preserving and rank-revealing QR-factorizations," *SIAM Journal on Scientific and Statistical Computing*, Vol. 12, No. 6, 1991, pp. 1332–1350.

[31] Hong, Y. P. and Pan, C. T., "Rank-revealing QR factorizations and the singular value decomposition," *Mathematics of Computation*, Vol. 58, 1992, pp. 213–232.

[32] Chandrasekaran, S. and Ipsen, I. C. F., "On rank-revealing factorizations," *SIAM Journal on Matrix Analysis and Applications*, Vol. 15, 1994, pp. 592–622.

[33] Gu, M. and Eisenstat, S., "Efficient Algorithms for Computing a Strong Rank-Revealing QR Factorization," *SIAM Journal on Scientific Computing*, Vol. 17, 1996, pp. 848–869.

[34] Bischof, C. H. and Quintana-Ortí, G., "Computing rank-revealing QR factorizations of dense matrices," *ACM Transactions on Mathematical Software*, Vol. 24, No. 2, 1998, pp. 226–253.

[35] Pan, C. T. and Tang, P. T. P., "Bounds on singular values revealed by QR factorizations," *BIT Numerical Mathematics*, Vol. 39, 1999, pp. 740–756.

[36] Pan, C.-T., "On the existence and computation of rank-revealing LU factorizations," *Linear Algebra and Its Applications*, Vol. 316, 2000, pp. 199–222.

[37] Drineas, P., Frieze, A., Kannan, R., Vempala, S., and Vinay, V., "Clustering large graphs via the singular value decomposition," *Machine Learning*, Vol. 56, No. 1-3, 2004, pp. 9–33.

[38] Drineas, P., Kannan, R., and Mahoney, M., "Fast Monte Carlo Algorithms for Matrices I: Approximating Matrix Multiplication," *SIAM Journal on Computing*, Vol. 36, 2006, pp. 132–157.

[39] Drineas, P., Kannan, R., and Mahoney, M., "Fast Monte Carlo Algorithms for Matrices II: Computing a Low-Rank Approximation to a Matrix," *SIAM Journal on Computing*, Vol. 36, 2006, pp. 158–183.

[40] Drineas, P., Kannan, R., and Mahoney, M., "Fast Monte Carlo Algorithms for Matrices III: Computing a Compressed Approximate Matrix Decomposition," *SIAM Journal on Computing*, Vol. 36, 2006, pp. 184–206.

[41] Rudelson, M. and Vershynin, R., "Sampling from large matrices: an approach through geometric functional analysis," *Journal of the ACM*, Vol. 54, No. 4, 2007, pp. Article 21.

[42] Drineas, P., Mahoney, M., and Muthukrishnan, S., "Relative-Error CUR Matrix Decompositions," *SIAM Journal on Matrix Analysis and Applications*, Vol. 30, 2008, pp. 844–881.

[43] Deshpande, A. and Vempala, S., "Adaptive sampling and fast low-rank matrix approximation," *Proceedings of the 10th International Workshop on Randomization and Computation*, 2006, pp. 292–303.

[44] Boutsidis, C., Mahoney, M., and Drineas, P., "An Improved Approximation Algorithm for the Column Subset Selection Problem," Tech. rep., Preprint: arXiv:0812.4293 (2008).

[45] Golub, G. and Loan, C. V., *Matrix Computations*, Johns Hopkins University Press, Baltimore, 1996.

[46] Chatterjee, S. and Hadi, A., *Sensitivity Analysis in Linear Regression*, John Wiley & Sons, New York, 1988.

[47] Hoaglin, D. and Welsch, R., "The Hat Matrix in Regression and ANOVA," *The American Statistician*, Vol. 32, No. 1, 1978, pp. 17–22.

[48] Chatterjee, S. and Hadi, A., "Influential Observations, High Leverage Points, and Outliers in Linear Regression," *Statistical Science*, Vol. 1, No. 3, 1986, pp. 379–393.

[49] Spielman, D. and Srivastava, N., "Graph Sparsification by Effective Resistances," *Proceedings of the 40th Annual ACM Symposium on Theory of Computing*, 2008, pp. 563–568.

[50] Candes, E. and Recht, B., "Exact Matrix Completion via Convex Optimization," Tech. rep., Preprint: arXiv:0805.4471 (2008).

[51] Draper, N. R. and Stoneman, D. M., "Testing for the Inclusion of Variables in Linear Regression by a Randomisation Technique," *Technometrics*, Vol. 8, No. 4, 1966, pp. 695–699.

[52] Velleman, P. and Welsch, R., "Efficient Computing of Regression Diagnostics," *The American Statistician*, Vol. 35, No. 4, 1981, pp. 234–242.

[53] Chatterjee, S., Hadi, A., and Price, B., *Regression Analysis by Example*, John Wiley & Sons, New York, 2000.

[54] Zachary, W., "An information flow model for conflict and fission in small groups," *Journal of Anthropological Research*, Vol. 33, 1977, pp. 452–473.

[55] Drineas, P., Mahoney, M., and Muthukrishnan, S., "Sampling Algorithms for ℓ_2 Regression and Applications," *Proceedings of the 17th Annual ACM-SIAM Symposium on Discrete Algorithms*, 2006, pp. 1127–1136.

[56] Drineas, P., Magdon-Ismail, M., Mahoney, M., and Woodruff, P. P. "Fast approximation of matrix coherence and statistical leverage," Technical report, Preprint: arxiv:1109.3843, 2011.

[57] Ailon, N. and Chazelle, B., "Approximate nearest neighbors and the fast Johnson-Lindenstrauss transform," *Proceedings of the 38th Annual ACM Symposium on Theory of Computing*, 2006, pp. 557–563.

[58] Matoušek, J., "On variants of the Johnson–Lindenstrauss lemma," *Random Structures and Algorithms*, Vol. 33, No. 2, 2008, pp. 142–156.

[59] Sarlós, T., "Improved Approximation Algorithms for Large Matrices via Random Projections," *Proceedings of the 47th Annual IEEE Symposium on Foundations of Computer Science*, 2006, pp. 143–152.

[60] Drineas, P., Mahoney, M., Muthukrishnan, S., and Sarlós, T., "Faster Least Squares Approximation," Tech. rep., Preprint: arXiv:0710.1435 (2007).

[61] Rokhlin, V. and Tygert, M., "A fast randomized algorithm for overdetermined linear least-squares regression," *Proc. Natl. Acad. Sci. USA*, Vol. 105, No. 36, 2008, pp. 13212–13217.

[62] Avron, H., Maymounkov, P., and Toledo, S., "Blendenpik: Supercharging LAPACK's least-squares solver," Manuscript. (2009).

[63] Boutsidis, C., Mahoney, M., and Drineas, P., "Unsupervised Feature Selection for Principal Components Analysis," *Proceedings of the 14th Annual ACM SIGKDD Conference*, 2008, pp. 61–69.

[64] Boutsidis, C., Mahoney, M., and Drineas, P., "Unsupervised Feature Selection for the *k*-means Clustering Problem," *Annual Advances in Neural Information Processing Systems 22: Proceedings of the 2009 Conference*, 2009.

[65] Newman, M., "A measure of betweenness centrality based on random walks," *Social Networks*, Vol. 27, 2005, pp. 39–54.

[66] Berry, M. and Browne, M., "Email Surveillance Using Non-negative Matrix Factorization," *Computational and Mathematical Organization Theory*, Vol. 11, No. 3, 2005, pp. 249–264.

[67] Deerwester, S., Dumais, S., Furnas, G., Landauer, T., and Harshman, R., "Indexing by latent semantic analysis," *Journal of the American Society for Information Science*, Vol. 41, No. 6, 1990, pp. 391–407.

[68] Chan, T. and Hansen, P., "Low-rank revealing QR factorizations," *Numerical Linear Algebra with Applications*, Vol. 1, 1994, pp. 33–44.

[69] Bischof, C. H. and Quintana-Ortí, G., "Algorithm 782: Codes for rank-revealing QR factorizations of dense matrices," *ACM Transactions on Mathematical Software*, Vol. 24, No. 2, 1998, pp. 254–257.

[70] Pothen, A., "Graph partitioning algorithms with applications to scientific computing," *Parallel Numerical Algorithms*, edited by D. E. Keyes, A. H. Sameh, and V. Venkatakrishnan, Kluwer Academic Press, 1996.

[71] Jain, A., Murty, M., and Flynn, P., "Data clustering: a review," *ACM Computing Surveys*, Vol. 31, 1999, pp. 264–323.

[72] Shi, J. and Malik, J., "Normalized Cuts and Image Segmentation," *IEEE Transcations of Pattern Analysis and Machine Intelligence*, Vol. 22, No. 8, 2000, pp. 888–905.

[73] Gaertler, M., "Clustering," *Network Analysis: Methodological Foundations*, edited by U. Brandes and T. Erlebach, Springer, 2005, pp. 178–215.

[74] von Luxburg, U., "A Tutorial on Spectral Clustering," Tech. Rep. 149, Max Plank Institute for Biological Cybernetics, August 2006.

[75] Newman, M., "Finding community structure in networks using the eigenvectors of matrices," *Physical Review E*, Vol. 74, 2006, pp. 036104.

[76] Schaeffer, S., "Graph Clustering," *Computer Science Review*, Vol. 1, No. 1, 2007, pp. 27–64.

[77] Spielman, D. and Teng, S.-H., "Spectral partitioning works: Planar graphs and finite element meshes," *FOCS '96: Proceedings of the 37th Annual IEEE Symposium on Foundations of Computer Science*, 1996, pp. 96–107.

[78] Karypis, G. and Kumar, V., "A Fast and High Quality Multilevel Scheme for Partitioning Irregular Graphs," *SIAM Journal on Scientific Computing*, Vol. 20, 1998, pp. 359–392.

[79] Karypis, G. and Kumar, V., "Multilevel k-way Partitioning Scheme for Irregular Graphs," *Journal of Parallel and Distributed Computing*, Vol. 48, 1998, pp. 96–129.

[80] Weiss, Y., "Segmentation using eigenvectors: a unifying view," *ICCV '99: Proceedings of the 7th IEEE International Conference on Computer Vision*, 1999, pp. 975–982.

[81] Ng, A., Jordan, M., and Weiss, Y., "On Spectral Clustering: Analysis and an algorithm," *NIPS '01: Proceedings of the 15th Annual Conference on Advances in Neural Information Processing Systems*, 2001.

[82] Kannan, R., Vempala, S., and Vetta, A., "On clusterings: Good, bad and spectral," *Journal of the ACM*, Vol. 51, No. 3, 2004, pp. 497–515.

[83] Cheeger, J., "A lower bound for the smallest eigenvalue of the Laplacian," *Problems in Analysis, Papers dedicated to Salomon Bochner*, Princeton University Press, 1969, pp. 195–199.

[84] Donath, W. and Hoffman, A., "Algorithms for partitioning graphs and computer logic based on eigenvectors of connection matrices," *IBM Technical Disclosure Bulletin*, Vol. 15, No. 3, 1972, pp. 938–944.

[85] Fiedler, M., "Algebraic connectivity of graphs," *Czechoslovak Mathematical Journal*, Vol. 23, No. 98, 1973, pp. 298–305.

[86] Mohar, B., "The Laplacian spectrum of graphs," *Graph Theory, Combinatorics, and Applications, Vol. 2*, edited by Y. Alavi, G. Chartrand, O. Oellermann, and A. Schwenk, Wiley, 1991, pp. 871–898.

[87] Chung, F., *Spectral graph theory*, Vol. 92 of *CBMS Regional Conference Series in Mathematics*, American Mathematical Society, 1997.

[88] Kernighan, B. and Lin, S., "An effective heuristic procedure for partitioning graphs," *The Bell System Technical Journal*, 1970, pp. 291–308.

[89] Fiduccia, C. and Mattheyses, R., "A linear-time heuristic for improving network partitions," *DAC '82: Proceedings of the 19th ACM/IEEE Conference on Design Automation*, 1982, pp. 175–181.

[90] Zhao, Y. and Karypis, G., "Empirical and Theoretical Comparisons of Selected Criterion Functions for Document Clustering," *Machine Learning*, Vol. 55, 2004, pp. 311–331.

[91] Dhillon, I., Guan, Y., and Kulis, B., "Weighted Graph Cuts without Eigenvectors: A Multilevel Approach," *IEEE Transactions on Pattern Analysis and Machine Intelligence*, Vol. 29, No. 11, 2007, pp. 1944–1957.

[92] Leighton, T. and Rao, S., "An Approximate Max-Flow Min-Cut Theorem for Uniform Multicommodity Flow Problems with Applications to Approximation Algorithms," *FOCS '88: Proceedings of the 28th Annual Symposium on Foundations of Computer Science*, 1988, pp. 422–431.

[93] Linial, N., London, E., and Rabinovich, Y., "The geometry of graphs and some of its algorithmic applications," *Combinatorica*, Vol. 15, No. 2, 1995, pp. 215–245.

[94] Shmoys, D. B., "Cut problems and their application to divide-and-conquer," *Approximation Algorithms for NP-Hard Problems*, edited by D. Hochbaum, PWS Publishing, Boston, 1996, pp. 192–235.

[95] Leighton, T. and Rao, S., "Multicommodity max-flow min-cut theorems and their use in designing approximation algorithms," *Journal of the ACM*, Vol. 46, No. 6, 1999, pp. 787–832.

[96] Arora, S., Rao, S., and Vazirani, U., "Expander flows, geometric embeddings and graph partitioning," *STOC '04: Proceedings of the 36th annual ACM Symposium on Theory of Computing*, 2004, pp. 222–231.

[97] Arora, S., Rao, S., and Vazirani, U., "Geometry, flows, and graph-partitioning algorithms," *Communications of the ACM*, Vol. 51, No. 10, 2008, pp. 96–105.

[98] Spielman, D. and Teng, S.-H., "Nearly-linear time algorithms for graph partitioning, graph sparsification, and solving linear systems," *STOC '04: Proceedings of the 36th annual ACM Symposium on Theory of Computing*, 2004, pp. 81–90.

[99] Andersen, R., Chung, F., and Lang, K., "Local Graph Partitioning using PageRank Vectors," *FOCS '06: Proceedings of the 47th Annual IEEE Symposium on Foundations of Computer Science*, 2006, pp. 475–486.

[100] Chung, F., "Four proofs of Cheeger inequality and graph partition algorithms," *Proceedings of ICCM*, 2007.

[101] Chung, F., "Random walks and local cuts in graphs," *Linear Algebra and its Applications*, Vol. 423, 2007, pp. 22–32.

[102] Chung, F., "The heat kernel as the pagerank of a graph," *Proceedings of the National Academy of Sciences of the United States of America*, Vol. 104, No. 50, 2007, pp. 19735–19740.

[103] Gallo, G., Grigoriadis, M., and Tarjan, R., "A fast parametric maximum flow algorithm and applications," *SIAM Journal on Computing*, Vol. 18, No. 1, 1989, pp. 30–55.

[104] Lang, K. and Rao, S., "A Flow-Based Method for Improving the Expansion or Conductance of Graph Cuts," *IPCO '04: Proceedings of the 10th International IPCO Conference on Integer Programming and Combinatorial Optimization*, 2004, pp. 325–337.

[105] Andersen, R. and Lang, K., "An algorithm for improving graph partitions," *SODA '08: Proceedings of the 19th ACM-SIAM Symposium on Discrete algorithms*, 2008, pp. 651–660.

[106] Mahoney, M. W., Orecchia, L., and Vishnoi, N. K., "A Spectral Algorithm for Improving Graph Partitions," Tech. rep., Preprint: arXiv:0912.0681 (2009).

[107] Arora, S., Hazan, E., and Kale, S., "$O(\sqrt{\log n})$ Approximation to SPARSEST CUT in $\tilde{O}(n^2)$ Time," *FOCS '04: Proceedings of the 45th Annual Symposium on Foundations of Computer Science*, 2004, pp. 238–247.

[108] Khandekar, R., Rao, S., and Vazirani, U., "Graph partitioning using single commodity flows," *STOC '06: Proceedings of the 38th annual ACM Symposium on Theory of Computing*, 2006, pp. 385–390.

[109] Arora, S. and Kale, S., "A combinatorial, primal-dual approach to semidefinite programs," *STOC '07: Proceedings of the 39th annual ACM Symposium on Theory of Computing*, 2007, pp. 227–236.

[110] Orecchia, L., Schulman, L., Vazirani, U., and Vishnoi, N., "On Partitioning Graphs via Single Commodity Flows," *Proceedings of the 40th Annual ACM Symposium on Theory of Computing*, 2008, pp. 461–470.

[111] Lang, K., Mahoney, M. W., and Orecchia, L., "Empirical Evaluation of Graph Partitioning Using Spectral Embeddings and Flow," *Proc. 8-th International SEA*, 2009, pp. 197–208.

[112] Guattery, S. and Miller, G., "On the Quality of Spectral Separators," *SIAM Journal on Matrix Analysis and Applications*, Vol. 19, 1998, pp. 701–719.

[113] Chung, F. and Lu, L., "The average distances in random graphs with given expected degrees," *Proceedings of the National Academy of Sciences of the United States of America*, Vol. 99, No. 25, 2002, pp. 15879–15882.

[114] Chung, F., Lu, L., and Vu, V., "The spectra of random graphs with given expected degrees," *Proceedings of the National Academy of Sciences of the United States of America*, Vol. 100, No. 11, 2003, pp. 6313–6318.

[115] Neumaier, A., "Solving Ill-Conditioned And Singular Linear Systems: A Tutorial On Regularization," *SIAM Review*, Vol. 40, 1998, pp. 636–666.

[116] Chen, Z. and Haykin, S., "On Different Facets of Regularization Theory," *Neural Computation*, Vol. 14, No. 12, 2002, pp. 2791–2846.

[117] Bickel, P. and Li, B., "Regularization in Statistics," *TEST*, Vol. 15, No. 2, 2006, pp. 271–344.

Chapter 17

Computational Challenges in Emerging Combinatorial Scientific Computing Applications

David A. Bader

Georgia Institute of Technology

Kamesh Madduri

Lawrence Berkeley National Laboratory

17.1 Introduction

In this chapter, we describe several novel uses of combinatorial techniques that have found widespread use in realms outside traditional scientific computing. The "emerging" scientific disciplines of *social and technological network analysis* and *computational molecular biology* serve as the backdrop for our discussion. As illustrative case studies, we present two key problems in each of these areas. Most typical formulations of our chosen case studies are $\mathcal{NP}$-hard optimization problems on graphs, and hence a great variety of new approximation algorithms and heuristics are being designed to solve them. We will provide a high-level overview of the problems, review some popular algorithmic strategies to solve them, and then focus our discussion on interesting combinatorial routines that these methods typically employ. Keeping with this book's recurring theme of efficient "large-scale computing," we highlight the state-of-the-art parallel approaches for solving these problems, and outline challenges that may hinder scalable implementations on current parallel platforms.

17.2   Analysis of Social and Technological Networks

The use of graph abstractions to analyze and understand social interaction data, complex engineered systems such as the power grid and the Internet, communication data such as email and phone networks, sensor networks, biological systems, and in general, all forms of relational data, has been gaining ever-increasing importance. In 2005, a committee of experts commissioned by the U.S. Army popularized the term "network science", which they define as "the study of networks using the scientific method" [1]. The simplest possible mapping of a real-world system to the discrete math "graph" structure would be as follows: actors or interacting entities are represented as vertices, and interactions between the entities are depicted as edges. These vertices and edges can be further assigned types, weights, or attributes to distinguish heterogeneous data. Watts was one the earliest to identify the spurt in networks-related research [2], Börner et al.'s primer [3] covers the use of graph abstractions and some common graph algorithms in a variety of application domains, and Alderson [4] reviews current research challenges in this broad domain of research from the perspective of the operations research community.

Network modeling (i.e., discovering or inventing mathematical formulations that quantitatively explain any underlying organizing principles behind a system of interacting entities) is a central problem and a key driver of network science research efforts. For instance, the "small-world" concept (first investigated by the social scientist Milgram [5]) and associated network models [6, 7] have been quite effective in explaining the phenomenon of low average path lengths seen in social and technological networks. In seminal research on empirical complex data modeling, Newman [8], Albert and Barabási [9], Faloutsos et al. [10], among others, extensively studied real-world data sets that are representative of Internet topologies, web crawls, socio-economic interactions, and biological networks. This work led to the development and subsequent refinement of a variety of graph-theoretic models to explain the statistical properties exhibited by these data sets. In addition to the small-world idea (which implies a low graph diameter), common themes in various models include skewed or "power law"-like vertex degree distribution, the property of self-similarity (or "scale-free" behavior), and the presence of dense subgraphs. A discussion of computational challenges involved in modeling data is outside the scope of this chapter, and we refer the reader to [11, 12, 13, 14, 15] for a sampling of popular models.

A report by Amaral et al. [16], summarizing discussions among leading complex network researchers in 2003, outlines a collection of research challenges that continue to be relevant to-date. In addition to modeling properties associated with static graphs, i.e., network structure at a particular instance of time, this report raises the need to study network evolution. Also, other modeling-related questions of interest include how dynamic processes on network abstractions affect the network topology, and how one characterizes relationships between interconnected systems.

We next discuss two problems that arise in the "analysis" of large-scale data sets. Example analytic queries include determining key interacting entities in a network, detecting anomalous entities or interaction patterns given a large data set, characterizing the network topology using well-understood graph-theoretic measures, and so on. Similar to the modeling challenge, graph analysis problems arise in a variety of domains: online network analysis [17], social networks [18, 19, 20, 21], biological networks [22, 23, 24], and intelligence and surveillance [25, 26]. Modeling and analysis are often inter-dependent and applications typically require a mixture of both to achieve efficient solutions. For instance, analysis problems such as determining whether a network is robust, or identifying links that are vulnerable to attacks, requires one to make accurate assumptions about the underlying network topology. We will only focus on the combinatorial aspects pertaining to straightforward graph formulations of the analytic routines in this chapter.

Concurrent with modeling and algorithmic research, there has been considerable effort focused on development of efficient software tools. Notable packages include GUI-based tools such as Network Workbench [27], libraries (e.g., igraph [28], JUNG [29]), and packages exclusively for social network analysis (e.g., Pajek [30], InFlow [31], UCINET [32]). SNAP [33, 34] and the Parallel BOOST graph library [35] are research software packages that implement efficient parallel versions of key network analysis routines. The DARPA High Productivity Computing Systems (HPCS) Scalable Synthetic Compact Application graph analysis benchmark [36, 37] also mimics some of the key kernels in social network analysis, operating on graphs exhibiting small-world properties and with a power-law degree distribution [38].

17.2.1 Community Identification

It has been observed that many real-world systems of interest, when modeled as networks, can be decomposed into variable-sized sub-systems with strong internal interactions and relatively weaker external connections. Thus a key problem in the analysis of networks is that of detecting these dense collections of vertices, also referred to as communities. We find very well-defined and direct formulations of this problem in several application domains (e.g., systems biology, web search, economics, physical infrastructure analysis). Other uses of community identification and its closely-related variants include the following:

- Identifying latent community structure in a network and boundaries between communities helps us classify vertices into smaller modules, and understand the relative importance and influence of vertices within the smaller group.

- Vertices that may belong to multiple communities typically have an important role to play in the spread of information through the network.

- The sizes and connectivity patterns exhibited by the communities are important global footprints of the network that may prove useful in comparative network analysis scenarios.

- Community identification also helps reveal hierarchical structure in the network, which plays an important role in explaining the evolution of the underlying system and the effect of dynamic processes on the network.

Traditional graph partitioning in scientific computing and community detection are related problems, but with important differences: the most commonly-used objective function in partitioning is minimization of edge cut, while trying to balance the number of vertices in each partition. The number of partitions is typically an input parameter for a partitioning algorithm. This partitioning approach is often used to split meshes onto parallel computers: the minimum edge cut educes communication and the equal number of vertices aim to balance the computational load among processors. Community identification, on the other hand, optimizes an appropriate application-dependent measure, and the number and sizes of modules are unknown beforehand. Multi-level partitioning algorithms, spectral techniques, flow-based approaches, hypergraph partitioning, and other techniques have been shown to be very effective for partitioning 2D and 3D mesh topologies. These approaches are discussed in more detail in other chapters of this book. There are also several open-source software packages that implement state-of-the-art graph partitioning algorithms, and these are quite efficient for tackling problems in scientific computing. However, empirical evidence suggests that these tools fail to provide a good partitioning for social, biological, and information network abstractions. Small-world networks, also typically "virtual" networks, encapsulate high-dimensional data and lack the topological regularity found in scientific meshes and physical networks, where the degree distribution is relatively constant and most connectivity is localized. Lang's experiments [39, 40] show that the cut quality varies inversely with cut balance for social graphs such as the Yahoo! IM network and the DBLP collaboration data set. We opine that the problem of partitioning small-world networks deserves more investigation, as there currently is no clear accepted strategy. Mahoney discusses partitioning and community identification in another chapter of this book, and offers his perspective via the analysis of "network community profiles" (NCPs) [41].

There are several different objective functions that one could utilize to resolve a network into communities (see [42, 43] for recent comparative studies of multiple objective functions for community detection). The notion of "modularity", first defined by Newman and Girvan [44, 45], is currently by far the most well-known method for assessing community qualities. Modularity measures internal connectivity of communities with respect to a randomized null model. The modularity objective function is formally defined as follows. Let $C = (C_1, ..., C_k)$ denote a partition of V such that $C_i \neq \phi$ and $C_i \cap C_j = \phi$. We call C a clustering of G and each C_i is defined to be a cluster. The cluster $G(C_i)$ is identified with the induced subgraph $G[C_i] := (C_i, E(C_i))$, where $E(C_i) := \{\langle u, v \rangle \in E : u, v \in C_i\}$. Then, $E(C) := \cup_{i=1}^{k} E(C_i)$ is the set of intra-cluster edges and $\tilde{E}(C) := E - E(C)$ is the set of inter-cluster edges. Let $m(C_i)$ denote the number of inter-cluster edges in C_i. Then, the modularity $q(C)$ of a clustering C is defined as

$$ q(C) = \sum_{i} \left[\frac{m(C_i)}{m} - \left(\frac{\sum_{v \in C_i} deg(v)}{2m} \right)^2 \right] $$

To maximize the first term, the number of intra-cluster edges should be high, whereas the second term is minimized by splitting the graph into multiple clusters with small total degrees. If a particular clustering gives no more intra-community edges than would be expected in a random graph with the same degree distribution, we will get $q = 0$. Values greater than 0 indicate deviation from randomness, and empirical results show that values greater than 0.3 indicate significant community structure.

Intuitively, modularity captures the idea that a good division of a network into communities is one in which the inter-community edge count is less than what is expected by random chance. This measure does not try to minimize the edge cut in isolation, nor does it explicitly favor a balanced community partitioning. Modularity has found widespread acceptance in the network analysis community, and there have been an array of heuristics, based on spectral analysis, simulated annealing, greedy agglomeration, extremal optimization, and other techniques proposed to optimize it. Brandes et al. [46] show that maximizing modularity is $\mathcal{NP}$-complete. For recent reviews of heuristics for this problem, see [47, 48].

Most of the known techniques for community identification fall into one of the following three broad categorizations: *divisive, agglomerative,* and *local* clustering. The algorithm by Newman and Girvan is a good example of the divisive approach. This approach is based on the notion of determining edges with high "betweenness centrality" (for the formal definition and algorithms, see [49, 50]). An edge has high betweenness if it lies on a high percentage of the shortest paths between vertices. Edge betweenness intuitively attempts to capture the impact on communication an edge has in the network, and can be used to identify critical entities whose removal may most disrupt the graph. In addition to use in community detection algorithms, betweenness centrality finds applicability in assessing lethality in biological networks [22, 24], study of sexual networks and AIDS [51], identifying key actors in terrorist networks [26], and supply chain management processes. Newman and Girvan's algorithm iteratively removes high betweenness edges, and recalculates scores for the rest of the graph. It deconstructs the initial graph into progressively smaller connected chunks until a final set of isolated nodes are obtained. The modularity score can be used as a stopping criteria in this divisive process. Thus, the key approach in divisive algorithms is to identify possible edges that connect vertices of different communities and iteratively remove them, so that the clusters get disconnected from each other. The algorithms differ in the property that allows for identification of inter-community edges. The hierarchical structure of community resolution and the divisions can be represented using a tree (referred to as a "dendrogram"). The final list of the communities is given by the leaves of the dendrogram, and internal nodes correspond to splits in the divisive approach. There are also several spectral heuristics for modularity maximizations that can also be thought of as divisive approaches. Instead of removing edge by edge, spectral heuristics use the Fiedler vector to eliminate multiple edges in one step and generate partitionings of the network.

In the bottom-up agglomerative scheme, each vertex initially belongs to a singleton community, and communities whose amalgamation produces an increase in a local or global measure of choice are merged together. The measure for ordering the merges is typically the increase in modularity [52] or some variant. Some approaches relax the global ordering of merges and instead grow communities around multiple central vertices. A dendrogram can be again constructed to capture the merge process.

Local clustering heuristics [47, 53] are based on the premise that a network is made up of many small communities which can be detected by local exploration. They are also motivated by the fact that modularity's null model implicitly assumes that each vertex could be attached to any other, whereas in real cases a cluster may be connected to only a few other clusters. These local approaches perform greedy search one or two hops from a vertex and attempt to extract small communities, and rely on local formulations of modularity for assessing quality. They also tend to favor extraction of smaller communities.

Maximizing modularity is also shown to be related to graph visualization techniques such as the Fruchterman-Reingold force-directed layout method [54]. This suggests that algorithms for community identification may also be applicable and lead to better techniques for visualizing large graphs, and vice versa. Modularity can also be used as an objective function for graph partitioning by adding the constraint that the communities obtained need to be roughly of the same size. In that case, there is a linear relationship between the edge cut and the final obtained modularity score.

The formulation of modularity has a known "resolution limit" issue. Fortunato and Barthelemy [55] demonstrated this problem using real and synthetic networks, showing how maximizing modularity results in missing communities that are smaller than a certain threshold size. The threshold depends on the size of the network and the extent of interconnectedness of its communities. Prior empirical work has shown that communities in real-world graphs show a diverse distribution, and hence many small communities may remain undetected with modularity maximization. There have been several fixes suggested to address this issue [56, 57], both to algorithms and to the actual modularity measure itself. It is also believed that modularity is very sensitive to individual edge linkages. In case of noisy data sets with missing or spurious relations between entities, the efficacy of modularity maximization is unclear. Further, enhancing the current topology-centric definition of modularity by including additional aspects (such as vertex or edge attributes, time dependence) remains an important open problem.

While there has been an explosion of research related to community detection algorithms in the past few years, work on parallelizing these heuristics is still in its infancy. There is certainly a need for faster parallel approaches, as most of the current open-source software packages implementing the state-of-the-art algorithms take hours to days for community identification on modest-sized graphs. Consider the three broad paradigms discussed above. The computational complexity of the divisive and agglomerative approaches depends on the measure that is used to split/merge communities. For instance, the divisive algorithm based on edge betweenness requires $O(mn)$ time (where n is the number of vertices and m is the number of edges in the graph) on each iteration to compute the edge betweenness scores, and this is a natural avenue for parallelization. In case of agglomerative clustering approaches, the main opportunities to exploit parallelism arise from the relaxed formulations: i.e., we can concurrently resolve and build communities that are sufficiently far away from each other. The local heuristics are conceptually much simpler to parallelize; one could concurrently work from different vertices to identify communities. Our parallel graph analysis framework SNAP [34, 33, 58] includes some of the first parallel implementations of modularity-maximizing community identification algorithms. Performance results on real-world networks indicate that the SNAP parallel community identification algorithms give substantial performance improvements on current multithreaded and multicore systems, while being comparable in quality to the best-performing serial approaches. We achieve this through a compact graph representation, highly-optimized underlying computational routines [59, 60], and judicious use of preprocessing steps before executing the community identification algorithms.

Next, we briefly discuss the key computational kernels and data representations that are required for efficient execution of serial and parallel community identification algorithms. Graph traversal and adjacency querying (primarily one or two hops from a vertex) are key components in many algorithms, and it is important that the graph representation is optimized for repeated execution of these routines. Maintaining the split and merge information for communities and tracking vertex-community updates mandates a separate data representation, independent of the graph, that can support fast and parallel insertions and deletions. A priority queue is also essential in many algorithms for ordering possible merges. In case of distributed-memory implementations, the challenge of partitioning the network for load balanced execution and inter-processor communication minimization arises. The partitioning scheme should be faster than the parallel community identification algorithm being implemented.

17.2.2   Graph Pattern Mining and Matching

In this section, we discuss the problems of "frequent pattern mining" and "pattern matching", two related and well-studied problems in the network analysis and data mining communities. Identifying frequently occurring patterns from collections of graphs, or graph databases, finds broad applicability in chemical informatics, and analytics related to XML documents, web logs, and online social networks. Pattern matching and its variants occur in anomaly detection, compression-related, and statistical classification-related applications.

Frequent graph pattern mining is a natural extension to the problem of frequent itemset mining [61] introduced by Agrawal et al. [62]. Consider a database of transactions D and a universe of items $I = i_1, i_2, \ldots, i_n$. Each transaction T_i is a subset of I, and let the cardinality of D be m. A k-itemset is defined a set of cardinality k that is a subset of I. A itemset is termed frequent if it occurs no fewer than θm times, where θ is a user-specified parameter called the "minimum support threshold" for the itemset. The problem of frequent pattern mining is to find all the itemsets which are above the minimum support threshold. Agrawal et al. formulated this problem to help with association rule mining. For instance, it can lead to conclusions such as "customers who purchased item A also purchased items B and C". It has been shown that the decision problem of whether "there exists a maximal frequent itemset with at least k items" is $\mathcal{NP}$-complete [63].

Agrawal and Srikant [64] designed an algorithm called "Apriori" to solve this problem. This algorithm utilizes an anti-monotone property among frequent k-itemsets: a k-itemset is frequent only if all of its sub-itemsets are frequent. This implies that frequent itemsets can be mined by first scanning the database to find the frequent 1-itemsets, then using this information to generate candidate frequent 2-itemsets, and check against the database to obtain the frequent 2-itemsets. This process iterates until no more frequent k-itemsets can be generated for some k. This "breadth-first" approach has received a lot of attention, improved, and extended [65, 66]. There are also several parallel algorithms based on this Apriori approach [67, 68].

The drawbacks with the Apriori approach include the cost of repeated candidate set generation, and repeated scans of the database to check whether the patterns are present. Han et al. [69] designed the FP-growth algorithm, a divide-and-conquer approach avoiding candidate generation. The first scan of the data is similar to the Apriori step, where a list of frequent items is derived and ordered by frequency-descending order. Now based on this list, the database is compressed into a compact tree representation called the FP-tree, which retains the itemset association information. The FP-tree is now mined by starting from each frequent length-1 pattern, constructing its conditional pattern base (a "sub-database" consisting of the set of prefix paths in the FP-tree co-occurring with the suffix pattern), then constructing its conditional FP-tree, and performing mining recursively. The pattern growth is achieved by the concatenation of the suffix pattern with the frequent patterns generated from a conditional FP-tree. Thus, the FP-growth algorithm transforms the problem of finding long frequent patterns to searching for shorter ones recursively, and then concatenating the suffix. It uses the least frequent items as a suffix, offering good selectivity. Prior performance studies demonstrate that this method is substantially faster than Apriori-based schemes.

In case of graph databases, the challenge is to detect frequently-occurring, interesting subgraphs (see [70] for a review of various algorithms). This is very compute-intensive, as graph and subgraph isomorphisms play a key role throughout the algorithm strategies. Apriori-based frequent subgraph mining algorithms (e.g., FSG [71]) share similar characteristics to the frequent itemset mining algorithms. The search for frequent subgraphs starts with graphs of a "small size", and proceeds in an iterative bottom-up manner. At each iteration, the size of newly-discovered frequent substructures is increased by one. These new substructures are generated by joining two similar but slightly different frequent subgraphs that were previously discovered, and the frequency of the newly-formed subgraphs is then checked using subgraph isomorphism queries.

Common optimizations to this approach include the use of graph invariants to reduce the subgraph isomorphism search space. If two graphs are topologically identical, i.e., isomorphic, they also have identical graph invariants. Examples of invariants include the number of vertices, the degree of each vertex, the number of cycles, and so on. Isomorphic graphs always have identical values of all graph invariants, while the identical values of given graph invariants does not imply the isomorphism of the graphs. Canonical labeling schemes are a common invariant strategy employed by various algorithms. A canonical labeling is a code that uniquely identifies a graph such that if two graphs are isomorphic to each other, they will have the same code. For instance, a simple method to define the canonical label of a graph is as the string obtained by concatenating the upper triangular entries of the graph's adjacency matrix, after symmetric permutation of the matrix so that this string becomes the lexicographically largest (or smallest) over the strings that can be obtained from all such permutations. However, the work complexity of this labeling scheme is $O(n!)$ (where n is the number of vertices), which makes it intractable even for moderate-sized graphs. The cost of canonical labeling can be reduced with various heuristics that may exploit special topological properties of a graph.

The NAUTY algorithm [72] is considered one of the fastest algorithms for subgraph isomorphism, and makes extensive use of graph invariants. It focuses on invariants associated with each vertex, and then partitions the set of vertices based on their invariant values. For comparing two graphs, it checks the graph isomorphism between the subsets having identical values of the graph invariants. If all subsets are isomorphic between the two graphs, then the full networks are also isomorphic. This divide and conquer approach based on graph invariants significantly enhances the computational efficiency of most practical problem formulations.

Holder et al. designed SUBDUE [73], a system for exact and approximate substructure pattern discovery using information-theoretic measures such as minimum description length. They also apply it to the problem of anomaly detection. Their algorithm is based on a heuristic search strategy called beam search. It begins with a subgraph comprising only a single vertex in the input graph and grows this subgraph incrementally. At each expansion, the total description length of the graph, which is defined as the sum of the subgraph and the input graph MDLs, is computed, and vertex additions that minimize this measure are greedily accepted.

Graph pattern matching queries also abound in the context of Resource Description Framework (RDF) data analysis. RDF is a simple data model that has emerged as the de-facto data import/export and exchange format for semantic web knowledge bases. In RDF, data items are represented in the form of (subject, predicate, object) triples. The data set in its entirety can be viewed as a directed multigraph, with vertices representing subjects and objects and edges representing the predicates. The SPARQL [74] query language is the official standard for searching and analyzing RDF repositories. It supports conjunctions and disjunctions of triple patterns with variables, which can be used to retrieve records of interest from the RDF data. Specialized RDF engines have been developed to store and index the RDF data, and also efficiently solve SPARQL-specified pattern queries. These queries are answered using multiple join computations on the data set, and the overall query optimization strategy is quite different from the graph database-centric approaches discussed above.

The literature contains prior work related to parallelization as well. Common challenges include the design of work-efficient approaches, load imbalance issues, and minimizing inter-processor communication that may arise due to expression of fine-grained parallelism in these problems. Buehrer et al. [75] discuss the parallelization and optimization of two different frequent subgraph mining algorithms, gSpan and GastonEL, on recent multicore architectures. They demonstrate the importance of dynamic load balancing in real data sets, where the frequent substructure frequency distribution may likely be skewed. Locality-improvement optimizations such as vertex reordering and compact representations of the search tree are also shown to be very important for achieving high performance.

17.3   Combinatorial Problems in Computational Biology

The field of computational molecular biology presents a rich collection of problems based on string and word combinatorics, as well as graph/tree manipulation and comparisons. A variety of discrete algorithms and combinatorial optimization strategies are applied for designing optimal solution strategies. We discuss algorithmic strategies for two important problems: "de novo" assembly of short-read genomic data and phylogeny inference.

17.3.1   Short-Read Genome Assembly

Assembly is a complex computational task, and is one of the first steps in the DNA sequencing process. Informally, it refers to the problem of piecing together a collection of fragments and reconstructing a long DNA sequence. A majority of the large-scale genome sequencing projects today employ the whole-genome shotgun sequencing strategy, where the genome of interest is randomly sheared to produce this fragment collection. The length of these fragments depends on experimental factors and can vary from 50 to 1000 base pairs. The problem is complicated by repeats (or near-identical DNA sections) and data quality variability in the sequencing data output. Further, the problem of de novo genome assembly, referring to reconstruction of a genome that is not similar to a known organism previously sequenced, is shown to be computationally $\mathcal{NP}$-hard [76]. Hence, a number of heuristic solutions are developed to tackle this problem. Fast and accurate de novo methods are essential to our efforts to characterize organisms and perform comparative studies.

The majority of de novo assemblers designed in the past decade have relied on greedy algorithmic strategies based on pairwise similarity evaluation of the DNA segments [77, 78]. However, these approaches have proven inadequate in handling the data volumes resulting from recent "high-throughput" sequencing technologies [79, 80, 81]. These new sequencers can produce billions of short fragments in a single run, but the fragment lengths are much smaller than prior-generation sequencers. An alternative approach, based on sub-quadratic time reconstruction of paths in a DNA string graph [82], has now emerged as a popular choice for assembly heuristics. In this method, overlaps between reads are implicitly captured by decomposing the collection of fragments into shorter words of k nucleotides (or k-mers), and then creating paths between reads that share k-mers. There are several open-source software packages (e.g., Velvet [83], ALLPATHS [84], YAGA [85, 86], and ABySS [87]) that implement variations of this heuristic. We next briefly discuss the main components of this assembly approach.

The idea of a graph of sequence fragments was first introduced by Idury and Waterman [88]. They presented a sequence assembly algorithm for the "sequencing by hybridization" strategy. This was extended by Pevzner et al. [82], who proposed a slightly different formalization of the sequence graph called the de Bruijn graph. In this graph, k-mers are represented as edges, and overlapping k-mers are joined at their tips. The assembly would be identification of an "Eulerian superpath" in this graph, which Medvedev et al. showed was $\mathcal{NP}$-hard by reduction from the shortest superstring problem [89]. The superstring problem, which asks to find the shortest string that has all input strings as substrings, has itself been proposed by Myers [90] as a model for the assembly problem, but it is unsuitable due to the abundance of repeats in genomes. Pevzner's graph formulation is inherently k-mer centric, and it is a robust method to deal with fragmentation of reads. A de Bruijn graph can also accommodate sequences of very different lengths, and this is particularly useful when attempting mixed-length sequencing or comparative genomics.

In the de Bruijn graph, there is a one-to-one mapping of sequences onto paths traversing the graph. Extracting the nucleotide sequence from a path is straightforward given the initial k-mer of the first vertex and the sequences of all the vertices in the path. Conversely, for every read there exists exactly one path that passes through the vertices corresponding to the sequence's k-mers. Note that there is a balance between sensitivity and specificity determined by the value of k. Smaller k-mers increase the connectivity of the graph due to a higher chance of overlap, but choosing larger k-mer values makes the assembler less vulnerable to ambiguities and errors in the reads. In the first step of the assembly process, a suitable value of k is chosen. A vertex is created for every sequence of length k, and vertices are joined by edges when they overlap by $k - 1$ bases. It is useful to construct a k-mer frequency profile (using tools such as tallymer [91]) to determine the value of k that may be appropriate for the data set.

The next step is to condense this graph. Some assemblers employ various cleaning strategies at this step to remove spurious vertices and edges created due to sequencing errors. For instance, dead-end branches and small bubbles in the graph can be removed. Graph condensation can be done by merging vertices linked by unambiguous edges. Ambiguous edges are removed from the graph, and the vertices are then merged and the corresponding k-mers concatenated.

While earlier approaches utilize a maximum flow-based or Chinese postman tour formulation for traversing the graph, most current assemblers resort to significantly faster near linear-time heuristic path discovery algorithms. After the path traversal, "contigs" (which are just long strings) are formed. Assemblers also utilize paired-end information when available to link contigs together. Paired-end reads correspond to fragments that are known to be separated from each other by a certain user-defined distance. Reads are aligned to the initial contigs to create a set of linked contigs, which is then filtered to remove erroneous links caused by mispaired or misaligned reads. Two contigs are considered to be linked if a sufficiently-large number of pairs join the contigs. A graph search is then performed to look for a possible single unique path that visits each contig. Heuristics are employed to limit the number of vertices visited in the search and hence bound the computational cost of this step. Paths that are ambiguous are removed, and this process is repeated until consistent paths are found to resolve the contigs of the final assembly.

The computational time and memory constraints limit the practical use of serial assemblers to genomes on the order of a megabase in size. The preliminary stages of the assembly process involving initial creation of the de Bruijn graph are particularly memory-intensive. There are several parallel assembly tools (for instance, see YAGA and ABySS) that attempt to solve this problem. They maintain a distributed representation of the de Bruijn graph and perform all steps of the assembly in parallel. Some techniques employed to improve locality and reduce inter-process communication include clustering and redistribution of reads, and periodic load balancing.

Once a reference sequence for a particular genome is available, it is then possible to perform comparative sequencing or resequencing to identify mutations, polymorphisms, and structural variations between organisms. Computational problems include fast alignment of reads to the reference genome, measuring the probability of error in the alignment, and discovery of inserts and deletions. These problems are also well-studied, and software packages such as MAQ [92] and SOAP [93] include efficient techniques to solve them.

17.3.2   Phylogeny Reconstruction

Phylogenetics [94] is the study of the evolutionary relationships among organisms, and trees are typically used to represent the evolutionary history in a compact manner. Phylogenetic tree reconstruction has applications in numerous practical domains, including epidemiological studies [95, 96], conservation biology [97], forensics, and drug discovery [98]. A multiple alignment of DNA and/or protein sequences is typically used as the input for phylogenetic inference. The basic assumption of process necessary for phylogenetic reconstruction is repeated divergence from ancestral populations or genes. In the output phylogenetic tree, the input taxa are depicted as the leaves, ancestral taxa occupy internal nodes, and the edges of the tree denote genetic, historical, and ancestor-descendant relationships among the taxa. The major computational aspects in phylogenetics involve estimation of the evolutionary tree (reconstruction problem) and using the estimated phylogeny as an analytic framework for further evolutionary study. There are several software packages for solving phylogeny-related problems (see [99] for a compendium). Popular tools for phylogeny inference include PAUP, PHYLIP, PAML, and MrBayes.

Almost every model of change proposed to underlie phylogenetic reconstruction has given rise to NP-hard optimization problems. Three major types of work have arisen: ad-hoc heuristics that run quickly, but offer no quality guarantees and may not even have a well defined optimization criterion (e.g., the popular neighbor-joining heuristic [100]); optimization problems based on a "parsimony" criterion, which seeks the phylogeny with the least total amount of change needed to explain modern data; and optimization problems based on a "maximum likelihood" criterion, which seeks the phylogeny that is the most likely (under some suitable statistical model) to have given rise to the modern data. Ad-hoc heuristics are fast and often rival the optimization methods in terms of accuracy; parsimony-based methods may take exponential time, but, at least for DNA data, can often be run to completion on datasets of moderate size. Methods based on maximum likelihood are comparatively slower and so restricted to very small instances, but appear capable of outperforming the others in terms of the quality of solutions.

In this section, we will focus on an approach for phylogeny reconstruction using the maximum parsimony (MP) criterion. In phylogeny inference problem formulations, species are described by their molecular sequences (nucleotide, DNA, amino acid, protein). The evolution of sequences is studied under a simplifying assumption that each site evolves independently, and the MP objective selects the tree with the smallest total evolutionary change. The edit distance between two species is the minimum number of evolutionary events through which one species evolves into the other. Given a tree in which each node is labeled by a species, the cost of this tree (tree length) is the sum of the costs of its edges. The cost of an edge is the edit distance between the species at the edge endpoints. The length of a tree T with all leaves labeled by taxa is the minimum cost over all possible labelings of the internal nodes. A brute-force approach examines all possible tree topologies to return one that shows the smallest amount of total evolutionary change. The number of unrooted binary trees on n leaves (representing the species or taxa) is $(2n-5)!! = (2n-5) \cdot (2n-7) \cdots 3$. For instance, this means that there are about 13 billion different trees for an input of $n = 13$ species. Hence it is very time-consuming to examine all trees to obtain the optimal tree. Most researchers focus on heuristic algorithms that examine a much smaller set of most promising topologies and choose the best one examined. The Branch and Bound strategy is particularly useful for this, as it provides instance-specific lower bounds showing how close a solution is to optimal.

Given a topology with leaf labels, the optimal internal labels for that topology can be computed in linear time per character. Consider a single character. In a leaf-to-root sweep, we compute for each internal node v a set of labels optimal for the subtree rooted at v (called the Farris Interval). Specifically, this is the intersection of its children's sets (connect children though v) or, if this intersection is empty, the union of its children's sets (agree with one child). At the root, we choose an optimal label and pass it down. Children agree with their parent if possible. Because we assume each site evolves independently, we can set all characters simultaneously. Thus for m character and n sequences, this takes $O(nm)$ time. Since most computers can perform efficient bitwise logical operations, a binary encoding of a state can be used in order to implement intersection and union efficiently using bitwise AND and bitwise OR. Even so, this operation dominates the parsimony B&B computation.

We outline the B&B strategy for MP that is used in the GRAPPA (Genome Rearrangement Analysis through Parsimony and other Phylogenetic Algorithms) toolkit [101, 102]. Each node in the B&B tree is associated with either a partial tree or a complete tree. A tree containing all n taxa is a complete tree. A tree on the first k ($k < n$) taxa is a partial tree. A complete tree is a candidate solution. Tree T is consistent with tree T' iff T' can be obtained from T by removing all the taxa in T that are not in T'. The subproblem for a node with partial tree T is to find the most parsimonious complete tree consistent with T.

The active nodes are partitioned into levels such that level k, for $3 \leq k \leq n$, contains all active nodes whose partial trees contain the first k taxa from the input. The root node contains the first three taxa (hence, indexed by level 3) since there is only one possible unrooted tree topology with three leaves. The branch function finds the immediate successors of a node associated with a partial tree T_k at level k by inserting the $(k+1)^{st}$ taxon at any of the $(2k-3)$ possible places. A new node (with this taxon attached by an edge) can join in the middle of any of the $(2k-3)$ edges in the unrooted tree. Depth-first search and a heuristic best-first search (BeFS) are used to break ties between nodes at the same depth. The search space explored by this approach depends on the addition order of taxa, which also influences the efficiency of the B&B algorithm.

A node v associated with tree T_k represents the subproblem to find the most parsimonious tree in the search space that is consistent with T_k. Assume T_k is a tree with leaves labeled by $S_1, \cdots, S_k$. The goal is to find a tight lower bound of the subproblem. However, one must balance the quality of the lower bound against the time required to compute it in order to gain the best performance of the overall B&B algorithm. Purdom et al. [103] use single-character discrepancies of the partial tree as the bound function. For each character, one computes a difference set, the set of character states that do not occur among the taxa in the partial tree and hence only occur among the remaining taxa. The single-character discrepancy is the sum over all characters of the number of the elements in these difference sets. The lower bound is therefore the sum of the single-character discrepancy plus the cost of the partial tree. An advantage of Purdom's approach is that given an addition order of taxa, there is only one single-character discrepancy calculation per level. The time needed to compute the bound function is negligible.

Next we discuss the candidate function and incumbent x_I. In phylogeny reconstruction, it is expensive to compute a meaningful feasible solution for each partial tree, so instead GRAPPA computes an upper bound on the input using a direct method such as neighbor-joining before starting the B&B search. This value the global upper bound, $f(x_I)$, the incumbent's objective function. In GRAPPA the first incumbent is the best returned by any of several heuristic methods.

The B&B approaches typically use DFS or BeFS as the search strategies. For phylogeny reconstruction with n taxa, the depth of the subproblems ranges from 3 to n. So an array is used to keep the open subproblems sorted by DFS depth. The array element at location i contains a priority queue (PQ) of the subproblems with depth i, and each item of the PQ contains an external pointer to stored subproblem information. The priority queues (PQs) support best-first-search tie breaking and allow efficient deletion of all dominated subproblems whenever a new incumbent is found. There are many ways to organize a PQ. However, in this case, most of the time is spent evaluating the tree length of a partial tree, and the choice of PQ data structure does not make a significant difference.

17.4   Summary and Concluding Remarks

We provide a brief overview of four different problems (community identification, subgraph mining, de novo assembly of short-read genomic data, and phylogeny reconstruction), and the key algorithmic strategies employed to solve them. Typical formulations of these problems are $\mathcal{NP}$-hard, and so the solution strategies are fast heuristic approaches and approximation algorithms. The common thread linking all these problems is that they are very intensive on graph/tree-based traversal and combinatorial optimization strategies. Many open problems exist, including better heuristics and performance improvements in all these problems, both in terms of solution quality and the time to solution. There is also an urgent need to design approaches that can handle massive data sets, exploiting the architectural features of emerging parallel systems.

Bibliography

[1] National Research Council Committee on Network Science for Future Army Applications, *Network Science*, The National Academies Press, 2006.

[2] Watts, D., "The "New" Science of Networks," *Annual Review of Sociology*, Vol. 30, 2004, pp. 243–270.

[3] Börner, K., Sanyal, S., and Vespignani, A., "Network Science," *Proc. Annual Review of Information Science and Technology*, 2007, pp. 537–607.

[4] Alderson, D., "Catching the "Network Science" Bug: Insight and Opportunity for the Operations Researcher," *Operations Research*, Vol. 56, No. 5, 2008, pp. 1047–1065.

[5] Milgram, S., "The small-world problem," *Psychology Today*, Vol. 1, No. 1, 1967, pp. 61–67.

[6] Watts, D. and Strogatz, S., "Collective dynamics of small world networks," *Nature*, Vol. 393, 1998, pp. 440–442.

[7] Amaral, L., Scala, A., Barthélémy, M., and Stanley, H., "Classes of small-world networks," *Proc. National Academy of Sciences USA*, Vol. 97, No. 21, 2000, pp. 11149–11152.

[8] Newman, M., "The structure and function of complex networks," *SIAM Review*, Vol. 45, No. 2, 2003, pp. 167–256.

[9] Albert, R. and Barabási, A.-L., "Statistical mechanics of complex networks," *Reviews of Modern Physics*, Vol. 74, 2002, pp. 47–97.

[10] Faloutsos, M., Faloutsos, P., and Faloutsos, C., "On power-law relationships of the Internet topology," *Proc. SIGCOMM '99*, 1999, pp. 251–262.

[11] Erdős, P. and Rényi, A., "On the evolution of random graphs," *Publ. Math. Inst. Hungar. Acad. Sci.*, Vol. 5, 1960, pp. 17–61.

[12] Aiello, W., Chung, F., and Lu, L., "A random graph model for power law graphs," *Experimental Mathematics*, Vol. 10, 2001, pp. 53–66.

[13] Newman, M., Strogatz, S., and Watts, D., "Random graphs with arbitrary degree distributions and their applications," *Physical Review E*, Vol. 64, 2001, pp. 026118.

[14] Newman, M., Watts, D., and Strogatz, S., "Random graph models of social networks," *Proc. National Academy of Sciences of the United States of America*, Vol. 99, No. 3, 2002, pp. 2566–2572.

[15] Leskovec, J., Chakrabarti, D., Kleinberg, J., and Faloutsos, C., "Realistic, Mathematically Tractable Graph Generation and Evolution, Using Kronecker Multiplication," *Proc. European International Conference on Principles and Practice of Knowledge Discovery in Databases (ECML/PKDD '05)*, 2005, pp. 133–145.

[16] Amaral, L., Barrat, A., Barabási, A.-L., and Caldarelli et al., G., "Virtual round table on ten leading questions for network research," *Eur. Phys. J. B*, Vol. 38, 2004, pp. 143–145.

[17] Brin, S. and Page, L., "The anatomy of a large-scale hypertextual Web search engine," *Computer Networks and ISDN Systems*, Vol. 30, No. 1–7, 1998, pp. 107–117.

[18] "International Network for Social Network Analysis," 2005, `http://www.insna.org`.

[19] Scott, J., *Social Network Analysis: A Handbook*, SAGE Publications, Newbury Park, CA, 2000.

[20] Wasserman, S. and Faust, K., *Social Network Analysis: Methods and Applications*, Cambridge University Press, Cambridge, UK, 1994.

[21] Freeman, L., *The development of social network analysis: a study in the sociology of science*, BookSurge Publishers, 2004.

[22] Jeong, H., Mason, S., Barabási, A.-L., and Oltvai, Z., "Lethality and centrality in protein networks," *Nature*, Vol. 411, 2001, pp. 41–42.

[23] Barabási, A.-L. and Oltvai, Z., "Network biology: understanding the cell's functional organization," *Nature Reviews Genetics*, Vol. 5, 2004, pp. 101–113.

[24] Pinney, J., McConkey, G., and Westhead, D., "Decomposition of Biological Networks using Betweenness Centrality," *Proc. 9th Ann. Int'l Conf. on Research in Computational Molecular Biology (RECOMB 2005)*, May 2005, Poster session.

[25] Krebs, V., "Mapping networks of terrorist cells," *Connections*, Vol. 24, No. 3, 2002, pp. 43–52.

[26] Coffman, T., Greenblatt, S., and Marcus, S., "Graph-based technologies for intelligence analysis," *Commun. ACM*, Vol. 47, No. 3, 2004, pp. 45–47.

[27] Team, N., "Network Workbench Tool," 2006, `http://nwb.slis.indiana.edu`.

[28] Csárdi, G. and Nepusz, T., "The igraph software package for complex network research," *InterJournal Complex Systems*, 2006, pp. 1695, `http://igraph.sf.net`.

[29] "JUNG: Java Universal Network/Graph Framework," 2010, `http://jung.sourceforge.net/`.

[30] Batagelj, V. and Mrvar, A., "Pajek – Program for large network analysis," *Connections*, Vol. 21, No. 2, 1998, pp. 47–57.

[31] Krebs, V., "InFlow 3.1 - Social Network Mapping Software," `http://www.orgnet.com`.

[32] Analytic Technologies, "UCINET 6 Social Network Analysis Software," `http://www.analytictech.com/ucinet`.

[33] Bader, D. and Madduri, K., "SNAP: Small-world network analysis and partitioning: an open-source parallel graph framework for the exploration of large-scale networks," *Proc. 22nd IEEE Int'l. Parallel and Distributed Processing Symposium (IPDPS 2008)*, IEEE, April 2008.

[34] Bader, D., Madduri, K., and Riedy, E., "SNAP: Small-world Network Analysis and Partitioning," 2010, `http://snap-graph.sf.net/`.

[35] Gregor, D. and Lumsdaine, A., "The Parallel BGL: A generic library for distributed graph computations," *Proc. Parallel Object-Oriented Scientific Computing (POOSC 2005)*, July 2005.

[36] Bader, D. and Madduri, K., "DARPA HPCS SSCA Graph Analysis benchmark v2.3," `http://www.graphanalysis.org`.

[37] Bader, D., Madduri, K., Gilbert, J., Shah, V., Kepner, J., Meuse, T., and Krishnamurthy, A., "Designing Scalable Synthetic Compact Applications for Benchmarking High Productivity Computing Systems," *CTWatch Quarterly*, Vol. 2, No. 4B, November 2006.

[38] Chakrabarti, D., Zhan, Y., and Faloutsos, C., "R-MAT: A Recursive Model for Graph Mining," *Proc. 4th SIAM Intl. Conf. on Data Mining (SDM)*, SIAM, April 2004.

[39] Lang, K., "Finding good nearly balanced cuts in power law graphs," Tech. rep., Yahoo! Research, 2004.

[40] Lang, K., "Fixing two weaknesses of the Spectral Method," *Proc. Advances in Neurals Information Proc. Systems 18 (NIPS)*, Vancouver, Canada, December 2005.

[41] Leskovec, J., Lang, K., Dasgupta, A., and Mahoney, M., "Community Structure in Large Networks: Natural Cluster Sizes and the Absence of Large Well-Defined Clusters," `http://arxiv.org/abs/0810.1355`.

[42] Brandes, U., Gaertler, M., and Wagner, D., "Engineering graph clustering: Models and experimental evaluation," *J. Exp. Algorithmics*, Vol. 12, 2008, pp. 1–26.

[43] Leskovec, J., Lang, K., and Mahoney, M., "Empirical Comparison of Algorithms for Network Community Detection," *Proc. 19th Int'l. Conference on the World Wide Web (WWW'10)*, 2010.

[44] Newman, M. and Girvan, M., "Finding and evaluating community structure in networks," *Physical Review E*, Vol. 69, 2004, pp. 026113.

[45] Newman, M., "Modularity and community structure in networks," *Proc. National Academy of Sciences*, Vol. 103, No. 23, 2006, pp. 8577–8582.

[46] Brandes, U., Delling, D., Gaertler, M., Görke, R., Höfer, M., Nikoloski, Z., and Wagner, D., "On finding graph clusterings with maximum modularity," *Proc. 33rd Intl. Workshop on Graph-Theoretic Concepts in CS (WG 2007)*, Dornburg, Germany, June 2007.

[47] Fortunato, S., "Community detection algorithms: a comparative analysis," *Phys. Rev. E*, Vol. 80, 2009, pp. 056117.

[48] Fortunato, S., "Community detection in graphs," *Physics Reports*, Vol. 486, 2010, pp. 75–174.

[49] Freeman, L., "A set of measures of centrality based on betweenness," *Sociometry*, Vol. 40, No. 1, 1977, pp. 35–41.

[50] Brandes, U., "On variants of shortest-path betweenness centrality and their generic computation," *Social Networks*, Vol. 30, No. 2, 2008, pp. 136–145.

[51] Liljeros, F., Edling, C., Amaral, L., Stanley, H., and Aberg, Y., "The Web of Human Sexual Contacts," *Nature*, Vol. 411, 2001, pp. 907–908.

[52] Clauset, A., Newman, M., and Moore, C., "Finding community structure in very large networks," *Physical Review E*, Vol. 70, 2004, pp. 066111.

[53] Andersen, R. and Lang, K., "Communities from seed sets," *Proc. 15th Int'l. Conference on World Wide Web (WWW '06)*, 2006, pp. 223–232.

[54] Noack, A., "Modularity clustering is force-directed layout," *Phys. Rev. E*, Vol. 79, 2009, pp. 026102.

[55] Fortunato, S. and Barthélemy, M., "Resolution limit in community detection," *Proc. National Academy of Sciences*, Vol. 104, No. 1, 2007, pp. 36–41.

[56] Lancichinetti, A., Fortunato, S., and Kértsz, J., "Detecting the overlapping and hierarchical community structure of complex networks," *New Journal of Physics*, Vol. 11, 2009, pp. 033015.

[57] Arenas, A., Fernández, A., and Gómez, S., "Multiple resolution of the modular structure of complex networks," *New Journal of Physics*, Vol. 10, No. 5, 2008, pp. 053039.

[58] Madduri, K., *A high-performance framework for analyzing massive complex networks*, Ph.D. thesis, Georgia Institute of Technology, July 2008.

[59] Bader, D. and Madduri, K., "Designing Multithreaded Algorithms for Breadth-First Search and st-connectivity on the Cray MTA-2," *Proc. 35th Int'l. Conf. on Parallel Processing (ICPP 2006)*, IEEE Computer Society, Aug. 2006, pp. 523–530.

[60] Bader, D. and Madduri, K., "Parallel Algorithms for Evaluating Centrality Indices in Real- world Networks," *Proc. 35th Int'l. Conf. on Parallel Processing (ICPP 2006)*, IEEE Computer Society, Aug. 2006, pp. 539–550.

[61] Han, J., Cheng, H., Xin, D., and Yan, X., "Frequent pattern mining: current status and future directions," *Data Mining and Knowledge Discovery*, Vol. 15, No. 1, 2007, pp. 55–86.

[62] Agrawal, R., Imielinski, T., and Swami, A., "Mining association rules between sets of items in large databases," *Proc. ACM Int'l. Conf. on Management of Data (SIGMOD'93)*, 1993, pp. 207–216.

[63] Yang, G., "The complexity of mining maximal frequent itemsets and maximal frequent patterns," *Proc. Int'l. Conf. on knowledge discovery in databases (KDD'04)*, 2004, pp. 344–353.

[64] Agrawal, R. and Srikant, R., "Fast algorithms for mining association rules," *Proc. Int'l. Conf. on very large data bases (VLDB'94)*, 1994, pp. 487–499.

[65] Park, J., Chen, M., and Yu, P., "Efficient parallel mining for association rules," *Proc. 4th Int'l. Conf. on Information and Knowledge Management (CIKM'95)*, 1995, pp. 31–36.

[66] Brin, S., Motwani, R., Ullman, J., and Tsur, S., "Dynamic itemset counting and implication rules for market basket analysis," *Proc. ACM Int'l. Conf. on Management of Data (SIGMOD'97)*, 1997, pp. 255–264.

[67] Cheung, D., Han, J., Ng, V., Fu, A., and Fu, Y., "A fast distributed algorithm for mining association rules," *Proc. 4th Int'l. Conf. on Parallel and Distributed Information Systems (DIS '96)*, 1996, pp. 31–43.

[68] Zaki, M., Parthasarathy, S., Ogihara, M., and Li, W., "Parallel algorithm for discovery of association rules," *Data Mining and Knowledge Disscovery*, Vol. 1, 1997, pp. 343–374.

[69] Han, J., Pei, J., and Yin, Y., "Mining frequent patterns without candidate generation," *Proc. ACM Int'l. Conf. on Management of Data (SIGMOD'00))*, 2000, pp. 1–12.

[70] Washio, T. and Motoda, H., "State of the art of graph-based data mining," *SIGKDD Explorations*, Vol. 5, 2003, pp. 59–68.

[71] Kuramochi, M. and Karypis, G., "GREW: a scalable frequent subgraph discovery algorithm," *Proc. Int'l. Conf. on Data Mining (ICDM)*, Nov. 2004, pp. 439–442.

[72] McKay, B., "Practical Graph Isomorphism," *Congressum Numerantium*, Vol. 30, 1981, pp. 45–87.

[73] Holder, L., Cook, D., and Djoko, S., "Substructure discovery in the SUBDUE system," *Proc. AAAI Workshop on Knowledge Discovery in Databases*, July 1994, pp. 169–180.

[74] Prud'hommeaux, E. and Seaborne, A., "SPARQL Query Language for RDF," Tech. rep., W3C, 2008, W3C Recommendation 15 January 2008, `http://www.w3.org/TR/rdf-sparql-query/`.

[75] Buehrer, G., Parthasarathy, S., and Chen, Y.-K., "Adaptive Parallel Graph Mining for CMP Architectures," *Proc. 6th Int'l. Conf. on Data Mining (ICDM'06))*, 2006, pp. 97–106.

[76] Myers, E., "Towards simplifying and accurately formulating fragment assembly," *Journal of Computational Biology*, Vol. 2, 1995, pp. 275–290.

[77] Pop, M., "Genome assembly reborn: recent computational challenges," *Briefings in Bioinformatics*, Vol. 10, No. 4, 2009, pp. 354–366.

[78] Myers, E., Sutton, G., Delcher, A., Dew, I., Fasulo, D., Flanigan, M., Kravitz, S., Mobarry, C., Reinert, K., and et al., "A Whole-Genome Assembly of Drosophila," *Science*, Vol. 287, No. 5461, 2000, pp. 2196–2204.

[79] Shendure, J. and Ji, H., "Next-Generation DNA sequencing," *Nature Biotechnology*, Vol. 26, 2008, pp. 1135–1145.

[80] Pop, M. and Salzberg, S., "Bioinformatics challenges of new sequencing technology," *Trends in Genetics*, Vol. 24, No. 3, 2008, pp. 142–149.

[81] Mardis, E., "The impact of next-generation sequencing technology on genetics," *Trends in Genetics*, Vol. 24, No. 3, 2008, pp. 133–141.

[82] Pevzner, P., Tang, H., and Waterman, M., "An Eulerian path approach to DNA fragment assembly," *Proc. National Academy of Sciences*, Vol. 98, No. 17, 2001, pp. 9748–9753.

[83] Zerbino, D. and Birney, E., "Velvet: algorithms for de novo short read assembly using de Bruijn graphs," *Genome Research*, Vol. 18, 2008, pp. 821–829.

[84] Butler, J., MacCallum, I., and Kleber, M., "ALLPATHS: de novo assembly of whole-genome shotgun microreads," *Genome Research*, Vol. 18, 2008, pp. 810–820.

[85] Jackson, B., *Parallel Methods for Short-read Genomic Assembly*, Ph.D. thesis, Iowa State University, 2009.

[86] Jackson, B., Regennitter, M., Yang, X., Schnable, P., and Aluru, S., "Parallel de novo assembly of large genomes from high-throughput short reads," *Proc. 24th Int'l. Parallel and Distributed Processing Symposium (IPDPS)*, April 2010.

[87] Simpson, J., Wong, K., Jackman, S., Schein, J., Jones, S., and Birol, I., "ABySS: A parallel assembler for short read sequence data," *Genome Research*, Vol. 19, 2009, pp. 1117–1123.

[88] Idury, R. and Waterman, M., "A New Algorithm for DNA Sequence Assembly," *Journal of Computational Biology*, Vol. 2, No. 2, 1995, pp. 291–306.

[89] Medvedev, P., Georgiou, K., Myers, G., and Brudno, M., "Computability of Models for Sequence Assembly," *Workshop on Algorithms in Bioinformatics*, edited by R. Giancarlo and S. Hannenhalli, Vol. 4645 of *Lecture Notes in Computer Science*, Springer Berlin/Heidelberg, 2007, pp. 289–301.

[90] Myers, E., "The fragment assembly string graph," *Bioinformatics*, Vol. 21, No. Suppl 2, 2005, pp. ii79–ii85.

[91] Kurtz, S., Narechania, A., Stein, J., and Ware, D., "A new method to compute K-mer frequencies and its application to annotate large repetitive plant genomes," *BMC Genomics*, Vol. 9, 2008, pp. 517.

[92] Li, H., Ruan, J., and Durbin, R., "Mapping short DNA sequencing reads and calling variants using mapping quality scores," *Genome Research*, Vol. 18, 2008, pp. 1851–1858.

[93] Li, R., Li, Y., Kristiansen, K., and Wang, J., "SOAP: short oligonucleotide alignment program," *Bioinformatics*, Vol. 24, 2008, pp. 713–714.

[94] Felsenstein, J., *Inferring phylogenies*, Sinauer Associates, 2004.

[95] Bush, R., Bender, C., Subbarao, K., Cox, N., and Fitch, W., "Predicting the evolution of human influenza," *Science*, Vol. 286, No. 5446, 1999, pp. 1921–1925.

[96] Ou, C., Ciesielski, C., Myers, G., Bandea, C., Luo, C., Korber, B., Mullins, J., Schochetman, G., Berkelman, R., and Economou, A., "Molecular epidemiology of HIV transmission in a dental practice," *Science*, Vol. 256, No. 5060, 1992, pp. 1165–1171.

[97] Faith, D., "Genetic diversity and taxonomic priorities for conservation," *Biol. Conserv.*, Vol. 68, 1992, pp. 69–74.

[98] Bader, D., Moret, B., and Vawter, L., "Industrial Applications of High-Performance Computing for Phylogeny Reconstruction," *Proc. SPIE Commercial Applications for High-Performance Computing*, Vol. 4528, SPIE, Denver, CO, Aug. 2001, pp. 159–168.

[99] Felsenstein, J., "Phylogeny Software," 2010, http://evolution.genetics.washington.edu/phylip/software.html.

[100] Saitou, N. and Nei, M., "The neighbor-joining method: a new method for reconstruction of phylogenetic trees," *Molec. Biol. Evol.*, Vol. 4, 1987, pp. 406–425.

[101] Moret, B., Wyman, S., Bader, D., Warnow, T., and Yan, M., "A new implementation and detailed study of breakpoint analysis," *Proc. 6th Pacific Symposium on Biocomputing*, 2001, pp. 583–594.

[102] Bader, D., Hart, W., and Phillips, C., "Parallel Algorithm Design for Branch and Bound," *Tutorials on Emerging Methodologies and Applications in Operations Research*, Vol. 76, chap. 5, Springer New York, 2005, pp. 1–44.

[103] Purdom, Jr., P., Bradford, P., Tamura, K., and Kumar, S., "Single column discrepancy and dynamic max-mini optimization for quickly finding the most parsimonious evolutionary trees," *Bioinformatics*, Vol. 2, No. 16, 2000, pp. 140–151.

Chapter 18

Spectral Graph Theory

Daniel Spielman

Yale University

18.1 Introduction

Spectral graph theory is the study and exploration of graphs through the eigenvalues and eigenvectors of matrices naturally associated with those graphs. It is intuitively related to attempts to understand graphs through the simulation of processes on graphs and through the consideration of physical systems related to graphs. Spectral graph theory provides many useful algorithms, as well as some that can be rigorously analyzed. We begin this chapter by providing intuition as to why interesting properties of graphs should be revealed by these eigenvalues and eigenvectors. We, then, survey a few applications of spectral graph theory.

The figures in this chapter are accompanied by the Matlab code used to generate them.

18.2 Preliminaries

We ordinarily view an undirected graph[1] G as a pair (V, E), where V denotes its set of vertices and E denotes its set of edges. Each edge in E is an unordered pair of distinct vertices, with the edge connecting vertices a and b written as (a, b). A weighted graph is a graph in which a weight (typically a real number) has been assigned to every edge. We denote a weighted graph by a triple (V, E, w), where (V, E) is the associated unweighted graph and w is a function from E to the real numbers. We restrict our attention to weight functions w that are strictly positive. We reserve the letter n for the number of vertices in a graph. The *degree* of a vertex in an unweighted graph is the number of edges in which it is involved. We say that a graph is *regular* if every vertex has the same degree, and d-regular if that degree is d.

We denote vectors by bold letters, and denote the ith component of a vector $\boldsymbol{x}$ by $\boldsymbol{x}(i)$. Similarly, we denote the entry in the ith row and jth column of a matrix M by $M(i, j)$.

If we are going to discuss the eigenvectors and eigenvalues of a matrix M, we should be sure that they exist. When considering undirected graphs, most of the matrices we consider are symmetric, and thus they have an orthonormal basis of eigenvectors and n eigenvalues, counted with multiplicity. The other matrices we associate with undirected graphs are *similar* to symmetric matrices, and thus also have n eigenvalues, counted by multiplicity, and possess a basis of eigenvectors. In particular, these matrices are of the form MD^{-1}, where M is symmetric and D is a non-singular diagonal matrix. In this case, $D^{-1/2}MD^{-1/2}$ is symmetric, and we have

$$D^{-1/2}MD^{-1/2}\boldsymbol{v}_i = \lambda_i \boldsymbol{v}_i \implies MD^{-1}(D^{1/2}\boldsymbol{v}_i) = \lambda_i \left(D^{1/2}\boldsymbol{v}_i\right).$$

So, if $\boldsymbol{v}_1, \ldots, \boldsymbol{v}_n$ form an orthonormal basis of eigenvectors of $D^{-1/2}MD^{-1/2}$, then we obtain a basis (not necessarily orthonormal) of eigenvectors of MD^{-1} by multiplying these vectors by $D^{1/2}$. Moreover, these matrices have the same eigenvalues.

---

[1]Strictly speaking, we are considering simple graphs. These are the graphs in which all edges go between distinct vertices and in which there is at most one edge between a given pair of vertices. Graphs that have multiple-edges or self-loops are often called multigraphs.

The matrices we associate with directed graphs will not necessarily be diagonalizable.

18.3 The Matrices Associated with a Graph

Many different matrices arise in the field of Spectral Graph Theory. In this section, we introduce the most prominent.

18.3.1 Operators on the Vertices

Eigenvalues and eigenvectors are used to understand what happens when one repeatedly applies an operator to a vector. If A is an n-by-n matrix having a basis of right-eigenvectors $\boldsymbol{v}_1, \ldots, \boldsymbol{v}_n$ with

$$A\boldsymbol{v}_i = \lambda_i \boldsymbol{v}_i,$$

then we can use these eigenvectors to understand the impact of multiplying a vector $\boldsymbol{x}$ by A. We first express $\boldsymbol{x}$ in the eigenbasis

$$\boldsymbol{x} = \sum_i c_i \boldsymbol{v}_i$$

and then compute

$$A^k \boldsymbol{x} = \sum_i c_i A^k \boldsymbol{v}_i = \sum_i c_i \lambda_i^k \boldsymbol{v}_i.$$

If we have an operator that is naturally associated with a graph G, then properties of this operator, and therefore of the graph, will be revealed by its eigenvalues and eigenvectors. The first operator one typically associates with a graph G is its *adjacency operator*, realized by its *adjacency matrix* A_G and defined by

$$A_G(i,j) = \begin{cases} 1 & \text{if } (i,j) \in E \\ 0 & \text{otherwise.} \end{cases}$$

To understand spectral graph theory, one must view vectors $\boldsymbol{x} \in \mathbb{R}^n$ as functions from the vertices to the Reals. That is, they should be understood as vectors in $\mathbb{R}^V$. When we apply the adjacency operator to such a function, the resulting value at a vertex a is the sum of the values of the function $\boldsymbol{x}$ over all neighbors b of a:

$$(A_G \boldsymbol{x})(a) = \sum_{b:(a,b)\in E} \boldsymbol{x}(b).$$

This is very close to one of the most natural operators on a graph: the *diffusion operator*. Intuitively, the diffusion operator represents a process in which "stuff" or "mass" moves from vertices to their neighbors. As mass should be conserved, the mass at a given vertex is distributed evenly among its neighbors. Formally, we define the *degree* of a vertex a to be the number of edges in which it participates. We naturally encode this in a vector, labeled d:

$$d(a) = |\{b : (a, b) \in E\}|,$$

where we write $|S|$ to indicate the number of elements in a set S. We then define the *degree matrix* D_G by

$$D_G(a, b) = \begin{cases} d(a) & \text{if } a = b \\ 0 & \text{otherwise.} \end{cases}$$

The *diffusion matrix* of G, also called the *walk matrix* of G, is then given by

$$W_G \overset{\text{def}}{=} A_G D_G^{-1}. \tag{18.1}$$

It acts on a vector x by

$$(W_G x)(a) = \sum_{b:(a,b)\in E} x(b)/d(b).$$

This matrix is called the *walk matrix* of G because it encodes the dynamics of a random walk on G. Recall that a random walk is a process that begins at some vertex, then moves to a random neighbor of that vertex, and then a random neighbor of that vertex, and so on. The walk matrix is used to study the evolution of the probability distribution of a random walk. If $p \in \mathbb{R}^n$ is a probability distribution on the vertices, then $W_G p$ is the probability distribution obtained by selecting a vertex according to p, and then selecting a random neighbor of that vertex. As the eigenvalues and eigenvectors of W_G provide information about the behavior of a random walk on G, they also provide information about the graph.

Of course, adjacency and walk matrices can also be defined for weighted graphs $G = (V, E, w)$. For a weighted graph G, we define

$$A_G(a, b) = \begin{cases} w(a, b) & \text{if } (a, b) \in E \\ 0 & \text{otherwise.} \end{cases}$$

When dealing with weighted graphs, we distinguish between the *weighted degree* of a vertex, which is defined to be the sum of the weights of its attached edges, and the *combinatorial degree* of a vertex, which is the number of such edges. We reserve the vector d for the weighted degree, so

$$d(a) = \sum_{b:(a,b)\in E} w(a, b).$$

The random walk on a weighted graph moves from a vertex a to a neighbor b with probability proportional to $w(a, b)$, so we still define its walk matrix by equation (18.1).

18.3.2 The Laplacian Quadratic Form

Matrices and spectral theory also arise in the study of quadratic forms. The most natural quadratic form to associate with a graph is the *Laplacian*, which is given by

$$x^T L_G x = \sum_{(a,b) \in E} w(a, b)(x(a) - x(b))^2. \tag{18.2}$$

This form measures the smoothness of the function x. It will be small if the function x does not jump too much over any edge. The matrix defining this form is the *Laplacian matrix* of the graph G,

$$L_G \overset{\text{def}}{=} D_G - A_G.$$

The Laplacian matrices of weighted graphs arise in many applications. For example, they appear when applying the certain discretization schemes to solve Laplace's equation with Neumann boundary conditions. They also arise when modeling networks of springs or resistors. As resistor networks provide a very useful physical model for graphs, we explain the analogy in more detail. We associate an edge of weight w with a resistor of resistance $1/w$, since higher weight corresponds to higher connectivity which corresponds to less resistance.

When we inject and withdraw current from a network of resistors, we let $i_{ext}(a)$ denote the amount of current we inject into node a. If this quantity is negative then we are removing current. As electrical flow is a potential flow, there is a vector $v \in \mathbb{R}^V$ so that the amount of current that flows across edge (a, b) is

$$i(a, b) = (v(a) - v(b))/r(a, b),$$

where $r(a, b)$ is the resistance of edge (a, b). The Laplacian matrix provides a system of linear equations that may be used to solve for v when given i_{ext}:

$$i_{ext} = L_G v. \tag{18.3}$$

We refer the reader to [1] or [2] for more information about the connections between resistor networks and graphs.

18.3.3 The Normalized Laplacian

When studying random walks on a graph, it often proves useful to normalize the Laplacian by its degrees. The *normalized Laplacian* of G is defined by

$$N_G = D_G^{-1/2} L_G D_G^{-1/2} = I - D_G^{-1/2} A_G D_G^{-1/2}.$$

It should be clear that normalized Laplacian is closely related to the walk matrix of a graph. Chung's monograph on spectral graph theory focuses on the normalized Laplacian [3].

18.3.4 Naming the Eigenvalues

When the graph G is understood, we will always let

$$\alpha_1 \geq \alpha_2 \geq \cdots \geq \alpha_n$$

denote the eigenvalues of the adjacency matrix. We order the eigenvalues of the Laplacian in the other direction:

$$0 = \lambda_1 \leq \lambda_2 \leq \cdots \leq \lambda_n.$$

We will always let

$$0 = \nu_1 \leq \nu_2 \leq \cdots \leq \nu_n$$

denote the eigenvalues of the normalized Laplacian. Even though ω is not a Greek variant of w, we use

$$1 = \omega_1 \geq \omega_2 \geq \cdots \geq \omega_n$$

to denote the eigenvalues of the walk matrix. It is easy to show that $\omega_i = 1 - \nu_i$.

For graphs in which every vertex has the same weighted degree, the degree matrix is a multiple of the identity; so, A_G and L_G have the same eigenvectors. For graphs that are not regular, the eigenvectors of A_G and L_G can behave very differently.

18.4 Some Examples

The most striking demonstration of the descriptive power of the eigenvectors of a graph comes from Hall's spectral approach to graph drawing [4]. To begin a demonstration of Hall's method, we generate the Delaunay graph of 200 randomly chosen points in the unit square.

```
xy = rand(200,2);
tri = delaunay(xy(:,1),xy(:,2));
elem = ones(3)-eye(3);
for i = 1:length(tri),
  A(tri(i,:),tri(i,:)) = elem;
end
A = double(A > 0);
gplot(A,xy)
```

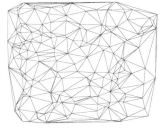

We will now discard the information we had about the coordinates of the vertices, and draw a picture of the graph using only the eigenvectors of its Laplacian matrix. We first compute the adjacency matrix A, the degree matrix D, and the Laplacian matrix L of the graph. We then compute the eigenvectors of the second and third smallest eigenvalues of L, v_2 and v_3. We then draw the same graph, using v_2 and v_3 to provide the coordinates of vertices. That is, we locate vertex a at position $(v_2(a), v_3(a))$, and draw the edges as straight lines between the vertices.

```
D = diag(sum(A));
L = D - A;
[v,e] = eigs(L, 3, 'sm');
gplot(A,v(:,[2 1]))
```

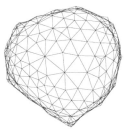

Amazingly, this process produces a very nice picture of the graph, in spite of the fact that the coordinates of the vertices were generated solely from the combinatorial structure of the graph. Note that the interior is almost planar. We could have obtained a similar, and possibly better, picture from the left-eigenvectors of the walk matrix of the graph.

```
W = A * inv(D);
[v,e] = eigs(W', 3);
gplot(A,v(:,[2 3]));
```

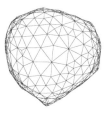

We defer the motivation for Hall's graph drawing technique to Section 18.7, so that we may first explore other examples.

One of the simplest graphs is the path graph. In the following figure, we plot the 2nd, 3rd, 4th, and 12th eigenvectors of the Laplacian of the path graph on 12 vertices. In each plot, the x-axis is the number of the vertex, and the y-axis is the value of the eigenvector at that vertex. We do not bother to plot the 1st eigenvector, as it is a constant vector.

```
A = diag(ones(1,11),1);
A = A + A';
D = diag(sum(A));
L = D - A;
[v,e] = eig(L);
plot(v(:,2),'o'); hold on;
plot(v(:,2));
plot(v(:,3),'o'); hold on;
plot(v(:,3));
. . . .
```

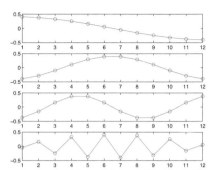

Observe that the 2nd eigenvector is monotonic along the path, that the second changes sign twice, and that the 12th alternates negative and positive. This can be explained by viewing these eigenvectors as the fundamental modes of vibration of a discretization of a string. We recommend [5] for a formal treatment.

By now, the reader should not be surprised to see that ring graphs have the obvious spectral drawings. In this case, we obtain the ring from the path by adding an edge between vertex 1 and 12.

```
A(1,12) = 1; A(12,1) = 1;
D = diag(sum(A));
L = D - A;
[v,e] = eig(L);
gplot(A,v(:,[2 3]))
hold on
gplot(A,v(:,[2 3]),'o')
```

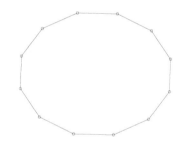

Our last example comes from the skeleton of the "Buckyball." This is the same as the graph between the corners of the Buckminster Fuller geodesic dome and of the seams on a standard Soccer ball.

```
A = full(bucky);
D = diag(sum(A));
L = D - A;
[v,e] = eig(L);
gplot(A,v(:,[2 3]))
hold on;
gplot(A,v(:,[2 3]),'o')
```

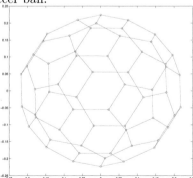

Note that the picture looks like a squashed Buckyball. The reason is that there is no canonical way to choose the eigenvectors v_2 and v_3. The smallest non-zero eigenvalue of the Laplacian has multiplicity three. This graph should really be drawn in three dimensions, using any set of orthonormal vectors

v_2, v_3, v_4 of the smallest non-zero eigenvalue of the Laplacian. As this picture hopefully shows, we obtain the standard embedding of the Buckyball in $\mathbb{R}^3$.

```
[x,y] = gplot(A,v(:,[2 3]));
[x,z] = gplot(A,v(:,[2 4]));
plot3(x,y,z)
```

The Platonic solids and all vertex-transitive convex polytopes in $\mathbb{R}^d$ display similar behavior. We refer the reader interested in learning more about this phenomenon to either Godsil's book [6] or to [7].

18.5 The Role of the Courant-Fischer Theorem

Recall that the Rayleigh quotient of a non-zero vector x with respect to a symmetric matrix A is

$$\frac{x^T A x}{x^T x}.$$

The Courant-Fischer characterization of the eigenvalues of a symmetric matrix A in terms of the maximizers and minimizers of the Rayleigh quotient (see [8]) plays a fundamental role in spectral graph theory.

Theorem 3 (Courant-Fischer) *Let A be a symmetric matrix with eigenvalues $\alpha_1 \geq \alpha_2 \geq \cdots \geq \alpha_n$. Then,*

$$\alpha_k = \max_{\substack{S \subseteq \mathbf{R}^n \\ \dim(S)=k}} \min_{\substack{x \in S \\ x \neq 0}} \frac{x^T A x}{x^T x} = \min_{\substack{T \subseteq \mathbf{R}^n \\ \dim(T)=n-k+1}} \max_{\substack{x \in T \\ x \neq 0}} \frac{x^T A x}{x^T x}.$$

The maximum in the first expression is taken over all subspaces of dimension k, and the minimum in the second is over all subspaces of dimension $n-k+1$.

Henceforth, whenever we minimize or maximize Rayleigh quotients we will only consider non-zero vectors, and thus will drop the quantifier "$x \neq 0$".

For example, the Courant-Fischer Theorem tells us that

$$\alpha_1 = \max_{x \in \mathbf{R}^n} \frac{x^T A x}{x^T x} \quad \text{and} \quad \alpha_n = \min_{x \in \mathbf{R}^n} \frac{x^T A x}{x^T x}.$$

We recall that a symmetric matrix A is *positive semidefinite*, written $A \succcurlyeq 0$, if all of its eigenvalues are non-negative. From (18.2) we see that the Laplacian is positive semidefinite. Adjacency matrices and walk matrices of non-empty graphs are not positive semidefinite as the sum of their eigenvalues equals their trace, which is 0. For this reason, one often considers the *lazy random walk* on a graph instead of the ordinary random walk. This walk stays put at each step with probability $1/2$. This means that the corresponding matrix is $(1/2)I + (1/2)W_G$, which can be shown to positive semidefinite.

18.5.1 Low-Rank Approximations

One explanation for the utility of the eigenvectors of extreme eigenvalues of matrices is that they provide low-rank approximations of a matrix. Recall that if A is a symmetric matrix with eigenvalues $\alpha_1 \geq \alpha_2 \geq \cdots \geq \alpha_n$ and a corresponding orthonormal basis of column eigenvectors $v_1, \ldots, v_n$, then

$$A = \sum_i \alpha_i v_i v_i^T.$$

We can measure how well a matrix B approximates a matrix A by either the operator norm $\|A - B\|$ or the Frobenius norm $\|A - B\|_F$, where we recall

$$\|M\| \overset{\text{def}}{=} \max_x \frac{\|Mx\|}{\|x\|} \quad \text{and} \quad \|M\|_F \overset{\text{def}}{=} \sqrt{\sum_{i,j} M(i,j)^2}.$$

Using the Courant-Fischer Theorem, one can prove that for every k, the best approximation of A by a rank-k matrix is given by summing the terms $\alpha_i v_i v_i^T$ over the k values of i for which $|\alpha_i|$ is largest. This holds regardless of whether we measure the quality of approximation in the operator or Frobenius norm.

When the difference between A and its best rank-k approximation is small, it explains why the eigenvectors of the largest k eigenvalues of A should provide a lot of information about A. However, one must be careful when applying this intuition as the analogous eigenvectors of the Laplacian correspond to its smallest eigenvalues. Perhpas the best way to explain the utility of these small eigenvectors is to observe that they provide the best low-rank approximation of the pseudoinverse of the Laplacian.

18.6 Elementary Facts

We list some elementary facts about the extreme eigenvalues of the Laplacian and adjacency matrices. We recommend deriving proofs yourself, or consulting the suggested references.

1. The all-1s vector is always an eigenvector of L_G of eigenvalue 0.

2. The largest eigenvalue of the adjacency matrix is at least the average degree of a vertex of G and at most the maximum degree of a vertex of G (see [9] or [10, Section 3.2]).

3. If G is connected, then $\alpha_1 > \alpha_2$ and the eigenvector of α_1 may be taken to be positive (this follows from the Perron-Frobenius theory; see [11]).

4. The all-1s vector is an eigenvector of A_G with eigenvalue α_1 if and only if G is an α_1-regular graph.

5. The multiplicity of 0 as an eigenvalue of L_G is equal to the number of connected components of L_G.

6. The largest eigenvalue of L_G is at most twice the maximum degree of a vertex in G.

7. $\alpha_n = -\alpha_1$ if and only if G is bipartite (see [12], or [10, Theorem 3.4]).

18.7   Spectral Graph Drawing

We can now explain the motivation behind Hall's spectral graph drawing technique [4]. Hall first considered the problem of assigning a real number $x(a)$ to each vertex a so that $(x(a) - x(b))^2$ is small for most edges (a, b). This led him to consider the problem of minimizing (18.2). So as to avoid the degenerate solutions in which every vertex is mapped to zero, or any other value, he introduces the restriction that x be orthogonal to $\mathbf{1}$. As the utility of the embedding does not really depend upon its scale, he suggested the normalization $\|x\| = 1$. By the Courant-Fischer Theorem, the solution to the resulting optimization problem is precisely an eigenvector of the second-smallest eigenvalue of the Laplacian.

But, what if we want to assign the vertices to points in $\mathbb{R}^2$? The natural minimization problem,

$$\min_{x,y\in\mathbb{R}^V} \sum_{(a,b)\in E} \|(x(a), y(a)) - (x(b), y(b))\|^2$$

such that

$$\sum_a (x(a), y(a)) = (0,0)$$

typically results in the degenerate solution $x = y = v_2$. To ensure that the two coordinates are different, Hall introduced the restriction that x be orthogonal to y. One can use the Courant-Fischer Theorem to show that the optimal solution is then given by setting $x = v_2$ and $y = v_3$, or by taking a rotation of this solution.

Hall observes that this embedding seems to cluster vertices that are close in the graph and separate vertices that are far in the graph. For more sophisticated approaches to drawing graphs, we refer the reader to Chapter 15.

18.8   Algebraic Connectivity and Graph Partitioning

Many useful ideas in spectral graph theory have arisen from efforts to find quantitative analogs of qualitative statements. For example, it is easy to show that $\lambda_2 > 0$ if and only if G is connected. This led Fiedler [13] to label λ_2 the *algebraic connectivity* of a graph, and to prove in various ways that better connected graphs have higher values of λ_2. This also led Fiedler to consider dividing the nodes of a graph into two pieces by choosing a real number t, and partitioning the nodes depending on whether or not $v_2(a) \geq t$. For $t = 0$, this corresponds to selecting all vertices in the right-half of the spectral embedding of the graph.

```
S = find(v(:,2) >= 0);
plot(v(S,2),v(S,1),'o')
```

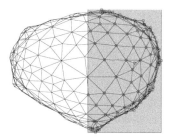

Fiedler proved [14] that for all $t \leq 0$, the set of nodes a for which $v_2(a) \geq t$ forms a connected component. This type of "nodal domain theorem" was extended by van der Holst [15] to the set of a such that $v(a) > 0$, when v is an eigenvector of λ_2 of minimal support.

The use of graph eigenvectors to partition graphs was also pioneered by Donath and Hoffman [16, 17] and Barnes [18]. It was popularized by experimental studies showing that it could give very good results [19, 20, 21, 22].

In many applications, one wants to partition the nodes of a graph into a few pieces of roughly equal size without removing too many edges (see Chapters 10 and 13). For simplicity, consider the problem of dividing the vertices of a graph into two pieces. In this case, we need merely identify one piece $S \subset V$. We then define $\partial(S)$ to be the set of edges with exactly one endpoint in S. We will also refer to S as a *cut*, as it implicitly divides the vertices into S and $V - S$, cutting all edges in $\partial(S)$. A tradeoff between the number of edges cut and the balance of the partition is obtained by dividing the first by a measure of the second, resulting in quantities called *cut ratio, sparsity, isoperimetric number,* and *conductance,* although these terms are sometimes used interchangeably. Wei and Cheng [23] suggested measuring the *ratio* of a cut, which they defined to be

$$R(S) \stackrel{\text{def}}{=} \frac{|\partial(S)|}{|S|\,|V-S|}.$$

Hagen and Kahng [24] observe that this quantity is always at least λ_2/n, and that v_2 can be described as a relaxation of the characteristic vector[2] of the set S that minimizes $R(S)$.

Let χ_S be the characteristic vector of a set S. For an unweighted graph G we have

$$\chi_S^T L_G \chi_S = |\partial(S)|\,,$$

and

$$\sum_{a<b} (\chi_S(a) - \chi_S(b))^2 = |S|\,|V-S|\,.$$

So,

$$R(S) = \frac{\chi_S^T L_G \chi_S}{\sum_{a<b}(\chi_S(a) - \chi_S(b))^2}.$$

On the other hand, Fiedler [14] proved that

$$\lambda_2 = n \min_{x \neq 0} \frac{x^T L_G x}{\sum_{a<b}(x(a) - x(b))^2}.$$

If we impose the restriction that x be a zero-one valued vector and then minimize this last expression, we obtain the characteristic vector of the set of minimum ratio. As we have imposed a constraint on the vector x, the ratio obtained must be larger than λ_2. Hagen and Kahng make this observation, and suggest using v_2 to try to find a set of low ratio by choosing some value t, and setting $S = \{a : v(a) \geq t\}$.

One may actually prove that the set obtained in this fashion does not have ratio too much worse than the minimum. Statements of this form follow from discrete versions of Cheeger's inequality [25]. The cleanest version relates to the *conductance* of a set S

$$\phi(S) \stackrel{\text{def}}{=} \frac{w(\partial(S))}{\min(d(S), d(V-S))},$$

where $d(S)$ denotes the sum of the degrees of the vertices in S and $w(\partial(S))$ denotes the sum of the weights of the edges in $\partial(S)$. The conductance of the graph G is defined by

$$\phi_G = \min_{\emptyset \subset S \subset V} \phi(S).$$

---

[2]Here, we define the characteristic vector of a set to be the vector that is one at vertices inside the set and zero elsewhere.

By a similar relaxation argument, one can show

$$2\phi_G \geq \nu_2.$$

Sinclair and Jerrum's discrete version of Cheeger's inequality [26] says that

$$\nu_2 \leq \phi_G^2/2.$$

Moreover, their proof reveals that if $\boldsymbol{v}_2$ is an eigenvector of ν_2, then there exists a t so that

$$\phi\left(\left\{a : \boldsymbol{d}^{-1/2}(a)\boldsymbol{v}_2(a) \geq t\right\}\right) \leq \sqrt{2\nu_2}.$$

Other discretizations of Cheeger's inequality were proved around the same time by a number of researchers. See [27, 28, 29, 30, 31]. We remark that Lawler and Sokal define conductance by

$$\frac{w(\partial(S))}{\boldsymbol{d}(S)\boldsymbol{d}(V-S)},$$

which is proportional to the *normalized cut* measure

$$\frac{w(\partial(S))}{\boldsymbol{d}(S)} + \frac{w(\partial(V-S))}{\boldsymbol{d}(V-S)}$$

popularized by Shi and Malik [21]. The advantage of this later formulation is that it has an obvious generalization to partitions into more than two pieces.

In general, the eigenvalues and entries of eigenvectors of Laplacian matrices will not be rational numbers; so, it is unreasonable to hope to compute them exactly. Mihail [32] proves that an approximation of the second-smallest eigenvector suffices. While her argument was stated for regular graphs, one can apply it to irregular, weighted graphs to show that for every vector $\boldsymbol{x}$ orthogonal to $\boldsymbol{d}^{1/2}$ there exists a t so that

$$\phi\left(\left\{a : \boldsymbol{d}^{-1/2}(a)\boldsymbol{x}(a) \geq t\right\}\right) \leq \sqrt{2\frac{\boldsymbol{x}^T N_G \boldsymbol{x}}{\boldsymbol{x}^T \boldsymbol{x}}}.$$

While spectral partitioning heuristics are easy to implement, they are neither the most effective in practice or in theory. Theoretically better algorithms have been obtained by linear programming [33] and by semi-definite programming [34]. Fast variants of these algorithms may be found in [35, 36, 37, 38, 39]. More practical algorithms are discussed in Chapters 10 and 13.

18.8.1 Convergence of Random Walks

If G is a connected, undirected graph, then the largest eigenvalue of W_G, ω_1, has multiplicity 1, equals 1, and has eigenvector $\boldsymbol{d}$. We may convert this eigenvector into a probability distribution $\boldsymbol{\pi}$ by setting

$$\boldsymbol{\pi} = \frac{\boldsymbol{d}}{\sum_a \boldsymbol{d}(a)}.$$

If $\omega_n \neq -1$, then the distribution of every random walk eventually converges to $\boldsymbol{\pi}$. The rate of this convergence is governed by how close $\max(\omega_2, -\omega_n)$ is to ω_1. For example, let $\boldsymbol{p}_t$ denote the distribution after t steps of a random walk that starts at vertex a. Then for every vertex b,

$$|\boldsymbol{p}_t(b) - \boldsymbol{\pi}(b)| \leq \sqrt{\frac{\boldsymbol{d}(b)}{\boldsymbol{d}(a)}} \, (1 - \max(\omega_2, -\omega_n))^t \, .$$

One intuition behind Cheeger's inequality is that sets of small conductance are precisely the obstacles to the convergence of random walks.

For more information about random walks on graphs, we recommend the survey of Lovàsz [40] and the book by Doyle and Snell [2].

18.8.2 Expander Graphs

Some of the most fascinating graphs are those on which random walks mix quickly and which have high conductance. These are called *expander graphs*, and may be defined as the d-regular graphs for which all non-zero Laplacian eigenvalues are bounded away from zero. In the better expander graphs, all the Laplacian eigenvalues are close to d. One typically considers infinite families of such graphs in which d and a lower bound on the distance of the non-zero eigenvalues from d remain constant. These are counter-examples to many naive conjectures about graphs, and should be kept in mind whenever one is thinking about graphs. They have many amazing properties, and have been used throughout Theoretical Computer Science. In addition to playing a prominent role in countless theorems, they are used in the design of pseudorandom generators [41, 42, 43], error-correcting codes [44, 45, 46, 47, 48], fault-tolerant circuits [49] and routing networks [50].

The reason such graphs are called *expanders* is that all small sets of vertices in these graphs have unusually large numbers of neighbors. That is, their neighborhoods expand. For $S \subset V$, let $N(S)$ denote the set of vertices that are neighbors of vertices in S. Tanner [51] provides a lower bound on the size of $N(S)$ in bipartite graphs. In general graphs, it becomes the following:

Theorem 4 *Let $G = (V, E)$ be a d-regular graph on n vertices and set*

$$\epsilon = \max\left(1 - \frac{\lambda_2}{d}, \frac{\lambda_n}{d} - 1\right)$$

Then, for all $S \subseteq V$,

$$|N(S)| \geq \frac{|S|}{\epsilon^2(1-\alpha) + \alpha},$$

where $|S| = \alpha n$.

The term ϵ is small when all of the eigenvalues are close to d. Note that when

α is much less than ϵ^2, the term on the right is approximately $|S|/\epsilon^2$, which can be much larger than $|S|$.

An example of the pseudo-random properties of expander graphs is the "Expander Mixing Lemma." To understand it, consider choosing two subsets of the vertices, S and T of sizes αn and βn, at random. Let $\vec{E}(S,T)$ denote the set of ordered pairs (a, b) with $a \in S$, $b \in T$ and $(a, b) \in E$. The expected size of $\vec{E}(S,T)$ is $\alpha\beta dn$. This theorem tells us that for *every* pair of large sets S and T, the number of such pairs is approximately this quantity. Alternatively, one may view an expander as an approximation of the complete graph. The fraction of edges in the complete graph going from S to T is $\alpha\beta$. The following theorem says that the same is approximately true for all sufficiently large sets S and T.

Theorem 5 (Expander Mixing Lemma) *Let $G = (V, E)$ be a d-regular graph and set*

$$\epsilon = \max\left(1 - \frac{\lambda_2}{d}, \frac{\lambda_n}{d} - 1\right)$$

Then, for every $S \subseteq V$ and $T \subseteq V$,

$$\left|\left|\vec{E}(S,T)\right| - \alpha\beta dn\right| \le \epsilon dn\sqrt{(\alpha - \alpha^2)(\beta - \beta^2)},$$

where $|S| = \alpha n$ and $|T| = \beta n$.

This bound is a slight extension by Beigel, Margulis and Spielman [52] of a bound originally proved by Alon and Chung [53]. Observe that when α and β are greater than ϵ, the term on the right is less than $\alpha\beta dn$. Theorem 4 may be derived from Theorem 5.

We refer readers who would like to learn more about expander graphs to the survey of Hoory, Linial, and Wigderson [54].

18.8.3   Ramanujan Graphs

Given the importance of λ_2, we should know how close it can be to d. Nilli [55] shows that it cannot be much closer than $2\sqrt{d-1}$.

Theorem 6 *Let G be an unweighted d-regular graph containing two edges (u_0, u_1) and (v_0, v_1) whose vertices are at distance at least $2k + 2$ from each other. Then*

$$\lambda_2 \le d - 2\sqrt{d-1} + \frac{2\sqrt{d-1}-1}{k+1}.$$

Amazingly, Margulis [56] and Lubotzky, Phillips and Sarnak [57] have constructed infinite families of d-regular graphs, called Ramanujan graphs, for which $\lambda_2 \ge d - 2\sqrt{d-1}$.

However, this is not the end of the story. Kahale [58] proves that vertex expansion by a factor greater than $d/2$ cannot be derived from bounds on λ_2. Expander graphs that have expansion greater than $d/2$ on small sets of vertices have been derived by Capalbo *et. al.* [59] through non-spectral arguments.

18.8.4 Bounding λ_2

I consider λ_2 to be the most interesting parameter of a connected graph. If it is large, the graph is an expander. If it is small, then the graph can be cut into two pieces without removing too many edges. Either way, we learn something about the graph. Thus, it is very interesting to find ways of estimating the value of λ_2 for families of graphs.

One way to explain the success of spectral partitioning heuristics is to prove that the graphs to which they are applied have small values of λ_2 or ν_2. A line of work in this direction was started by Spielman and Teng [60], who proved upper bounds on λ_2 for planar graphs and well-shaped finite element meshes.

Theorem 7 ([60]) *Let G be a planar graph with n vertices of maximum degree d, and let λ_2 be the second-smallest eigenvalue of its Laplacian. Then,*

$$\lambda_2 \leq \frac{8d}{n}.$$

This theorem has been extended to graphs of bounded genus by Kelner [61]. Entirely new techniques were developed by Biswal, Lee and Rao [62] to extend this bound to graphs excluding bounded minors. Bounds on higher Laplacian eigenvalues have been obtained by Kelner, Lee, Price and Teng [63].

Theorem 8 ([63]) *Let G be a graph with n vertices and constant maximum degree. If G is planar, has constant genus, or has a constant-sized forbidden minor, then*

$$\lambda_k \leq O(k/n).$$

Proving lower bounds on λ_2 is a more difficult problem. The dominant approach is to relate the graph under consideration to a graph with known eigenvalues, such as the complete graph. Write

$$L_G \succcurlyeq cL_H$$

if $L_G - cL_H \succcurlyeq 0$. In this case, we know that

$$\lambda_i(G) \geq c\lambda_i(H),$$

for all i. Inequalities of this form may be proved by identifying each edge of the graph H with a path in G. The resulting bounds are called Poincaré inequalities, and are closely related to the bounds used in the analysis of preconditioners in Chapter 12 and in related works [64, 65, 66, 67]. For examples of such arguments, we refer the reader to one of [68, 69, 70].

18.9   Coloring and Independent Sets

In the graph coloring problem one is asked to assign a color to every vertex of a graph so that every edge connects vertices of different colors, while using as few colors as possible. Replacing colors with numbers, we define a k-coloring of a graph $G = (V, E)$ to be a function $c : V \to \{1, \dots, k\}$ such that

$$c(i) \neq c(j), \text{ for all } (i, j) \in E.$$

The *chromatic number* of a graph G, written $\chi(G)$, is the least k for which G has a k-coloring. Wilf [71] proved that the chromatic number of a graph may be bounded above by its largest adjacency eigenvalue.

Theorem 9 ([71])
$$\chi(G) \leq \alpha_1 + 1.$$

On the other hand, Hoffman [72] proved a lower bound on the chromatic number in terms of the adjacency matrix eigenvalues. When reading this theorem, recall that α_n is negative.

Theorem 10 *If G is a graph with at least one edge, then*

$$\chi(G) \geq \frac{\alpha_1 - \alpha_n}{-\alpha_n} = 1 + \frac{\alpha_1}{-\alpha_n}.$$

In fact, this theorem holds for arbitrary weighted graphs. Thus, one may prove lower bounds on the chromatic number of a graph by assigning a weight to every edge, and then computing the resulting ratio.

It follows from Theorem 10 that G is not bipartite if $|\alpha_n| < \alpha_1$. Moreover, as $|\alpha_n|$ becomes closer to 0, more colors are needed to properly color the graph. Another way to argue that graphs with small $|\alpha_n|$ are far from being bipartite was found by Trevisan [73]. To be precise, Trevisan proves a bound, analogous to Cheeger's inequality, relating $|E| - \max_{S \subseteq V} |\partial(S)|$ to the smallest eigenvalue of the *signless Laplacian matrix*, $D_G + A_G$.

An *independent set* of vertices in a graph G is a set $S \subseteq V$ such that no edge connects two vertices of S. The size of the largest independent set in a graph is called its *independence number*, and is denoted $\alpha(G)$. As all the nodes of one color in a coloring of G are independent, we know

$$\alpha(G) \geq n/\chi(G).$$

For regular graphs, Hoffman derived the following upper bound on the size of an independent set.

Theorem 11 *Let $G = (V, E)$ be a d-regular graph. Then*

$$\alpha(G) \leq n \frac{-\alpha_n}{d - \alpha_n}.$$

This implies Theorem 10 for regular graphs.

18.10   Perturbation Theory and Random Graphs

McSherry [74] observes that the spectral partitioning heuristics and the related spectral heuristics for graph coloring can be understood through matrix perturbation theory. For example, let G be a graph and let S be a subset of the vertices of G. Without loss of generality, assume that S is the set of the first $|S|$ vertices of G. Then, we can write the adjacency matrix of G as

$$\begin{bmatrix} A(S) & 0 \\ 0 & A(V-S) \end{bmatrix} + \begin{bmatrix} 0 & A(S, V-S) \\ A(V-S, S) & 0 \end{bmatrix},$$

where we write $A(S)$ to denote the restriction of the adjacency matrix to the vertices in S, and $A(S, V-S)$ to capture the entries in rows indexed by S and columns indexed by $V-S$. The set S can be discovered from an examination of the eigenvectors of the left-hand matrix: it has one eigenvector that is positive on S and zero elsewhere, and another that is positive on $V - S$ and zero elsewhere. If the right-hand matrix is a "small" perturbation of the left-hand matrix, then we expect similar eigenvectors to exist in A. It seems reasonable that the right-hand matrix should be small if it contains few edges. Whether or not this may be made rigorous depends on the locations of the edges. We will explain McSherry's analysis, which makes this rigorous in certain random models.

We first recall the basics perturbation theory for matrices. Let A and B be symmetric matrices with eigenvalues $\alpha_1 \geq \alpha_2 \geq \cdots \geq \alpha_n$ and $\beta_1 \geq \beta_2 \geq \cdots \geq \beta_n$, respectively. Let $M = A - B$. Weyl's Theorem, which follows from the Courant-Fischer Theorem, tells us that

$$|\alpha_i - \beta_i| \leq \|M\|$$

for all i. As M is symmetric, $\|M\|$ is merely the largest absolute value of an eigenvalue of M.

When some eigenvalue α_i is well-separated from the others, one can show that a small perturbation does not change the corresponding eigenvector too much. Demmel [75, Theorem 5.2] proves the following bound.

Theorem 12 *Let $v_1, \ldots, v_n$ be an orthonormal basis of eigenvectors of A corresponding to $\alpha_1, \ldots, \alpha_n$ and let $u_1, \ldots, u_n$ be an orthonormal basis of eigenvectors of B corresponding to $\beta_1, \ldots, \beta_n$. Let θ_i be the angle between v_i and w_i. Then,*

$$\frac{1}{2} \sin 2\theta_i \leq \frac{\|M\|}{\min_{j \neq i} |\alpha_i - \alpha_j|}.$$

McSherry applies these ideas from perturbation theory to analyze the behavior of spectral partitioning heuristics on random graphs that are generated to have good partitions. For example, he considered the planted partition model of Boppana [76]. This is defined by a weighted complete graph H determined by a $S \subset V$ in which

$$
w(a,b) = \begin{cases} p & \text{if both or neither of } a \text{ and } b \text{ are in } S, \text{ and} \\ q & \text{if exactly one of } a \text{ and } b \text{ are in } S, \end{cases}
$$

for $q < p$. A random unweighted graph G is then constructed by including edge (a,b) in G with probability $w(a,b)$. For appropriate values of q and p, the cut determined by S is very likely to be the sparsest. If q is not too close to p, then the largest two eigenvalues of H are far from the rest, and correspond to the all-1s vector and a vector that is uniform and positive on S and uniform and negative on $V - S$. Using results from random matrix theory of Füredi and Komlós [77], Vu [78], and Alon, Krievlevich and Vu [79], McSherry proves that G is a slight perturbation of H, and that the eigenvectors of G can be used to recover the set S, with high probability.

Both McSherry [74] and Alon and Kahale [80] have shown that the eigenvectors of the smallest adjacency matrix eigenvalues may be used to k-color randomly generated k-colorable graphs. These graphs are generated by first partitioning the vertices into k sets, $S_1, \ldots, S_k$, and then adding edges between vertices in different sets with probability p, for some small p.

For more information on these and related results, we suggest the book by Kannan and Vempala [81].

18.11   Relative Spectral Graph Theory

Preconditioning (see Chapter 12) has inspired the study of the *relative eigenvalues of graphs*. These are the eigenvalues of $L_G L_H^+$, where L_G is the Laplacian of a graph G and L_H^+ is the pseudo-inverse of the Laplacian of a graph H. We recall that the pseudo-inverse of a symmetric matrix L is given by

$$
\sum_{i:\lambda_i \neq 0} \frac{1}{\lambda_i} \boldsymbol{v}_i \boldsymbol{v}_i^T,
$$

where the λ_i and $\boldsymbol{v}_i$ are the eigenvalues and eigenvectors of the matrix L. The eigenvalues of $L_G L_H^+$ reveal how well H approximates G.

Let K_n denote the complete graph on n vertices. All of the non-trivial eigenvalues of the Laplacian of K_n equal n. So, L_{K_n} acts as n times the identity on the space orthogonal to $\mathbf{1}$. Thus, for every G the eigenvalues of $L_G L_{K_n}^+$ are just the eigenvalues of L_G divided by n, and the eigenvectors are the same. Many results on expander graphs, including those in Section 18.8.2, can be derived by using this perspective to treat an expander as an approximation of the complete graph (see [82]).

Recall that when L_G and L_H have the same range, $\kappa_f(L_G, L_H)$ is defined to be the largest non-zero eigenvalue of $L_G L_H^+$ divided by the smallest. The Ramanujan graphs are d-regular graphs G for which

$$\kappa_f(L_G, L_{K_n}) \leq \frac{d + 2\sqrt{d-1}}{d - 2\sqrt{d-1}}.$$

Batson, Spielman and Srivastava [82] prove that every graph H can be approximated by a sparse graph almost as well as this.

Theorem 13 *For every weighted graph G on n vertices and every $d > 1$, there exists a weighted graph H with at most $\lceil d(n-1) \rceil$ edges such that*

$$\kappa_f(L_G, L_H) \leq \frac{d + 1 + 2\sqrt{d}}{d + 1 - 2\sqrt{d}}.$$

Spielman and Srivastava [83] show that if one forms a graph H by sampling $O(n \log n / \epsilon^2)$ edges of G with probability proportional to their effective resistance and rescaling their weights, then with high probability $\kappa_f(L_G, L_H) \leq 1 + \epsilon$.

Spielman and Woo [84] have found a characterization of the well-studied *stretch* of a spanning tree with respect to a graph in terms of relative graph spectra. For simplicity, we just define it for unweighted graphs. If T is a spanning tree of a graph $G = (V, E)$, then for every $(a, b) \in E$ there is a unique path in T connecting a to b. The stretch of (a, b) with respect to T, written $\mathsf{st}_T(a, b)$, is the number of edges in that path in T. The stretch of G with respect to T is then defined to be

$$\mathsf{st}_T(G) \overset{\text{def}}{=} \sum_{(a,b) \in E} \mathsf{st}_T(a, b).$$

Theorem 14 ([84])

$$\mathsf{st}_T(G) = \text{trace}\left(L_G L_T^+\right).$$

See Chapter 12 for a proof.

18.12 Directed Graphs

There has been much less success in the study of the spectra of directed graphs, perhaps because the nonsymmetric matrices naturally associated with directed graphs are not necessarily diagonalizable. One naturally defines the adjacency matrix of a directed graph G by

$$A_G(a, b) = \begin{cases} 1 & \text{if } G \text{ has a directed edge from } b \text{ to } a \\ 0 & \text{otherwise.} \end{cases}$$

Similarly, if we let $d(a)$ denote the number of edges leaving vertex a and define D as before, then the matrix realizing the random walk on G is

$$W_G = A_G D_G^{-1}.$$

The Perron-Frobenius Theorem (see [11, 8]) tells us that if G is strongly connected, then A_G has a unique positive eigenvector v with a positive eigenvalue λ such that every other eigenvalue μ of A satisfies $|\mu| \leq \lambda$. The same holds for W_G. When $|\mu| < \lambda$ for all other eigenvalues μ, this vector is proportional to the unique limiting distribution of the random walk on G.

These Perron-Frobenius eigenvectors have proved incredibly useful in a number of situations. For instance, they are at the heart of Google's PageRank algorithm for answering web search queries (see [85, 86]). This algorithm constructs a directed graph by associating vertices with web pages, and creating a directed edge for each link. It also adds a large number of low-weight edges by allowing the random walk to move to a random vertex with some small probability at each step. The PageRank score of a web page is then precisely the value of the Perron-Frobenius vector at the associated vertex. Interestingly, this idea was actually proposed by Bonacich [87, 88, 89] in the 1970's as a way of measuring the centrality of nodes in a social network. An analogous measure, using the adjacency matrix, was proposed by Berge [90, Chapter 4, Section 5] for ranking teams in sporting events. Palacios-Huerta and Volij [91] and Altman and Tennenholtz [92] have given abstract, axiomatic descriptions of the rankings produced by these vectors.

A related approach to obtaining rankings from directed graphs was proposed by Kleinberg [93]. He suggested using singular vectors of the directed adjacency matrix. Surprising, we are unaware of other combinatorially interesting uses of the singular values or vectors of matrices associated with directed graphs.

To avoid the complications of non-diagonalizable matrices, Chung [94] has defined a symmetric Laplacian matrix for directed graphs. Her definition is inspired by the observation that the degree matrix D used in the definition of the undirected Laplacian is the diagonal matrix of d, which is proportional to the limiting distribution of a random walk on an undirected graph. Chung's Laplacian for directed graphs is constructed by replacing d by the Perron-Frobenius vector for the random walk on the graph. Using this Laplacian, she derives analogs of Cheeger's inequality, defining conductance by counting edges by the probability they appear in a random walk [95].

18.13   Concluding Remarks

Many fascinating and useful results in Spectral Graph Theory are omitted in this survey. For those who want to learn more, the following books and survey papers take an approach in the spirit of this Chapter: [96, 97, 98, 81, 3, 40]. I also recommend [10, 99, 6, 100, 101].

Anyone contemplating Spectral Graph Theory should be aware that there are graphs with very pathological spectra. Expanders could be considered examples. But, Strongly Regular Graphs (which only have 3 distinct eigenvalues) and Distance Regular Graphs should also be considered. Excellent treatments of these appear in some of the aforementioned works, and also in [6, 102].

Bibliography

[1] Bollobás, B., *Modern graph theory*, Springer-Verlag, New York, 1998.

[2] Doyle, P. G. and Snell, J. L., *Random Walks and Electric Networks*, Vol. 22 of *Carus Mathematical Monographs*, Mathematical Association of America, 1984.

[3] Chung, F. R. K., *Spectral Graph Theory*, American Mathematical Society, 1997.

[4] Hall, K. M., "An r-dimensional quadratic placement algorithm," *Management Science*, Vol. 17, 1970, pp. 219–229.

[5] Gould, S. H., *Variational Methods for Eigenvalue Problems*, Dover, New York, 1995.

[6] Godsil, C., *Algebraic Combinatorics*, Chapman & Hall, London, UK, 1993.

[7] van der Holst, H., Lovàsz, L., and Schrijver, A., "The Colin de Verdière Graph Parameter," *Bolyai Soc. Math. Stud.*, Vol. 7, 1999, pp. 29–85.

[8] Horn, R. A. and Johnson, C. R., *Matrix Analysis*, Cambridge University Press, 1985.

[9] Collatz, L. and Sinogowitz, U., "Spektren endlicher Grafen," *Abh. Math. Sem. Univ. Hamburg*, Vol. 21, 1957, pp. 63–77.

[10] Cvetković, D. M., Doob, M., and Sachs, H., *Spectra of Graphs*, Academic Press, New York, 1978.

[11] Bapat, R. B. and Raghavan, T. E. S., *Nonnegative Matrices and Applications*, No. 64 in Encyclopedia of Mathematics and its Applications, Cambridge University Press, 1997.

[12] Hoffman, A. J., "On the Polynomial of a Graph," *The American Mathematical Monthly*, Vol. 70, No. 1, 1963, pp. 30–36.

[13] Fiedler., M., "Algebraic connectivity of graphs," *Czechoslovak Mathematical Journal*, Vol. 23, No. 98, 1973, pp. 298–305.

[14] Fiedler, M., "A property of eigenvectors of nonnegative symmetric matrices and its applications to graph theory," *Czechoslovak Mathematical Journal*, Vol. 25, No. 100, 1975, pp. 618–633.

[15] van der Holst, H., "A Short Proof of the Planarity Characterization of Colin de Verdière," *Journal of Combinatorial Theory, Series B*, Vol. 65, No. 2, 1995, pp. 269 – 272.

[16] Donath, W. E. and Hoffman, A. J., "Algorithms for Partitioning Graphs and Computer Logic Based on Eigenvectors of Connection Matrices," *IBM Technical Disclosure Bulletin*, Vol. 15, No. 3, 1972, pp. 938–944.

[17] Donath, W. E. and Hoffman, A. J., "Lower Bounds for the Partitioning of Graphs," *IBM Journal of Research and Development*, Vol. 17, No. 5, Sept. 1973, pp. 420–425.

[18] Barnes, E. R., "An Algorithm for Partitioning the Nodes of a Graph," *SIAM Journal on Algebraic and Discrete Methods*, Vol. 3, No. 4, 1982, pp. 541–550.

[19] Simon, H. D., "Partitioning of Unstructured Problems for Parallel Processing," *Computing Systems in Engineering*, Vol. 2, 1991, pp. 135–148.

[20] Pothen, A., Simon, H. D., and Liou, K.-P., "Partitioning Sparse Matrices with Eigenvectors of Graphs," *SIAM Journal on Matrix Analysis and Applications*, Vol. 11, No. 3, 1990, pp. 430–452.

[21] Shi, J. B. and Malik, J., "Normalized Cuts and Image Segmentation," *IEEE Trans. Pattern Analysis and Machine Intelligence*, Vol. 22, No. 8, Aug. 2000, pp. 888–905.

[22] Ng, A. Y., Jordan, M. I., and Weiss, Y., "On Spectral Clustering: Analysis and an algorithm," *Adv. in Neural Inf. Proc. Sys. 14*, 2001, pp. 849–856.

[23] Wei, Y.-C. and Cheng, C.-K., "Ratio cut partitioning for hierarchical designs," *IEEE Transactions on Computer-Aided Design of Integrated Circuits and Systems*, Vol. 10, No. 7, Jul 1991, pp. 911–921.

[24] Hagen, L. and Kahng, A. B., "New Spectral Methods for Ratio Cut Partitioning and Clustering," *IEEE Transactions on Computer-Aided Design of Integrated Circuits and Systems*, Vol. 11, 1992, pp. 1074–1085.

[25] Cheeger, J., "A lower bound for smallest eigenvalue of the Laplacian," *Problems in Analysis*, Princeton University Press, 1970, pp. 195–199.

[26] Sinclair, A. and Jerrum, M., "Approximate Counting, Uniform Generation and Rapidly Mixing Markov Chains," *Information and Computation*, Vol. 82, No. 1, July 1989, pp. 93–133.

[27] Lawler, G. F. and Sokal, A. D., "Bounds on the L^2 Spectrum for Markov Chains and Markov Processes: A Generalization of Cheeger's Inequality," *Transactions of the American Mathematical Society*, Vol. 309, No. 2, 1988, pp. 557–580.

[28] Alon, N. and Milman, V. D., "λ_1, Isoperimetric inequalities for graphs, and superconcentrators," *J. Comb. Theory, Ser. B*, Vol. 38, No. 1, 1985, pp. 73–88.

[29] Alon, N., "Eigenvalues and expanders," *Combinatorica*, Vol. 6, No. 2, 1986, pp. 83–96.

[30] Dodziuk, J., "Difference Equations, Isoperimetric Inequality and Transience of Certain Random Walks," *Transactions of the American Mathematical Society*, Vol. 284, No. 2, 1984, pp. 787–794.

[31] Varopoulos, N. T., "Isoperimetric inequalities and Markov chains," *Journal of Functional Analysis*, Vol. 63, No. 2, 1985, pp. 215–239.

[32] Mihail, M., "Conductance and Convergence of Markov Chains—A Combinatorial Treatment of Expanders," *30th Annual IEEE Symposium on Foundations of Computer Science*, 1989, pp. 526–531.

[33] Leighton, T. and Rao, S., "Multicommodity max-flow min-cut theorems and their use in designing approximation algorithms," *Journal of the ACM*, Vol. 46, No. 6, Nov. 1999, pp. 787–832.

[34] Arora, S., Rao, S., and Vazirani, U., "Expander flows, geometric embeddings and graph partitioning," *J. ACM*, Vol. 56, No. 2, 2009, pp. 1–37.

[35] Khandekar, R., Rao, S., and Vazirani, U., "Graph partitioning using single commodity flows," *J. ACM*, Vol. 56, No. 4, 2009, pp. 1–15.

[36] Sherman, J., "Breaking the Multicommodity Flow Barrier for O(sqrt(log n))-approximations to Sparsest Cut," *Proceedings of the 50th IEEE Symposium on Foundations of Computer Science*, 2009, pp. 363–372.

[37] Arora, S., Hazan, E., and Kale, S., "O(sqrt (log n)) Approximation to Sparsest Cut in $\tilde{O}(n^2)$ Time," *45th IEEE Symposium on Foundations of Computer Science*, 2004, pp. 238–247.

[38] Arora, S. and Kale, S., "A combinatorial, primal-dual approach to semidefinite programs," *Proceedings of the 39th Annual ACM Symposium on Theory of Computing*, 2007, pp. 227–236.

[39] Orecchia, L., Schulman, L. J., Vazirani, U. V., and Vishnoi, N. K., "On partitioning graphs via single commodity flows," *Proceedings of the 40th Annual ACM Symposium on Theory of Computing*, 2008, pp. 461–470.

[40] Lovász, L., "Random walks on graphs: a survey," *Combinatorics, Paul Erdös is Eighty, Vol. 2*, edited by T. S. D. Miklos, V. T. Sos, Janos Bólyai Mathematical Society, Budapest, 1996, pp. 353–398.

[41] Impagliazzo, R. and Zuckerman, D., "How to recycle random bits," *30th annual IEEE Symposium on Foundations of Computer Science*, 1989, pp. 248–253.

[42] Karp, R. M., Pippenger, N., and Sipser, M., "A time randomness trade-off," *AMS Conf. on Probabilistic Computational Complexity*, Durham, NH, 1985.

[43] Ajtai, M., Komlós, J., and Szemerédi, E., "Deterministic Simulation in LOGSPACE," *Proceedings of the Nineteenth Annual ACM Symposium on Theory of Computing*, 1987, pp. 132–140.

[44] Alon, N., Bruck, J., Naor, J., Naor, M., and Roth, R. M., "Construction of Asymptotically Good Low-Rate Error-Correcting Codes through Pseudo-Random Graphs," *IEEE Transactions on Information Theory*, Vol. 38, No. 2, March 1992, pp. 509–516.

[45] Sipser, M. and Spielman, D., "Expander codes," *IEEE Transactions on Information Theory*, Vol. 42, No. 6, Nov 1996, pp. 1710–1722.

[46] Zemor, G., "On expander codes," *IEEE Transactions on Information Theory*, Vol. 47, No. 2, Feb 2001, pp. 835–837.

[47] Barg, A. and Zemor, G., "Error exponents of expander codes," *IEEE Transactions on Information Theory*, Vol. 48, No. 6, Jun 2002, pp. 1725–1729.

[48] Guruswami, V., "Guest column: error-correcting codes and expander graphs," *SIGACT News*, Vol. 35, No. 3, 2004, pp. 25–41.

[49] Pippenger, N., "On networks of noisy gates," *Proceedings of the 26th Ann. IEEE Symposium on Foundations of Computer Science*, 1985, pp. 30–38.

[50] Pippenger, N., "Self-Routing Superconcentrators," *Journal of Computer and System Sciences*, Vol. 52, No. 1, Feb. 1996, pp. 53–60.

[51] Tanner, R. M., "Explicit construction of concentrators from generalized n-gons," *SIAM J. Algebraic Discrete Methods*, Vol. 5, 1984, pp. 287–293.

[52] Beigel, R., Margulis, G., and Spielman, D. A., "Fault diagnosis in a small constant number of parallel testing rounds," *SPAA '93: Proceedings of the fifth annual ACM symposium on Parallel algorithms and architectures*, ACM, New York, NY, USA, 1993, pp. 21–29.

[53] Alon, N. and Chung, F., "Explicit Construction of Linear Sized Tolerant Networks," *Discrete Mathematics*, Vol. 72, 1988, pp. 15–19.

[54] Hoory, S., Linial, N., and Wigderson, A., "Expander Graphs and Their Applications," *Bulletin of the American Mathematical Society*, Vol. 43, No. 4, 2006, pp. 439–561.

[55] Nilli, A., "On the second eigenvalue of a graph," *Discrete Math*, Vol. 91, 1991, pp. 207–210.

[56] Margulis, G. A., "Explicit group theoretical constructions of combinatorial schemes and their application to the design of expanders and concentrators," *Problems of Information Transmission*, Vol. 24, No. 1, July 1988, pp. 39–46.

[57] Lubotzky, A., Phillips, R., and Sarnak, P., "Ramanujan Graphs," *Combinatorica*, Vol. 8, No. 3, 1988, pp. 261–277.

[58] Kahale, N., "Eigenvalues and expansion of regular graphs," *J. ACM*, Vol. 42, No. 5, 1995, pp. 1091–1106.

[59] Capalbo, M., Reingold, O., Vadhan, S., and Wigderson, A., "Randomness conductors and constant-degree lossless expanders," *Proceedings of the 34th Annual ACM Symposium on Theory of Computing*, 2002, pp. 659–668.

[60] Spielman, D. A. and Teng, S.-H., "Spectral partitioning works: Planar graphs and finite element meshes," *Linear Algebra and its Applications*, Vol. 421, No. 2-3, 2007, pp. 284–305, Special Issue in honor of Miroslav Fiedler.

[61] Kelner, J. A., "Spectral partitioning, eigenvalue bounds, and circle packings for graphs of bounded genus," *SIAM J. Comput.*, Vol. 35, No. 4, 2006, pp. 882–902.

[62] Biswal, P., Lee, J., and Rao, S., "Eigenvalue bounds, spectral partitioning, and metrical deformations via flows," *Journal of the ACM*, 2010, to appear.

[63] Kelner, J. A., Lee, J., Price, G., and Teng, S.-H., "Higher Eigenvalues of Graphs," *Proceedings of the 50th IEEE Symposium on Foundations of Computer Science*, 2009, pp. 735–744.

[64] Vaidya, P. M., "Solving linear equations with symmetric diagonally dominant matrices by constructing good preconditioners." Unpublished manuscript UIUC 1990. A talk based on the manuscript was presented at the IMA Workshop on Graph Theory and Sparse Matrix Computation, October 1991, Minneapolis.

[65] Bern, M., Gilbert, J., Hendrickson, B., Nguyen, N., and Toledo, S., "Support-graph preconditioners," *SIAM Journal on Matrix Analysis and Applications*, Vol. 27, No. 4, 2006, pp. 930–951.

[66] Boman, E. G. and Hendrickson, B., "Support Theory for Preconditioning," *SIAM Journal on Matrix Analysis and Applications*, Vol. 25, No. 3, 2003, pp. 694–717.

[67] Spielman, D. A. and Teng, S.-H., "Nearly-Linear Time Algorithms for Preconditioning and Solving Symmetric, Diagonally Dominant Linear Systems," *CoRR*, Vol. abs/cs/0607105, 2009, Available at http://www.arxiv.org/abs/cs.NA/0607105.

[68] Diaconis, P. and Stroock, D., "Geometric Bounds for Eigenvalues of Markov Chains," *The Annals of Applied Probability*, Vol. 1, No. 1, 1991, pp. 36–61.

[69] Guattery, S., Leighton, T., and Miller, G. L., "The Path Resistance Method for Bounding the Smallest Nontrivial Eigenvalue of a Laplacian," *Combinatorics, Probability and Computing*, Vol. 8, 1999, pp. 441–460.

[70] Guattery, S. and Miller, G. L., "Graph Embeddings and Laplacian Eigenvalues," *SIAM Journal on Matrix Analysis and Applications*, Vol. 21, No. 3, 2000, pp. 703–723.

[71] Wilf, H. S., "The Eigenvalues of a Graph and its Chromatic Number," *J. London math. Soc.*, Vol. 42, 1967, pp. 330–332.

[72] Hoffman, A. J., "On eigenvalues and colorings of graphs," *Graph Theory and Its Applications*, Academic Press, New York, 1970, pp. 79–92.

[73] Trevisan, L., "Max cut and the smallest eigenvalue," *Proceedings of the 41st Annual ACM Symposium on Theory of Computing*, 2009, pp. 263–272.

[74] McSherry, F., "Spectral Partitioning of Random Graphs," *Proceedings of the 42nd IEEE Symposium on Foundations of Computer Science*, 2001, pp. 529–537.

[75] Demmel, J., *Applied Numerical Linear Algebra*, SIAM, 1997.

[76] Boppana, R. B., "Eigenvalues and graph bisection: an average-case analysis," *Proc. 28th IEEE Symposium on Foundations of Computer Science*, 1987, pp. 280–285.

[77] Füredi, Z. and Komlós, J., "The eigenvalues of random symmetric matrices," *Combinatorica*, Vol. 1, No. 3, 1981, pp. 233–241.

[78] Vu, V., "Spectral Norm of Random Matrices," *Combinatorica*, Vol. 27, No. 6, 2007, pp. 721–736.

[79] Alon, N., Krivelevich, M., and Vu, V. H., "On the concentration of eigenvalues of random symmetric matrices," *Israel Journal of Mathematics*, Vol. 131, No. 1, 2002, pp. 259–267.

[80] Alon, N. and Kahale, N., "A Spectral Technique for Coloring Random 3-Colorable Graphs," *SIAM Journal on Computing*, Vol. 26, 1997, pp. 1733–1748.

[81] Kannan, R. and Vempala, S., "Spectral Algorithms," *Foundations and Trends in Theoretical Computer Science*, Vol. 4, No. 3-4, 2008, pp. 132–288.

[82] Batson, J. D., Spielman, D. A., and Srivastava, N., "Twice-Ramanujan sparsifiers," *Proceedings of the 41st Annual ACM Symposium on Theory of computing*, 2009, pp. 255–262.

[83] Spielman, D. A. and Srivastava, N., "Graph sparsification by effective resistances," *SIAM Journal on Computing*, 2010, To appear. Preliminary version appeared in the Proceedings of the 40th Annual ACM Symposium on Theory of Computing.

[84] Spielman, D. A. and Woo, J., "A Note on Preconditioning by Low-Stretch Spanning Trees," *CoRR*, Vol. abs/0903.2816, 2009, Available at http://arxiv.org/abs/0903.2816.

[85] Page, L., Brin, S., Motwani, R., and Winograd, T., "The PageRank Citation Ranking: Bringing Order to the Web," Tech. rep., Computer Science Department, Stanford University, 1998.

[86] Langville, A. N. and Meyer, C. D., *Google's PageRank and Beyond: The Science of Search Engine Rankings*, Princeton University Press, Princeton, NJ, USA, 2006.

[87] Bonacich, P., "Technique for Analyzing Overlapping Memberships," *Sociological Methodology*, Vol. 4, 1972, pp. 176–185.

[88] Bonacich, P., "Factoring and Weighting Approaches to Status Scores and Clique Identification," *Journal of Mathematical Sociology*, Vol. 2, 1972, pp. 113–120.

[89] Bonacich, P., "Power and Centrality: A family of Measures," *American Journal of Sociology*, Vol. 92, 1987, pp. 1170–1182.

[90] Berge, C., *Graphs*, North-Holland, 1985.

[91] Palacios-Huerta, I. and Volij, O., "The Measurement of Intellectual Influence," *Econometrica*, Vol. 72, No. 3, 2004, pp. 963–977.

[92] Altman, A. and Tennenholtz, M., "Ranking systems: the PageRank axioms," *Proceedings of the 6th ACM Conference on Electronic Commerce*, 2005, pp. 1–8.

[93] Kleinberg, J., "Authoratitive sources in a hyperlinked environment," *Journal of the ACM*, Vol. 48, 1999, pp. 604–632.

[94] Chung, F., "The Diameter and Laplacian Eigenvalues of Directed Graphs," *The Electronic Journal of Combinatorics*, Vol. 13, No. 1, 2006.

[95] Chung, F., "Laplacians and the Cheeger Inequality for Directed Graphs," *Annals of Combinatorics*, Vol. 9, No. 1, 2005, pp. 1–19.

[96] Mohar, B. and Poljak, S., "Eigenvalues in combinatorial optimization," *Combinatorial and graph-theoretical problems in linear algebra*, IMA Volumes in Mathematics and Its Applications, Springer-Verlag, 1993, pp. 107–151.

[97] Mohar, B., "The Laplacian spectrum of graphs," *Graph Theory, Combinatorics, and Applications*, Wiley, 1991, pp. 871–898.

[98] Brouwer, A. E. and Haemers, W. H., "Spectra of Graphs," Electronic Book. Available at http://www.win.tue.nl/~aeb/ipm.pdf.

[99] Godsil, C. and Royle, G., *Algebraic Graph Theory*, Graduate Texts in Mathematics, Springer, 2001.

[100] Biggs, N. L., *Algebraic graph theory*, Cambridge Tracts in Math., Cambridge University Press, London, New York, 1974.

[101] Brualdi, R. A. and Ryser, H. J., *Combinatorial Matrix Theory*, Cambridge University Press, New York, 1991.

[102] Brouwer, A. E., Cohen, A. M., and Neumaier, A., *Distance-Regular Graphs.*, Springer Verlag, New York, 1989.

Chapter 19

Algorithms for Visualizing Large Networks

Yifan Hu

AT&T Labs Research

19.1 Introduction

Graphs are often used to encapsulate relationship between objects. Graph drawing enables visualization of such relationships. The usefulness of this visual representation is dependent on whether the drawing is aesthetic. While there are no strict criteria for aesthetics of a drawing, it is generally agreed, for example, that such a drawing has minimal edge crossing, with vertices evenly distributed in the space, connected vertices close to each other, and symmetry that may exist in the graph preserved.

One of the earliest example of graph drawing is probably the Mill (also known as Morris) game boards, found in the thirteenth century *The Book of*

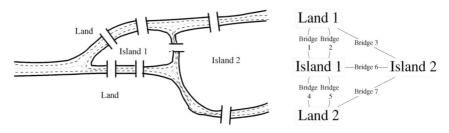

FIGURE 19.1: An illustration of the seven bridges of Königsberg (left) and the graph representation (right).

Games, produced under the direction of Alfonso X (1221-1284), King of Castile (Spain). Other examples of early graph drawing are drawings of family trees, and trees of virtues and vices.

In 1735, Euler presented and later published his famous paper [1] on the Seven Bridges of Königsberg. The problem was to find a walk through the city of Königsberg, which include two islands connected to the mainland by seven bridges (Figure 19.1). The walk must cross each bridge once and only once. Euler proved that such a walk is not possible. Euler's paper laid the foundation of graph theory. Interestingly, however, Euler did not provide a drawing of the graph representation of the problem itself in the paper.

In the eighteenth century, the Irish physicist, astronomer, and mathematician, William Hamilton, invented the Iconsian game. To play the game, one has to visit every node of a dodecahedral graph once, and back to the starting node. Every edge can be passed at most once. In 1857 he sold the game to a London game dealer for 25 Pounds (about 3000 US dollars in today's money after taking inflation into account). The dealer made two versions; unfortunately, neither sold well, probably because the game was so simple that even a child could succeed. Around the same time, British mathematician Arthur Cayley wrote a paper on the enumeration of tree diagrams. In the paper he pioneered the concept of a tree, which is now fundamental to computer sciences. The paper also contains drawings of trees.

Between the eighteenth and nineteenth centuries, graph drawing also appeared in areas outside of mathematics. For example, in crystallography and chemistry, graph drawings were used to illustrate molecular structures. In 1878, the English mathematician Sylvester introduced the concept of a *graph* for the first time.

From the mid-twentieth century, with the demands from the military, transportation and communication, and science and technology, a great number of graph theory problems emerged. Often it is helpful to be able to visualize a graph. However drawing a complex graph by hand is time consuming, if not impossible, for large graphs. Therefore automatic generation of graph drawings became of interest, and were facilitated by the ever increasing computing power.

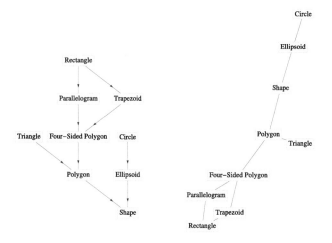

FIGURE 19.2: A graph drawn in hierarchical (left) and undirected (right) styles.

In 1963, Tutte [2] proposed an algorithm to draw planar graphs by fixing nodes on a face and placing the rest of the nodes at the barycenters of their neighbors. In 1981, Sugiyama et al. [3] proposed a method for the hierarchical drawing of directed graphs. Eades [4] proposed a force-directed algorithm in 1984, and Kamada and Kawai [5] the spring embedding algorithm in 1989. These algorithms were further improved [6, 7], and some of them applied to large graphs [8, 9, 10, 11, 12, 13] later.

Depending on the applications and properties of the graph to be visualized, there are many styles of graph drawing. For example, for planar graphs, a planar embedding is more appropriate. For general graphs, there are also two broad styles of drawing: straight line drawing, where each edge is represented by a straight line, and orthogonal drawing, where each edge is represented by lines segments that are either horizontal or vertical. Finally, for directed graphs where it is important to show the direction of the edges, a hierarchical drawing style can be used. Figure 19.2 shows a graph drawn using the hierarchical style (left) and the undirected straight-edged style (right).

In this chapter, we look at algorithms for visualizing large networks. The term large is relative to the computing power and memory available. In this paper, by large, we mean graphs of more than a few thousand vertices. Such large graphs occur in areas such as Internet mapping, social networks, biological pathways and genealogy. Large networks bring unique issues. For example, the complexity of the algorithm becomes very important. The ability of the algorithm in escaping from local minimum and achieving a globally optimal drawing is also vital. We present some of the algorithms suitable for large graphs. We shall limit our attention to straight edge drawing of undirected graphs, mostly because scalable algorithms for these have been more intensely investigated. Our emphasis is on algorithms, with a tilt toward issues relating to combinatorial computing. We do not attempt to give a comprehensive review of the literature and history, nor of software and visualization systems; for these the reader is referred to [14, 15].

We note that not all graph can be aesthetically embedded into two or three-dimensional space. For example, many Internet graphs are of "small-world" nature [16]. Such graphs are difficult to layout in Euclidean space, and often require an interactive system to comprehend and explore. A number of techniques can be brought to bear on such graphs, including visual abstraction [17, 18], fisheye-like distortion and hyperbolic layout [18, 19, 20, 21, 22]. These topics are not within the scope of this chapter, but are nevertheless very important parts of large graph visualization.

19.2   Algorithms for Drawing Large Graphs

We use $G = \{V, E\}$ to denote an undirected graph, with V the set of vertices and E the set of edges. Denote by $|V|$ and $|E|$ the number of vertices and edges, respectively. If vertices i and j form an edge, we denote that $i \leftrightarrow j$, and call i and j neighboring vertices. We denote by x_i the current coordinates of vertex i in d-dimensional Euclidean space. Typically $d = 2$ or 3.

The aim of graph drawing is to find x_i for all $i \in V$ so that the resulting drawing gives a good visual representation of the connectivity information between vertices. One way to solve this problem is to turn the graph drawing problem into that of finding a minimal energy configuration of a physical system. Within this framework, two popular methods, the spring-electrical model [4, 7], and the stress model [5], are the most well-known.

We first discuss the spring-electrical model, and the multilevel approach and force approximation techniques that make this model suitable for large graphs. We note that the multilevel approach is not limited to the spring-electrical model, but for convenience we introduce it in the context of this model. We then discuss the stress model and classical MDS (strain model), and look at attempts to apply these models to large graphs. Finally we present high-dimensional embeddings and Hall's algorithm for large graph layout.

Within the graph drawing literature, the name "force-directed algorithm" has often been used with the both spring-electrical and the spring models. Strictly speaking, the name "force-directed algorithm" is appropriate when we talk about a procedure that involves calculating forces exerted on vertices, moving them along the direction of the forces, and repeating until the system comes to an equilibrium state. Therefore we shall only use the term "force-directed algorithm" when referring to this iterative procedure.

19.2.1 Spring-Electrical Model

The spring-electrical model [4] represents the graph drawing problem by a system of electrically charged vertices attracted to each other by springs; vertices also repel each other via electrical forces. Here, following [7], the attractive spring force exerted on vertex i from its neighbor j is proportional to the squared distance between these two vertices,

$$F_a(i,j) = -\frac{\|x_i - x_j\|^2}{K} \frac{x_i - x_j}{\|x_i - x_j\|}, \quad i \leftrightarrow j,$$

where K is a parameter related to the nominal edge length of the final layout. The repulsive electrical force exerted on vertex i from any vertex j is inversely proportional to the distance between these two vertices,

$$F_r(i,j) = \frac{K^2}{\|x_i - x_j\|} \frac{x_i - x_j}{\|x_i - x_j\|}, \quad i \neq j.$$

The energy of this physical system [23] is

$$E(x) = \sum_{i \leftrightarrow j} \|x_i - x_j\|^3 / (3K) - \sum_{i \neq j} K^2 \ln \left(\|x_i - x_j\| \right),$$

with its derivatives a combination of the attractive and repulsive forces. We note that there are other variations of this force model. See, e.g., [4, 24].

The spring-electrical model can be solved with the aforementioned force-directed algorithm by starting from a random layout, calculating the combined attractive and repulsive forces on each vertex, and moving the vertices along the direction of the force for a certain step length. This process is repeated, with the step length decreasing every iteration, until the layout stabilizes. The following algorithm starts with a random, or user supplied, initial layout x.

Algorithm 1: Force-directed algorithm

- ForceDirectedAlgorithm(G, x, tol) {
    - *converged* = $FALSE$;
    - *step* = initial step length;
    - while (*converged* equals $FALSE$) {
        * $x^0 = x$;
        * for $i \in V$ {
            · $f = 0$;
            · for $(j \leftrightarrow i)$ $f := f + F_a(i,j)$;
            · for $(j \neq i, \ j \in V)$ $f := f + F_r(i,j)$;
            · $x_i := x_i + step * (f/\|f\|)$;
        * }
        * *step* := *update_steplength*($step, x, x^0$);

          * if $(||x - x^0|| < tol * K)$ *converged = TRUE*;

  – }

  – return x;

• }

This procedure can be enhanced by an adaptive step length updating scheme [10, 25], and usually works well for small graphs.

For large graphs, this simple iterative procedure is not sufficient to overcome the many local minima that often exist in the space of all possible layouts. Instead, a multilevel approach has to be used (Section 19.2.1.2). Furthermore, a nested space partitioning data structure is needed to approximate the all-to-all electrical force so as to reduce the quadratic complexity to $O(|V|\log|V| + |E|)$ (Section 19.2.1.1). Combining these two powerful techniques results in efficient implementations of the spring-electrical model [9, 10] that are capable of handling graphs of millions of vertices and edges [26].

19.2.1.1 Fast Force Approximation

Each iteration of the force-directed algorithm (Algorithm 1) involves two loops. The outer loop iterates over each vertex. Of the two inner loops, the latter involves calculation of all-to-all repulsive forces, and $O(|V|)$ force calculations are needed for every vertex. Thus the overall complexity is $O(|V|^2)$.

The repulsive force calculation resembles the n-body problem in physics, which is well-studied. One of the widely used techniques to approximate the repulsive forces in $O(n \log n)$ time with good accuracy, but without ignoring long range forces, is to treat groups of far away vertices as supernodes, using a suitable data structure [27]. This idea was adopted by Tunkelang [12] and Quigley [11]. They both used an quadtree (2D) or octree (3D) data structure.

For simplicity, hereafter we use the term quadtree exclusively, which should be understood as octree for 3D. A quadtree data structure is constructed by first forming a square that encloses all vertices. This is the level 0 square. This square is subdivided into 4 squares if it contains more than 1 vertex, and forms the level 1 squares. This process is repeated until each square contains no more than 1 vertex. Figure 19.3 (left) shows a quadtree on the `jagmesh1` graph.

The quadtree forms a recursive grouping of vertices, and can be used to efficiently approximate the repulsive force in the spring-electrical model. The idea is that in calculating the repulsive force on a vertex i, if a group of vertices, S, lies in a square that is sufficiently "far" from i, the whole group can be treated as a supernode. Otherwise we traverse down the hierarchy and examine the four sibling squares. The supernode is assumed to be situated at the center of gravity of the group, $x_S = (\sum_{j \in S} x_j)/|S|$. The repulsive force on vertex i from this supernode is

$$f_r(i, S) = \frac{K^2|S|}{\|x_i - x_S\|} \frac{x_i - x_S}{\|x_i - x_S\|}.$$

It remains to define what "far" means. Following [11, 12], the supernode S is far away from vertex i, if the width d_S of the square that contains the supernode is small, compared with the distance between the supernode and the vertex i,

$$\frac{d_S}{\|x_i - x_S\|} \leq \theta. \tag{19.1}$$

This inequality is called the Barnes-Hut opening criterion, and was originally formulated by Barnes and Hut [27]. Here $\theta \geq 0$ is a parameter. The smaller the value of θ, the more accurate the approximation to the repulsive force, and the larger the number of force calculations. A typical value that works well in practice is $\theta = 1.2$.

The quadtree data structure allows efficient identification of all the supernodes that satisfy (19.1). The process starts from the level 0 square. Each square is checked, and recursively opened, until the inequality (19.1) is satisfied. Figure 19.3 (right) shows all the supernodes (the squares) and the vertices these supernodes consist of, with reference to vertex i located at the top-middle part of the graph. In this case there are 936 vertices, and 32 supernodes.

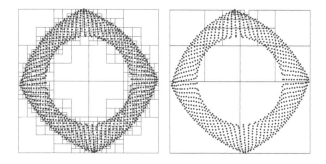

FIGURE 19.3: An illustration of the quadtree data structure. Left: the overall quadtree. Right: supernodes with reference to a vertex at the top middle part of the graph, with $\theta = 1$.

Under a reasonable assumption [27, 28] of the distribution of vertices, it can be proved that building the quadtree takes a time complexity of $O(|V| \log |V|)$. Finding all the supernodes with reference to a vertex i can be done in a time complexity of $O(log|V|)$. Overall, using a quadtree structure to approximate the repulsive force, the complexity for each iteration of the force-directed Algorithm 1 is reduced from $O(|V|^2)$ to $O(|V| \log |V|)$. This force approximation scheme can be further improved by considering force approximation at supernode-supernode level instead of vertex-supernode level [29].

A force approximation algorithm with the same $O(|V| \log |V|)$ complexity but that is independent of the distribution of vertices is the multipole method [30], Hachul and Jünger applied this force approximation to graph drawing [9].

19.2.1.2 Multilevel Approach

While fast approximation of long range forces resolves the quadratic complexity of the force-directed algorithm for the spring-electrical model, it does not change the fact that the algorithm repositions one vertex at a time, without a "global view" of the layout. Due to the fact that this physical system of springs and electrical charges has many local minimum configurations, applying the force directed algorithm directly to a random initial layout is unlikely to give an optimal final layout. A multilevel approach can overcome this limitation. In this approach, a sequence of successively smaller graphs are generated, each captures the essential connectivity information of its parent. Global optimal layout can be found much more easily on a small graph, which are then used as a starting layout for its parent. From this initial layout, further refinement is carried out to achieve the optimal layout for the parent.

A multilevel approach has been used in many large-scale combinatorial optimization problems, such as graph partitioning [31, 32, 33], matrix ordering [34, 35], the traveling salesman problem [36], and was proved to be a very useful meta-heuristic tool [37]. A multilevel approach was later used in graph drawing [13, 38, 39, 40]. Note that a multilevel approach is not limited to the spring-electrical model, but for convenience we are introducing it in the context of this model.

A multilevel approach has three distinctive phases: coarsening, coarsest graph layout, and prolongation and refinement. In the coarsening phase, a series of coarser and coarser graphs, $G^0 = G, G^1, \ldots, G^l$, are generated, each coarser graph G^{k+1} encapsulates the information needed to layout its "parent" G^k, while containing fewer vertices and edges. The coarsening continues until a graph with only a small number of vertices is reached. The optimal layout for the coarsest graph can be found cheaply. The layout on the coarser graphs are recursively prolonged to the finer graphs, with further refinement at each level.

Graph coarsening and initial layout is the first phase in the multilevel approach. There are a number of ways to coarsen an undirected graph. One often used method is based on edge collapsing (EC) [31, 32, 33]. In this scheme, a maximal independent edge set (MIES) is selected. This is a maximal set of edges, with no edges incident to the same vertex. The vertices correspond to this edge set form a maximal matching. Each edge, and its corresponding pair of vertices, are coalesced into a new vertex. Figure 19.4 (a) illustrates MIES and the result of coarsening using edge collapsing.

Alternatively, coarsening can be performed based on a maximal independent vertex set (MIVS) [41]. This is a maximal set of vertices such that no

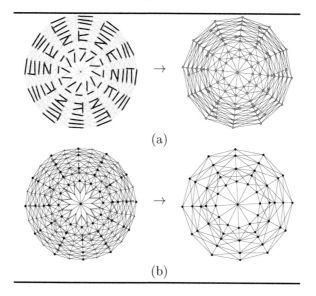

(a)

(b)

FIGURE 19.4: An illustration of graph coarsening: (a) Left: original graph with 229 vertices. Edges in a maximal independent edge set are thickened. Right: a coarser graph with 115 vertices resulted from coalescing thickened edges; (b) Left: original graph with 229 vertices. Vertices in a maximal independent vertex set are darkened. Right: a coarser graph with 55 vertices resulted from the maximal independent vertex set.

two vertices in the set are connected by an edge in the graph. Edges of the coarser graph are formed through linking two vertices in the maximal independent vertex set by an edge if their distance apart is no greater than three. Figure 19.4 (b) illustrates MIVS and the result of coarsening using maximal independent vertex set.

Coarsest graph layout is carried out at the end of the recursive coarsening process. Coarsening is performed repeatedly until the graph is very small; at that point we can layout the graph using a suitable algorithm, for example, the force-directed Algorithm 1. Because the graph on the coarsest level is very small, it is likely that it can be laid out optimally.

The prolongation and refinement step is the third phase in a multilevel procedure. The layout on the coarser graphs are recursively interpolated to the finer graphs, with further refinement at each level.

Row "spring electrical" of Figure 19.5 shows drawings of two graphs using this multilevel force-directed algorithm [10]. The drawings are of good quality for both `jagmesh1`, a mesh-like graph, and `1138_bus`, a sparser graph. The spring-electrical model does suffer slightly from "warping effect" [42]. For example, for the `jagmesh1`, vertices are closer together near the boundary, compared to the interior of the mesh. This effect can be mitigated using post-processing techniques [42].

19.2.1.3 An Open Problem: More Robust Coarsening Schemes

The multilevel approach can work efficiently, provided that the coarsening scheme is able to generate a coarsened graph with many fewer vertices than its parent graph. The aforementioned multilevel algorithm was found to work well for a lot of graphs from the graph drawing literature [10]. However, when applied to the University of Florida Sparse Matrix Collection [43], we found that, for some matrices, the coarsening scheme could not coarsen sufficiently, and the multilevel scheme has to be terminated prematurely, which results in poor drawings. An example is shown in Figure 19.6. On the left of the figure is the sparsity pattern of the matrix gupta1. From this plot we can conclude that this matrix describes a graph of three groups of vertices: those represented by the top 1/3 of the rows in the matrix plot, the middle 1/3 of the rows, and the rest. Vertices in each group are all connected to a selected few in that group; these links are seen as dense horizontal and vertical bars in the matrix plot. At the same time, vertices in the top group are connected to those in middle group, which in turn are connected to those in the bottom group, as represented by the off-diagonal lines parallel to the diagonal. However, the graph drawing in the middle of Figure 19.6 shows none of these structures.

This graph exemplifies many of the problematic graphs: they contain star-graph like substructures, with a lot of vertices all connected to a few vertices. Such structures pose a challenge to the usual graph coarsening schemes. These schemes are not able to coarsen graphs like that adequately. For example, Figure 19.7 (left) shows such a graph, with $k = 10$ vertices on the outskirts all connected to two vertices at the center. Because of this structure, any MIES can only contain two edges (the two thick edges in Figure 19.7). When the end vertices of the MIES are merged at the center of each edge, the resulting graph has only two fewer vertices. Therefore if k is large enough, the coarsening can be arbitrarily slow ($k/(k + 2) \rightarrow 1$ as $k \rightarrow \infty$).

One solution proposed [43] is to find vertices that share the same neighbors. These vertices are matched in pairs. The usual MIES scheme is then used to match the remaining unmatched vertices. Finally the matched vertices are merged to get the coarsened graph. The scheme is able to overcome the slow coarsening problem associated with graphs having star-graph likes substructures. Applying this scheme to the graph in Figure 19.8 (left) resulted in a graph with 1/2 the number of vertices (Figure 19.8 right). With this new coarsening scheme, we are able to layout many more graphs aesthetically. For example, when applied to the gupta1 matrix, the drawing at Figure 19.6 (right) reveals the correct visual structures as we expected, with three groups of vertices, each connected to a few within the group, and a linear connectivity relation among the groups. This new coarsening scheme is able to handle many of the graphs MIES or MIVS based coarsening schemes

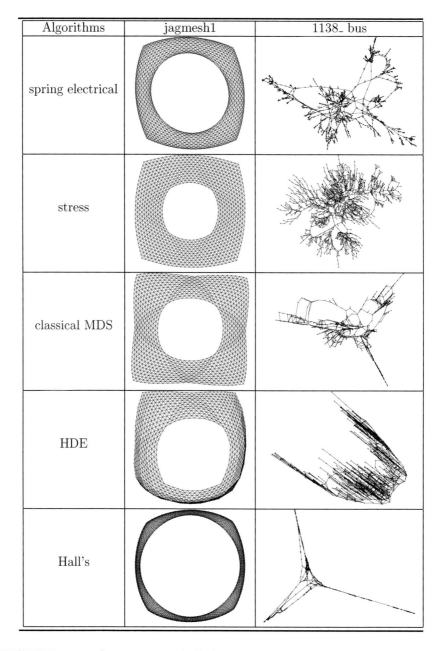

FIGURE 19.5: An overview of all algorithms described in the chapter, applied to two graphs, `jagmesh1` and `1138_bus`.

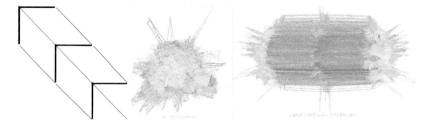

FIGURE 19.6: Matrix plot of `gupta1` matrix (left) and the initial graph drawing (middle). After applying the new coarsening scheme, the graph drawing reflects the structure of the matrix much better (right).

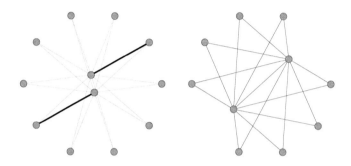

FIGURE 19.7: A maximal independent edge set based coarsening scheme fails to coarsen sufficiently a star-graph like structure: a maximal independent edge set (thick edges) (left); when merging the end vertices of the edge set at the middle of these edges, the resulting coarsened graph only has 2 fewer vertices (right).

fail to. But a more general and robust coarsening scheme that works for all graphs is still an open problem. A promising route for investigation may be the coarsening schemes associated with algebraic multigrid (AMG) [35].

19.2.2 Stress and Strain Models

While the spring-electrical model is scalable and can layout graphs of millions of nodes in minutes, it does have the limitation of not coping well when edges have predefined lengths. It is possible to assign weaker attractive force and stronger repulsive force for longer edges, but such treatment is not as principled and direct as the following stress model.

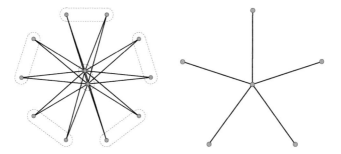

FIGURE 19.8: Matching and merging vertices with the same neighborhood structure (left, with dashed line enclosing matching vertex pairs) resulted in a new graph (right) with 1/2 the number of vertices.

19.2.2.1 Stress Model

The stress model assumes that there are springs connecting all pairs of vertices of the graph, with the ideal spring length equal to the predefined edge length. The energy of this spring system is

$$\sum_{i \neq j} w_{ij} \left(\|x_i - x_j\| - d_{ij} \right)^2, \tag{19.2}$$

where d_{ij} is the ideal distance between vertices i and j, and w_{ij} is a weight factor, typically $1/d_{ij}^2$. The layout that minimizes the above stress energy is an optimal layout of the graph according to this model.

The stress model has its root in Multidimensional Scaling (MDS) [44, 45]. Note that typically we only know the ideal distance between vertices that share an edge, which is usually taken to be one for graphs without predefined edge length. Alternatively, it has been proposed to set the edge length equal the total number of non-common neighbors of the two end vertices [46]. For other vertex pairs, one way to define d_{ij} is to take it as the shortest distance between vertex i and j. The practice of taking the shortest graph distance as the ideal edge length date back at least to 1980 in social network layout [47], and in graph drawing using classical MDS [45], but is often attributed to Kamada and Kawai [5].

There are several ways to minimize (19.2). A force-directed algorithm (Algorithm 1) can be used. The repulsive/attractive force exerted on vertex i from the spring between vertices i and j is

$$F(i, j) = -w_{ij}(\|x_i - x_j\| - d_{ij}) \frac{x_i - x_j}{\|x_i - x_j\|}, \quad i \neq j. \tag{19.3}$$

Stress-majorization: a stress-majorization technique [46] can be employed to solve the stress model. Consider the cost function (19.2),

$$\sum_{i \neq j} w_{ij} \left(\|x_i - x_j\| - d_{ij} \right)^2 = \sum_{i \neq j} \left(w_{ij} \|x_i - x_j\|^2 - 2 d_{ij} w_{ij} \|x_i - x_j\| + w_{ij} d_{ij}^2 \right)$$

On the right hand side, the first and third terms are either constant or quadratic with regard to x, except the second one. Using the Cauchy-Schwartz inequality, $(x_i - x_j)^T (y_i - y_j) \leq \|x_i - x_j\| \|y_i - y_j\|$, we can bound the cost function by

$$g(x, y) = \sum_{i \neq j} \left(w_{ij} \|x_i - x_j\|^2 - 2 d_{ij} w_{ij} \frac{(x_i - x_j)^T (y_i - y_j)}{\|y_i - y_j\|} + w_{ij} d_{ij}^2 \right),$$

with the bound tight when $y = x$. The idea of stress majorization is to minimize a sequence of quadratic function $g(x, y^k)$, with $y^0 = x^0$ the initial layout, and subsequent y^k the result of minimizing $g(x, y^{k-1})$, $k = 1, 2, \dots$.

The minimum of the quadratic function $g(x, y)$ is derived by setting $\partial_{x_i} g(x, y) = 0$, giving

$$L_w x = L_{w,d}\, y \tag{19.4}$$

where the weighted Laplacian matrix L_w has elements

$$(L_w)_{ij} = \begin{cases} \sum_{i \neq l} w_{il}, & i = j \\ -w_{ij}, & i \neq j \end{cases}$$

and matrix $L_{w,d}$ has elements

$$(L_{w,d})_{ij} = \begin{cases} \sum_{i \neq l} w_{il}\, d_{il}/ \|y_i - y_l\|, & i = j \\ -w_{ij}\, d_{ij}/ \|y_i - y_j\|, & i \neq j \end{cases}$$

In summary, the process of finding a minima of (19.2) becomes that of solving a series of linear systems (19.4), with the solution x served as y in the next iteration. This iterative process is found to be quite robust, although for large graphs, it still benefits from a good initial layout. Row "stress" of Figure 19.5 gives drawings of the stress model. It performed very well on both graphs.

19.2.2.2   Strain Model (Classical MDS)

The strain model, also known as classical MDS [48], predates the stress model. Classical MDS tries to fit the inner product of positions, instead of the distance between points. Specifically, assume that the final embedding is centered around the origin:

$$\sum_{i=1}^{|V|} x_i = 0.$$

Furthermore, assume that in the ideal case, the embedding fits the distance exactly:

$$\|x_i - x_j\| = d_{ij}. \tag{19.5}$$

It is then easy to prove [49] that the product of the positions, $b_{ij} = x_i^T x_j$, can be expressed as the squared and double centered distance,

$$b_{ij} = x_i^T x_j = -1/2 \left(d_{ij}^2 - \frac{1}{|V|} \sum_{k=1}^{|V|} d_{kj}^2 - \frac{1}{|V|} \sum_{k=1}^{|V|} d_{ik}^2 + \frac{1}{|V|^2} \sum_{k=1}^{|V|} \sum_{l=1}^{|V|} d_{kl}^2 \right). \tag{19.6}$$

In real data, it is unlikely that we can find an embedding that fits the distances perfectly, hence assumption (19.5) does not stand. But we would still expect that b_{ij} is a good approximation of $x_i^T x_j$. Therefore we try to find an embedding that minimizes the difference between the two,

$$\min_X \|X^T X - B\|_F, \tag{19.7}$$

where X is the $|V| \times d$ dimensional matrix of x_i's, B is the $|V| \times |V|$ symmetric matrix of b_{ij}'s, and $\|.\|_F$ is the Frobenius norm. If the eigen-decomposition of B is $B = Q^T \Lambda Q$, then the solution to (19.7) is $X = \Lambda_d^{1/2} Q$, where Λ_d is the diagonal matrix of Λ, with all but the d largest eigenvalues on the diagonal set to zero. Because the strain model does not fit the distance directly, graph drawings given by solving this model are not as satisfactory as those using the stress model [47], but it can be used as a good starting point for the stress model. Row "classical MDS" of Figure 19.5 gives drawings using classical MDS. It performed well on the mesh like `jagmesh1` graph, but not so well on the sparser `1138_bus` graph, where vertices cling close to each other, making some details of the graph unclear.

19.2.2.3 MDS for Large Graphs

In a typical usage of the stress or strain models, the ideal distance between all pairs of vertices has to be calculated, which requires an all-pair shortest path calculation. Using Johnson's algorithm, this needs $O(|V|^2 \log |V| + |V||E|)$ computation, and a storage of $O(|V|^2)$. Therefore for very large graphs, this formulation is computationally expensive and memory prohibitive. A number of strategies [8, 46, 50] have been proposed to approximately minimize (19.2) or (19.7).

A **multiscale algorithm** [38, 51] applies the multilevel approach in solving the stress model. In the GRIP algorithm [38], graph coarsening is carried out through vertex filtration, an idea similar to the maximal independent vertex set. A sequence of vertex sets, $V^0 = V \subset V^1 \subset V^2, \ldots, \subset V_L$, is generated. However, coarser graphs are not constructed explicitly. Instead, a vertex set

V^k at level k of the vertex set hierarchy is constructed so that distance between vertices is at least $2^{k-1} + 1$. On each level k, the stress model is solved by a force-directed procedure. The spring force (19.3) on each vertex $i \in V^k$ is calculated by considering a neighborhood $N^k(i)$ of this vertex, with $N^k(i)$ the set of vertices in level k, chosen so that the total number of vertices in this set is $O(|E|/|V^k|)$. Thus the force calculation on each level can be done in time $O(|E|)$. It was proved that with this multilevel procedure and the localized force calculation algorithm, for a graph of bounded degree, the algorithm has close to linear computational and memory complexity.

LandmarkMDS [50] approximates the result of classical MDS by choosing $k << |V|$ vertices as landmarks, and calculating a layout of these vertices using the classical MDS, based on distances among these vertices. The k landmarks are chosen to be well dispersed in the graph. One possibility is to use a MaxMin strategy where the first landmark is randomly chosen, then each subsequent landmark is chosen to be furthest away from the previous landmarks, which is a well-known 2-approximation to the k-center problem. The positions of the rest of vertices are then calculated by placing these vertices at the weighted barycenter, with weights based on distances to the landmarks. So essentially classical MDS is applied to a $k \times k$ submatrix of the $|V| \times |V|$ matrix B. The complexity of this algorithm is $O(k|V| + k^2)$, and only $O(k|V|)$ distances need to be stored.

PivotMDS [8], on the other hand, takes a $|V| \times k$ submatrix C of B. The two eigenvectors corresponding to the largest eigenvalues of the $|V| \times |V|$ matrix CC^T are then calculated using power iterations and used as the $x-$ and $y-$ coordinates. By an algebraic argument, if v is an eigenvector of the $k \times k$ matrix $C^T C$, then

$$CC^T(Cv) = C(C^T Cv) = \lambda(Cv),$$

hence Cv is an eigenvector of CC^T. Therefore PivotMDS proceeds by finding the largest eigenvectors of the smaller $k \times k$ matrix $C^T C$, then projects back to $|V|$-dimensional space by multiplying them with C. Using this technique, the overall complexity is similar to LandmarkMDS, but unlike LandmarkMDS, PivotMDS utilizes distances between landmark vertices and other vertices in the matrix product. In practice PivotMDS was found to give drawings that are closer to the classical MDS than LandmarkMDS, and both are very efficient when used with a small number (e.g., $k \approx 50$) of pivots/landmarks.

It is worth pointing out that, for sparse graphs, there is a limitation in both algorithms. For example, if the graph is a tree, and pivots/landmarks are chosen to be non-leaves, then two leaf nodes that have the same parent will have exact the same distances to any of the pivots/landmarks, consequently their final positions based on these algorithms will also be the same. This problem may be alleviated by utilizing the layout given by these algorithms as an initial placement for a stress model based algorithm, but taking only a sparse set of terms in the stress function [46, 47].

19.2.3 High-Dimensional Embedding

The high-dimensional embedding (HDE) algorithm [52] finds coordinates of vertices in a k-dimensional space, then projects back to two or three dimensional space.

First, a k-dimensional coordinate system is created based on k-centers, where k-centers are chosen as in LandmarkMDS/PivotMDS (Section 19.2.2.3). The graph distances from each vertex to the k-centers form a k-dimensional coordinate system. The $|V|$ coordinate vectors form an $|V| \times k$ matrix Y, where the i-th row of Y is the k-dimensional coordinates for vertex i.

Since it is only possible to visualize in two or three dimensions, and since the coordinates may be correlated, the coordinates are projected back to two or three dimensions by suitable linear combinations that minimize correlations. To make this projection shift-invariant, Y is first normalized so that the center of gravity of the vertices is at the origin, i.e.,

$$Y := Y - \frac{1}{|V|} e e^T Y,$$

where e is the $|V|$-dimensional vector of all 1's.

Assume we want to project the k dimensional coordinate system into 2-dimensions. Let v_1 and v_2 be two k-dimensional linear combination vectors, so that $Y v_1$ and $Y v_2$ form the $x-$ and $y-$ coordinates. The two linear combinations should be uncorrelated, so we take them to be orthogonal to each other:

$$v_1^T Y^T Y v_2 = 0.$$

Furthermore, each should be as far away from 0 as possible, so we want to maximize

$$\frac{v_l^T Y^T Y v_l}{||v_l||}, \; l = 0, \; 1.$$

These can be achieved by taking v_1 and v_2 to be the two eigenvectors that correspond to the two largest eigenvalues of the $k \times k$ symmetric matrix $Y^T Y$. This process of choosing highly uncorrelated vectors out of high dimensional data is known as principal component analysis. The final $x-$ and $y-$ coordinates are given by $Y v_1$ and $Y v_2$.

HDE clearly has many commonalities to PivotMDS. Each utilizes k-centers, and each finds the largest eigenvectors of a $k \times k$ dimensional matrix derived by multiplying the transpose of a $|V| \times k$ matrix with itself. The main difference is that in high-dimensional embedding this $|V| \times k$ matrix consists of distances to the k-centers, while in PivotMDS this matrix consists of distances squared and double centered (see (19.6)). High-dimensional embedding suffers from the same limitation as PivotMDS and LandmarkMDS, in that it is not able to layout sparse graphs as well as mesh-like graphs. Row "HDE" of Figure 19.5 gives drawings using HDE. It performs particularly badly on 1138_bus graph, with vertices close to each other, obscuring many details.

19.2.4 Hall's Algorithm

In 1970, Hall [53] remarked that many sequencing and placement problems could be characterized as finding locations of points which minimize the weighted sum of square distances. In our notation, what he proposed was to minimize

$$\sum_{i \leftrightarrow j} w_{ij} \left\| x_i - x_j \right\|^2, \text{ subject to } \sum_{k=1}^{|V|} x_k^2 = 1$$

where x_i is the 1-dimensional coordinate value for vertex i. The objective function can be written as

$$\sum_{i \leftrightarrow j} w_{ij} \left\| x_i - x_j \right\|^2 = x^T L_w x,$$

with $x = \{x_1, x_2, ..., x_{|V|}\}$. Here L_w is the weighted Laplacian matrix with elements

$$(L_w)_{ij} = \begin{cases} \sum_{(i,l) \in E} w_{il}, & i = j \\ -w_{ij}, & i \neq j \end{cases}$$

For a connected graph, the Laplacian is positive semi-definite with one eigenvalue of 0 corresponding to the trivial eigenvector of all 1's. The solution x of the minimization problem is the eigenvector corresponding to the smallest positive eigenvalue of the weighted Laplacian L_w. We can achieve a 2-dimensional layout by taking the two eigenvectors corresponding to the two smallest positive eigenvalues. Row "Hall's" of Figure 19.5 gives drawings employing Hall's algorithm. It performed reasonably well on the mesh like jagmesh1 graph, but is almost useless on the sparser 1138_bus graph.

Koren et al. [54] proposed an extremely fast algorithm for calculating the two extreme eigenvectors using a multilevel algorithm. The algorithm is called ACE (Algebraic multigrid Computation of Eigenvectors). Using this algorithm, they were able to layout graphs of millions of nodes in less than a minute. However, the fundamental weakness of Hall's algorithm on sparse graphs remains.

19.3   Examples of Large Graph Drawings

In this section we give example drawings of some large graphs. These graphs are drawn using a multilevel spring-electrical model based code, sfdp [10], available from graphviz [55].

One rich source of large graphs is the University of Florida Sparse Matrix Collection [43]. This is a collection of 2272 (as of January, 2010) sparse matrices. The largest matrix has 27 million rows and columns. Graph visualization provides a way to visualize these matrices and get a glimpse of the application underneath these matrices. Figure 19.9 gives the drawing of four graphs in this collection. For instance, by its name, the boneS10 graph could be from a model of the bone. The porous structure can be seen clearly from the drawing. More drawings for all the matrices in the collection can be found at [26].

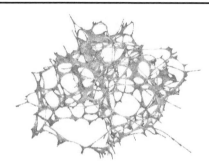

boneS10. $|V| = 914898, |E| = 27276762$ cvxbqp1. $|V| = 40000, |E| = 120000.$

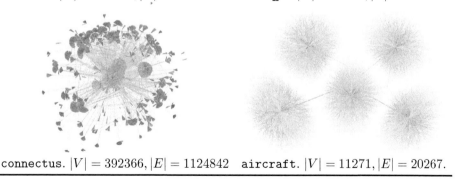

connectus. $|V| = 392366, |E| = 1124842$ aircraft. $|V| = 11271, |E| = 20267.$

FIGURE 19.9: (See color insert.) Drawing of some graphs from the University of Florida Sparse Matrix Collection [43]. More drawings can be found at [26].

Figure 19.10 gives a drawing of the tree of life. The data came from the Tree of Life Project [56]. This project documents phylogeny of organisms, i.e., the history of organismal lineages as they change through time. It implies that different species arise from previous forms via descent, and that all organisms are connected by the passage of genes along the branches of the phylogenetic tree. The data, as collected in February, 2009, contains 76425 species, representing a tiny fraction of the estimated 5 to 100 million species on Earth today. Figure 19.10 clearly shows the tree structure. The root vertex "Life on Earth" (see closeup) sits near the northwest of the tree; to its west are "Green plants" and its descendants; to the east are animals. Fungi are just north of it. While it is difficult to tell due to the size of the figure, there are many more animals than plants, probably a reflection of the data itself rather than reality.

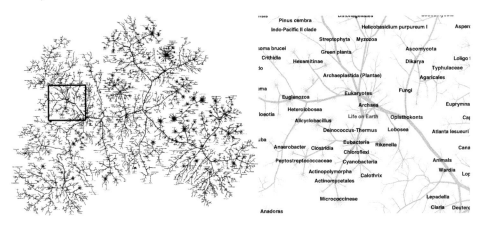

FIGURE 19.10: The tree of life drawing. Left: overview with 76425 species. Right: a closeup view.

19.4   Conclusions

In this chapter, we looked at algorithms for drawing large graphs. Since the 1980's, when the field of graph drawing became very active, much progress has been made in visualizing very large graphs. The key enabling ingredients include a multilevel approach, force approximations by space decomposition, and algebraic techniques for the robust solution of the stress model and in the sparsification of it. Many of the graph drawing algorithms presented have strong links with topics in algebraic graph theory, and a number of open problems remain. One of these is the need for a robust coarsening scheme (Section 19.2.1.3). There are many topics we could not possibly cover in the limited space available. For example, laying out and visualizing a dynamically changing graph remains a challenge [57]. Another challenge is how to draw and explore very complex real world graphs of a "small-world" nature [16]. This often requires not just a good layout algorithm, but also visualization techniques such as edge-bundling [58], interactive exploration systems [18, 59], and additional visual aids such as maps and bubble-sets [60, 61] – all fertile grounds for problems of a combinatorial nature.

Acknowledgments

The author would like to thank Emden Gansner and Stephen North, as well as anonymous referees, for valuable comments.

Bibliography

[1] Euler, L., "Commentarii academiae scientiarum Petropolitanae," *Solutio problematis ad geometriam situs pertinentis*, Vol. 8, 1741, pp. 128–140.

[2] Tutte, W., "How to draw a graph," *Proceedings of the London Mathematical Society*, Vol. 13, 1963, pp. 743–768.

[3] Sugiyama, K., Tagawa, S., and Toda, M., "Methods for Visual Understanding of Hierarchical Systems," *IEEE Trans. Systems, Man and Cybernetics*, Vol. SMC-11, No. 2, 1981, pp. 109–125.

[4] Eades, P., "A heuristic for graph drawing," *Congressus Numerantium*, Vol. 42, 1984, pp. 149–160.

[5] Kamada, T. and Kawai, S., "An algorithm for drawing general undirected graphs," *Information Processing Letters*, Vol. 31, 1989, pp. 7–15.

[6] Gansner, E. R., Koutsofios, E., North, S. C., and Vo, K. P., "A Technique for Drawing Directed Graphs," *IEEE Trans. Softw. Eng.*, Vol. 19, 1993, pp. 214–230.

[7] Fruchterman, T. M. J. and Reingold, E. M., "Graph drawing by force directed placement," *Software - Practice and Experience*, Vol. 21, 1991, pp. 1129–1164.

[8] Brandes, U. and Pich, C., "Eigensolver methods for progressive multidimensional scaling of large data," *Proc. 14th Intl. Symp. Graph Drawing (GD '06)*, Vol. 4372 of *LNCS*, 2007, pp. 42–53.

[9] Hachul, S. and Jünger, M., "Drawing large graphs with a potential field based multilevel algorithm," *Proc. 12th Intl. Symp. Graph Drawing (GD '04)*, Vol. 3383 of *LNCS*, Springer, 2004, pp. 285–295.

[10] Hu, Y. F., "Efficient and High Quality Force-Directed Graph Drawing," *Mathematica Journal*, Vol. 10, 2005, pp. 37–71.

[11] Quigley, A., *Large scale relational information visualization, clustering, and abstraction*, Ph.D. thesis, Department of Computer Science and Software Engineering, University of Newcastle, Australia, 2001.

[12] Tunkelang, D., *A Numerical Optimization Approach to General Graph Drawing*, Ph.D. thesis, Carnegie Mellow University, 1999.

[13] Walshaw, C., "A Multilevel Algorithm for Force-Directed Graph Drawing," *J. Graph Algorithms and Applications*, Vol. 7, 2003, pp. 253–285.

[14] Batagelj, V., "Visualization of large networks," *Encyclopedia of Complexity and Systems Science*, edited by R. A. Meyers, Springer, New York, 2009.

[15] Krujaa, E., Marks, J., Blair, A., and Waters, R., "A Short Note on the History of Graph Drawing," *Proc. 9th Intl. Symp. Graph Drawing (GD '01)*, Springer-Verlag, London, UK, 2002, pp. 272–286.

[16] Watts, D. and Strogate, S., "Collective dynamics of "small-world" networks," *Nature*, Vol. 393, 1998, pp. 440–442.

[17] Quigley, A. and Eades, P., "FADE: Graph Drawing, Clustering, and Visual Abstraction," *LNCS*, Vol. 1984, 2000, pp. 183–196.

[18] Gansner, E. R., Koren, Y., and North, S., "Topological fisheye views for visualizing large graphs," *IEEE Transactions on Visualization and Computer Graphics*, Vol. 11, 2005, pp. 457–468.

[19] Herman, I., Melancon, G., and Marshall, M. S., "Graph visualization and navigation in information visualization: A survey," *IEEE TRANSACTIONS ON VISUALIZATION AND COMPUTER GRAPHICS*, Vol. 6, 2000, pp. 24–43.

[20] Lamping, J., Rao, R., and Pirolli, P., "A Focus+Context Technique Based on Hyperbolic Geometry for Visualizing Large Hierarchies," *SIGCHI CONFERENCE ON HUMAN FACTORS IN COMPUTING SYSTEMS (CHI '95)*, ACM, 1995, pp. 401–408.

[21] Munzner, T. and Burchard, P., "Visualizing the structure of the World Wide Web in 3D hyperbolic space," *VRML '95: Proceedings of the first symposium on Virtual reality modeling language*, ACM, New York, USA, 1995, pp. 33–38.

[22] Munzner, T., "Exploring Large Graphs in 3D Hyperbolic Space," *IEEE Comput. Graph. Appl.*, Vol. 18, 1998, pp. 18–23.

[23] Noack, A., "An Energy Model for Visual Graph Clustering," *Proceedings of the 11th International Symposium on Graph Drawing (GD 2003)*, Vol. 2912 of *LNCS*, Springer, 2004, pp. 425–436.

[24] Battista, G. D., Eades, P., Tamassia, R., and Tollis, I. G., *Algorithms for the visualization of Graphs*, Prentice-Hall, 1999.

[25] Bruss, I. and Frick, A., "Fast Interactive 3-D Graph Visualization," *LNCS*, Vol. 1027, 1995, pp. 99–11.

[26] Hu, Y. F., "A gallery of large graphs," http://research.att.com/~yifanhu/GALLERY/GRAPHS/index.html.

[27] Barnes, J. and Hut, P., "A hierarchical O(NlogN) force-calculation algorithm," *Nature*, Vol. 324, 1986, pp. 446–449.

[28] Pfalzner, S. and Gibbon, P., *Many-Body Tree Methods in Physics*, Cambridge University Press, Cambridge, UK, 1996.

[29] Burton, A., Field, A. J., and To, H. W., "A cell-cell Barnes Hut algorithm for fast particle simulation," *Australian Computer Science Communications*, Vol. 20, 1998, pp. 267–278.

[30] Greengard, L. F., *The rapid evaluation of potential fields in particle systems*, The MIT Press, Cambridge, Massachusetts, 1988.

[31] Gupta, A., Karypis, G., and Kumar, V., "Highly scalable parallel algorithms for sparse matrix factorization," *IEEE Transactions on Parallel and Distributed Systems*, Vol. 5, 1997, pp. 502–520.

[32] Hendrickson, B. and Leland, R., "A multilevel algorithm for partitioning graphs," Technical Report SAND93-1301, Sandia National Laboratories, Allbuquerque, NM, 1993, Also in Proceeding of Supercomputing'95 (http://www.supercomp.org/sc95/proceedings/509_BHEN/SC95.HTM).

[33] Walshaw, C., Cross, M., and Everett, M. G., "Parallel dynamic graph partitioning for adaptive unstructured meshes," *Journal of Parallel and Distributed Computing*, Vol. 47, 1997, pp. 102–108.

[34] Hu, Y. F. and Scott, J. A., "A multilevel algorithm for wavefront reduction," *SIAM Journal on Scientific Computing*, Vol. 23, 2001, pp. 1352–1375.

[35] Safro, I., Ron, D., and Brandt, A., "Multilevel algorithms for linear ordering problems," *J. Exp. Algorithmics*, Vol. 13, 2009, pp. 1.4–1.20.

[36] Walshaw, C., "A Multilevel Approach to the Travelling Salesman Problem," *Oper. Res.*, Vol. 50, 2002, pp. 862–877.

[37] Walshaw, C., "Multilevel refinement for combinatorial optimisation problems," Annals of Operations Research, Vol. 131, pp. 325–372, 2004; originally published as Univ. Greenwich Tech. Rep. 01/IM/73.

[38] Gajer, P., Goodrich, M. T., and Kobourov, S. G., "A fast multidimensional algorithm for drawing large graphs," *LNCS*, Vol. 1984, 2000, pp. 211 – 221.

[39] Hadany, R. and Harel, D., "A multi-scale algorithm for drawing graphs nicely," *Discrete Applied Mathematics*, Vol. 113, 2001, pp. 3–21.

[40] Harel, D. and Koren, Y., "A Fast Multi-Scale Method for Drawing Large Graphs," *J. Graph Algorithms and Applications*, Vol. 6, 2002, pp. 179–202.

[41] Barnard, S. T. and Simon, H. D., "Fast multilevel implementation of recursive spectral bisection for partitioning unstructured problems," *Concurrency: Practice and Experience*, Vol. 6, 1994, pp. 101–117.

[42] Hu, Y. F. and Koren, Y., "Extending the spring-electrical model to overcome warping effects," *Proceedings of IEEE Pacific Visualization Symposium*, IEEE Computer Society, 2009, pp. 129–136.

[43] Davis, T. A. and Hu, Y. F., "University of Florida sparse matrix collection," *in preparation*, 2009.

[44] Kruskal, J. B., "Multidimensioal scaling by optimizing goodness of fit to a nonmetric hypothesis," *Psychometrika*, Vol. 29, 1964, pp. 1–27.

[45] Kruskal, J. B. and Seery, J. B., "Designing network diagrams," *Proceedings of the First General Conference on Social Graphics*, U. S. Department of the Census, Washington, D.C., July 1980, pp. 22–50, Bell Laboratories Technical Report No. 49.

[46] Gansner, E. R., Koren, Y., and North, S. C., "Graph Drawing by Stress Majorization," *Proc. 12th Intl. Symp. Graph Drawing (GD '04)*, Vol. 3383 of *LNCS*, Springer, 2004, pp. 239–250.

[47] Brandes, U. and Pich, C., "An experimental study on distance based graph drawing," *Proc. 16th Intl. Symp. Graph Drawing (GD '08)*, Vol. 5417 of *LNCS*, Springer-Verlag, 2009, pp. 218–229.

[48] Torgerson, W. S., "Multidimensional scaling: I. Theory and method," *Psychometrika*, Vol. 17, 1952, pp. 401–419.

[49] Young, G. and Householder, A. S., "Discussion of a set of points in terms of their mutual distances," *Psychometrica*, Vol. 3, 1938, pp. 19–22.

[50] de Solva, V. and Tenenbaum, J. B., "Global versus local methods in non-linear dimensionality reduction," *Advances in Neural Information Processing Systems 15*, MIT Press, 2003, pp. 721–728.

[51] Harel, D. and Koren, Y., "A Fast Multi-Scale Method for Drawing Large Graphs," *Journal of graph algorithms and applications*, Vol. 6, 2002, pp. 179–202.

[52] Harel, D. and Koren, Y., "Graph Drawing by High-Dimensional Embedding," *lncs*, 2002, pp. 207–219.

[53] Hall, K. M., "An *r*-dimensional quadratic placement algorithm," *Management Science*, Vol. 17, 1970, pp. 219–229.

[54] Koren, Y., Carmel, L., and Harel, D., "ACE: A Fast Multiscale Eigenvectors Computation for Drawing Huge Graphs," *INFOVIS '02: Proceedings of the IEEE Symposium on Information Visualization (InfoVis'02)*, IEEE Computer Society, Washington, DC, USA, 2002, pp. 137–144.

[55] Gansner, E. R. and North, S., "An Open Graph Visualization System and its Applications to Software Engineering," *Software—Practice & Experience*, Vol. 30, 2000, pp. 1203–1233.

[56] Maddison, D. R. and (eds.), K.-S. S., "The Tree of Life Web Project," http://tolweb.org, 2007.

[57] Frishman, Y. and Tal, A., "Online Dynamic Graph Drawing," *proceeding of Eurographics/IEEE VGTC Symposium on Visualization (EuroVis)*, 2007, pp. 75–82.

[58] Holten, D. and van Wijk, J. J., "Force-Directed Edge Bundling for Graph Visualization," *Computer Graphics Forum*, Vol. 28, 2009, pp. 983–990.

[59] Moscovich, T., Chevalier, F., Henry, N., Pietriga, E., and Fekete, J., "Topology-aware navigation in large networks," *CHI '09: Proceedings of the 27th international conference on Human factors in computing systems*, ACM, New York, USA, 2009, pp. 2319–2328.

[60] Collins, C., Penn, G., and Carpendale, S., "Bubble Sets: Revealing Set Relations with Isocontours over Existing Visualizations," *IEEE Transactions on Visualization and Computer Graphics*, Vol. 15, No. 6, 2009, pp. 1009–1016.

[61] Gansner, E. R., Hu, Y. F., and Kobourov, S. G., "GMap: Drawing Graphs and Clusters as Map," *IEEE Pacific Visualization Symposium*, 2010, pp. 201 – 208.

Index

T - #0152 - 171019 - C8 - 234/156/26 - PB - 9780367381752